AF897838

Textile Science and Clothing Technology

Series editor

Subramanian Senthilkannan Muthu, Hong Kong, Hong Kong SAR

More information about this series at http://www.springer.com/series/13111

Sohel Rana · Raul Fangueiro
Editors

Fibrous and Textile Materials for Composite Applications

Springer

Editors
Sohel Rana
University of Minho
Guimarães
Portugal

Raul Fangueiro
School of Engineering
University of Minho
Guimarães
Portugal

ISSN 2197-9863 ISSN 2197-9871 (electronic)
Textile Science and Clothing Technology
ISBN 978-981-10-0232-8 ISBN 978-981-10-0234-2 (eBook)
DOI 10.1007/978-981-10-0234-2

Library of Congress Control Number: 2015957090

© Springer Science+Business Media Singapore 2016
This work is subject to copyright. All rights are reserved by the Publisher, whether the whole or part of the material is concerned, specifically the rights of translation, reprinting, reuse of illustrations, recitation, broadcasting, reproduction on microfilms or in any other physical way, and transmission or information storage and retrieval, electronic adaptation, computer software, or by similar or dissimilar methodology now known or hereafter developed.
The use of general descriptive names, registered names, trademarks, service marks, etc. in this publication does not imply, even in the absence of a specific statement, that such names are exempt from the relevant protective laws and regulations and therefore free for general use.
The publisher, the authors and the editors are safe to assume that the advice and information in this book are believed to be true and accurate at the date of publication. Neither the publisher nor the authors or the editors give a warranty, express or implied, with respect to the material contained herein or for any errors or omissions that may have been made.

Printed on acid-free paper

This Springer imprint is published by SpringerNature
The registered company is Springer Science+Business Media Singapore Pte Ltd.

Dedicated to my mentor and dearest grandfather, Md. Abdur Razzaque, who has always loved, guided and inspired me and without whose endless efforts and inspiration this book was not possible

—Sohel Rana

To my father

—Raul Fangueiro

Preface

The possibility to achieve any targeted sets of properties has promoted the wide application of composite materials in various industrial sectors. Composites and nanocomposites are being considered as superior materials for aerospace, transportation, sports, civil construction, medical and in many other technical sectors. Fibrous and textile materials have been extensively used in composite materials in different forms (sort fibres, tows, fabrics, mats, nanofibre webs, etc.) for reinforcing purpose or for adding various functionalities. Huge flexibility in terms of material selection, structure designing and achievable properties led to their extensive utilization in composite materials.

This book focuses on the different types of fibrous and textile materials applied in composite industries. An introductory discussion on the definition and fundamental aspects of composite materials, types of composites, reinforcements, matrices and applications is presented in the chapter "Introduction to Composite Materials". Chapter "Essential Properties of Fibres for Composite Applications" presents various essential properties of fibres for their successful application to composite materials. As fibres are used in various forms in composite materials such as short fibres, unidirectional tows, directionally oriented structures or advanced 2D and 3D textile structures, these different forms and architectures of fibres have been presented in the chapter "Fiber Architectures for Composite Applications". The subsequent chapters, i.e. "Synthetic Fibres for Composite Applications", Natural Fibers for Composite Applications, "Metallic Fibers for Composite Applications" cover various synthetic, natural, as well as metallic fibres used for reinforcement of polymeric, cementitious, metallic and other matrices. Properties of these fibres, their manufacturing process, processing and properties of composites are discussed in detail. Looking at the tremendous growth in the nanofibre market in recent times, the properties, processing and composite application of carbon nanofibres (also nanotubes) and natural nanofibres are presented in the chapters "Carbon Nanofibres and Nanotubes for Composite Applications" and "Natural Nanofibres for Composite Applications", respectively. In the chapter "Surface Preparation of Fibres for Composite Applications", different surface treatments and finishes which are applied to improve the fibre/matrix interface and other essential properties of composites are covered. Detailed discussions of some special properties of fibres and composites such as piezoresistivity, self-sensing, self-healing, electromagnetic shielding, etc., which are essential requirements for advanced multifunctional composites, have been included in the chapter. "Reinforcements and Composites with Special Properties". The concluding chapter "Comparison of Performance, Cost-Effectiveness and Sustainability" presents the comparison of performance, cost-effectiveness and sustainability aspects of different fibres used in composite industries. In all of these chapters, both the existing technologies used in commercial applications and the recently explored advanced research and developments are presented.

This book is a complete collection of information on all types of fibres and textile structures used to reinforce a wide variety of materials including polymers, cement, metal, soils and so on to improve their general performances as well as multifunctional behaviours. Therefore, we believe that this book will serve as a highly useful reference for a wide range of readers, including engineers, bachelor, master and Ph.D. students, teachers as well as advanced researchers both from academics and industry, covering different disciplines such as textile engineering, fibre science and technology, materials science, mechanical engineering, chemical engineering, nanotechnology, medical sciences, environmental science and so on.

We would like to express our sincere thanks and gratitude to all authors, who have contributed to different chapters, for their excellent efforts. Sincere thanks are also due to the members of the Fibrous materials research group (Fibrenamics) of University of Minho for their kind help and support throughout the preparation of the book.

Sohel Rana
Raul Fangueiro

Contents

About the Editors

Dr. Sohel Rana is currently a Senior Scientist at Fibrous Materials Research Group, University of Minho, Portugal. He obtained his bachelor's degree in Textile Technology from University of Calcutta, India, and master's degree and Ph.D. in Fiber Science and Technology from Indian Institute of Technology (IIT, Delhi), India. His current research areas include advanced fibrous and composite materials, natural fibers, nanocomposites, electrospinning, multifunctional and biocomposite materials and so on. He is the author of 1 book, has edited four books and contributed to 14 book chapters, six keynote and invited papers, and about 100 publications in various refereed journals and international conferences. He participates on the editorial board of several scientific journals and is a potential reviewer for numerous scientific journals including Composite Science and Technology, Composites Part A, Composite Interfaces, Journal of Composite Materials, Journal of Reinforced Plastics and Composites, Powder Technology, Journal of Nanomaterials, Journal of Applied Polymer Science and so on.

Prof. Raul Fangueiro is currently professor and senior researcher in the School of Engineering at the University of Minho, Portugal. He is the Head of the Fibrous Materials Research Group of the same university with expertise in advanced materials (nano, smart, composites) and structures (3D, auxetic, multiscale) with 25 researchers. He is the mentor and coordinator of the FIBRENAMICS International Platform (www.fibrenamics.com), including 200 partners developing promotion, dissemination, technology transfer and research activities on fibre-based advanced materials. He has more than 110 published papers in international reputed scientific journals, 320 conference publications, 36 books and 14 patents. He is the scientific coordinator of several national and international research projects on advanced fibrous and composite materials, mainly for building, architectural and healthcare applications. He has supervised various Ph.D. and post-doc scientific works and is an expert in the European Technological Textile Platform and member of the editorial board of several leading international scientific journals on composite and fibrous materials.

About the Editors

Introduction to Composite Materials

M. Balasubramanian

Abstract This chapter gives an overview of the important aspects of composite materials. The need for composite materials and the advantages over conventional materials are discussed. The definition of composite materials is given with suitable examples. The classification of composites based on the matrix and reinforcement is explained in this chapter. The fundamental aspects like, load transfer and rule of mixtures are discussed in detail. The important characteristics of different fibres and matrices are explained briefly. Manufacturing methods for polymer, metal and ceramic matrix composites are briefly explained in this chapter. Some of the important applications of composites are also highlighted here.

Materials with unusual combination of properties are required for modern technologies. Since the properties of conventional materials are limited, they cannot meet the requirements. In general, the high strength materials are heavy or brittle. The material property combinations have been extended with composite materials. It is possible to produce a composite material with high strength and light weight. Like this, materials with unusual combination of properties can be made with composite materials. For example, cemented carbide, the composite of tungsten carbide and cobalt, has very high hardness and appreciable toughness. Similarly light weight composites can be made with good electrical properties, chemical resistance or optical properties.

Composite materials are a combination of two or more materials in such a way that there are certain desired properties or improved properties. The nature itself is making many composites because of their high performance. Wood and bone are the common examples of natural composites. The constituents of wood are cellulose fibre and lignin matrix, whereas in bone, soft and strong collagen surrounds hard and brittle apatite. The principle of combined action is utilised in composites to get better properties. For example, the polymer composite gets flexibility and light weight from polymer matrix and high strength from the fibre reinforcement.

M. Balasubramanian (✉)
Department of Metallurgical & Materials Engineering, Indian Institute of Technology Madras, Chennai 600036, India
e-mail: mbala@iitm.ac.in

© Springer Science+Business Media Singapore 2016

S. Rana and R. Fangueiro (eds.), *Fibrous and Textile Materials for Composite Applications*, Textile Science and Clothing Technology,
DOI 10.1007/978-981-10-0234-2_1

The composite materials have many advantages over conventional materials. Most of the composites are made with light-weight materials. However, their mechanical properties are significantly high. Hence, the specific strength and modulus values of composites are very high. They have very good chemical resistance and weather resistance. There are many reinforcements and a variety of matrix materials available. By a suitable selection of reinforcement and matrix material, it is possible to get the desired properties from the final composite. When the fibres are used as reinforcements, a composite with directional properties can be obtained by aligning the fibre reinforcement in the required direction. This type of composites is very useful for structures, which require different properties in different directions. A variety of manufacturing techniques are available to produce polymer composites. Hence, it is very easy to manufacture composites with complex shapes. Because of their high specific strength and stiffness, composites can lead to a significant weight reduction. In the present scenario of energy crisis and environmental consciousness, this is very important to increase the fuel efficiency of transport vehicles.

In a composite material, the discontinuous constituent is called reinforcement and the continuous constituent surrounding the reinforcement is called matrix. Composites are generally heterogeneous materials, consisting of two or more materials. There are some composites made with the same material, but in different forms. A few examples are silicon carbide fibre reinforced silicon carbide and carbon fibre reinforced carbon composites.

There are many factors, which control the final properties of a composite material. The most important factor is the properties of constituents, viz. the reinforcement and the matrix. The next important factor is the concentration of constituents, i.e. the quantity of reinforcement and matrix. Apart from these factors, the size, shape, orientation and the nature of the distribution of reinforcement also play a major role in controlling the properties of the composites. These factors are schematically illustrated in Fig. 1. It is desirable to use very fine fibrous reinforcements, which are uniformly distributed throughout the matrix to achieve good and reliable properties in the composites. The orientation of the reinforcement controls the directional properties. It is possible to achieve very high mechanical properties by aligning the reinforcement fibres along the loading direction. In general, the reinforcements have better mechanical properties than the matrix materials. Hence, the composites with higher quantity of reinforcements have better mechanical properties than that of composites with lesser quantity of reinforcements.

The approximate properties of composites can be estimated by using the rule of mixtures, which is given below:

$$X_c = V_r X_r + V_m X_m \tag{1}$$

where, X_c is the property of composites and X_r and X_m are the properties of reinforcement and matrix respectively. V_r and V_m are the volume fractions of

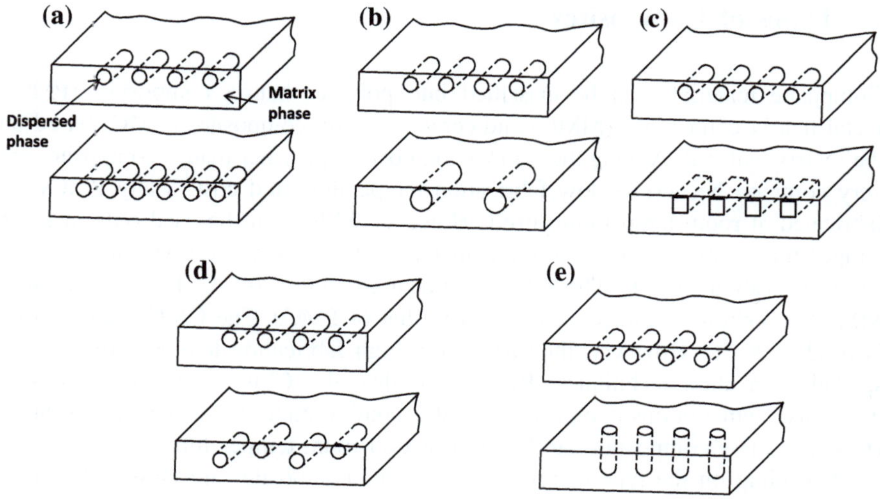

Fig. 1 Schematic of geometrical and spatial characteristics of reinforcements in composites: **a** concentration, **b** size, **c** shape, **d** distribution, and **e** orientation [1]

reinforcement and matrix, respectively. For example, the modulus of a composite material is,

$$E_c = V_r E_r + V_m E_m \tag{2}$$

In practice, it is very difficult to take the reinforcement and matrix on volume basis. The volume and weight of materials are related to each other through density. Once the density values of the constituents are available, it is possible to find out the weight from the volume, or vice versa using the following equations.

$$W_r = \frac{V_r \rho_r}{V_r \rho_r + V_m \rho_m} \tag{3}$$

$$V_r = \frac{W_r/\rho_r}{W_r/\rho_r + W_m/\rho_m} \tag{4}$$

where, ρ is density and W is weight fraction. When the porosity of a composite is negligible, $1-V_f$ can give the volume fraction of matrix, V_m.

The rule of mixtures can predict the structure insensitive properties with reasonable accuracy, whereas there will be very large deviations for the structure sensitive properties. Some of the structure insensitive properties are elastic modulus, thermal conductivity and electrical conductivity. An example of structure sensitive properties is strength.

1 Types of Composites

Composite materials can be classified into polymer matrix composites (PMC), metal matrix composites (MMC) and ceramic matrix composites (CMC) based on the matrix material. Among these three composites, polymer matrix composites are very common because of low fabrication temperature and cost. Many PMCs are fabricated at room temperature itself. However, PMCs can be used only at room temperature or slightly above room temperature. The maximum service temperature for PMC is about 300 °C. The MMCs can be used up to 1200 °C. The fabrication of MMCs is somewhat difficult compared to that of PMCs. The CMCs can be used beyond 1200 °C. However, they need very high fabrication temperature or some special fabrication techniques. Because of this, the CMCs are very expensive materials. Cementitious composites are also coming under the category of CMCs. However, these composites can be fabricated at room temperature.

Depending on the type of reinforcement, the composites are classified as particulate composites and fibre reinforced composites. The performance level of fibre reinforced composites is very high compared to particulate composites. The fibre reinforced composites can be further classified as short fibre reinforced composites and continuous fibre reinforced composites. There are two categories of short fibre reinforced composites based on the alignment of fibres. They are aligned and randomly oriented short fibre reinforced composites. Whiskers can also be considered as short fibres, but each fibre is a single crystal. The continuous fibre reinforced composites can be further classified into unidirectional, bidirectional and multidirectional composites depending on the orientation of fibres.

There are two types of structural composites, which are laminates and sandwich composites. The laminates are multilayered composites, in which the fibres are oriented in the same or different directions in each layer. The sandwich composites consist of composite skin layers with a light weight core material. The core can be foam or a honeycomb structure, depending on the application. The sandwich composites are light in weight with high stiffness. When more than one type of reinforcements is used, then the composite is known as hybrid composite.

The composites can also be classified based on the size of reinforcement into macro, micro and nanocomposites. The macro composites are made with more than millimetre size reinforcements. Micrometre size reinforcements are used in microcomposites. The glass and carbon fibre reinforced composites are the examples of microcomposites. In nanocomposites, at least one of the dimensions of the reinforcement should be less than 100 nm. Many nanocomposites are becoming popular now because of significant improvement in properties with less quantity of reinforcements. Another advantage is that the same processing equipment used for the matrix material can also be used to produce the nanocomposites. The nanocomposites can be further divided into zero, one, two and three dimensional nanocomposites depending on the dimensionality of reinforcements.

2 Fundamental Aspects

Most of the composites produced today are fibre reinforced composites. Fibres are excellent reinforcements compared to particulates because of the following characteristics.

1. Very high strength
2. High aspect ratio (length to diameter ratio)
3. High flexibility

According to Griffith's theory, the fracture strength and flaw size are inversely related.

$$\sigma_f = \sqrt{\frac{2E\gamma}{\pi c}} \tag{5}$$

where, σ_f is the fracture strength, E is Young's modulus, γ is surface energy and c is surface crack length. Most of the fibre reinforcements used in composites have diameter in the range of 6–15 μm. The flaw size is restricted by the diameter of the fibre, since it may not be possible to have a flaw, which is larger than the diameter of fibre. Larger crack along the fibre direction may be possible, but they will not propagate during loading, since they are parallel to the loading direction. The smaller flaw size is responsible for the high strength of fibre reinforcements.

The flexibility of any material is controlled by its elastic modulus and the moment of inertia of its cross-section. For any specific material, the modulus value is constant and the flexibility can be modified by changing the dimensions. The inverse of product of bending moment and radius of curvature is a measure of flexibility. According to the bending theory of circular rod, the flexibility can be written as:

$$\frac{1}{MR} = \frac{64}{E\pi d^4} \tag{6}$$

where, M is bending moment, R is the radius of curvature, E is modulus and d is diameter. Since the fibres can be considered as fine rod like material, this theory is applicable to them also. From this equation, it is evident that the flexibility is a sensitive function of diameter. Hence, a flexible fibre can be obtained from any brittle material by reducing the diameter. The glass and other ceramic fibres are flexible because of this reason. The diameter for various fibre forming materials to have equal flexibility as that of 25 μm polyamide fibre is shown in Fig. 2. The flexibility is very essential during the manufacturing of composites with complex shapes.

In a composite material, the reinforcements have better mechanical properties compared to the matrix. Hence, they are the load bearing members. For the effective transfer of load to the reinforcements, the aspect ratio should be very high. Fibrous materials generally have high aspect ratio because of smaller diameter and higher length. When a tensile load is applied to the composite material, both the fibre and

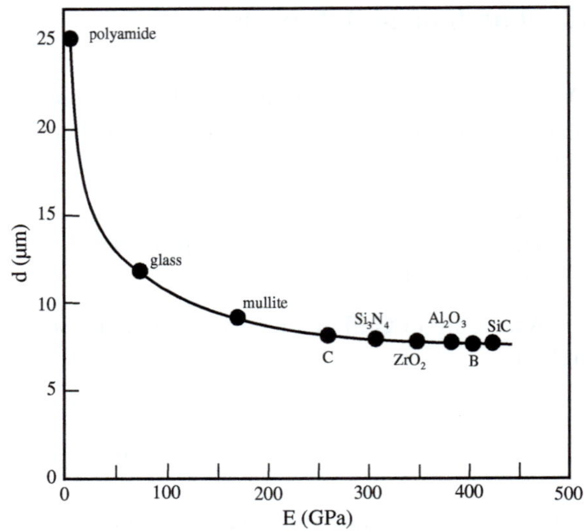

Fig. 2 Diameter of fibre from different materials to have flexibility equal to that of a 25 μm diameter polyamide fibre [2]

matrix will try to deform. Since the modulus of fibre is generally high, there will be less deformation in the fibre. When the bonding between fibre and matrix is good, the fibre will prevent the excessive deformation of matrix. That means the load is now transferred to the fibre. The load transfer is not effective at fibre ends, since the fibre will try to separate from the matrix. The load transfer to the fibre will increase from the fibre end towards the centre of fibre. The applied load transferred is reaching a maximum after certain length. The length at which the maximum applied load is transferred to the fibre is called load transfer length. This length varies with the applied load. A load transfer length, which is independent of applied load, is defined as critical fibre length. It is based on the applied load equivalent to the failure load of fibre. When the fibre length is equal to critical fibre length, the maximum load transfer occurs at the fibre midpoint. This load transfer behaviour for various fibre lengths is illustrated in Fig. 3. To have maximum load transfer at large area of fibre, the fibre length should be at least 15 times greater than the critical fibre length.

Even with continuous fibres, it may not be possible to get the reinforcement effect when the fibre content is less than a minimum. Only above certain minimum fibre volume fraction, the strength of a composite starts increasing with increasing fibre content. Above this minimum fibre volume fraction, the composite failure is controlled by the fibre failure. The composite will fail immediately after the failure of fibre. Thus, the composite strength can be written as:

$$\sigma_{cu} = \sigma_{fu} V_f + \sigma_m^* \left(1 - V_f\right) \quad (7)$$

where, σ_{fu} is the failure strength of fibres and σ_m^* is the stress in the matrix at the fibre failure strain. When the fibre content is less, the failure of the composite is controlled by the matrix failure. Even after the failure of all fibres, the matrix can

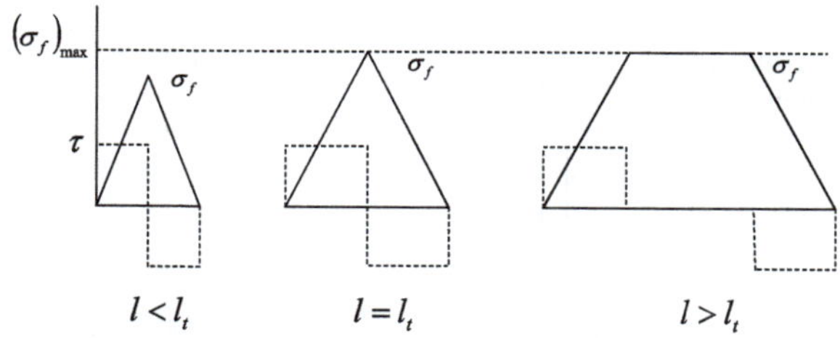

Fig. 3 Variations of fibre stress and interface shear stress for different fibre lengths [3]

take the load and continue to deform. The composite will fail after the failure of the matrix. At this condition, the failure strength of the composite is given as:

$$\sigma_{cu} = \sigma_{mu}(1 - V_f) \tag{8}$$

The minimum fibre content, at which the failure of composite is controlled by fibre failure is called the minimum fibre volume fraction. It can be given as:

$$\sigma_{cu} = \sigma_{fu}V_f + \sigma_m^*(1 - V_f) \geq \sigma_{mu}(1 - V_f) \tag{9}$$

Hence,

$$V_f,\ \text{i.e. } V_{\min} = \frac{\sigma_{mu} - \sigma_m^*}{\sigma_{fu} + \sigma_{mu} - \sigma_m^*} \tag{10}$$

The critical fibre volume fraction can be defined as the volume fraction above which the composite failure strength is greater than matrix ultimate strength. That is:

$$\sigma_{cu} = \sigma_{fu}V_f + \sigma_m^*(1 - V_f) \geq \sigma_{mu} \tag{11}$$

$$V_f,\ \text{i.e. } V_{crit} = \frac{\sigma_{mu} - \sigma_m^*}{\sigma_{fu} - \sigma_m^*} \tag{12}$$

The minimum and critical fibre volume fractions are illustrated in Fig. 4. As mentioned earlier, the properties of composites are mainly controlled by the relative proportions of the constituents. Once the concentrations are known, it is possible to predict the property of composite based on the properties of constituents. The prediction of Young's modulus of a fibre reinforced composite is described below.

Consider a fibre reinforced composite material, in which all the fibres are aligned in a single direction. A tensile load of F_c is applied to this composite in the fibre direction. This load will be shared by the fibre and matrix.

$$F_c = F_f + F_m \tag{13}$$

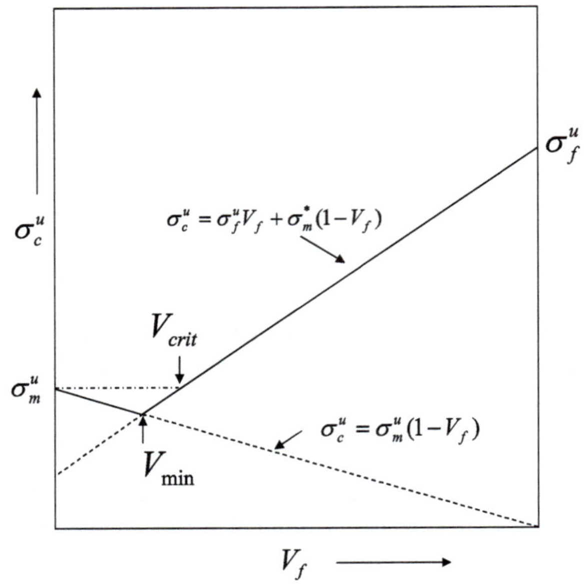

Fig. 4 Tensile strength of unidirectional composite as a function of fibre volume fraction [4]

Since stress, $\sigma = F/A$, hence $F = \sigma A$

$$\sigma_c A_c = \sigma_f A_f + \sigma_m A_m \tag{14}$$

where, σ_c, σ_f and σ_m are stresses in the composite, fibre and matrix, respectively and A_c, A_f and A_m are the cross-sectional areas of the composite, fibre and matrix, respectively.

On dividing the Eq. 14 by the cross-sectional area of the composite, then

$$\sigma_c = \sigma_f \frac{A_f}{A_c} + \sigma_m \frac{A_m}{A_c} \tag{15}$$

For a given length, the area fraction is equal to the volume fraction. Therefore,

$$\sigma_c = \sigma_f V_f + \sigma_m V_m \tag{16}$$

Assuming that the constituents are undergoing elastic deformation and obey Hooke's law, i.e.

$$\frac{\sigma}{\varepsilon} = E \tag{17}$$

Then,

$$E_c \varepsilon_c = E_f \varepsilon_f V_f + E_m \varepsilon_m V_m \tag{18}$$

When the bonding between fibre and matrix is perfect, then $\varepsilon_c = \varepsilon_f = \varepsilon_m$ (iso-strain condition)

So,

$$E_c = E_f V_f + E_m V_m \tag{19}$$

When the porosity of the composite is negligible, then $V_m = 1 - V_f$
Now,

$$E_c = E_f V_f + E_m (1 - V_f) \tag{20}$$

This is the rule of mixtures to predict the modulus of a composite in the fibre direction.

When the load is applied transverse to the fibre direction, the stresses in the composite, fibre and matrix are equal for a given length, i.e. $\sigma_c = \sigma_f = \sigma_m$ (iso-stress condition). The deformation in the composite is the summation of deformations in the fibre and matrix.

$$\delta_c = \delta_f + \delta_m \tag{21}$$

Since, $\varepsilon = \frac{\delta}{t}$, where t is thickness.

$$\varepsilon_c t_c = \varepsilon_f t_f + \varepsilon_m t_m \tag{22}$$

Divide by t_c, then

$$\varepsilon_c = \varepsilon_f \frac{t_f}{t_c} + \varepsilon_m \frac{t_m}{t_c} \tag{23}$$

For a given cross-sectional area of the composite, $\frac{t_f}{t_c} = V_f$ and $\frac{t_m}{t_c} = V_m$
Therefore,

$$\varepsilon_c = \varepsilon_f V_f + \varepsilon_m V_m \tag{24}$$

Again by assuming elastic deformation of constituents, $\varepsilon = \frac{\sigma}{E}$, then

$$\frac{\sigma_c}{E_c} = \frac{\sigma_f V_f}{E_f} + \frac{\sigma_m V_m}{E_m} \tag{25}$$

Since, the stresses are equal,

$$\frac{1}{E_c} = \frac{V_f}{E_f} + \frac{V_m}{E_m} \quad or \quad \frac{1}{E_c} = \frac{V_f}{E_f} + \frac{(1 - V_f)}{E_m} \tag{26}$$

This is the equation to predict the modulus in the transverse direction. The properties of real composites lie in between the values predicted by Eqs. 20 and 26.

3 Types of Matrix

As mentioned earlier, there are three major classes of matrix materials. They are polymeric, metallic and ceramic matrix materials. The majority of the composites produced today are based on polymer matrices, since they are relatively cheap and easily processable.

3.1 Polymer Matrix

Polymer is a macromolecule consisting of one or more units repeating in large numbers. For example, the very common polymer, polyethylene is made-up of hundreds of ethylene ($–CH_2–CH_2–$) repeating units. A polymeric material is the collection of many polymeric molecules of similar chemical structure. The polymeric chains are either randomly arranged or randomly arranged with some chain orientation in a polymeric material. The former is an amorphous polymer and the latter is a semicrystalline polymer. The same polymer can be obtained in the amorphous state or semi-crystalline state depending on the processing conditions. The polymers are further classified into thermosetting and thermoplastic polymers. In a thermosetting polymer, the polymer molecules form a three dimensional network structure with strong covalent bonds between the chains. Once this type of network structure is formed, it is very difficult to deform the material. It is similar to cement setting. In the context of composites, thermoset polymers have many advantages and hence they are widely used compared to thermoplastics. Some of the advantages are:

- easy fabrication of composites, since they are available in liquid form
- good wetting with the reinforcements
- high thermal stability
- good creep resistance
- good chemical resistance

However, there are a few disadvantages also; which are:

- high brittle nature
- long fabrication time
- relatively short storage life

In a thermoplastic material, the polymer molecules are held together by weak van der Waals forces. Hence, by applying heat and presence, it is easy to deform a thermoplastic material. That means, it is possible to reuse it. Hence, it is not necessary to fabricate the final shape by a single stage processing. Some regular shapes can be made in bulk and then they can be deformed to the final shape at a later date. Apart from this advantage, there are some additional advantages also. The fabrication time with thermoplastics is less and they have good toughness. The thermoplastic materials can be stored for longer time. Even with these advantages, the thermoplastics are not commonly used because of the requirement of high processing temperatures, poor wettability to reinforcements and high melt viscosity. These problems are addressed to some extent and thermoplastic composites are now becoming popular. The details of commonly used polymer matrix materials are given in the following sections.

3.1.1 Polyester Resin

This is the most commonly used polymeric material in composites. It is a thermosetting polymer. This is unsaturated polyester with a number of double bonds in the chain. An ester is an organic compound formed between an acid and alcohol. When there are two reactive groups in each constituent, they can form polyester. These resins are generally prepared using an unsaturated acid and alcohol. The number of unsaturated groups present in the polyester resin can be modified by introducing a saturated acid. A variety of polyester resins is available depending on the type of unsaturated acid, saturated acid and alcohol. The unsaturated polyester resin is generally dissolved in a reactive diluent, such as styrene to reduce the viscosity and facilitate the easy fabrication of composites. The common varieties of polyester resins are orthophthalic, isophthalic, and terephthalic resins.

The liquid resin transforms to a solid after the curing reaction. These resins can be cured at room temperature or high temperature by suitably selecting a catalyst. An accelerator is also added to increase the speed of reaction. During curing, the double bond is attacked by the catalyst and form a reactive group. This reactive group attacks the double bond present in styrene, which in turn attack the double bond of another polyester molecule. In this way cross-links are established between polyester molecules through styrene. This curing reaction is illustrated in Fig. 5. The properties of polyester resin depend on the cross-link density, which is controlled by the number of double bonds. A brittle product is obtained at higher cross-link density, whereas a flexible product is obtained at lower cross-link density. UV resistant additives and fire retardants are also added to the resin, depending on the applications. The required additives are added into the resin, just before the fabrication of composites and then the composite part is fabricated. In the presence of catalyst and accelerator, the resin will start solidify after a few minutes. The fabrication of the composite should be completed before the gelation of the resin. Polyester resin based composites are widely used in boat hulls, instrument covers, wind-mill blades, roofing, storage tanks, etc.

3.1.2 Epoxy Resins

Epoxy resins have better adhesive strength and chemical resistance than polyester resin. However, they are more expensive. Polymers containing epoxy groups are called epoxy resins. An epoxy group is a cyclic group formed by two or more carbon atoms and an oxygen atom. The reactivity of epoxy increases with

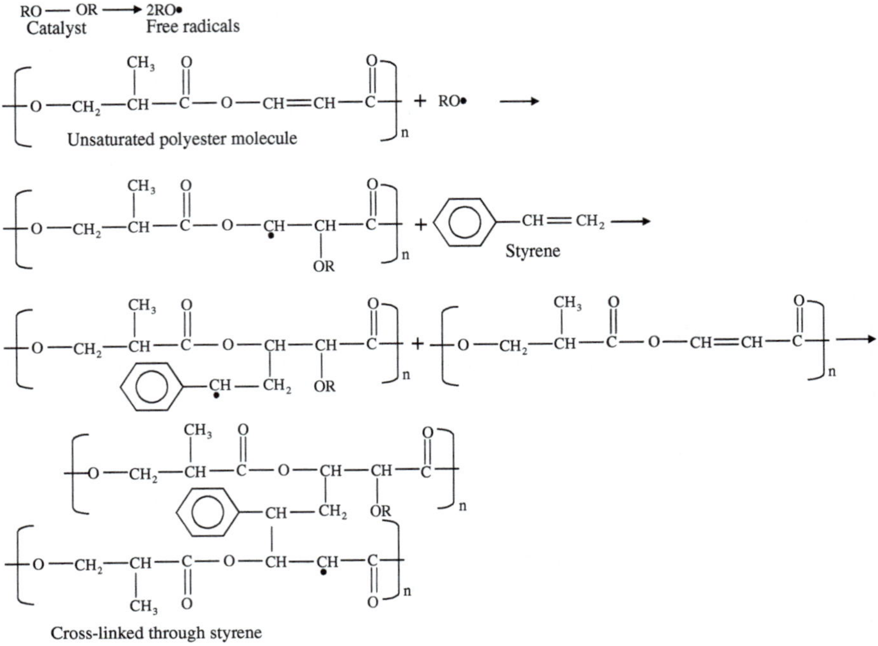

Fig. 5 Curing reaction of thermoset polyester resin [3]

decreasing number of carbon atoms in the epoxy group. The epoxy group consisting of two carbon atoms and one oxygen atom is more reactive. Like polyester resin, there are a variety of epoxy resins available in the market depending on the remaining organic groups. One of the very common epoxy resins is diglycidylether of bisphenol A (DGEBA) resin. There are aliphatic epoxy resins as well as aromatic epoxy resins. Generally, the aromatic epoxy resins have better mechanical and thermal properties than aliphatic epoxy resins.

Unlike polyester resins, hardeners are used to cure the epoxy resins. The hardeners become the part of network structure after curing. Hence, the properties of epoxy resins are also controlled by the nature of hardeners. There are a variety of hardeners available to cure the epoxy resin at room temperature or at high temperature. The amine curing agents are very commonly used with epoxy resin. Apart from these, acid anhydride curing agents are also available. The curing of epoxy resin proceeds in two stages. In the first stage, linear polymer chains are produced by the reaction of epoxy groups and amine groups. The cross-links are then formed by the reaction of remaining epoxy groups and secondary amine groups. The curing behaviour of epoxy resin is illustrated in Fig. 6. It is possible to stop the curing of epoxy resin at the intermediate stage. Epoxy resin based prepregs (pre-impregnated fibres) are made by stopping the curing at the intermediate stage (B-stage). The prepregs are generally made with high temperature curing systems. After certain curing, the curing reaction is stopped by lowering the temperature. To avoid the

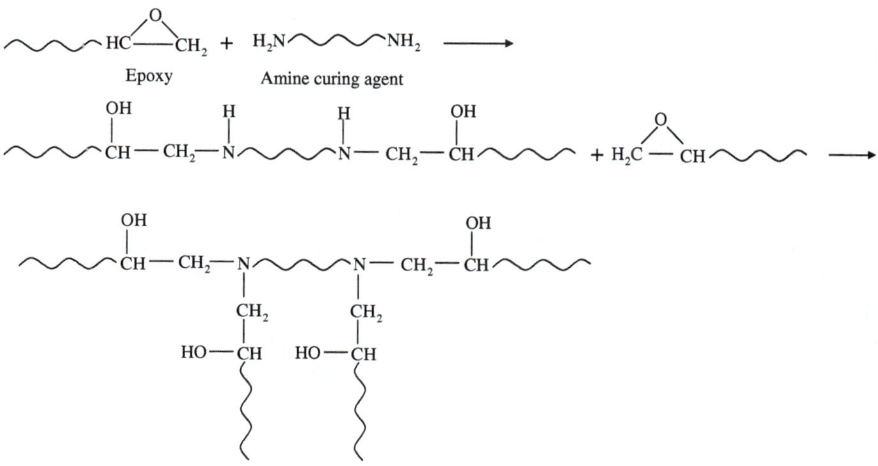

Fig. 6 Curing reaction of epoxy resin [5]

progression of curing reaction, the prepregs need to be stored at −18 °C. At this temperature, it is possible to store the prepregs up to 6 months. Prepregs are made in the form of thin tapes. The composite part is fabricated with this prepreg and then cured at high temperature, generally in an autoclave. The required fibre content and orientation are easily achieved with these prepregs and hence composites with the desired properties can be fabricated. Since the epoxy resins are expensive, they are mainly used in aerospace and sporting goods applications.

3.1.3 Vinyl Ester Resins

They have intermediate properties between epoxy and polyester resins. The curing behaviour is similar to polyester resins. Since they have many –OH groups along the chain, the adhesion with glass fibre is good.

3.1.4 Phenolic Resins

They are based on phenol formaldehyde. The reaction between phenol and formaldehyde leads to the formation of phenol formaldehyde. These resins show low fire spread and smoke emission. Hence, the composites based on phenolic resins are widely used where fire resistance is very important, such as under-ground railway lines and panels for low cost housing. The main problem with these resins is the volatile emission during curing. Hence, it is essential to cure the resin under pressure to avoid pore formation. There are two types of phenolic resins, depending on the phenol-formaldehyde ratio. When the ratio is less than 1, it is called resole and more than 1, it is called novolac resin. The high temperature properties of novolac resin are better than that of resole resin. The ablative composites used in re-entry vehicles are also made with phenolic resin. On high temperature ablation under inert conditions, this resin transforms into carbon matrix.

3.1.5 Polyimide Resins

Polyimide resins can be used up to 230 °C for long periods and up to 315 °C for short durations. Apart from high temperature resistance, they also offer high chemical and solvent resistance. However, these resins are inherently brittle and their composites are prone to microcracking. This problem can be solved to some extent by the incorporation of a thermoplastic polyimide. They form a semi-interpenetrating network, which leads to better toughness. Some of the examples of polyimide resins are bismaleimides, PMR-15 (polymerisation of monomer reactants) and acetylene terminated polyimide.

3.1.6 Thermoplastics

Thermoplastics have good toughness and resistance to damage from low velocity impacts due to their high ductility. Most of the thermoplastics are used in commodity applications, where high mechanical performance is not needed. However, there are a few thermoplastics with good mechanical properties and high temperature resistance. High performance thermoplastics contain aromatic rings that impart rigidity and high temperature stability to these materials. The incorporation of reinforcements significantly improves the mechanical properties and creep resistance. Some of the thermoplastic matrix materials used in composites meant for room temperature applications are polyamides, polypropylene and polyethylene terephthalate. Polyether ether ketone, polyphenylene sulphide, polysulphone and thermoplastic polyimides are the examples of high performance thermoplastic matrices.

3.2 Metal Matrix

Metal matrix includes metals and alloys. Light weight metals and alloys based on aluminium, magnesium and titanium are generally used in composites. Metals have more toughness and hardness than polymeric materials. Moreover, they can be used at relatively higher temperatures. However, the fabrication temperatures of metal matrix composite are high (more than 600 °C). Because of this, the fabrication of MMCs with continuous fibres is very difficult or more expensive. It is not possible to use the natural and synthetic organic fibres with metal matrix. High temperature resistant fibres, such as carbon, silicon carbide and alumina fibres can only be used, but they are very expensive. Even the glass fibres cannot be used with metal matrix, since they will deform at the processing temperature. Ceramic particulates and short fibres, including whiskers are generally used in MMCs. The metal matrix should have good wettability with the reinforcements, but there should not be any adverse reaction between them. This is very critical during the fabrication of composites. Many MMCs are commercially produced now and they are replacing many metallic parts and also some PMCs. Most of the MMCs are based on aluminium, magnesium and titanium alloys and some intermetallic materials. The details of these matrices are given in the following sections.

3.2.1 Aluminium Alloys

They are the most commonly used metal matrices. These alloys have very good corrosion resistance and mechanical properties. They are light in weight and relatively inexpensive compared to other matrices used in MMCs. Aluminium-silicon alloys are used when the composite is fabricated by melt route and aluminium-copper or aluminium-magnesium alloys are used when the composite is fabricated by powder metallurgy technique. Alumina, silicon carbide and fly ash particles are generally used for the fabrication of composites with these alloys. Many in situ composites with TiC/TiB_2 particles are also processed with these alloys.

3.2.2 Magnesium Alloys

This is one of the lightest metals with the density of 1.74 g/cm^3. Since magnesium has hexagonal close packed structure, it is very difficult to deform this material at room temperature. Great care should be taken during the fabrication of composites with these alloys because they undergo vigorous reaction with oxygen and sometimes trigger explosion.

3.2.3 Titanium Alloys

The aluminium and magnesium based composites are useful for room temperature or slightly above room temperature applications. When the service temperature is more than 300 °C, these matrices may not be suitable. Since the melting point of titanium is 1672 °C, it can be used up to 1000 °C. For example, the skin temperature of supersonic aircraft may go beyond 300 °C. Aluminium alloy based composites are not suitable for this application, whereas titanium alloy based composites can be used. Titanium exists in two crystalline forms, viz. α and β forms. The α-form has hexagonal structure and is stable up to 885 °C, whereas the β–form has cubic structure, which is stable above 885 °C. The alloying elements can modify this transformation temperature. Aluminium stabilises the α-form and hence the transformation temperature is increased beyond 885 °C. Elements like V stabilise β-form and thus the transformation temperature is reduced. A very common titanium alloy is Ti-6Al-4V alloy. This alloy has a mixture of α and β forms. Although the corrosion resistance of Ti alloys is good, they have great affinity to gases like O_2 and H_2 at high temperatures. Hence, the processing of composites should be carried out under highly inert conditions.

3.2.4 Intermetallic Compounds

They are the compounds formed between metals. The main difference between an alloy and an intermetallic compound is that the alloy can have variable composition and the intermetallic compound should have a fixed composition. Like alloys they have very good thermal and electrical conductivity, but their ductility is very poor. The ductility can be improved by forming glassy intermetallics through rapid cooling or by adding elements like boron. Another way of improving toughness is to form composites. The intermetallics can retain the strength at high temperatures and in some cases an improvement of strength at high temperature is also observed. Some of the intermetallics used as matrices are NiAl, TiAl, etc.

3.3 Ceramic Matrix

Ceramic materials are the compounds formed between metals and non-metals. However, only those compounds, which have better mechanical properties and high temperature resistance, are considered as ceramic materials. Generally, they are the oxides, carbides and nitrides of metals. The melting points of ceramic materials are generally high and they can retain the mechanical properties at high temperatures. They have very good oxidation and chemical resistance. Since the modulus and strength values of ceramic materials are high, the main purpose of forming composites is to improve the fracture toughness.

Unlike the metallic materials, the ceramic materials are iono-covalent compounds, i.e. the ions or atoms are bonded with strong ionic and covalent bonds. Hence, the number of slip systems is restricted, and that is responsible for the low ductility of ceramic materials. Since the ceramic materials are processed at high temperatures, the formation of flaws is unavoidable. During loading those cracks propagate and lead to catastrophic failure. The presence of reinforcements in a ceramic material can prevent the propagation of cracks in many ways and thus increases the fracture toughness. Some of the toughening mechanisms operate in CMCs are crack deflection, crack bridging, fibre pull-out and transformation toughening. When the reinforcements have higher fracture toughness, they will not allow the crack to propagate through them. Either the crack bows out or is deflected. In this way, the crack propagation is delayed. When fibrous reinforcements are present in a ceramic matrix, they can bridge the crack faces and thus preventing the easy crack propagation. A tailored interface will allow the pull-out of fibrous reinforcements. Hence, a part of energy utilised for this process and the crack propagation is delayed. The transformation toughening mechanism operates in zirconia toughened ceramics. Zirconia exists in three crystalline forms, viz., cubic, tetragonal and monoclinic forms. The transformation from tetragonal to monoclinic form involves a volume expansion of about 4 %. When the tetragonal zirconia particles are present in a ceramic matrix, they can transform to monoclinic form during loading. This generally happens near the crack tip, since the stress concentration is high in the region. The transformation associated with volume expansion will try to close the crack, thus preventing easy propagation.

The interface engineering is very important in ceramic matrix composites. The interface should not be very strong or very weak. To have a tailored interface, many surface modification procedures are adopted. Some of the matrix materials used in CMCs are alumina, silicon carbide, silicon nitride, glass and cement, and the details about these materials are given in the following sections.

3.3.1 Alumina

Alumina or aluminium oxide is one of the most widely used advanced ceramic materials. It exists in various forms like γ, θ, δ and α. The α-alumina is the stable form and the other forms are transitional aluminas. The other forms of alumina can be transformed into α–alumina by heating at high temperature. Alumina has a melting point of 2020 °C. The α-form has hexagonal structure and the other forms have cubic structure. It also has high hardness and strength. It is generally produced from bauxite ore. It has very good chemical resistance even at high temperature. The fracture toughness of this material is around 3 $\text{MPa}\sqrt{\text{m}}$. Silicon carbide fibres and whiskers and zirconia particles are generally used to improve the fracture toughness of this material.

3.3.2 Silicon Carbide

SiC exists in two crystalline forms. The α–SiC is hexagonal, whereas the β-SiC is cubic. This material also has high hardness and strength. The thermal conductivity of SiC is also high. Depending on the processing conditions, it exists in either α-form or β-form. Bulk quantity of SiC is produced by the carbothermal reduction of silica using carbon in an arc furnace. This is one of the common abrasive materials used in grinding wheels and abrasive papers. This material also has low fracture toughness. SiC fibres and whiskers are generally used as reinforcements for this material.

3.3.3 Silicon Nitride

Apart from high hardness and strength, this material also has low thermal expansion coefficient. It also has good oxidation resistance at high temperatures. Because of its high thermal stability, many engine components are made with this material. SiC fibres and whiskers are generally used as reinforcements. Various methods like compaction and sintering, hot pressing, hot isostatic pressing are used to product silicon nitride ceramic components. Reaction bonding technique using silicon is another attractive method to prepare ceramic parts with this material.

3.3.4 Glass

Glass matrix can also be used to form CMCs. The brittleness of glass can be reduced by forming a composite. Carbon, SiC, alumina fibres are generally used as reinforcements to glass. The low temperature requirement for the processing of composite is the main attraction for glass matrix. Various glasses can be prepared by varying the composition of oxides.

3.3.5 Cements

The main advantage of cement matrix is that the composite can be formed at room temperature. Hence, glass fibres, organic fibres and plant fibres can be used as reinforcements. After the ban of asbestos fibres, cement sheets are currently produced using natural fibres in many countries.

4 Types of Reinforcement

There are a variety of reinforcements available for the fabrication of composites. Depending on the application and cost, one can select a suitable reinforcement. The reinforcements can be broadly classified into natural and synthetic reinforcements. The natural reinforcements can be further divided into inorganic and organic reinforcements. Many mineral fibres come under the category of inorganic reinforcements. The plant and animal based fibres are organic reinforcements. The natural reinforcements are relatively cheap. The plant based fibres are also environmental friendly. However, there are many disadvantages with natural fibres. Some of the mineral fibres are carcinogenic. The plant fibres absorb moisture and degrade over a period of time. Natural fibres are not available as continuous fibres. The mechanical properties of natural fibres are generally inferior to the synthetic fibres and also there is wide variation. Some of the inorganic fibrous materials are asbestos and wollastonite. The usage of asbestos is banned in many countries. The other mineral fibres are not effective as the synthetic fibres.

Animal hair is another category of natural fibres. Although they are used in protective clothing, they are not commonly used in composites. There are a variety of plant fibres available throughout the world. Some of the common plant fibres are jute, coir, sisal, cotton, etc.

The plant based natural fibres are one among the fastest growing types of reinforcement for polymeric materials. They are finding increasing use in composites for automotive applications, since the cost is 30 % less than glass fibres. Moreover, the natural fibres weigh only about 50 % of glass fibres for the same volume and the composites made with natural fibres are easier to dispose. The interior parts of automobiles, furnitures, doors and flower pots are some of the common products made with natural fibre composites.

Among the plant based natural fibres, wood fibres are widely used in North America. The cost of wood fibre is less than that of widely used fillers, such as calcium carbonate. Natural fibres can also be obtained from the stem, leaf or seed of certain plants. The fibres obtained from stem are called 'bast' fibres and a few examples of bast fibres are flax, hemp and jute. Flax fibre is produced largely in Canada and China, hemp is mainly produced in Philippines and the largest producers of jute are India, China and Bangladesh. Sisal and abaca are the common leaf fibres. Brazil and Tanzania are the two largest producers of sisal. Abaca fibre is also produced in large quantities by Philippines. The commonly used seed fibres are cotton and coir. China, India and USA are the leading producers of cotton. The major producers of coir fibre are India and Sri Lanka. More information on the natural fibres can be found in the Chap. 5.

The synthetic fibres are high performance fibres with very high mechanical properties. They're available in various forms for the fabrication of composites. Some of the forms are continuous fibre rovings, chopped strand mat and woven rovings. There are a variety of synthetic fibres available in the market. Depending on the property requirement one can select the appropriate fibre. The commonly used synthetic fibres are glass, carbon, aramid, polyethylene, alumina and silicon carbide. Among the synthetic fibres, the glass fibre is widely used in composites because of its relatively lower cost.

4.1 Glass Fibres

Glass is an amorphous material. It is normally based on silica. Other oxides are added to silica to modify the properties. Depending on the composition, the properties will vary. The common glass fibre types are E-glass, S-glass, C-glass, A-glass and Z-glass. The E–glass is electrically insulating glass and originally used for electrical insulation. This is the very commonly used glass fibre in composites. The S-glass is high strength glass fibre. The C-glass has high chemical resistance and A-glass has higher alkali resistance.

The chemical compositions of these glass fibres are given in Table 1. Glass fibres are available in various forms for the fabrication of composites [6]. They are continuous roving, woven roving, chopped strand mat, chopped strand, yarn and fabrics. During the production of glass fibres, the fibres are collected as strands consisting of around 200 individual filaments with size coating. The rovings are a collection of continuous strands. Woven roving mats are produced by weaving the rovings.

Table 1 Composition of various glass fibres

Constituents	Composition (wt%)				
	E glass	S glass	C glass	A glass	Z glass
SiO_2	52.0–53.0	64.0–65.0	65.0	72.5	60.0
Al_2O_3	14.0–15.0	25.0–26.0	4.0	0.7–1.5	–
B_2O_3	8.0–10.0	–	6.0	–	–
MgO	4.5	10.0	3.0	2.5	–
CaO	17.5	–	14.0	10.0	–
Na_2O/K_2O	0.5	–	8.0	13.5–14.0	20.0
Fe_2O_3	0.4	–	–	–	–
SO_3	–	–	–	0.7	–
TiO_2	–	–	–	–	5.0
ZrO_2	–	–	–	–	15.0

Chopped strands are produced by cutting the strands to 5–50 mm length and the chopped strand mats are produced by spraying these short strands with a binder over a flat conveyor. Yarns are twisted fibres with better integrity and the fabrics are made by the conventional weaving of these yarns.

4.2 Carbon Fibres

Carbon is a light element with the density of 2.2 g/cm^3. It exists in three crystalline forms, viz. diamond, graphite and fullerene. The diamond form has cubic structure with strong covalent bonds in all three directions. This is the hardest material on earth. Graphite has hexagonal structure with covalent bonds in the plane and weak van der Waals forces between the planes. Because of these different kinds of bonding, it has anisotropic properties. The modulus value along the plane is about 1000 GPa, whereas it is only 35 GPa in the perpendicular direction. Fullerene has bulky ball structure with 60 or 70 carbon atoms. The graphitic form of carbon is present in carbon fibre. Hence, it is necessary to align all the planes along the fibre direction to achieve better properties. Carbon fibre is prepared from polymeric precursor fibres, which are special textile fibres that can be carbonised without melting. A commonly used precursor fibre is polyacrylonitrile fibre and the other precursor fibres are rayon and the fibres obtained from pitch and phenolics. After spinning the precursor fibre, it is stabilised by heating in air at 250 °C. The stabilised fibres are carbonised at 1000–1400 °C under inert atmosphere. During carbonisation all other elements, except carbon leave the fibre. The resulting carbon fibre may not have good mechanical properties. To achieve better mechanical properties, a graphitization treatment is carried out, which involves heating at 2000–2200 °C with stress under argon atmosphere. The regular arrangement of lamellar planes along the fibre direction happens during this treatment.

Although the preparation of carbon fibre looks simple, each and every parameter during each step of processing should be controlled to the optimum level. The processing details are kept as closely guarded secret and only a few companies mastered the art of preparation of carbon fibres. However, the properties of carbon fibre vary from one manufacturer to another. Hence, a variety of carbon fibres with different properties is available in the market. Like glass fibres, carbon fibres are also available in various forms for the fabrication of composites. Carbon fibres have very high modulus value compared to glass fibres. Instead of converting to carbon fibre, some people tried to improve the properties of polymeric fibre itself by playing with the molecular structure. The result is the birth of polyethylene and aramid fibres.

4.3 Polyethylene Fibre

Polymer molecules in a polymeric material are neither aligned nor stretched. They assume random coil configuration. Although the atoms in a polymer molecule are covalently bonded, the individual molecules are held together only by weak van der Waals forces. This is responsible for the poor mechanical properties of polymeric materials. A small force is enough to overcome the intermolecular bonding forces, thus leading to stretching and alignment of polymer molecules. Once the polymer molecules are completely stretched and aligned, it is very difficult to deform further. This is the basis for the development of high performance polymeric fibres.

The stretching and orientation of polymer molecules depend on the draw ratio (the ratio of original diameter to final diameter). It may not be possible to apply any draw ratio to polymeric materials. The maximum draw ratio depends on molecular structure and drawing conditions, like temperature and strain rate. The polymer molecules should stretch and align without the formation of defects during the drawing process.

Polyethylene is a very common polymer with poor mechanical properties. The modulus value is less than 10 GPa and the strength hardly exceeds 50 MPa. Very high mechanical properties can be realized in this polymer by stretching and aligning. Polyethylene fibres with the modulus value of 120 GPa and strength of more than 2000 MPa are available in the market. High performance polyethylene fibres are prepared by gel-spinning process. Gels are swollen network, in which the crystalline regions present at the junctions. Polyethylene gel fibre is prepared from a dilute solution of polyethylene using the appropriate spinning process. This gel fibre is subjected to drawing at high temperature to form the polyethylene fibre. The crystalline regions present at the junction facilitate the full stretching of polymer chains. The specific gravity of polyethylene fibre is less than 1. Apart from very high modulus and strength, this fibre also has very good damping properties, which is useful for shock absorption.

4.4 Aramid Fibre

This is also a polymeric fibre. Aramid is the short form of aromatic polyamide. These aromatic polyamides form rigid rod like structure. Stretching may not be needed for these molecules, but molecular alignment is needed to achieve better mechanical properties. Under certain conditions these aramids form liquid crystals in an appropriate solvent. In the liquid crystal state, a group of polymer rods align in the same direction. However, different groups are oriented in different directions. The formation of liquid crystal state can be observed using polarised light microscope.

Because of the rigid rod like structure, these polymers are very difficult to deform and also difficult to dissolve in common solvents. The solvent suitable for dissolving aramid is 100 % sulphuric acid. The viscosity of this solution shows an abnormal behaviour after certain concentration. The viscosity of this solution initially increases with increasing concentration of aramid. After certain concentration, there is a sudden drop in viscosity. This is due to the formation of liquid crystal state. The liquid crystal regions act like a dispersoid. The solution at the liquid crystal state is used for the preparation of fibres, by dry-jet wet spinning process. It is necessary to maintain the liquid bath at low temperatures (−4 °C). It is a well-known fact the concentration of solution increases with temperature. Hence, by keeping the spinneret 1 cm above the liquid bath a higher temperature can be maintained in the solution. The higher the concentration, the higher will be the yield. Because of the air gap, this process is called dry-jet wet spinning process. The aramid fibres are also produced by a few companies and commercially available in the name of Kevlar, Technora, Nomex, etc. Apart from high modulus and strength, these fibres also have good vibration damping properties. Bullet proof clothes for low enforcing agencies throughout the world are made using these fibres. The major disadvantage of these polymeric fibres is their poor high temperature stability. It is not possible to use these fibres in composites intended for high temperature applications.

4.5 Ceramic Fibres

For the composites used above 1000 °C, only ceramic fibres are suitable. They have high temperature stability and can retain their mechanical properties at high temperatures. The commonly used ceramic fibres are silicon carbide and alumina based fibres. Silicon carbide fibres are prepared by chemical vapour deposition (CVD) and polymer pyrolysis processes. The chemical vapour deposition of SiC is carried out on carbon or tungsten substrate fibres in a CVD reactor. Methyl tricholorosilane (MTS) is an ideal raw material to deposit SiC, since it contains only one carbon and one silicon in its structure. This MTS undergoes reduction reaction in the presence of hydrogen gas. The flow of hydrogen should be optimum for the formation of stoichiometric SiC. The fibres produced by CVD process have less flexibility because of large diameter ($\sim$150 μm). However, these fibres have good oxidation and creep resistance. The SiC fibres having good flexibility are produced by polymer pyrolysis of polymeric precursor fibres. The process is almost similar to that used for carbon fibre production and the major difference is in the precursor fibre. A polymer containing silicon in its backbone structure is used as the starting material. This polymer on pyrolysis at 1400 °C forms SiC. The oxidation resistance of this fibre is poor compared to CVD derived fibre. This is due to oxygen pick-up during stabilisation treatment and the presence of residual carbon and silicon. The oxidation resistance of present SiC fibres is significantly high because of the better control of processing parameters.

Alumina based ceramic fibres are generally produced by sol-gel process. Sol is one of the colloidal systems, in which fine solid particles are dispersed in a liquid medium. A gel is the interpenetrating network of solid particles with the liquid present within the network. The processes involving sol and gel stages are called sol-gel process. A sol can be converted to gel by the evaporation of liquid or by controlling the pH. Metallo-organic compounds are used as the starting materials to prepare these oxide fibres. For example, aluminium isopropoxide or secondary butoxide is the starting material for alumina and tetraethyl orthosilicate is the starting material for silica. The required amount of the respective metallo-organic compounds are dissolved in the appropriate solvents separately and hydrolysed in the presence of catalysts to form sols. The required quantities of sols are mixed depending on the final composition. Alumina based fibres are prepared from 100 % alumina to 40 % alumina with other oxides like silica. Either the sol or the gel can be used to prepare the fibre. The gel fibre is converted to ceramic fibre by controlled drying and calcination. Since the solid content of sol/gel is very less, there is very large shrinkage during the conversion. Hence, it is essential to heat the gel fibre at a very slow heating rate to avoid fibre breakage. These oxide fibres are mainly used in metal matrix composites. More information on the synthetic fibres can be found in Chap. 4.

4.6 Whiskers

Whiskers are elongated single crystals. They are available as short fibres. Since the whiskers are almost defect free, it is possible to realise the mechanical properties close to theoretical values. Although carbon and silicon nitride whiskers are available, silicon carbide whiskers are the commonly used reinforcements in composites. The SiC whiskers are generally prepared by the carbothermal reduction of rice husk or by vapour-liquid-solid (VLS) process. Rice husk is the waste material obtained during polishing of rice grains. It contains an intimate mixture of silica and organic compounds. The cleaned rice husk is coked at 700 °C under inert conditions to convert the organic compounds into carbon. The temperature is then raised to 1400 °C. At this temperature, the carbon reacts with silica and form SiC particulates and whiskers. The residual carbon and silicon carbide particles are removed from the SiC whiskers by various treatments. The aspect ratio of whiskers formed by this process is low and also there is wide variation in properties.

Good quality SiC whiskers with high aspect ratio are prepared by the VLS process. The VLS stands for the vapour feed gases, liquid catalyst and solid whiskers. Transition metal particles of size $\sim$30 μm are used as catalyst. These metallic particles are taken on a carbon substrate and heated to 1400 °C in a reaction chamber. SiO and carbonaceous gases are fed into the chamber in a controlled flow rate. The catalyst particles melt at this temperature and absorb SiO and C gases. These react and form SiC. Once it is supersaturated, SiC is precipitated out as whiskers. These whiskers can have strength as high as 20 GPa.

SiC whiskers can be used as reinforcements in MMCs and CMCs. However there are many problems in using whiskers. High mechanical properties in the composites can be realised only when the whiskers are aligned in the loading direction. The alignment of whiskers is a major problem. Another problem is the wide variation of properties. Since the whisker diameter is at the submicron level, they have the tendency to form agglomerates. The uniform dispersion is another major hurdle. SiC whiskers are carcinogenic in nature. Since they are very fine, whiskers can go into the lungs and deposit there, which can cause cancer. Hence, it is necessary to take appropriate precautions while handling the whiskers.

4.7 Nanofibres

When the diameter of a fibre is less than 100 nm, then it is called nanofibre. Nanofibres based on alumina, silicon carbide and carbon are available now. Even the carbon nanotubes (CNT) can be considered as nanofibrous materials. A method commonly used in the recent years to produce nanofibres is electrospinning process. An appropriate solution coming out of a nozzle is drawn into nanofibre by applying very high voltage between the nozzle and a metallic substrate. To prepare ceramic nanofibres, the corresponding polymeric precursor fibres are produced first and then they are converted to ceramic fibres by pyrolysis. CNTs are prepared by arc discharge and chemical vapour deposition methods. Although the yield is high in arc discharge method, CVD process is preferred to get high quality CNTs. There is no doubt that the nanofibres are better than micron size fibres, but there are many challenges with the nanofibres. At present the nanofibres are more expensive. The commercial production of nanofibres is yet to start. Although long nano fibres can be produced, continuous nanofibres are a distant dream. There are many issues on the uniform distribution of nanofibres in the matrix material, since the nanofibres have high tendency to form agglomerates. The compatibility between the nanofibre and matrix material is another issue to be addressed. More information on the nanofibres is available in Chap. 7.

Plant based natural fibres in general are suitable for plastics to improve their mechanical properties. Nevertheless, the wide variation in properties of natural fibres is a major drawback. The variation is mainly due to the fibre structure, which depends on the overall environmental conditions during growth. Hence, the natural fibres are preferred only for uncritical applications. Synthetic fibres are produced to specific

Table 2 Mechanical properties of natural fibres and synthetic fibres [7]

Fibre	Density (g/cm^3)	Elongation (%)	Tensile strength (MPa)	Young's modulus (GPa)
Cotton	1.5–1.6	7.0–8.0	287–597	5.5–12.6
Jute	1.3	1.5–1.8	393–773	26.5
Flax	1.5	2.7–3.2	345–1035	27.6
Hemp	–	1.6	690	–
Ramie	–	3.6–3.8	400–938	61.4–128.0
Sisal	1.5	2.0–2.5	511–635	9.4–22.0
Coir	1.2	30.0	175	4.0–6.0
Viscose (cord)	–	11.4	593	11.0
Soft wood kraft	1.5	–	1000	40.0
E-glass	2.5	2.5	2000–3500	70.0
S-glass	2.5	2.8	4570	86.0
Aramid (normal)	1.4	3.3–3.7	3000–3150	63.0–67.0
Carbon (standard)	1.4	1.4–1.8	4000	230.0–240.0

properties and also the variation is very less. The advanced composites, for which the properties are very critical, are only made with synthetic fibres. The properties of common plant based natural fibres and synthetic fibres are given in Table 2.

5 Manufacturing Techniques

There are a variety of techniques available to produce composites and some of the techniques are amenable for mass production. It may not be possible to produce complex shapes by using some techniques. Some of the techniques are more expensive to use. Depending on the application, cost and properties, an appropriate method can be selected. Generally, the MMCs and CMCs need higher temperatures for processing. Hence, the methods used to produce PMCs may not be suitable for MMCs and CMCs. The commonly used manufacturing methods for these three types of composites are explained briefly in the following sections.

5.1 Manufacturing of PMCs

The manufacturing methods of PMCs can be broadly classified into thermoset resin based and thermoplastics based methods. In the thermoset resin based methods, the composites are fabricated using the liquid resin. Hence, the fabrication can be carried out at room temperature. Thermoplastic composites are fabricated above the melting temperature of matrix material. The commonly used manufacturing methods with thermoset resins are hand lay-up, resin transfer moulding (RTM), resin infusion, autoclave processing, compression molding, filament winding and pultrusion. Injection moulding, reinforced reaction injection moulding, and thermoforming are the common methods used with thermoplastics. These manufacturing methods are briefly explained below.

5.1.1 Hand Lay-up

This is the most widely used method for the manufacturing of composites because of its simplicity. No expensive equipment is needed for this process. Even the mould will not cost much, since it can be made from wood, plastics and fibre reinforced plastics. Any complex shape can be made easily with this process. In this process, a mould is made according to the final shape of product. It can be a single piece mould or multi-segment mould depending on the complexity of the product. The mould surface should have good surface finish, since the surface finish on the product depends on this. Moreover, the release of the product after curing will be easy, when the surface finish is good. After that a thin release film is applied over the mould surface. Polyvinyl alcohol solution can be used to form the release film. A water solution of polyvinyl alcohol (PVA) forms a solid film over the surface on drying at room temperature. It may stick to the product, but it can be easily removed by putting the product in hot water for some time.

Once the release film is formed, a gel-coat layer is generally applied first. The gel-coat is made with the same resin used for the fabrication of composite. Some inert fillers, colouring agents and other special additives are added to the resin to prepare the gel-coat. The gel-coat layer is about 0.5 mm thick. Before the gel-coat is completely solidified (thumb impression should form while touching), the reinforcement mat is placed over the gel-coat. The resin with the curing agent is applied over the reinforcement using a paint brush. To ensure complete wetting of the reinforcement and to remove entrapped air a roller is rolled over the surface. Special kind of rollers for the fabrication of composites is available in the market. The required number of layers is laid one by one like this. The schematic of this process is shown in Fig. 7. This is a manual process. Hence, the quality of the product

Fig. 7 Schematic of hand lay-up process [8]

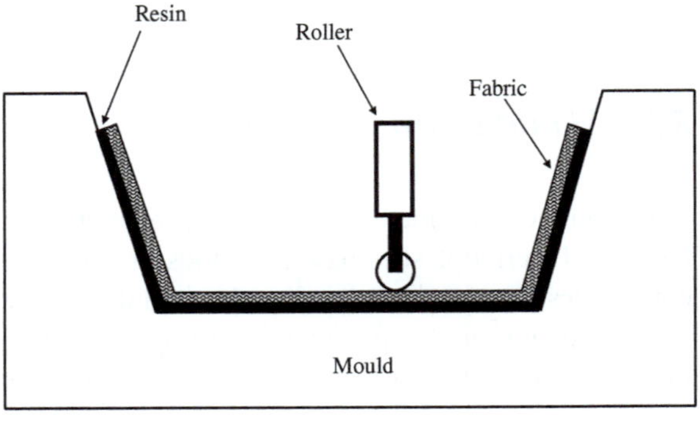

depends on the skill of the fabricator. A product with good surface finish, uniform distribution and thickness of resin and, negligible pores is considered as a good quality product. There are certain disadvantages with this process. The product can have good surface finish only on one side, i.e. the surface in contact with the mould during fabrication. It is very difficult to produce products with consistent quality. It is a time-consuming and labour-intensive process.

5.1.2 Resin Transfer Moulding (RTM)

This is a semi-automated process. Some of the short-comings of hand lay-up process are overcome in this process. Consistent quality products with good surface finish on both the sides can be produced by this process. For this process, the moulds are made with fibre reinforced plastics or metals. The required amount of reinforcement is stacked over the female mould and then closed with the male mould. The resin is mixed with curing agent in a mixing head and then pumped into the mould cavity. Vents are provided at the farther ends of mould. After completely filling the mould cavity, the resin starts coming out through the vents. The resin injection is stopped at this stage and the resin is allowed to cure. The mould is opened to remove the product after curing. This is a faster process than hand lay-up. When the size of the product is large, multiple injection parts should be provided. The fibre wash-out can happen near the injection parts, if chopped strand mat is used. The stitching of the mat can reduce this problem to some extent. The mould design is a critical element for this process to achieve complete filling of the mould. A variant of this process is vacuum assisted RTM, in which vacuum is applied through the vents to remove entrapped air and to facilitate easy flow of resin.

5.1.3 Resin Infusion/Vacuum Bag Moulding

This process is similar to RTM, but the resin is flowing into the mould cavity by the application of vacuum. The top mould can be rigid mould or flexible. Resin trap should be provided in the vacuum line; otherwise the resin can enter the vacuum pump after filling the mould. Very large products such as boat hulls are made with this process.

5.1.4 Autoclave Process

Autoclave is a high temperature pressure vessel. The composites fabricated using prepregs are generally cured in autoclave. A high quality composite product can be produced by this process. A composite product fabricated over a mould is covered with a vacuum bag. By applying vacuum, the air entrapped within the product is removed. The pressure and temperature within the autoclave are then increased slowly. The temperature initiates the curing reaction and the pressure facilitates better consolidation. By introducing bleeder fabric layers the resin content can be accurately controlled. By theoretical analysis, one can determine the orientation of fibres and its content to meet the required properties. The same properties can be achieved by this process. The composites fabricated by the hand lay-up process can also be consolidated by this process.

5.1.5 Compression Moulding

Moulding compounds are generally used in the compression moulding process. The moulding compounds are prepared by mixing the fibres and fillers in a thermoset resin. Hot curing catalysts are used in the moulding compounds. Hence, the curing reaction will not start after mixing. Further, to avoid any chance of curing, the moulding compounds are generally stored at 18 °C. The moulding compounds can be made in the form of sheet or bulk. Very large products are generally made using sheet moulding compounds and small products are made using bulk moulding compounds. The required quantity of moulding compound is placed on a compression moulding die. The die and the platens of the press are heated to the molding temperature. Pressure is applied to the moulding compound by moving the top or bottom platen. The moulding compound flows and completely fills the die cavity. The curing reaction will be over in a few minutes. The die is opened by moving the platens apart and then the product is removed from the die. The die is now ready for another cycle. The pressure applied during the compression moulding process is very high. The moulding temperature is also of the order of 120 °C. Hence, metallic dies are used in this process. The products with good surface finish can be made by this process. Thermoplastic composites can also be fabricated by compression moulding using thermoplastic composite sheets.

5.1.6 Filament Winding

This process is generally used to produce axisymmetric parts such as, cylinders, spheres, etc. It is possible to incorporate very high quantity of fibres in this process. A fibre content of more than 70 vol % is very common in this process. Fibre rovings and prepreg tapes are generally used. A collection of rovings impregnated with resin is wound on rotating mandrel. The winding angle is controlled by the rotating speed of mandrel and the speed of movement of fibre feeder. The schematic of this process is shown in Fig. 8. The final properties of the composite can be controlled by the fibre winding angle and it is possible to fabricate the composite with the required fibre angle in this process. Very long pipes for the transportation of different liquids can be made by this process.

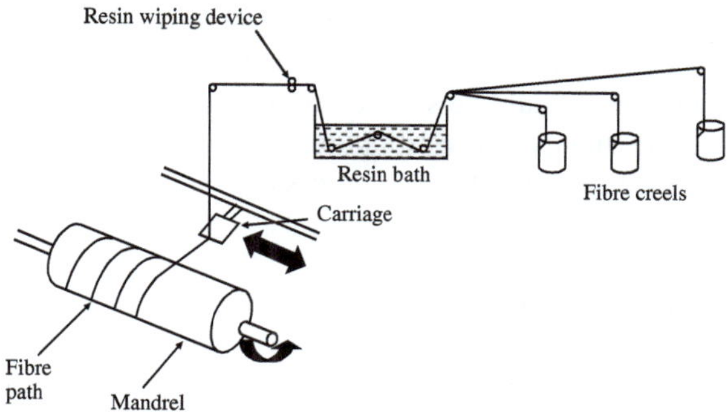

Fig. 8 Schematic of a filament-winding process [5]

5.1.7 Pultrusion

Very long products with constant cross-section are made by this process. Very high fibre content can be realised in this process also. Long tubes and rods are some of the products. These regular shapes are used for the construction of walk-ways, side grills, ladders, etc. These pultruded polymer composite products are more preferred for off-shore structures because of their good corrosion resistance. In this process resin impregnated fibres are allowed to pass through a hot die, where consolidation and curing of resin take place. The cured composite product is pulled through the other end of the die. A cut-off saw cuts the product to the required length. Metallic dies with hard surfaces are used in this process to withstand the high temperature and abrasion. Improper wetting and fibre breakage are some of the problems with this process.

5.1.8 Injection Moulding

It is one of the common processing methods for thermoplastics. The same equipment can be used to produce thermoplastic composites with short fibres or particulates. Since hard reinforcements are used, the dies should have high abrasion resistance. A thermoplastic material is mixed with the reinforcement in an extruder above the melting temperature of the thermoplastics. The molten matrix with reinforcement is injected into the die. After sufficient cooling, the die is opened to remove the product. Only fibres with low aspect ratio are used in this process, since the long fibres will increase the viscosity and affect the flow.

5.1.9 Reinforced Reaction Injection Moulding (RRIM)

It is similar to the RTM process used with thermoset resin. Instead of liquid thermoset resin, liquid monomer is injected into the mould cavity containing fibre reinforcement. The monomer undergoes polymerisation reaction within the mould and forms the final thermoplastic matrix. Since the melt viscosity of thermoplastics is high, it is very difficult to get easy flow. This problem is overcome by the use of monomer, which has lower viscosity. There is a limitation with this process, i.e. the monomer should be selected in such a way that it should not produce a by-product during the polymerisation reaction. This restricts the number of thermoplastic matrices suitable for this process.

5.1.10 Thermoforming

Metal forming process is used to produce metallic parts by the deformation of metal sheets. In the thermoforming process, thermoplastic composite sheets are used to produce composite products. The thermoplastic composite sheet is heated with infrared lamp above the glass transition temperature of the thermoplastic matrix. It is then transferred to a die kept in a press. The hot thermoplastic composite sheet undergoes deformation within the die and forms the product according to the shape of die.

5.2 *Manufacturing of MMCs*

Metal matrix composites are being produced by melt processing and powder metallurgy techniques. Deposition techniques like electrodeposition, physical vapour deposition and chemical vapour deposition can also be used. Although the product quality is good with these methods, they are time-consuming and expensive methods. Recently, MMCs are also produced by in situ processing methods.

5.2.1 Stir Casting

This is the simplest method to produce MMCs. In this process, the required alloy is melted above its melting temperature. The reinforcements in the form of particulates or short fibres are slowly added to the melt. To ensure the uniform distribution of reinforcements, the melt is stirred well during the reinforcement addition. Even then the segregation and settling of reinforcements are the major problems with this process. The chemical nature and size of reinforcements play a major role for the uniform distribution. Sometimes the reinforcements are chemically modified to improve wetting. When the size of reinforcement is very small, it will drastically increase the viscosity of melt. Moreover, the fine reinforcements stay at the surface of melt. Larger reinforcements try to settle at the bottom of melt, when their density is higher than melt density. Since this processing is carried out above the melting temperature of metal matrix, there is a possibility of unwanted reactions between reinforcement and matrix materials. These reactions may lead to the formation of detrimental compounds, which can affect the properties of MMC. Once the required amount of reinforcement is added, the melt is stirred for a few minutes and then poured into a mould. A better control of processing parameter is needed to get a good quality composite with this process.

5.2.2 Melt Infiltration

A porous preform is made with particulates or short fibres. This preform is infiltrated with molten metal. Generally, it is difficult to get complete infiltration without the application of pressure. The pressure is applied to the melt by mechanical means on by using high pressure gas. The former is called squeeze infiltration process and its schematic is shown in Fig. 9. In this process, the preform is kept in a pre-heated die, the required quantity of molten metal is poured over that and then pressure is applied to the melt by lowering the top punch. The precise quantity of reinforcement is ensured in this process. The probability of reactions is also less due to the shorter duration of the process. Even without any chemical modification, complete wetting of reinforcement can be achieved in this process. The only condition is that the preform should have continuous porous network.

5.2.3 Powder Metallurgy Technique

Reinforcements in the form of particulates or short-fibres are generally used in this process. The reinforcement is thoroughly mixed with the matrix powder in a mixer. The mixture is compacted in a die and then sintered at high temperature. The sintering temperature is generally 0.7–0.8 of the melting point of matrix material.

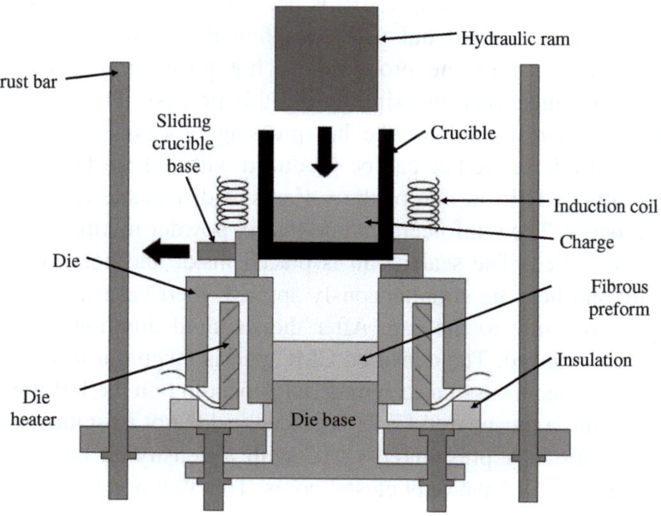

Fig. 9 Schematic of squeeze infiltration process [9]

Since the metallic materials undergo rapid oxidation reactions at high temperature, the sintering process should be carried out under inert atmospheric conditions. Most of the times, a thin oxide layer is already present over the metallic particles. Hence, it is necessary to use a reducing atmosphere during the initial periods of sintering. The densification of composite during sintering occurs by the diffusion of metallic atoms. The presence of reinforcements may prevent the diffusion. Hence, it is very difficult to achieve good density by normal sintering process. It is necessary to use sintering additives or high pressure to facilitate easy diffusion.

5.3 *Manufacturing of CMCs*

The manufacturing of CMCs needs still higher temperatures. Hence, the melt processing routes are not generally preferred. The powder based methods are widely used to produce CMCs. Apart from these methods; the deposition methods are also common with CMCs. Some special techniques, such as reaction bonding, directed oxidation and polymer infiltration and pyrolysis are also used. In any case, the manufacturing of CMC is more expensive than that of PMC and MMC.

5.3.1 Compaction and Sintering

The reinforcements in the form of short-fibres or particulates are thoroughly mixed with matrix powder. This mixture is compacted in a die using a press and then sintered at high temperature. As in MMC, the composite is densified by the diffusion process, which is affected by the presence of reinforcements. Hence, hot pressing or hot isostatic pressing (HIP) is necessary to achieve good density. Hot pressing is generally carried out using graphite dies with a typical pressure of 50 MPa. The final composite produced by hot pressing may have anisotropic properties due to directional pressing during this process. The complexity of product shape is also restricted for the hot pressing process. CMCs with complex shapes and isotropic properties can be produced with hot isostatic pressing. There are two variants in HIP; in one of the variants, a disposable can material is used during the process. The reinforcement and matrix powder mixture is taken in a can, evacuated and sealed. The sealed can is placed inside the hot isostatic press and then pressure and heat are simultaneously applied. Inert gases at high pressure are used to apply pressure to the can. After the required duration, the pressure and temperature are reduced. The densified CMC part is taken out from the can. Either glass or stainless steel is used as can material depending on the HIPing temperature. In another variant, a presintered CMC is placed inside hot isostatic press to achieve further densification. A presintered CMC with a density of more than 92 % of theoretical density will have only closed pores. This will not allow the pressurising gases to enter the composite; hence further densification can be achieved.

5.3.2 Reaction Bonding

This is a relatively low temperature process. In this process, silicon based composites are first made by the powder metallurgy technique. These composites are subjected to nitridation at 1300°–1400 °C. The silicon particles react with nitrogen and form silicon nitride matrix. This reaction closes the pores. The main advantage of this process is that the composite will not undergo any shrinkage during the CMC formation. However, the presence of unreacted silicon and residual pores are unavoidable.

5.3.3 Directed Oxidation

This process was invented by the Lanxide Corporation. A reinforcement preform is placed over molten metal. The molten metal infiltrates into the preform by capillary action. During the infiltration, the metal reacts with the surrounding gases and forms the respective ceramic matrix. For example, alumina based composites can be produced by the infiltration of aluminium metal and subsequent oxidation using oxygen. In this case, the CMCs are formed at the processing temperatures of MMC. Hence, the cost of production of CMC will be low. However, the presence of residual metal is a major problem, especially for the composites used at high temperatures. The presence of residual metal may be beneficial for the CMCs used at room temperature, since it will increase the toughness.

5.3.4 Polymer Infiltration and Pyrolysis (PIP)

In this process, a reinforcement preform is infiltrated with liquid/molten polymer and then the polymer is converted to a ceramic material by high temperature pyrolysis. Since the polymer to ceramic conversion takes place with the liberation of volatile materials from the polymer, it may not be possible to get a dense ceramic matrix in a single cycle. The infiltration and pyrolysis cycle should be repeated for 3–4 times, to get appreciable amount of ceramic matrix. Even after 3–4 cycles, the presence of pores is unavoidable. Hence, the mechanical properties of the CMC may not be high. SiC and Si_3N_4 based composites are generally produced by this method.

5.3.5 Chemical Vapour Infiltration

When the chemical vapour deposition happens inside a porous preform, then the process is called chemical vapour infiltration. Although this process is more expensive, it is justified by the high performance of CMCs. In this process, the appropriate gases react at high temperature and form the ceramic material, which is deposited over the reinforcement surface within the pores. The schematic of this

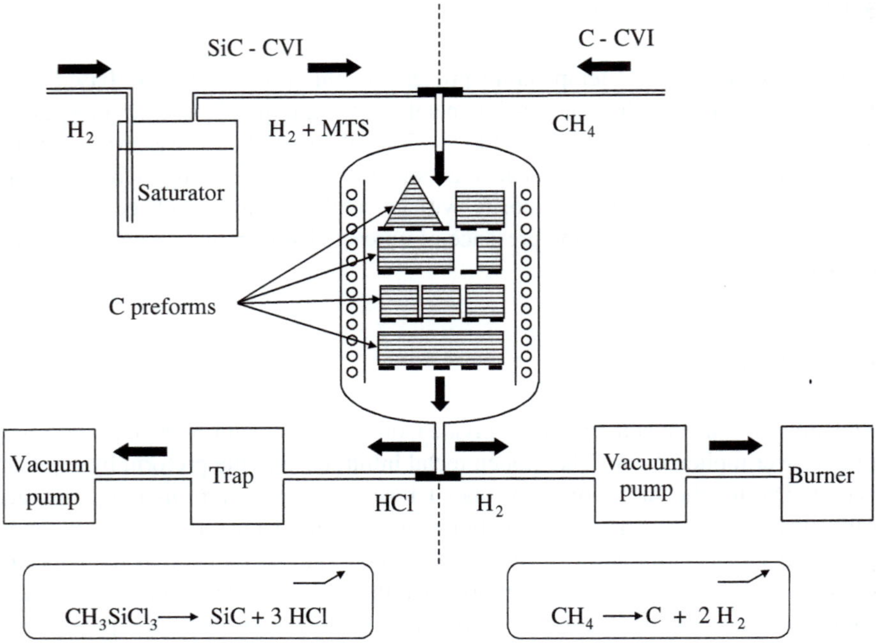

Fig. 10 Schematic of isothermal CVI process [10]

process is shown in Fig. 10. Under normal heating process in a furnace, the surface of preform will be at a higher temperature than inside. Hence, the deposition is more predominate at the surface, which will close the surface pores. Once the surface pores are closed, the interior surfaces of the preform may not have access to the gases for the formation of ceramic matrix. Hence, the preform is removed from the reactor frequently and the surface layer is removed by grinding. Instead of this, a temperature gradient can be maintained from top to the bottom of preform which can facilitate the directional deposition. However, this will restricts the number of products formed at the same time and also the complexity of product shape. For example, SiC_f/SiC composites are produced by chemical vapour infiltration of SiC_f preform with methyl trichlorosilane (MTS) and hydrogen gases. As mentioned earlier, this is a very time-consuming process.

6 Applications

At present composites are finding applications in all engineering fields. A study by DuPont indicates that about 50 % of the engineering materials used in the year 2020 will be made-up of composite materials. The composites are finding applications from the house-hold items to aerospace parts. The main driving force for the usage of composite material is weight reduction. Apart from this, a tailorable property is another major advantage of composites. The polymer composites are the most widely used composite materials, followed by MMCs and then CMCs. Some of the important applications of these three types of composites are given in the following sections.

6.1 Applications of PMCs

The applications of PMCs started with aerospace applications during the Second World War. After the war, people started using PMCs for making boat hulls. During this time, many glass fibre production facilities have been started all over the world. This has driven the expansion of application of composites to many areas. Most of the structural parts of military aircrafts are currently being made with PMCs. Rotor blades of all helicopters are made with PMCs because of their extremely high fatigue resistance. The usage of PMCs in commercial aircrafts is also increasing steadily. The recently developed Boeing 787 aircraft has 50 % PMCs on its total weight. In the ground transport also PMCs are widely used. Bus bodies, car bodies, bumpers, doors, seat-rest, brake-pad are some of the applications of PMCs in automobiles. Truck-cabin, elliptical spring and containers are other applications.

Even now the cost of PMCs is higher than that of cement based building materials. However, the wood based parts, such as window frames and doors, bath room doors, partitions and furnitures can be replaced with PMCs. The PMC parts will have better performance than wooden parts. The roofing for parking areas, domes and translucent roofs are also made with PMCs. Repairing of concrete structures is another important application of PMCs. The damaged part of bridges and multi-storey buildings can be strengthened with fibre reinforced plastics. Very large pipes for the transportation of gases and liquids are being made with PMCs (Fig. 11).

Fig. 11 HOBAS 3 m FRP pipes (*Courtesy*: HOBAS Engineering GmbH, Pischeldorferstraße 128, 9020 Klagenfurt)

Because of their high electrical insulating properties, PMCs are the preferred choice for electrical and electronics applications. High voltage insulators, printed circuit boards, etc. are currently being made with PMCs. In the energy sector, wind mill blades are the important applications of PMCs. Antenna dishes, which will not undergo any dimensional change due to the variation in temperature are being made with carbon fibre reinforced epoxy.

Many sporting goods are currently being made with PMCs. Tennis rackets, Golf clubs, fishing rods, ski boards and sports cycle frames are some of the applications of PMCs. Light weight, good impact properties, and high strength of PMCs are attractive properties for these applications.

6.2 Applications of MMCs

Many PMCs and metallic parts are currently being replaced by MMCs. The abrasion/scratch resistance of MMCs is better than PMCs. With improved strength and modulus, the MMC part can perform better than the metallic parts. In some applications the high wear resistance of MMCs is utilized. In aerospace applications, fan exit guide vans, stator vans, fuel tank covers are some of the parts made with MMCs. Either the short fibres or the particulates are generally used in MMCs.

Automobile pistons with selective reinforcement in the regions, where high wear resistance required, can be made with MMCs. The critical speed of drive shaft depends on the stiffness. High stiffness, hence high critical speed can be achieved with MMC shafts. Brake pads/discs for many transport vehicles are being made with MMCs (Fig. 12). They have high wear resistance coupled with good thermal conductivity. Roller coaster brake fins are another interesting application of MMCs.

Cemented carbides are the commonly used cutting tool materials now. They are made with tungsten carbide particles and cobalt. Substrate carriers of electronic circuit boards need very high thermal conductivity. The carbon fibre reinforced aluminium composite is a suitable choice for this application.

Fig. 12 An air-cooled brake disc made of A359 Al alloy reinforced with 20 vol % of SiC particles [11]

6.3 Applications of CMCs

The CMCs are generally preferred for high temperature applications. Many engine components can be made with CMCs. Gas turbine engines with CMCs are under development. Automobile engines made with CMCs can give high fuel efficiency. These engines are adiabatic engines, since there is no need of cooling system to control the temperature of engine. Already the prototype has been made in Japan.

Another important application is cutting tools. Ceramics are generally hard, but they are very brittle. By forming a ceramic composite, the fracture toughness of ceramics can be improved. A CMC cutting tool will have high hardness and appreciable toughness. High speed machining is possible with these CMC cutting tools.

Brake systems for aircrafts are made with CMCs. Carbon fibre reinforced carbon composites are used because of their high wear resistance and good thermal conductivity. These composites are also used in re-entry vehicles. Since these composites are biocompatible, many medical implants can be made.

Cement based CMCs can be used in building applications. The natural fibre reinforced cement sheets are used for roofing in many developing countries. The incorporation of glass or polymeric fibres can improve the performance of concrete.

7 Conclusion

The composite materials have many advantages and the important advantage is high specific strength/stiffness. Hence, it is possible to make light-weight and high performance components with composite materials. The properties of composites are controlled by the type and quantity of reinforcement and matrix, and size, shape and the nature of distribution of reinforcement. The composites can be classified into various types based on the matrix, the type and arrangement of reinforcement and the size of reinforcement. The fibre reinforced composites are very common because of the many advantages of fibres compared to particulates. There are natural fibres as well as synthetic fibres. Although the natural fibres are inexpensive and environmental friendly, their mechanical properties may not be suitable to produce high performance composites. The synthetic fibres are high performance fibres and there are a variety of fibres available to meet the requirements for various applications. Similarly, a variety of matrix materials is also available. Depending on the service temperature, a suitable matrix material can be selected. The processing of polymer composites is matured enough and there are plenty of methods like, hand lay-up, filament winding, pultrusion, etc. available to make the composites of any shape and size. The processing methods of metal and ceramic matrix composites are evolving and there are a few methods available to fabricate these composites on a commercial scale. Wherever possible, the components made with conventional materials are being replaced with composite materials because of weight advantage and high performance.

References

1. Balasubramaniam R (2010) Callister's Materials science and engineering, Wiley India, New Delhi, p 541
2. Chawla KK (1998) Composite materials: science and engineering, 2nd edn. Springer, New York, p 8
3. Balasubramanian M (2013) Composite materials and processing. CRC Press, Boca Raton, p 20, 109
4. Agarwal BD, Broutman LJ, Chandrashekhara K (2006) Analysis and performance of fiber composites, 3rd edn. Wiley, New York, p 76
5. Mallick PK (2008) Fiber-reinforced composites, 3rd edn. CRC Press, Boca Raton, p 72, 409
6. ASM Handbook (2001) Composites, vol 21. ASM International, Materials Park, Ohio, p 31
7. Bledzki AK, Gassan J (1999) Composites reinforced with cellulose based fibres. Prog Polym Sci 24:221–274 (Table page no. 225)
8. Mazumdar SK (2002) Composites manufacturing. CRC Press, Boca Raton, p 131
9. Clyne TW, Withers PJ (1993) An introduction to metal matrix composites. Cambridge University Press, Cambridge, p 325
10. Naslain R, Langlais F (1986) Proceedings twenty-first university conference on ceramic science. Plenum, New York, p 145
11. Anonym (1996) Materials at the 1996 SAE international congress and exposition. Adv Mater Process 7:33–36

Essential Properties of Fibres for Composite Applications

Wilhelm Steinmann and Anna-Katharina Saelhoff

Abstract In this chapter, essential properties of fibres required for fabricating good performance composites are presented. Fundamental aspects of these properties and principles of their characterization techniques are discussed. Firstly, the geometrical aspects of fibres (for short and endless fibres) are described. The second part deals with the different types of structures in fibres with respect to molecular orientation mostly responsible for anisotropy of fibres. In the third part, the mechanical properties (especially tensile properties) and failure mechanisms for different types of fibres are discussed and correlated to the structure of the fibres. The following part is concerned with the surface of fibres, which is responsible for the interaction of the fibres with the matrix material in composites and has a large influence on the wetting behavior and adhesion to matrix materials. In the last parts, further physical properties (heat capacity, thermal conductivity, thermomechanical properties and electrical conductivity) and the durability of fibres are described.

1 Introduction

Fibers are a unique class of materials because of their anisotropy. Usually, fibers have a high length compared to their diameter. This so called high aspect ratio is responsible for the unique properties compared to a bulk material. Beside the geometrical aspect of anisotropy, the most important fiber types also have an anisotropic structure. Both aspects together result in anisotropic physical properties, which are described in this book chapter.

W. Steinmann (✉) · A.-K. Saelhoff
Institut Für Texiltechnik (ITA) der RWTH Aachen University, Otto-Blumenthal-Straße 1, 52074 Aachen, Germany
e-mail: wilhelm.steinmann@ita.rwth-aachen.de

A.-K. Saelhoff
e-mail: anna-katharina.saelhoff@ita.rwth-aachen.de

© Springer Science+Business Media Singapore 2016

S. Rana and R. Fangueiro (eds.), *Fibrous and Textile Materials for Composite Applications*, Textile Science and Clothing Technology,
DOI 10.1007/978-981-10-0234-2_2

First of all, the geometric and structural aspects of anisotropy result in extraordinary mechanical properties. Typical fibers for composite applications have high strength and stiffness in fiber direction, while they are weak and flexible perpendicular to it. Also other physical properties, like electrical or thermal conductivity may be totally different when measured along or perpendicular to a fiber. When embedded into a matrix, these anisotropic properties can also be transferred to the fiber reinforced composite.

In this case, the mechanical and other physical properties can be tailored in the composite part depending on the placement of fibers. For designing composites with tailored properties, it is essential to understand the anisotropic properties of fibrous materials.

Another unique property is the high surface area of fibers, caused by their small diameter compared to their length. The fiber surface with its topography and chemistry is the interface to the matrix material. First of all, the fiber surface determines the wetting behavior with matrix material, where surface tension of the materials and capillary effects have to be taken into account. Secondly, the interface between fiber and matrix is responsible for the transfer of load to the fiber. By controlling the adhesion between the two components, the failure of a composite part can be adjusted to specific needs. A detailed understanding of the interaction of fiber and matrix is therefore necessary to design tailored fiber reinforced composites.

The geometrical properties, the fiber structure, the most important physical properties of fibers and the fiber's surface are described in this chapter with special respect to the application of these materials in fiber reinforced composites. Furthermore, suitable methods for analyzing these properties are presented. These fundamentals will help the reader of this book to understand the specific properties of fibers and composites in the following chapters.

2 Geometrical Properties

The geometrical aspects of fibers are mainly defined by their length L and diameter D, resulting in the so called aspect ratio L/D. Concerning the fiber's diameter, different shapes of fibers beside a round shape are possible. Furthermore, the so called fineness or titer T, which is the mass per length unit, is an important factor to describe the fibers geometry, since it takes into account differences in density of the specific fiber types [1].

2.1 Cross-Section

The most common synthetic reinforcement fibers have a round cross-section, since it provides a uniform mechanical loading of the fiber material. This leads to a higher specific strength of the materials compared to other fiber shapes [1, 2].

However, different cross-sections can be found in fibrous materials for composite applications

(1) *Process-based deviations from round cross-section*: Depending on the spinning process of the reinforcement fibers, non-round filaments can be formed due to inhomogeneous processing conditions. A typical examples are carbon fibers with the so-called "kidney-shaped" cross-section. This cross-section is formed in the wet-spinning process depending on the coagulation conditions (see Fig. 1) [3, 4].
(2) *Intended deviations from round cross-section*: By changing the shape of the spinneret during fiber production or specific post-treatment of the fibers, different fiber cross-sections can be achieved. Examples are star-shaped or triangular fibers [5]. Furthermore, hollow fibers are common for creating lighter materials or materials with special transport properties (e.g. in membranes) [6]. Beside the mechanical properties of the filaments, rovings and composites, the fiber cross-section can have an influence on the processing of the fibers (e.g. tribology or wetting behavior).

Beside the outer (geometrical) shape of the fiber, the material distribution within a filament can differ from a homogeneous distribution. This deviation can also be caused by the process (e.g. core-sheath structures in wet spinning or carbon fiber conversion) or be intended (e.g. bicomponent fibers made of two polymers) [3, 7].

The (geometrical) cross-section and the material distribution is usually analyzed by light or electron microscopy, depending on the necessary spatial resolution [8]. By combining microscopy with spectroscopic methods (e.g. Energy-dispersive X-ray spectroscopy), information about the chemical composition over the cross-section can be obtained.

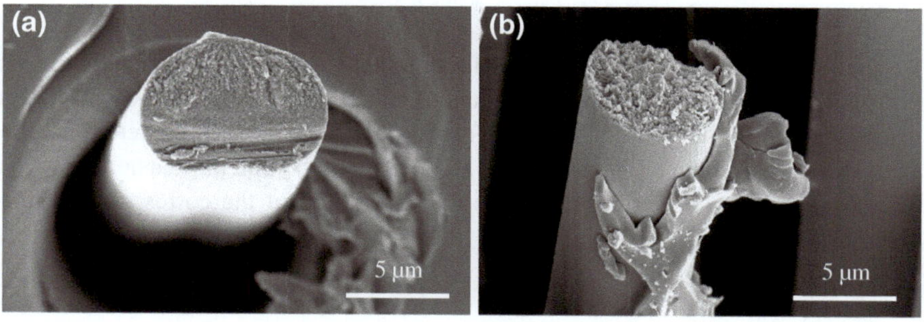

Fig. 1 Scanning electron microscopy (SEM) pictures of **a** "kidney-shaped", **b** round carbon fibers

Table 1 Classification of fiber length for textiles and plastics processing

Fiber type	Length's range (textiles) (mm) [1]	Length's range (plastics processing) (mm) [9]
Infinite length	>200.0	>50.0
Long staple fibers	40.0–200.0	1.0–50.0
Short staple fibers	6.0–40.0	
Short fibers/pulp	<6	0.1–1.0

2.2 *Fiber Length*

For the fiber's length, short or staple fibers ($L/D < \infty$) and fibers with infinite length ($L/D = \infty$) have to be taken into account [9, 10]. In the following Table 1, the classification of fibers length for textiles and plastics processing is displayed.

The length and its distribution of fibers with finite length is usually analyzed by microscopy [11]. Because of the small number of fibers analyzed by one microscopic image, novel methods for analyzing length and diameter of a large number of fibers were developed.

2.3 *Fiber Fineness*

The fineness (or titer T) of a fiber describes the mass per unit length. Therefore, it depends on the diameter d of the fiber and its density ρ (see section below). The unit for fiber fineness is tex (grams per 1,000 m) or den (grams per 9,000 m). The following equation can be used to calculate the fineness from filament diameter or vice versa [1]:

$$D = \sqrt{\frac{4}{\pi \cdot \rho} \cdot T} \tag{1}$$

The fineness can be measured either for complete rovings or for single filaments. For rovings, a reel is used to prepare fibers of a fixed length (usually 100 m) for the determination of their mass. The fineness of single filaments can be determined by analyzing their resonance frequency f at a defined length l and pre-tensioning with a force F. The following equation allows the calculation of the corresponding fineness [12]:

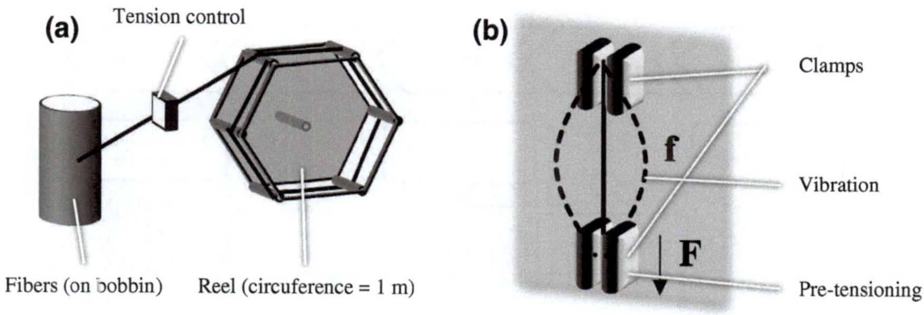

Fig. 2 Illustration of methods for determining the fineness of fibers (according to [12]). **a** Determination of weight per length by reeling. **b** Measuring single filament diameter by analyzing the resonance frequency during vibration

$$T = \frac{F}{4 \cdot f^2 \cdot l^2} \tag{2}$$

The methods for the determination of fiber fineness are displayed in Fig. 2.

3 Fiber Structure

The structure of a fibre is responsible for all of the properties described in this chapter. Therefore, a basic overview on the structures present in a fibre shall be given. The structure of the common reinforcement fibres can be considered as hierarchical, in which a certain structural element is used to build the next larger part of structure, so that the size of structures in a single filament range over several orders of magnitude (from 10^{-10} m, the distance between two atoms, to 10^{-5} m, the diameter of a filament). Due to the anisotropy of fibres, special attention has to be paid also to the anisotropy of structure. Several structural elements can be oriented in a filament, depending on the material [13–15]. The different types of structures are illustrated in Fig. 3 and described in the following sections.

3.1 Description of Structures

The *atomic or molecular structure* is being formed by the interaction of single atoms and the resulting type of chemical bonding. Depending on the type of chemical bonding, anisotropic structures can be formed on the molecular level. Usually, covalent bonding between atoms causes such an anisotropy (such as graphene layers in carbon fibres and the backbone in polymeric fibres), whereas metallic or ionic bonds are isotropic.

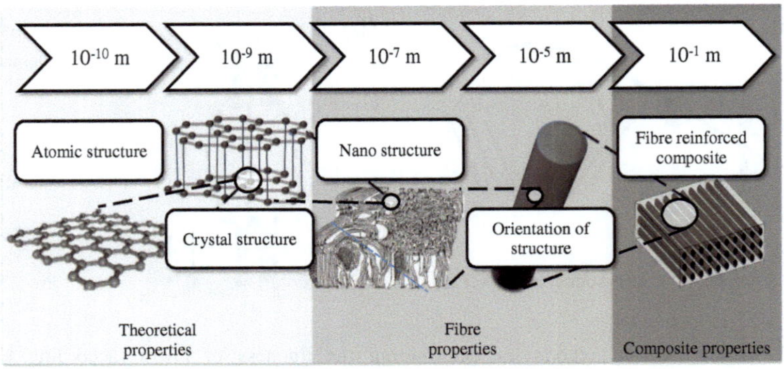

Fig. 3 Types of structures in a reinforcement fibre with their size indicated (structure illustrated exemplarily for carbon fibres) [16]

Depending on a regular or non-regular atomic or molecular structure, the atoms respectively molecules can be arranged to a *crystalline structure* or non-crystalline (amorphous) structure. In crystalline structures, atoms or molecules have fixed distances to each other in all space directions over a long distance, which are described by the unit cell. Amorphous structure are contrarily characterized by a medium (most probable) distance to the next atom or molecule. However, also the size of crystalline structures is limited and crystallites of different sizes and shapes can be formed in a fibre. If the molecular structure is anisotropic, usually anisotropic crystalline structures are formed, which may have an anisotropic (needle, disk- or platelet-like) shape. The crystalline or amorphous structure of a material usually defines the theoretical properties of a material [13, 14, 17].

Different crystallites or amorphous structures are combined to the *nano structure* of a fibre. It is characterized by the arrangement of the single elements, which may differ over the diameter of a filament [18]. Depending on the molecular architecture, crystalline and amorphous structures as well as nano-sized pores coexist in the nanostructure of a filament [19]. The nano structure is usually anisotropic, and can even be anisotropic if the single elements are isotropic or amorphous, such as in glass fibres.

The different elements of the nano structure as well as the nano structure itself may have a distinguished *orientation* in the filament. Usually, anisotropic crystalline structures such as in polymeric or carbon fibres can be highly oriented in the direction of the fibre [3, 18, 20]. The assembly of the single elements to the nano structure as well as their orientation define the properties of the material in the fibrous form.

The most outer parts of the nano structure form the *surface* of a filament. Therefore, the arrangement and orientation of structures in the outer parts of a filament can have an effect on the surface morphology (and roughness) as well as the possibilities of adding chemical functionalities to the surface. This part of the structure is then responsible for the interaction with the matrix material and has an influence on the composite properties such as wetting behavior and adhesion.

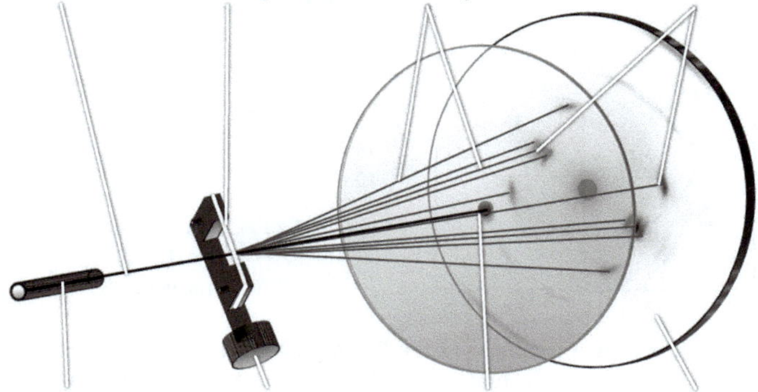

Fig. 4 Illustration of WAXS setup for analyzing structure and orientation of fiber samples

3.2 Structure Analysis by X-Ray Diffraction

The structural features of a fiber are usually analyzed by diffraction and scattering methods. The most important method for structure analysis is the diffraction or scattering of X-rays, while Wide Angle X-ray Diffraction (WAXD, or Wide Angle X-ray Scattering—WAXS) is used for the evaluation of the atomic and crystalline structure of a material. A typical setup for analyzing fibers by WAXS is displayed in Fig. 4. By using two dimensional detectors, both the structural features as well as their orientation within the fiber can be analyzed [15].

Most of the primary X-ray beam with the wavelength λ is transmitted through the fiber sample. However, the X-rays are partly scattered elastically. Depending on the distance of scattering structures and their orientation in the fiber, positive interference takes place at certain angles to the primary beam (diffraction angle 2θ) and relatively to the fiber axis (azimuthal angle φ) [21].

Structural features of a fiber are analyzed by the evaluation of intensity distribution as a function of the diffraction angle 2θ. As indicated in Fig. 5, the intensity has to integrated first over the azimuthal angle for taking into account all structural features independent from their orientation. The resulting intensity diagram can then be used to fit peaks to the data. Typically, sharp peaks from crystalline structures and/or broad peaks from amorphous structures can be found in the intensity distribution [15].

The center position of a peak is connected to the distance within the crystalline lattice by Bragg's equation, while d_{hkl} is the distance for arbitrary lattice planes [21]:

$$\lambda = 2 \cdot d_{hkl} \cdot \sin(\theta) \tag{3}$$

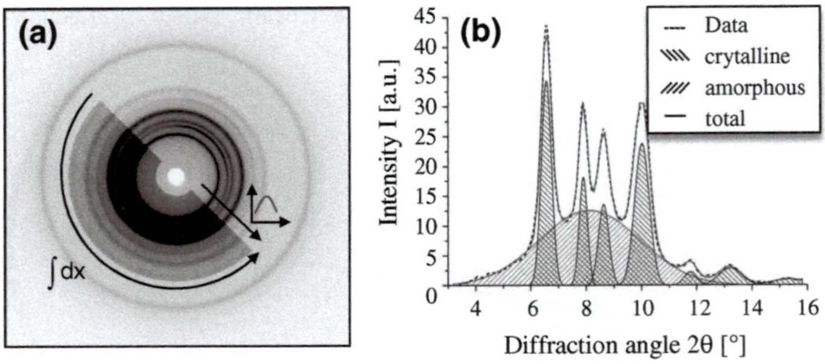

Fig. 5 Analysis of structural parameters from WAXS 2D images. **a** Integration of intensity over azimuthal angle. **b** Intensity as a function of diffraction angle for data fitting (crystalline and amorphous peaks)

The width of the peaks is determined by the finite size of the structures (or the size of domains within a structure, which are correlated to each other). The peak width $\Delta 2\theta$ can therefore be used for the determination of the expansion D_{hkl} of structures in the direction of an arbitrary lattice plane by using Scherrer's equation (K is a constant factor determined by the crystalline structure) [15, 21]:

$$D = \frac{K}{\Delta 2\theta \cdot \cos(\theta)} \tag{4}$$

The number of scattering structures determines the total intensity of X-ray beam which is scattered. If two or more different phases (e.g. crystalline and amorphous) are present in a fiber, the phase content X can be determined by determining the relation of the intensity I_x of a specific phase relatively to the overall intensity I_{total} scattered by the fiber [15]:

$$X = \frac{I_x}{I_{total}} \tag{5}$$

By analyzing the azimuthal distribution of intensity, the orientation of structures can be determined. The peak intensity in the direction of diffraction angle has to be integrated and evaluated as a function of azimuthal angle (as displayed in Fig. 6).

By analyzing the center position of peaks φ_c, the orientation of specific lattice planes relatively to the fiber axis is analyzed. The standard deviation $\langle \sin^2(\varphi - \varphi_c) \rangle$ from this position indicates the degree of orientation, which is expressed by Herrmann's orientation factor f for uniaxially oriented structures [14, 15]:

$$f = 1 - \frac{3}{2} \cdot \langle \sin^2(\varphi - \varphi_c) \rangle \tag{6}$$

Contrarily to WAXS, the method of Small Angle X-ray Scattering (SAXS) can be used to analyze the nanostructure of fibers. By analyzing intensity distribution at

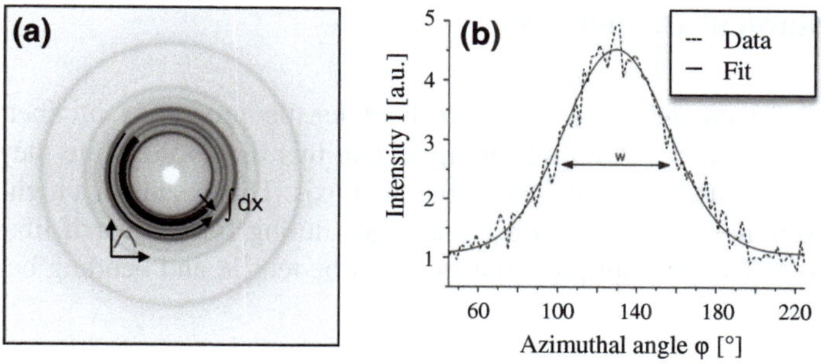

Fig. 6 Analysis of orientation parameters from WAXS 2D images. **a** Integration of intensity over diffraction angle. **b** Intensity as a function of azimuthal angle for data fitting (determination of width of crystalline peaks)

lower diffraction angles, scattering from objects with larger distances can be analyzed according to Bragg's equation. SAXS can be used to identify fibrils or other types of secondary structure formed in a fiber [14, 18].

3.3 Other Techniques for Structure Analysis

Beside X-rays, other types of radiation (e.g. neutrons or electrons) can be used to perform scattering experiments. Furthermore, other analysis techniques can be used to evaluate the structure of fibers:

- A direct visualization of structures is possible by microscopic methods such as electron microscopy. Especially with transmission electron microscopy (TEM), structures can be visualized with nearly atomic resolution. Because of high electron intensity necessary for that spatial resolution, only fibres with satisfying temperature stability and electrical conductivity (e.g. carbon fibres) can be analyzed in that way [22]. Furthermore, the area to be analyzed is very small (usually in a range of <100 nm), resulting in poor statistics of the analyzed data.
- Spectroscopic methods (such das IR, Raman or NMR spectroscopy) can be used to obtain information about the chemical bonds in a fiber, which allows the identification of the structures which are present. A quantitative description of structures as known from X-ray diffraction is however only possible if the method is calibrated with samples with well-known structural features [14].

4 Mechanical Properties

Typical reinforcement fibers have excellent tensile properties in fiber direction allowing high load transfer from the matrix in this direction, while they are relatively weak if loaded perpendicular to the fiber axis. This results in a high flexibility during processing, but also to fiber breakages during bending or if fibers are not well oriented. The relevant properties to describe tensile and bending behavior are described here.

4.1 Tensile Properties

The relevant measures to describe the tensile properties of reinforcement fibres are there tenacity, maximum elongation and Young's modulus [1, 3].

A high tenacity σ (or strength, stress at maximum elongation) is necessary for acting as a reinforcement of the matrix material. The strength of the fibres is usually one or two orders of magnitude higher than the strength of the matrix material. If mechanical load is transferred completely from the matrix to uniaxially oriented fibers (see section surface properties), the composite strength can reach the fiber strength multiplied by the fiber volume content. For reinforcement fibres, tenacity is usually measured in the unit of MPa or GPa. For lightweight construction, the specific tenacity (tenacity divided by the fiber's density) is indicated additionally to take into account the density of different material classes [3, 9].

The maximum elongation ε_{max} of reinforcement fibres is usually low compared to the possible elongation of the matrix material. In case of good load transfer, composite failure takes place at the level of maximum fibre elongation due to fibre breakages. The failure behavior of the composite is usually brittle. By adjusting fiber matrix adhesion (see section surface properties), a more ductile failure behavior can be achieved if cracks propagate slowly between the single filaments [3, 9, 10].

Young's modulus E describes the stiffness of the fibre measured in the region of linear elastic deformation and is indicated usually in GPa. Fibres with high Young's modulus have a positive impact on the stiffness of the composite part if load is transferred sufficiently from the matrix [9].

4.2 Tensile Test Methods

Tensile properties of fibres can measured on single filaments or complete rovings. For single filaments, standard test methods [23] and special testing machines (e.g. Favimat by Textechno GmbH, Mönchengladbach, Germany) can be used [24] (see Fig. 7a). For standard test methods, filaments are fixed on a paper frame, which is

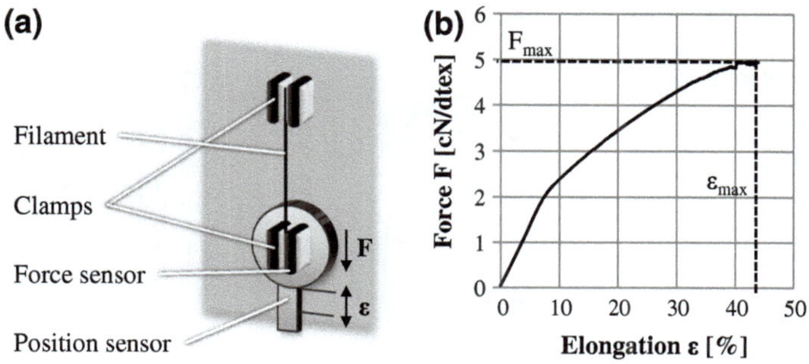

Fig. 7 **a** Experimental setup for single filament testing (Favimat method). **b** Resulting stress-strain diagram of a polymeric fiber

transferred to the testing machine and then cut before the tensile test. Stress-strain diagrams should at least be measured at 50 filaments [23]. A typical stress-strain diagram of a single filament is displayed in Fig. 7b. Maximum force (or tenacity) and maximum elongation can be easily extracted from the diagram. For the evaluation of Young's modulus, the linear elastic region (usually below 1 % of elongation) has to be taken into account [23].

For tensile tests on rovings, the following test methods are available:

1. The ends of the fiber roving are placed between two clamps, which can be fed automatically (e.g. by special testing machines like Statimat from Textechno GmbH, Mönchengladbach, Germany). This method is suitable if a sufficient load transfer between all filaments and the clamps can be achieved and the filaments are not to brittle to break between the clamps. It is therefore usually applied for polymeric fibres with low fineness.
2. Fibre ends are impregnated with a suitable resin at a defined geometry, which fits into a standard testing machine. The method can be applied to avoid filament breakages at the clamps. However, it is only valid if the tenacity is not to high, since fibre pull-out can occur at the resin forms (see section surface properties).
3. The standard method for glass and carbon fibres requires a complete impregnation of the rovings with a suitable resin [25, 26]. Only by a complete impregnation, a complete load transfer to all filaments without fibre pull-out can be achieved. A valid setup for the coating of rovings is displayed in Fig. 8. The coated rovings can then be placed into a standard testing machine.

4.3 *Effect of Fibre Length and Diameter*

Both the fibre gage length in mechanical tests and the filament diameter have an influence on the measured strength of a fibre. Defects in the fibre structure are mainly responsible for their failure. The effect of length and diameter depend on the

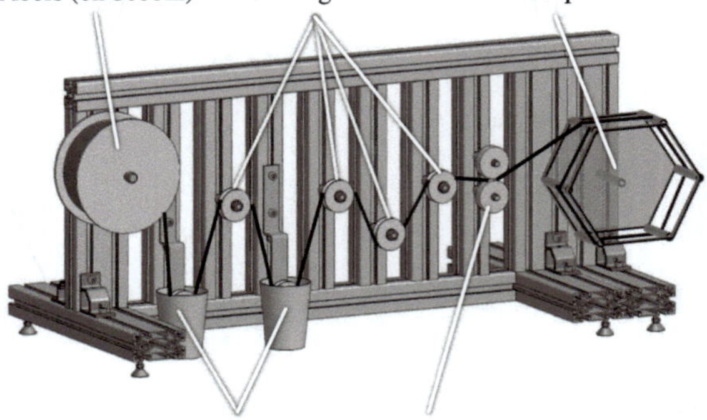

Fig. 8 Experimental setup for coating glass or carbon rovings according to [25, 26]

probability to find a critical defect in the test volume, which is responsible for the failure of the fibre [2].

The probability to find a critical defect in a fibre of a length l is determined by the Weibull statistics. The resulting mean strength of $\langle\sigma\rangle$ is then a function of fibre length with $\ln \langle\sigma\rangle \sim -\ln(l)$ [27].

The relation for the fiber diameter is similar. Also here the probability of finding a defect within the fibre cross-section area A has to be taken into account. The impact of diameter on the measured mean fibre strength depends on the type of material used in the tests, e.g. $\langle\sigma\rangle \sim 1/d^{1/2}$ for glass fibres [2].

4.4 Bending Properties

While the bending properties of a composite can be easily determined by standard test methods, the flexural strength and bending stiffness of single fibres are only accessible by microscopic test methods at filament level.

Three-point bending tests can be implemented on the filament level by using very precise testing machines to measure the flexural strength (see Fig. 9a), since maximum forces in bending mode do not exceed the range of several mN. Fibres have to be fixed with a length of several 100 μm and bending has to be performed with a probe of a very small diameter, usually not exceeding 1 μm [28]. Another test method is based on applying a compression load in the fibre direction (see Fig. 9b). This leads to a bending deformation, so that flexural strength and bending modulus can be calculated [29].

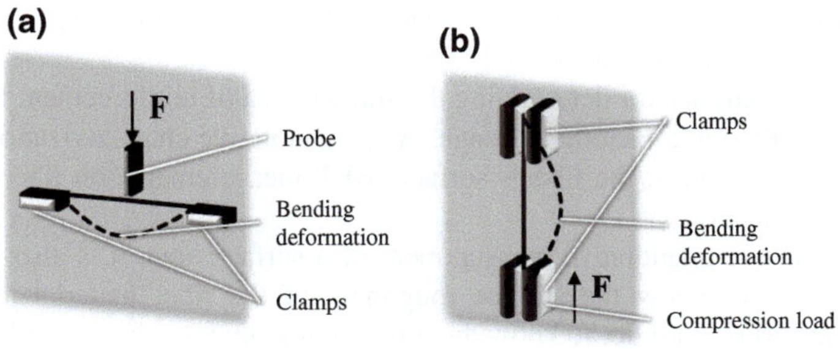

Fig. 9 Methods for measuring bending properties of filaments. **a** Three-point bending. **b** Application of compression load

5 Surface Properties

As the interface between fiber and matrix material dramatically influences a composite's performance, the fibers' surface properties are an important part of the reinforcing materials.

The parameter which characterizes the binding strength of fiber and matrix is called interlaminar shear strength (ILSS); its unit is [N/mm^2] [3]. The value of the ILSS between fiber and matrix is determined by mainly by two influencing factors: the mechanical and the chemical interlock [30].

In order to modify the degree of adhesion between fiber and matrix, most reinforcement fibers are not directly embedded into the composite matrix. They will undergo a surface treatment (leading to chemical and structural changes on the surface) as well as a coating procedure called sizing application (the sizing has the double function to protect the fiber during textile processing and promote adhesion in the composite).

Apart from the resulting adhesion properties in the composite, also wettability of the fiber with matrix materials is important during impregnation.

In order to characterize the properties mentioned above, analytical methods for their determination are presented in the following section.

5.1 Surface Area and Surface Roughness

The most common synthetic reinforcement fibers have a diameter in the range of less than 50 μm, while pores and fibrils in the range of nm contribute to the surface which is later bound to the matrix material.

Apart from the adhesion properties, the surface area and roughness have an impact on the fibers' tribological properties [31]. These properties are important when it comes to fiber-fiber-interaction and friction with machinery elements during textile processing. If the surface is too rough and a protection (e.g. in form of a sizing) is missing, a lot of fiber breakage and hence a weakening of the reinforcement fiber material will occur [32].

Therefore, analytics for determining the surface area of reinforcement fibers also has to be sensitive in a nanometer range. As a quantitative characterization method for the determination of the fiber's surface, BET measurements on fibers are presented in this section.

Apart from the quantitative measurement of a surface area, it is also important for composites to know the surface roughness of the fiber material as well as imaging the surface in order to know the different roughness values in different fiber directions.

5.1.1 BET Surface Area Measurements (Named After Stephen Brunauer, Paul Hugh Emmet and Edward Teller)

In this measurement method, the phenomenon of physical adsorption of gases on solid surfaces is used to determine the surface of the material analyzed.

For the measurement, the fibers are put into a sample holder (usually a cavity similar to a glass test tube) which is connected to a gas supply. From the cooled gas supply, the gases (nitrogen, argon or krypton at their boiling temperature) can float to the cavity with the samples (see Fig. 10a). The sample holder is cooled down to temperatures lower than the triplex point of the measurement gas, leading to a volumetric measurement with a precision of 0.5 % or better [33].

The pressure in the sample cavity is constantly held on the level of the equilibrium pressure p_0, where as many adsorbtions as desorptions take place (see Fig. 10b). When the pressure value goes below p_0 due to adsorption of the gas on the sample's surface, more gas is guided into the sample cavity. The amount of gas which floats into the cavity is characterized by the according pressure p. The relation p/p_0 is plotted against the number of particles adsorbed. By taking into

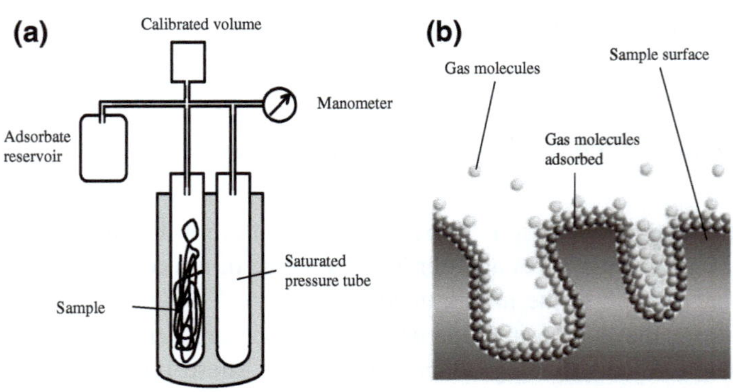

Fig. 10 **a** Measurement setup for BET surface area measurements (static volumetric method); **b** gas adsorption on a solid surface: the phenomenon used in BET surface area measurements

account the cross-section of one gas molecule adsorbed to the surface and the number of layers on the surface (BET theory [34]), the specific surface area per weight unit (m^2/g) is calculated.

5.1.2 Atomic Force Microscopy (AFM)

With atomic force microscopy (AFM), the topography of surfaces is characterized by measuring the inter-atomic forces between the sample surface and a measurement tip. The tip end ideally consists of only very few atoms. The tip is mounted on a cantilever which is oscillating with a frequency close to it resonance frequency. A piezo motor approaches the tip to the surface until the amplitude is lowered by the inter-atomic forces. The height information of the piezo element and the difference in the amplitude are recorded while the measurement process is repeated for every point of the sample to be measured [35]. The experimental setup is shown in Fig. 11a).

As the measurement is time-consuming, only parts of the surface can be characterized by this method. Typically, the size of AFM scans lies in the range of 2.5 or 5 μm². The number of points to be measured per line are always powers of two (256, 512 or 1024) due to digitalization. With the data taken, height information is available all over the scan area. Therefore, roughness values in different directions can be analyzed. Moreover, a three-dimensional image of the surface area scanned can be calculated (see Fig. 11b).

5.2 Chemical Functionalities

In order to evaluate the possibilities for chemical interlock between fiber and matrix material it is necessary to analyze the types of functional groups on the surface. This

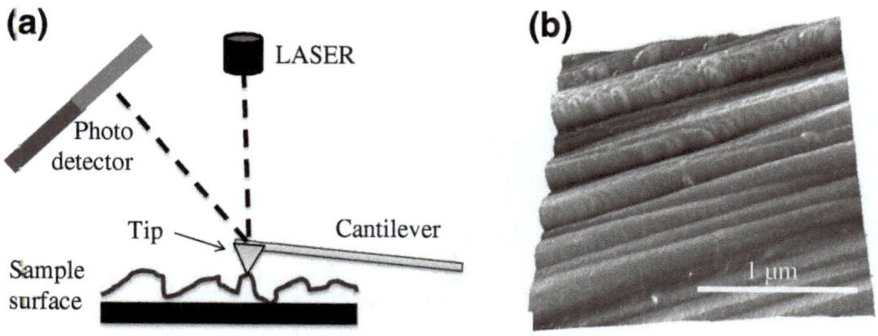

Fig. 11 **a** Measurement setup for AFM. **b** AFM image of a carbon fiber surface before surface treatment

can be done by different spectroscopic methods. Two spectroscopy methods often used for determining functional groups on fiber surfaces are presented here.

5.2.1 X-Ray Photoelectron Spectroscopy (XPS)

XPS is a method for the quantitative analysis of the elements on a solid surface. The measurement method the phenomenon that electrons are separated out of a solid by electromagnetic radiation. When the solid is exposed to X-ray source, electrons are hit by photons from the X-ray radiation with the energy $\hbar\omega$. If the energy transferred to the electron is sufficient to overcome the binding energy E_b and the electric potential Φ, the electrons will leave the solid with their rest energy E_{kin}. The energy balance equation is

$$\hbar \cdot \omega = \Phi + E_{kin} + E_b \tag{7}$$

As ω is known from the X-ray source applied, while the other factors are constants, the binding energy E_b can be calculated when the kinetic energy is measured. The binding energy E_b is characteristic for the atomic or molecular orbital where the electron comes from. Hence, the information on the binding energy of the separated electrons lead to the knowledge of elements on the specimen surface investigated [36].

The measurement setup for XPS measurements is shown in Fig. 12. The x-ray source provides the photons which excite the electrons on the sample's surface (depth of the electrons up to 3 nm). The electrons with sufficient kinetic energy will reach the hemisphere electron energy analyzer where the voltage applied between to concentric electrodes can select the electrons which will reach the detector. An electron multiplier transforms the weak electron signal into a measurable signal. The resulting spectrum is a graph with the counts per second versus the binding energies E_b. A Comparison of the binding energies with a database leads to the information of the elements and molecules on the surface [37].

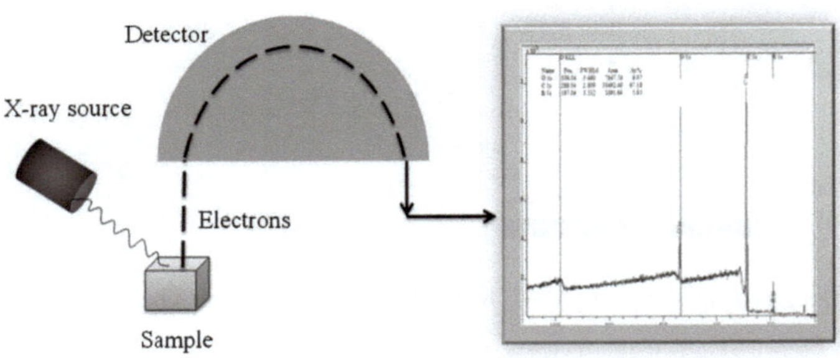

Fig. 12 Measurement setup for XPS

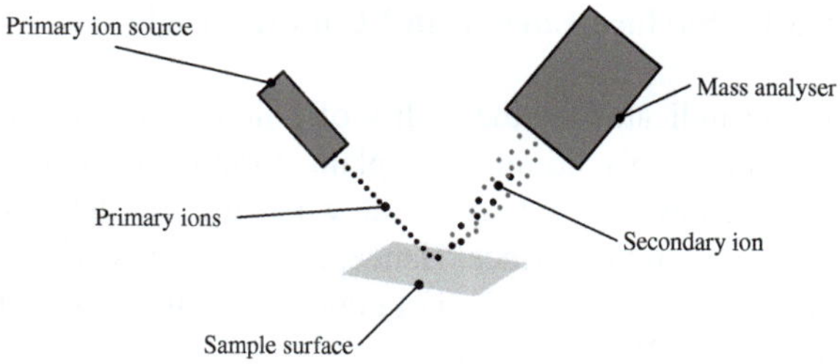

Fig. 13 Measurement setup for ToF-SIMS

5.2.2 ToF-SIMS (Time-of-Flight Secondary Ion Mass Spectroscopy)

In this measurement method, a sample's surface is exposed to ions (e.g. O_2^+, Cs^+, Ga^+, Ar^+, Bi^+) which are accelerated by a voltage of 0.2–30 keV. During the interaction of the accelerated ions, new ions (charge: neutral, negative, positive) are emitted. The measurement setup is set under a high vacuum (approx. 10^{-7} mbar). When the newly formed ions are leaving the sample's surface, they all have the same energy. Hence, light ions have a higher velocity than heavier ions. This velocity difference is used when a time-of-flight-analyzer is employed for the measurement: this analyzer consists of a vacuum tube with a fast detector at its end [38]. The experimental setup is displayed in Fig. 13.

The method is very sensitive and can separate ions as well as ion clusters up to $\frac{m}{\Delta m} > 10.000$. Therefore, also ion clusters with the same mass number can be separated. The measurement data for the ion masses are compared to a database. With the knowledge on the ions which were accelerated towards the surface (primary ions) and the comparison with the newly formed ions measured in the analyzer (secondary ions), an evaluation of the molecular composition of the sample's surface is possible [39].

5.3 *Surface Energy and Wettability*

When fiber materials shall reinforce matrix materials, the textile structures have to be impregnated with the matrix material still in a liquid condition. The processes for impregnating textiles with resins vary according to the matrix materials used and according to the application of the composite [10]. All processes have in common that the degree and quality of impregnation strongly influence the properties of the resulting composites. The wettability of the fiber material is crucial for the set-up of process parameters, as the wetting behavior has a strong impact on the cycle time for the impregnation process.

In this section, at first the impact of surface energy on the wettability are explained. Then, measurement methods for the wettability of fibers with matrix materials are explained.

5.3.1 Surface Energy and Contact Angle

As an indicator for the ability of a liquid to wet a material, the contact angle of a droplet of this liquid on a plain horizontal solid surface can be measured. The contact angle α is very small when the droplet immediately loses its form after contact with the surface. Hence, the wettability of the surface with the liquid is very high. If the wettability is not good, the contact angle α lies in the range between 180 and 90° [40] (see Fig. 14).

The wettability and hence the contact angle α depends on surface energies of the materials involved in the wetting process. The surface energy σ is defined as the amount of energy ΔW which has to be overcome in order to increase the surface by ΔA:

$$\sigma = \Delta W / \Delta A \tag{8}$$

The work ΔW which has to be overcome is due to the intermolecular cohesion within the liquid. If a liquid is brought on a surface, then the tension of the boundary layer σ_{LS} becomes important (see Fig. 15).

Following the equation $\sigma_L \cdot \cos \alpha = \sigma_s - \sigma_{LS}$, good wettability is achieved when the surface energy of the solid (fibers) is higher than the surface energy of the wetting liquid (e.g. matrix material).

Surface energies can be separated into a polar part σ_p (due to polar chemical bonds, e.g. hydrogen bridge linkage) and an unpolar part σ_u (due to non-polar chemical bonds, e.g. van-der-Waals forces) [41]:

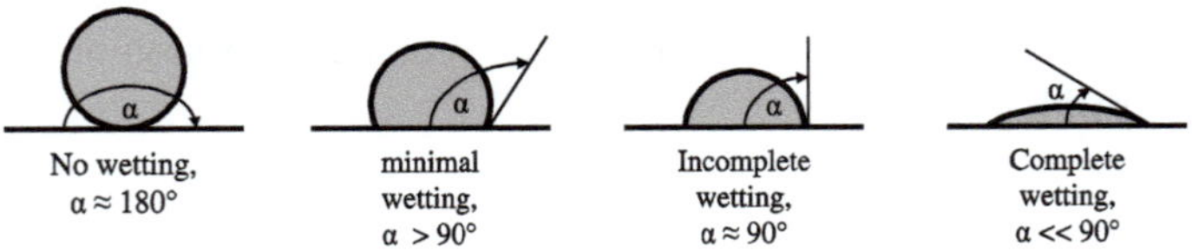

Fig. 14 Liquid droplets showing a different wetting behavior

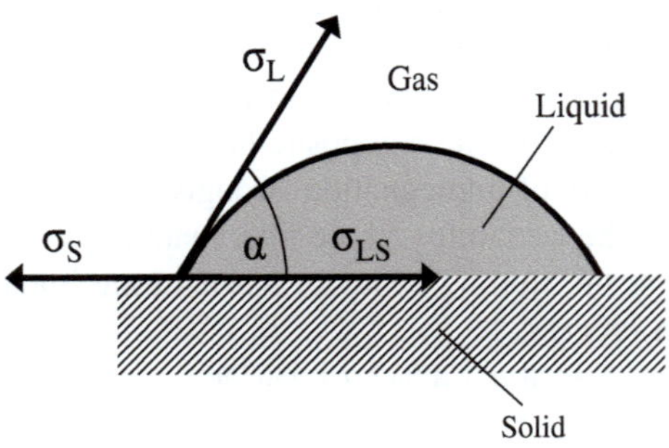

Fig. 15 Relations between surface energies of liquid and solid determine the contact angle and the resulting wettabilty

$$\sigma = \sigma_p + \sigma_u \tag{9}$$

5.3.2 Measurement of Surface Energies and Wetting Behavior

The surface energy of liquids can be measured by the Wilhelmy method, employing an inert testing body which is dipped into the liquid [40]. The measurement of the surface energy of fibres is performed by using testing liquids with a defined surface energy. Using testing liquids with defined polarity leads to discovering the polarity properties of the solid (fibres) [42].

These measurements are performed with a tensiometer (see Fig. 16a), which measures the weight increase of the specimens dipped into the testing liquid [43]. If single filaments (Fig. 16b) are measured, a sufficient number of samples has to be measured in order to provide statistical significance.

The wetting behavior of fiber bundles (Fig. 16c) is described by the Washburn equation [44], while the measurement is carried in a set-up developed for powder materials [45]. The method takes into account capillary effects (expressed with the capillary constant C) and properties of the testing liquid (viscosity η, density ρ_L and surface tension σ_L). With the following equation, the contact angle can be determined from the measurement of the weight increase per time:

$$\cos\theta = \frac{2 \cdot \eta}{C \cdot \rho_L^2 \cdot \sigma_L} \cdot \frac{m^2}{t} \tag{10}$$

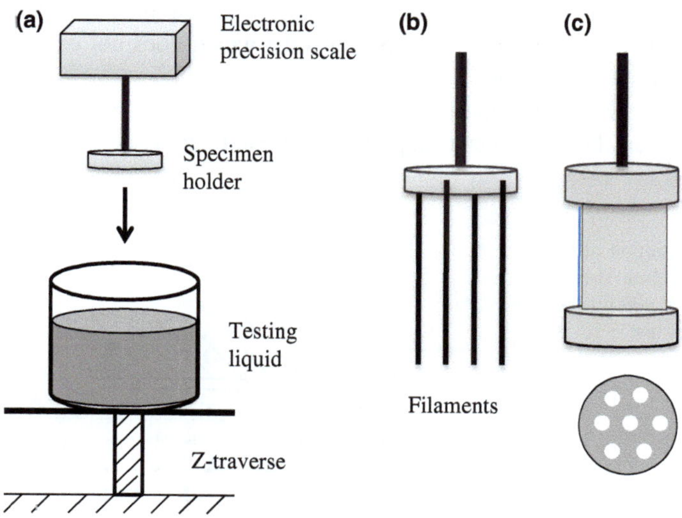

Fig. 16 **a** Wettability measurements setup using a tensiometer. **b** Sample holder for surface tension measurements on single filaments. **c** Sample holder for fiber rovings: the liquid can flow into the sample holder through the holes in the bottom

The method presented above also allows direct measurements of the wetting behavior between fibres and impregnation materials (e.g. sizings or matrix materials), replacing the testing liquid by the impregnation material.

5.4 Adhesion to Matrix

When it comes to reinforcement applications of the fibers, the fiber matrix adhesion is the composite's property which decides on its performance. In this section, at first the principle of load transfer from a matrix to a single fiber is explained. After that, different methods for measuring the fiber matrix adhesion are presented.

5.4.1 Load Transfer and Critical Fiber Length

A parameter which allows the quantification of the fibre-matrix-adhesion (FMA) is the interlaminar shear strength τ_{FMA}. The interlaminar shear strength (ILSS) is the maximum force per surface area with the unit [N/mm^2]. In other words: when the shear stress between fiber and matrix exceeds the interlaminar shear strength, a delamination of the interlayers will take place [46] (see Fig. 17).

When stress is transferred from the matrix into a fiber, to general failure behaviours can occur:

(1) Pull-Out of the fiber: the interlaminar shear strength is lower than the tensile strength of the fiber. Therefore, a delamination takes place and, due to debonding, no more stress can be transferred into the fiber. The reinforcement of the fibers is only possible up to a force which does not exceed the interlaminar shear strength.
(2) Fiber breakage: the matrix can conduct more stress into the fiber than the fiber's tensile strength can support. Hence, the fiber breaks into smaller pieces

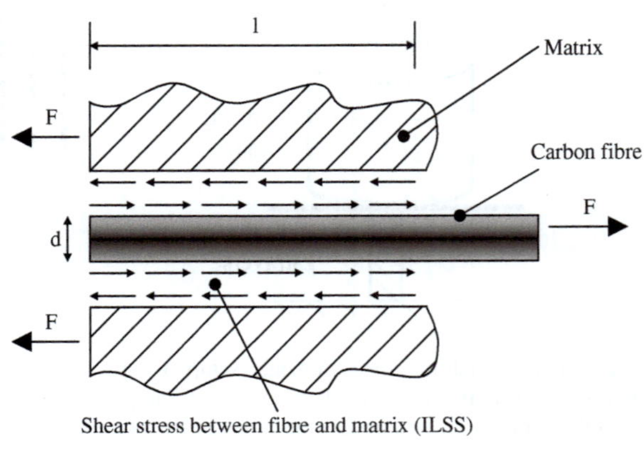

Fig. 17 Definition of the interlaminar shear strength (ILSS), illustrated for the case of a pull-out test

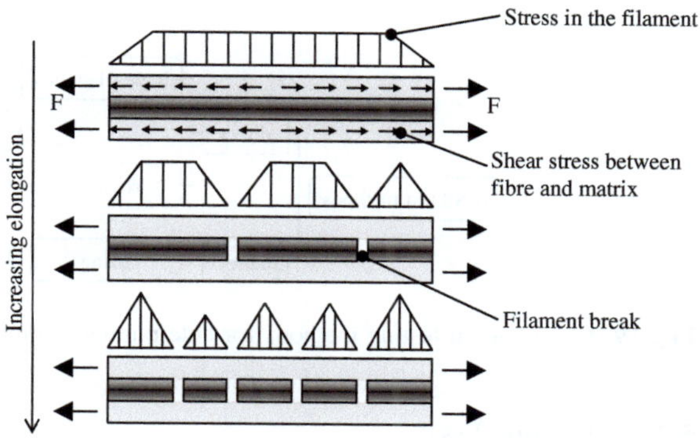

Fig. 18 Stress and fracture development for the case of a composite's elongation: load is transfered to the fibers until the critical fiber length is reached by filament breakage

and the fiber's reinforcement is ideally used until the critical fiber length (length of the fiber where no more stress can be transferred from the matrix to the fibers) is reached (see Fig. 18).

No more reinforcement is possible when the shearing force

$$F_{shear} = A_{surface} \cdot \tau_{FMA} = \pi \cdot d \cdot \frac{l_c}{2} \cdot \tau_{FMA} \tag{11}$$

is at the same level as the maximum tensile force the fiber can withstand:

$$F_{max,fiber} = A_{fiber} \cdot \sigma = \pi \cdot \left(\frac{d}{2}\right)^2 \cdot \sigma \tag{12}$$

Equalizing the two equations leads to the critical fiber length l_c:

$$l_c = \frac{\sigma}{2\tau_{FMA}} \cdot d \tag{13}$$

5.4.2 Overview on Measurement Methods for Fiber Matrix Adhesion

In order to determine the interlaminar shear strength between fibre and matrix, different testing methods (both direct and indirect) have been developed. In real samples, the interlaminar shear stress always occurs in a combination with other stresses, such as tensile stress, pressure and torsion. Nevertheless, the capacitance of fibre reinforced plastics depends significantly on the interlaminar shear strength.

In general, testing methods for the determination of the ILSS can be divided into two groups: direct and indirect testing methods (see Fig. 19).

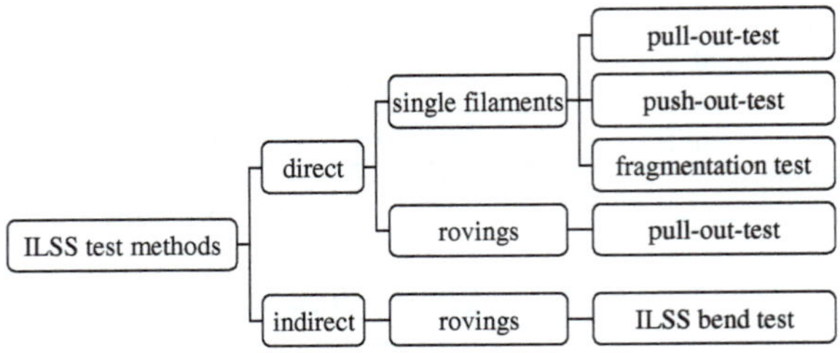

Fig. 19 Overview on testing methods for determining the interlaminar shear strength

5.4.3 Roving Tests

The industrially most commonly used test for characterizing the fiber matrix adhesion is the tree-point-bending test according to ISO 14130:1997 "Fibre-reinforced plastic composites—Determination of apparent interlaminar shear strength by short-beam method" [47]. The sample and measurement setup is shown in Fig. 20.

In the setup shown for the three-point-bending test, a specimen with length l, height h and depth d is set on two supporting axis. In the middle between the two axis, a force F is applied to the specimen until the sample breaks at the force F_{max}. The interlaminar shear strength can be calculated with the data obtained from the test by using the equation:

$$\tau \cong \frac{3 \cdot F_{max}}{4 \cdot d \cdot h} \quad (14)$$

The advantages of this method are that it can be carried out quickly with standard testing machines. The constraints of this method are that the results obtained are no absolute values. Pure shear stress is only applied in the middle of the sample during testing and the results dramatically change according to the textile structure, the fiber-volume content and the impregnation method used. Hence, the value for τ obtained in this test is called the "apparent" interlaminar shear strength and can be only compared with results obtained with the same setup and the same sample preparation (textile structure, impregnation method).

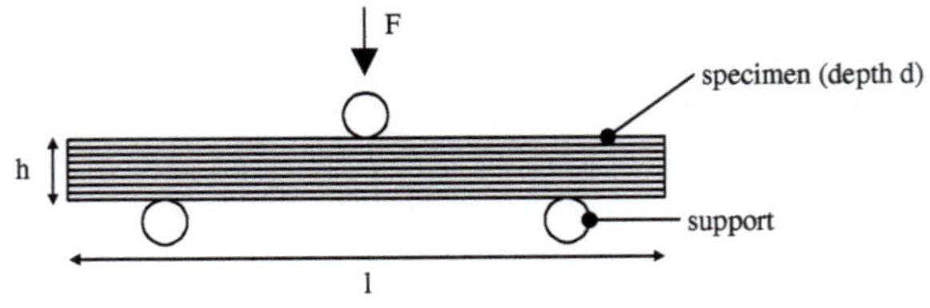

Fig. 20 Measurement setup for three-point-bending test according to ISO 14130:1997

Other common and direct testing method for the interlaminar shear strength is the pull-out test on a roving scale [48]. Therefore, the roving is embedded into resin block on both sides. Between the blocks, the roving is not embedded. One of the resin blocks is shorter than the other one so that the contact area is smaller. When a constantly increasing force pulls the ends away from each other, the roving end in the smaller resin block will delaminate at a certain force level. As the embedded length l_{emb} of the fibers in the shorter resin block, the number of filaments n in the roving and the filament diameter D is known, the interlaminar shear strength can be calculated with the equation:

$$\tau = \frac{F_{max}}{n \cdot \pi \cdot D \cdot l_{emb}} \tag{15}$$

The constraint of this method is that the impregnation has to be perfect i.e. every filament in the roving is surrounded by matrix material on the whole embedded length. Additionally, the roving has to be embedded perfectly straight in the matrix block; otherwise the real embedded length will be different from l_{emb} which was used for the calculation.

5.4.4 Single Filament Tests

Other tests for determining fiber matrix adhesion are employing single filament samples: the single filament pull-out test, the fragmentation test and the single-filament-push-out test.

The single filament pull-out-test and its sample preparation is carried out in the same way as the pull-out-test on roving scale [49].

Within the single filament fragmentation test, a filament is completely embedded into a transparent matrix material. With a special apparatus, the whole specimen is stretched with a constant rate of elongation [50]. Therefore, in the interface between matrix material and fibre surface, a constant shear stress is occurring. If the adhesion between fiber and matrix is weak, it will directly come to delamination. If the fiber matrix adhesion is strong, the shear stress in the boundary layer can induce tensile load into the fiber. If the tensile stress exceeds the tenacity σ of the fibre, the filament breaks into two parts. With further elongation, the effect is reproduced within all fracture pieces remaining until the length of the fiber pieces are shorter than the critical fiber length l_c and no more load can be transferred [51].

The testing procedure is carried out under a light microscope with polarized light while photos of the specimen are taken. The load transfer which has been taking place between fiber and matrix material can be observed by micro cracks (see Fig. 21). The micro cracks appear in white colour. Also the fractures of the fiber can be observed. The microscope images of the elongated samples allow a quantitative comparison of specimen with different grades of fiber matrix adhesion by taking into account the form of the micro crack area.

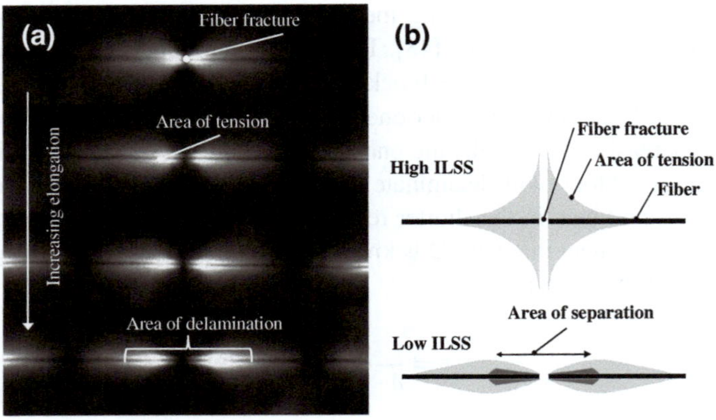

Fig. 21 **a** Microscope images of delamination. **b** Schematic illustration of tension distribution in fiber reinforced composites with different values of ILSS

By counting the number of fractures at different elongation stages, the critical fiber length can be determined and the interlaminar shear strength can be calculated according to Eq. (13).

Both the single filament pull-out test as well as the single filament fragmentation test require an effort in sample preparation as in both cases, single filaments have to be prepared out of a roving. As the effects due to the presence of other filaments is neglected in theses testing methods, the values obtained for the interlaminar shear strength can be taken as a characterization of the fibers' surfaces, but not directly intend the performance of the material as reinforcement material for composite applications.

The single filament push-out test allows the characterization of a single filament's adhesion to a matrix system in the presence of other filaments. Hence, the conditions of a real composite material are maintained. Moreover, even a detailed investigation of the composite's mechanical properties and crack propagation is possible [52]. The sample preparation (thin sections of approx. 100 μm) is very complex, but on one sample, a nearly unlimited number of single filament push-outs can be carried out.

The principle of the single filament push-out can be seen in Fig. 22. A thin section of a fiber reinforced composite is layed above a substrate with a cavity. The set-up is put into the testing chamber of a nanomechanical testing device. The nanoindenter approaches the sample's surface and the indenter tip pushes out the single filaments into the cavity. The force applied for the push-out is measured as a function of displacement of the indenter tip [53].

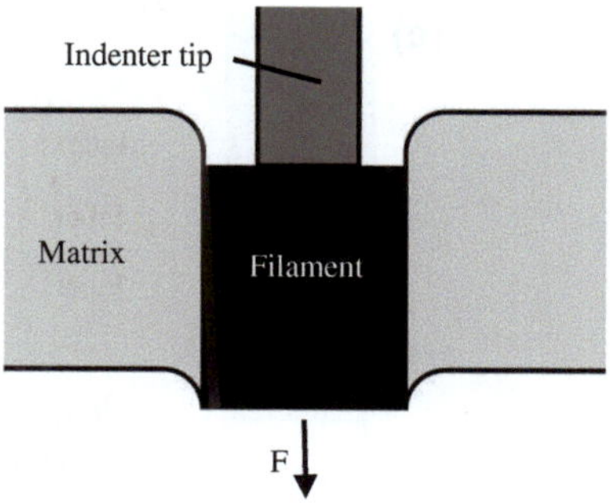

Fig. 22 Measurement principle for a single filament push-out-test

6 Physical Properties

In this section, further physical properties of fibres are described together with suitable analytical methods for measuring them.

6.1 Density

The density of a reinforcement fibre is an important measure, because it is the main factor determining the fibre's potential for lightweight construction beside its mechanical properties [9]. Furthermore, the measurement of density is an important tool for quality control in fibre production, because it can be used to determine whether defects are present (e.g. pores leading to a lower density) or production processes are completed (e.g. density increase in carbon fibre conversion) [3].

For the density measurement of fibres, several methods can be applied as indicated in Fig. 23. The following methods are suitable:

(a) *Density gradient column*: The method is suitable for a direct measurement of the density. Two miscible liquids with different densities are mixed in a column. Due to gravitational effects, a density gradient is found in the column. By putting floats with defined density into the column, a density index can be marked on the column. If fibers with a certain density are put into the column, they sink to the place with corresponding density and start to float, to that their density can be obtained from the scale [54]. The method is very precise if high columns and the right liquids are used for a specific fiber material. However, it can mainly be used for polymeric fibers or other fibers with low density because of the availability of suitable liquids [55].

(b) Archmedes' principle: The density is determined by measuring mass and volume of the fibres. Fibres with a defined mass m are put into a sample holder mounted on a precise scale, which is lowered to a liquid with a defined density ρ_l (lower than the fiber's density). By measuring the mass in the liquid m'

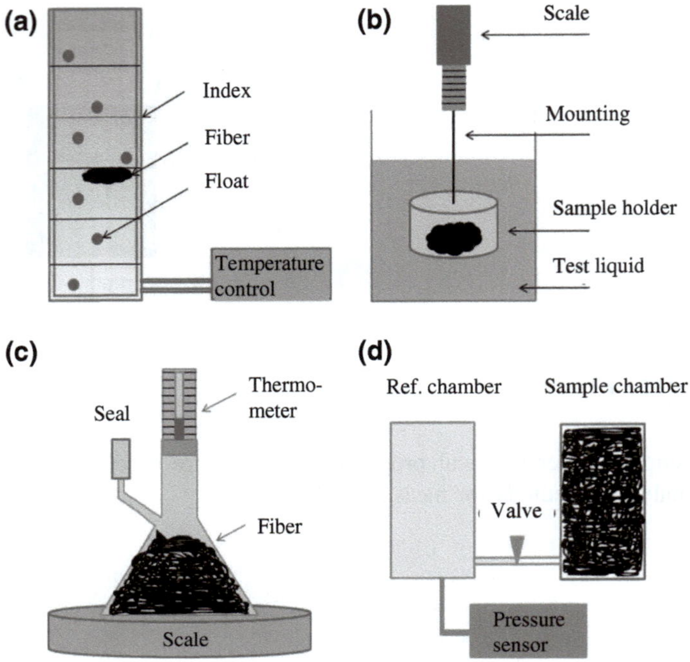

Fig. 23 Methods for measuring the density of fibers. **a** Density gradient column method [54]. **b** Archimedes' principle. **c** Pycnometry (with liquids). **d** Gas pycrometry

(mass change due to the buoyancy force), the fiber's density can be calculated [54, 55]:

$$\rho = \frac{m}{m - m'} \cdot \rho_l \tag{16}$$

(c) Pycnometry: Fibers with a defined mass m are put into a pycnometer with a precisely defined volume V, which is then filled with a liquid with known density ρ_l. By using a thermometer, the liquid's density can precisely be determined from corresponding tables as a function of temperature. By measuring the total mass of fibers and liquid M in the totally filled pycnometer, the density of the fibers can then be calculated from the mass difference to the pycnometer filled only with liquid [54]:

$$\rho = \frac{m}{V \cdot \rho_l - (M - m)} \rho_l \tag{17}$$

(d) Gas pycnometry: Gas pycnometry is a very precise method for measuring the volume V of fibres. Together with a precise scale, the density can be obtained. The fibers are filled into a sample chamber with volume V′, which is then evacuated. After evacuation of the chamber, a gaseous medium (usually helium) is filled into the sample chamber from a reference chamber of the size V″ by opening a valve. By measuring the pressure in the chambers (p before expansion, p′ after expansion) and taking into account the ideal gas equation for the test gas, the remaining volume of the test chamber can be obtained and then be used to calculate the fibers' volume [56, 57]:

$$V = V' - \frac{V''}{\frac{p}{p'} - 1} \tag{18}$$

6.2 *Thermal Properties*

Important thermal properties of reinforcement fibers are described in this section. These are temperatures for phase transitions, heat capacity and thermal conductivity.

6.2.1 Phase Transitions

Phase transitions like glass transition or melting of materials mainly determine the temperature range for their application (e.g. decreasing Young's modulus and tenacity above the glass transition temperature) as well as their processing range (e.g. above their melting temperature for thermoplastic materials) [58].

Phase transitions in fibers are measured by Differential Scanning Calorimetry (DSC). The fiber sample is put into a crucible with good heat contact. The sample and an empty crucible acting as a reference are put onto a sensor with thermocouples (see Fig. 24a). The surrounding chamber is then heated or cooled with a specific temperature profile. By measuring the temperature difference between sample and reference, latent heat in the fiber can be identified (see Fig. 24b). Since the temperature difference is proportional to the heat flow, it can be determined quantitatively by calibrating the method with substances with known melting enthalpy [58].

According to [59], a standard method is available for the determination of transition temperatures. For more specific analysis, special methods with adjusted temperature profiles have to be developed to measure the desired properties. A direct measurement of heat flow is also possible by using so-called power compensating DSC, where two different furnaces (sample and reference) are heated or cooled and differences in power consumption are measured.

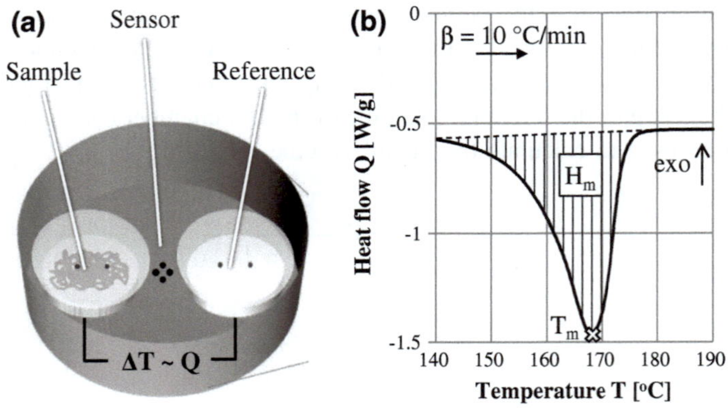

Fig. 24 Differential scanning calorimetry. **a** Sensor for determination of heat flow in sample and reference crucible. **b** Heat flow during the melting of a polymeric fiber

6.2.2 Heat Capacity

The heat capacity c_p describes the amount of energy needed to heat the material by one °C. It can be also determined by DSC, whereas two methods can be applied.

1. The reference crucible is filled with a substance with the same mass and known heat capacity (usually sapphire). By measuring differences in heat flow at temperature ranges with no phase transitions, the heat capacity can be calculated [60].
2. By applying temperature modulation to the standard temperature profile and measuring the time-dependent response of the sample, the reversing part of the heat flow can be identified, which is correlated to the heat capacity [58, 61].

6.2.3 Thermal Conductivity

The thermal conductivity κ of a fiber material describes its property to transfer heat. Due to anisotropic structures, the heat capacity can differ in the directions parallel or perpendicular to the fiber axis (e.g. in carbon fibers, where oriented graphite structures along the fiber increase the thermal conductivity in that direction). Furthermore, the thermal conductivity of textiles or composites is usually better in the direction of oriented fibers, since heat can be conducted along the fibers without heat transfer to other filaments or filament bundles. By adjusting fiber materials and their orientation within the composite, materials with good heat conductivity or thermal insulation in specific directions can be designed.

Single fibres can be tested on their thermal conductivity with special analytical methods. An individual fibre is mounted on a hot wire and temperature changes at the end of the fibre are measured as a function of time [62].

6.3 Thermomechanical Properties

Thermomechanical properties of fibers are important to analyze differences in fiber properties in their application's temperature range. Beside conducting mechanical tests (as described above) at defined temperatures, especially the measurement of thermal expansion (or shrinkage) of the fibers is important. Shrinkage of a fiber can have an impact during the processing of fibers (e.g. in carbon fiber conversion, where draw ratios have to be adjusted) or composites (e.g. by decreasing the fibers' orientation in a textile, which influences the mechanical properties of the composite). Furthermore, internal stress or delamination in the composite part can take place if the thermal expansion of fibers and matrix do not match [1, 9, 10].

The shrinkage behavior of fibers can be measured by standard shrinkage tests or more precisely by thermo-mechanical analysis (TMA). A possible setup is displayed in Fig. 25. Single filaments are fixed at a constant length or force between two clamps and the resulting shrinkage forces or length changes are measured as a function of temperature, while the sample is heated or cooled with a specific temperature profile [63].

6.4 Electrical Properties

The electrical properties of a material can be described with the help of its electrical conductivity σ and dielectrical constant ε. Like other physical properties (e.g. thermal conductivity), anisotropic electrical properties can be found in fiber materials due to oriented structures. Depending on the orientation and distribution of fibers, anisotropic electrical properties can also be found in the composite material. The following applications are of great importance:

1. Conductive fibers (especially short or long fibers) are placed in a non-conducting matrix material and can form conductive paths within the material, if a so-called percolating network is formed from the single filaments.

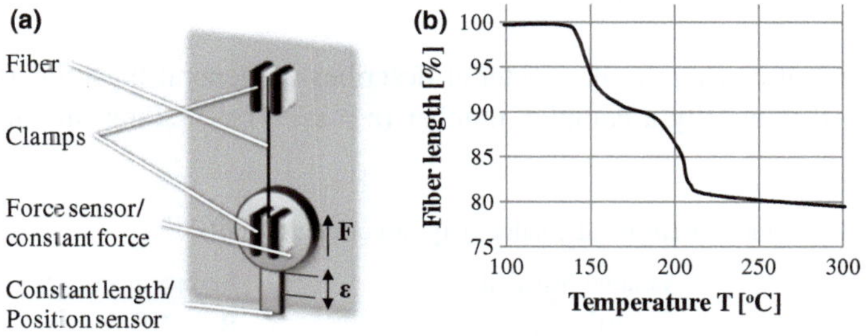

Fig. 25 Thermo-mechanical analysis. **a** Experimental setup. **b** Specific length as a function of temperature for a PAN fiber

If the fiber volume content passes the percolation threshold, a spontaneous increase of conductivity can be found at this point [64].

2. Anisotropic textile structures (e.g. with uni- or bidirectional fiber placement) made from conductive fibers increase the composite's conductivity in fiber direction. Such structures can be placed in non-conductive composites to remove electrical charges from their surface [65].

6.4.1 Electrical Conductivity

Usually, materials are classified regarding their specific conductivity or their specific resistivity. Resistivity can be measured in the volume (ρ in $\Omega \cdot m$) or on their surface (R_o in Ω). As displayed in Table 2, the following classes of materials are defined according to [66, 67].

Surface and volume conductivity can be easily measured in composites according to the standards. Anisotropic properties have to be taken into account by measuring in different directions of the material.

For fiber materials, the test procedure has to be adapted. The volume conductivity can only be measured easily in the direction of fibers. Fibers are clamped between two connectors, so that the resistivity of the connections and contacts have to be taken into account. The contact resistivity should be decreased by applying highly conductive material (e.g. silver-based coatings) on the fiber. This procedure is especially important to ensure the contacting of all filaments if fiber bundles are measured. The resistivity values have to be measured at different fiber lengths to allow the calculation of material's conductivity by linear regression [68]. A possible setup for conductivity measurements and the equivalent circuit diagram (taking into account the contact resistivity) are displayed in Fig. 26.

Conductivity tests can be either performed with direct current (DC) or alternating current (AC) at different frequencies. The variation of current frequency allows the investigation if complete conductive paths or single conductive domains only contributing to AC conductivity are present in the fibre [69].

6.4.2 Dielectrical Constant

The dielectrical constant ε of a material describes its general behavior in an electrical field. It is usually a complex number ($\varepsilon = \varepsilon_1 - i \cdot \varepsilon_2$) taking into account the

Table 2 Classification of electrical conductivity (according to [66])

Material class	Volume resistivity	Surface resistivity
Conductive	$\rho < 10^4\ \Omega \cdot m$	$R_o < 10^4\ \Omega$
Discharging	$10^4\ \Omega \cdot m < \rho < 10^{11}\ \Omega \cdot m$	$10^4\ \Omega < R_o < 10^{11}\ \Omega$
Insulating	$10^{11}\ \Omega \cdot m < \rho$	$10^{11}\ \Omega < R_o$

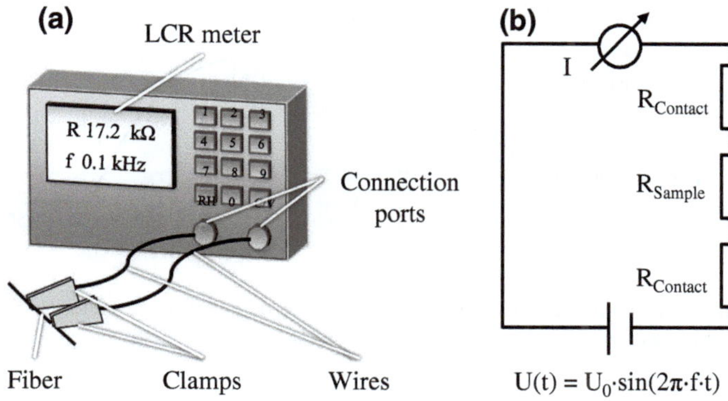

Fig. 26 **a** Experimental setup for measurement of AC resistivity of fiber samples. **b** Equivalent circuit diagram for resistivity measurement of fibers

polarization of the material in an electrical field (real part ε_1) and the dielectrical losses (imaginary part ε_2). Dielectrical losses are an important measure for special applications, e.g. for microwave absorption during composite manufacturing or shielding of electromangnetical fields [70].

Dielectrical properties of a fiber material can be measured by dielectrical spectroscopy. Fibres with a defined geometry are placed between two condensator plates, so that the complex capacity of the condensator with the mounted fibres can be measured as a function of frequency. Taking into account the geometry of the fibres acting as a dielectric between the plates, the dielectrical constant can be calculated.

7 Durability

The durability of fibres has to be taken into account both for the manufacturing of composites as well as during their period of application (even if the fibres are covered with matrix material in the composite, smaller cracks or pores allow the interaction of surrounding media with the fibre). The most suitable test method for all types of durability is to expose to the fibres (or composites) to the specific environmental conditions and test the relevant fibre properties after exposition. For certain applications such as automotive, standardized test methods are available describing the specific environmental conditions. The following types of durability should be taken into account:

1. Thermal stability and degradation: The influence of temperature is present during the curing of the matrix material (especially in thermoplastic matrices or high-temperature matrices) and also in specific applications. Specific temperature ranges to be tested are defined for certain fields of application such as

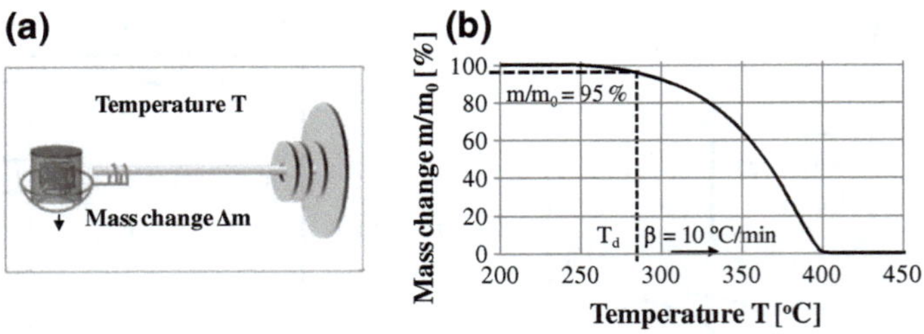

Fig. 27 Thermogravimetic analysis for determination of degradation behavior. **a** Experimental setup. **b** Specific mass as a function of temperature for a polymeric fiber

aerospace or automotive. Especially, the influence on tensile properties should be taken into account. Materials can shrink, melt or degrade at high temperatures or become brittle at low temperatures. Periodic changes in temperature, even if temperature ranges do not exceed certain limits, can lead to structural changes in the fibres which influence the tensile properties.

Temperature ranges to be examined can be identified using DSC (see above, thermal properties) for the identification of phase changes or thermogravimetric analysis (TGA, see Fig. 27a). TGA allows the determination of thermal degradation temperature T_d, which is defined at a relative mass loss of 5 % of the material (see Fig. 27b) [71].

2. Chemical stability: During the curing of matrix systems (e.g. in special systems like concrete), pH values can drastically change from a neutral value. Furthermore, fibre materials may get in contact with alkaline solutions or acids in their fields of application. The stability can be tested by exposing the material to the specific pH value for a period of time and, if necessary, at elevated temperatures. For assessing long-term effects, which may occur after years of application, more harsh conditions can be chosen to study possible effects qualitatively. After exposition, the specific fibre properties can be tested. Degradation due to exposure to chemicals can be tested by determining the fibres weight before and after testing.
3. Mechanical stability: Even if fibres or composites are loaded far below their maximum strength in the linear elastic region, fatigue can occur in the material in the form of plastic deformation, creep or cracks. The behavior can be examined by fatigue tests, where a certain load is applied periodically. Effects on the fibre material and composites can then be studied by the test methods described above [72].

References

1. Wulfhorst B, Gries T, Veit D (2006) Textile technology. Hanser, Munich
2. Elices M, Llorca J (2002) Fiber fracture. Elsevier, Kidlington

3. Morgan P (2005) Carbon fibers and their composites. CRC Press, Boca Raton
4. Pakravan HR, Jamshidi M, Latif M, Pacheco-Torgal F (2012) Influence of acrylic fibers geometry on the mechanical performance of fiber-cement composites. J App Pol Sci 125 (4):3050–3057
5. Liu X, Wang R, Wu Z, Liu W (2012) The effect of triangle-shape carbon fiber on the flexural properties of the carbon fiber reinforced plastics. Mater Let 73:21–23
6. Liu Y, Chae HG, Choi YH (2015) Preparation of low density hollow carbon fibers by bi-component gel-spinning method. J Mater Sci 50:3614–3621
7. Houis S, Schreiber F, Gries T (2008) Fiber-Table according to P.-A. Koch. In: Bicomponent fibres, Shaker, Aachen
8. ASTM D5103—07 (2012) Standard test method for length and length distribution of manufactured staple fibers (Single-Fiber Test)
9. Schürmann H (2005) Konstruieren mit Faser-Kunststoff-Verbunden. Springer, Berlin
10. Ehrenstein GW (2006) Faserverbund-Kunststoffe. Hanser, Munich
11. DIN 53811: 1970–07. Prüfung von Textilien—Faserdurchmesser-Messung in Mikroprojektion der Längsansicht
12. DIN EN ISO 1973: 1995–12. Textilien—Fasern—Bestimmung der Feinheit—Gravimetrisches Verfahren und Schwingungsverfahren
13. Eichhorn S, Hearle JWS, Jaffe M, Kikutani T (2009) Handbook of textile fibre structure: fundamentals and manufactured polymer fibres, vol 1. CRC Press, Boca Raton
14. Salem DR (2000) Structure formation in polymeric fibers. Hanser, Munich
15. Steinmann W, Vad T, Seide G, Gries T, Roth G (2012) Simultaneous analysis of structure and orientation in polymeric fibers by wide-angle X-ray diffraction. Paper presented at the MSE2012 materials science and engineering, Darmstadt, 25–29 Sept 2012
16. Steinmann W, Saelhoff AK, Seide G, Gries T (2014) Analysis of structure formation in carbon fibers: a new method for process development. Paper presented at the ISF 2014 international symposium on fiber science and technology, Tokyo, 28 Sept–1 Oct 2014
17. Steinmann W, Seide G, Gries T (2014) How carbon fibers can get stronger: structure based process development. Paper presented at AUTEX 2014: 14th World textile conference, Bursa, 26–28 May 2014
18. Prevorsek D, Oswald H (1990) Melt-spinning of PET and nylon fibers. In: Schultz JM, Fakirov S (eds) Solid state behavior of linear polyesters and polyamides. Prentice Hall, Englewood Cliffs
19. Bennett SC, Johnson DJ (1978) Strength structure relationships in PAN-based carbon fibres. Paper presented at 5th London international carbon and graphite conference, 1978
20. Ziabicki A (1976) Fundamentals of fiber formation. Wiley, London
21. Borchardt-Ott W (2009) Kristallographie: Eine Einführung für Naturwissenschaftler. Springer, Berlin
22. Schafer et al (1990) High-resolution electron microscopy observations of carbon fibre structures. Acta Polym 41:515–518
23. ASTM D3822/D3822M—14: Standard test method for tensile properties of single textile fibers
24. Textechno GmbH (2015) FAVIMAT+ ROBOT2 and AIROBOT2. http://www.textechno.com/index.php/en/fibre-testing-products-65/favimat-robot-products-145. Accessed 16 Jun 2015
25. ASTM D2343—09 Standard test method for tensile properties of glass fiber strands, yarns, and rovings used in reinforced plastics
26. ASTM D4018—11 Standard test methods for properties of continuous filament carbon and graphite fiber tows
27. Anderson J (2015) Tensile strength of single fibers: test methods and data analysis. http://cost-fp0802.tuwien.ac.at/fileadmin/mediapool-cost/Diverse/Stockholm_Workshop/Andersons.pdf. Accessed 16 Jun 2015

28. Asad RAM, Yu W, Zheng Y, He Y (2015) Characterization of prickle tactile discomfort properties of different textile single fibers using an axial fiber-compression-bending analyzer. Text Res J 85(5):512–523
29. Naito K, Tanaka Y, Yang JM, Kagawa Y (2009) Tensile and flexural properties of single carbon fibres. Paper presented at ICCM 17 international committee on composite materials, Edinburgh, 27–31 July 2009
30. Saelhoff AK, Jäger M, Steinmann W, Gries T (2014) Surface treatment of carbon fibers—increasing the interlaminar shear strength in CFRP. Paper presented at the ADITC2014 Aachen-Dresdner international textile conference, Dresden, 27–28 Nov 2015
31. Steinmann W, Wulfhorst J, Walter S, Seide G, Gries T (2012) Nanoparticles in polymeric fibers: novel possibilities for the modification of surface, mechanical and electrical properties. Paper presented at the nanoscience conference, Yaiza, 14–17 Feb 2012
32. Rosiepen C, Beck T, Hehl A, Gries T (2012) Tribology of textiles: challenging carbon fibre processing. Paper presented at the MSE2012 materials science and engineering, Darmstadt, 25–29 Sept 2012
33. ISO 9277:2010: Determination of the specific surface area of solids by gas adsorption—BET method
34. Braunauer S, Emmet PH, Teller E (1938) Adsorption of gases in mulitmolecular layers. Am Chem Soc 60(2):309–319
35. Binnig G, Quate CF, Gerber C (1986) Atomic force microscope. Phys Rev Lett 56(9):930–933
36. Cardona M, Ley L (eds) (1979) Photoemission in solids I, II. Topics Appl Phy 1:26–27 (Springer, Berlin)
37. Hüfner S (1996) Photoelectron spectroscopy, principles and applications. Solid-State Sciences, vol 82. Springer, Berlin
38. Benninghoven A, Rüdenauer FG, Werner HW (1987) Secondary ion mass spectrometry: basic concepts, instrumental aspects, applications, and trends. Wiley, New York
39. Liebl H (1967) Ion microprobe mass analyzer. J Appl Phys 38:5277–5280
40. Adamson AW, Gast AP (1997) Physical chemistry of surfaces. Wiley, Hoboken
41. Ström G, Fredriksson M, Stenius P (1987) Contact angles, work of adhesion, and interfacial tensions at a dissolving Hydrocarbon surface. J Colloid Interface Sci 119:352–361
42. Fowkes FM (1964) Predicting attractive forces at interfaces. Ind Eng Chem Res 56:40–53
43. Hoecker F, Karger-Kocsis J (1996) Surface energetics of carbon fibers and its effects on the mechanical performance of CF/CP composites. J Appl Polym Sci 59:139–153
44. Washburn EW (1921) The dynamics of capillary flow. Phys Rev 17:273
45. Bruil HG, Aartsen JJ (1974) The determination of contact angles of aqueous surfactant solutions on powders. Colloid Polym Sci 252:32–38
46. Drazal L, Madhukar M (1993) Fibre-matrix adhesion and its relationship to composite mechanical properties. J Mater Sci 28:569–610
47. ISO 14130:1997—Fibre-reinforced plastic composites—determination of apparent interlaminar shear strength by short-beam method
48. Michaeli W, Fölster T, Klink R, Kocker K (1993) Faserbündel-Pull-Out-Versuch. Kunststoffe 83:65–69
49. Bannister DJ, Andrews MC, Cervenka AJ, Young RJ (1995) Analysis of the single-fibre pull-out test by means of Raman spectroscopy: part II. Compos Sci Technol 53:411–421
50. Feih S, Wonsyld K, Minzari D, Westermann P, Lilholt H (2004) Testing procedure for the single fiber fragmentation test. Risø National Laboratory, Roskilde
51. Netravali AN, Henstenburg RB, Phoenix SL, Schwartz P (1989) Interfacial shear strength studies using the single-filament-composite test. Polym Compos 10(4):226–241
52. Jäger J, Sause MGR, Burkert F, Moosburger-Will J, Greisel M, Horn S (2015) Influence of plastic deformation on single-fiber push-out tests of carbon fiber reinforced epoxy resin. Compos A 71:157–167
53. Greisel M, Jäger J, Moosburger-Will J, Sause MGR, Mueller WM, Horn S (2014) Influence of residual thermal stress in carbon fiber-reinforced thermoplastic composites on interfacial fracture toughness evaluated by cyclic single-fiber push-out tests. Compos A 66:117–127

54. ISO 10119-2 Carbon fibre—determination of density
55. Truong M, Zhong W (2009) A comparative study on natural fibre density measurement. J Text Inst 100(6):525–529
56. GmbH Micromeritics (2010) Highly adaptable density determinations—the AccyPyc II 1340 gas displacement pycnometrie system. Micromeritics GmbH, Mönchengladbach
57. DIN 51913 Bestimmung der Dichte mit einem Gaspyknometer (volumetrisch) unter Verwendung von Helium als Messgas
58. Steinmann W, Walter S, Beckers M, Seide G, Gries T (2013) Thermal analysis of phase transitions and crystallization in polymeric fibers. In: Elkordy AA (ed) Applications of calorimetry in a wide contet: differential scanning calorimetry, isothermal titration calorimetry and minicalorimetry. InTech Europe, Rieka, pp 277–306
59. ASTM D3418 Standard test method for transition temperatures and enthalpies of fusion and crystallization of polymers by differential scanning calorimetry
60. Takahashi Y (1985) Latent heat measurement by DSC with sapphire as standard material. Thermochim Acta 88(1):199–204
61. Shawe JEK, Hütter T, Heitz C, Alig I, Lellinger D (2006) Thermochim Acta 446(1,2): 147–155
62. Wang JL, Gu M, Zhang X, Song Y (2009) Thermal conductivity measurement of an individual fibre using a T type probe method. J Phys D 42:105502–105509
63. Heine M (1988) Optimierung der Reaktionsbedingungen von thermoplastischen Polymer-Fasern zur Kohlenstoffaser-Herstellung am Beispiel von Polyacrylnitril. Universität Karlsruhe, Dissertation
64. Steinmann W, Vad T, Weise B, Wulfhorst J, Seide G, Gries T, Heidelmann M, Weirich T (2013) Extrusion of CNT-modified polymers with low viscosity—influence of crystallization and CNT orientation on the electrical properties. J Polym Polym Compos 21(8):473–482
65. Kilbride M, Pethrick RA (2012) Enhancement of the surface electrical conductivity of thermoplastic composite matrices. J Mater Des Appl 226(3):252–264
66. Bundesanstalt für Arbeitsschutz und Arbeitsmedizin (BAuA) (2009) Vermeidung von Zündgefahren infolge elektrostatischer Aufladungen. BAuA, Dortmund
67. ANSI/ESD S541-2008 For the protection of electrostatic discharge susceptible items: packaging materials for ESD sensitive items
68. Steinmann W (2014) Elektrisch leitfähige Fasern aus Polymer-Nanoverbundwerkstoffen. RWTH Aachen, Dissertation
69. Glauß B, Steinmann W, Walter S, Beckers M, Seide G, Gries T, Roth G (2013) Spinnability and characteristics of polyvinylidene fluoride (PVDF)-based bicomponent fibers with a carbon nanotube (CNT) modified polypropylene core for piezoelectric applications. Materials 6 (7):2642–2661
70. Elimat ZM, Hamideen MS, Schulte KI, Wittich H, de la Vega A, Wichmann M, Buschhorn S (2010) Dielectric properties of epoxy/short carbon fiber composites. J Mater Sci 45: 5196–5203
71. ASTM D3850 Test method for rapid thermal degradation of solid electrical insulating materials by thermogravimetric method (TGA)
72. Gude M, Hufenbach W, Koch I, Koschichow R, Schulte K, Knoll J (2013) Fatigue testing of carbon fibre reinforced polymers under VHCF loading. Procedia Mater Sci 2:18–24

Fiber Architectures for Composite Applications

Kadir Bilisik, Nesrin Sahbaz Karaduman and Nedim Erman Bilisik

Abstract Two dimensional (2D) and three dimensional (3D) fabric architectures, methods, processes and some of their properties have been reviewed in this chapter. 2D woven, braided, knitted and nonwoven fabrics are used for structural composites but they exhibit delamination between the layers. Triaxial fabrics, on the other hand, have an open structure and low fabric volume fraction. But in-plane properties of triaxial fabric are homogeneous due to ±bias yarns. 3D woven fabrics show no delamination due to the z-fibers. But, possess low in-plane mechanical properties. 3D braided fabrics also show no delamination but they have low transverse properties due to lack of transverse (filling) yarns. Multiaxis 3D knitted fabrics exhibit no delamination and, the in-plane properties are enhanced due to the ±bias yarn layers but, it has limitations for the number of layers. Similarly, multiaxis 3D woven fabrics also have multiple layers and suffer no delamination due to the z-fibers and in-plane properties are enhanced due to ±bias yarn layers. However, multiaxis 3D technique is still in its early stages of development and this is the future technological challenge in multiaxis 3D preform fabrication for composite application.

K. Bilisik (✉)
Department of Textile Engineering, Faculty of Engineering, Erciyes University, 38039 Talas-Kayseri, Turkey
e-mail: kadirbilisik@gmail.com

N.S. Karaduman
Akdagmadeni Vocational High School, Bozok University, 66300 Akdagmadeni-Yozgat, Turkey
e-mail: nesrin38@gmail.com

N.E. Bilisik
Department of Electronic Engineering, Faculty of Engineering, Istanbul Kultur University, 34156 Bakirkoy-Istanbul, Turkey
e-mail: ermanbilisik66@gmail.com

© Springer Science+Business Media Singapore 2016

S. Rana and R. Fangueiro (eds.), *Fibrous and Textile Materials for Composite Applications*, Textile Science and Clothing Technology,
DOI 10.1007/978-981-10-0234-2_3

1 Introduction

Textile structural composites are widely used in various industrial applications, such as civil and defense [1, 2] as they have better specific properties compared to basic materials such as metal and ceramics [3, 4]. Research conducted on textile structural composites indicated that they can be considered as alternative materials since they are delamination-free and damage-tolerant [3, 5]. From a textile processing viewpoint, they are readily available, cheap, and not labor intensive [1]. The textile preform fabrication is made by weaving, braiding, knitting, stitching, and by using nonwoven techniques, and they can be chosen generally based on end-use requirements. Simple three dimensional (3D) preform consists of two dimensional (2D) fabrics that are stitched together on the basis of a determined stacking sequence. 3D woven preforms are fabricated by using specially designed automated looms and manufactured to near-net shape in order to reduce scrap [6, 7]. However, it was indicated that they have low in-plane properties due to the through-the-thickness fiber reinforcement [1, 2, 8]. Simple 3D braided preform consists of 2D biaxial fabrics that are stitched together depending on a certain stacking sequence. 3D braiding is a preform production technique used in the multidirectional near-net shape manufacturing of high damage tolerant structural composites [9–12]. 3D braiding is highly automated and readily available. Generally, 3D braided preforms are fabricated by traditional maypole braiding (slotted horn gear matrix) or innovative 4-step and 2-step braiding (track and column) or more recently by 3D rotary braiding and multi-step braiding [11, 13–17]. The fabrication of small sectional 3D braided preforms is cheap, and is not labor intensive [1]. However, the fabrication of large sections of 3D braided preform may not be feasible due to position displacement of the yarn carriers. Simple 3D knitted preform includes 2D warp and weft knitted fabrics. These fabrics can be fabricated in various shapes such as spheres, cones, ellipsoids and T-pipe junctions. However, 2D knitted composites generally show low mechanical properties due to their looped architecture and low fiber volume fraction. 3D nonwoven preform is a web or felt structure which is composed of short fibers. It is formed by needle-punching, stitch-bonding, high-frequency welding, ultrasound or laminating. However, it demonstrates low mechanical properties due to lack of fiber continuity. Multiaxis knitted preform has four fiber sets such as ±bias, warp (0°) and weft (90°) together with stitching fibers which enhance the in-plane properties [18]. It was explained that multiaxis knitted preforms suffer from limitations in fiber architecture, the through-thickness reinforcement due to thermoplastic stitching thread and three dimensional shaping during molding [3]. Multiaxis 3D woven preform enhances the in-plane properties of the composites by positioning the fibers in the preform off-axis direction [19, 20]. Multi-step braiding is a relatively new concept and with this technique it is possible to make multidirectional 3D braided preforms by orienting the fibers in various directions in the preform [21].

2 Types of Advanced Architectures

2.1 2D Architectures

Woven Fabric

2D woven fabric is the most widely used material in the composite industry. It has two yarn sets as warp (0°) and filling (90°) which are interlaced to each other to form the woven fabric surface. Basic weave types are plain, twill and satin. Various derivative weave patterns can be made for specific applications. Some of them are shown in Fig. 1 [22]. In plain weave, each warp passes alternately under and over each weft. The plain woven fabric is symmetrical, with good directional stability but it has high crimp and is difficult to shape during molding. Basket weave is a derivative of the plain weave which has two or more warp yarns alternately interlacing with two or more weft yarns. It has less crimp and good drapability but it is unstable. In twill weave, one or more warp yarns alternately weave over and under two or more weft yarns in a regularly repeated manner. The twill woven fabric has a smoother surface and low crimp. It has a good wettability and drapability. However, it shows low stability compared to the plain weave. In satin weave, one or more warp yarns alternately weave over and under two or more weft yarns to make fewer intersections. Therefore, it has a smooth surface; good wettability and a high degree of drapability. It has also low crimp. However, it has low stability and an asymmetrical structure. Another 2D woven architecture is a leno weave in which adjacent warp yarn is twisted around consecutive weft yarn. One of the derivatives of the leno weave is called mock leno in which occasional warp deviate from the alternate under-over interlacing and interlaces every two or more weft. This results in a thick and rough surface with high porosity [23–25].

2D woven fabrics suffer from poor impact resistance because of high degree of crimp as well as low delamination strength due to lack of binder fibers (z-fibers) in the thickness direction. They also have low in-plane shear properties due to absence of off-axis fiber orientation other than material principal directions [4]. Although the through-the-thickness reinforcement eliminates the delamination weakness, it reduces the in-plane properties [1, 2]. Uniweave is a comparatively new fabric structure that has only one yarn set i.e. warp (0°) where multiple warp yarns were locked by stitching yarns [26]. Biaxial noncrimped fabric was developed to replace the unidirectional cross-ply lamina structure [24]. This fabric has basically two sets of fibers such as warp and filling along with locking fibers. Warp yarns are positioned in 0° direction whereas filling yarns are positioned under the warp layer in the cross-direction (90°). Two sets of yarns are locked by two sets of stitching yarns one in 0° direction and another in 90° direction. Another biaxial fabric structure has two fiber sets oriented in 0/90°. The yarns are all positioned in the in-plane direction of the fabric structure without any interlacement with each other as shown in Fig. 1. Integrated 2D shaped woven connector fabric was developed to join the sandwiched structures together for the aircraft applications [27]. 2D integrated woven connector has warp and filling yarns. Basically, two yarn sets are interlaced with each other.

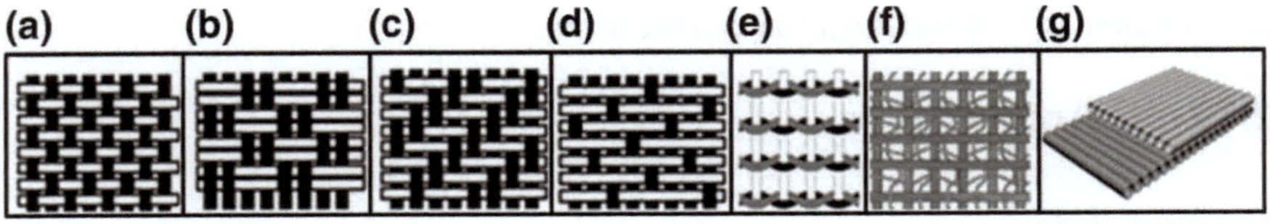

Fig. 1 2D various woven fabrics. **a** Uniform plain. **b** Basket (2/2). **c** Twill (2/2). **d** Satin (4/1). **e** Leno (1/1). **f** Non-interlace woven fabric with stitching yarn. **g** Non-interlace woven fabric without stitching yarn [23–25]

Z-fibers can be used based on the connector thickness. The connector can be woven into **Π, Y, H** shapes according to joint types as shown in Fig. 2. Rib or spars as the form of sandwiched structures are joined with connector by gluing. T-shaped 3D orthogonal woven connector with a hole was also developed as shown in Fig. 2.

Triaxial Woven Fabric
Triaxial weave has basically three sets of yarns as ±bias (±warp) and filling [28]. They are interlaced to each other at about 60° angle to make the fabric shown in Fig. 3. The interlacement is similar to that of a traditional fabric: one set of yarn is passed over and under another and this process is repeated across the fabric width and length. The fabric generally has large open areas between interlacements. Altough, dense fabrics can be made in this way, it is not possible to weave a fabric as dense as a traditional woven fabric. This is mainly because of open-reed process. Triaxial fabrics have been developed as two variants i.e. loose-weave and tight-weave. It was found that loose-weave triaxial fabric has certain stability and higher shear stiffness in ±45° directions compared to the biaxial fabrics. It also has a more isotropic structure [29]. Another type of fabric called "quart-axial" has four sets of yarns such as the ±bias, the warp and the filling as shown in Fig. 3. All fibers are interlaced to each other to form the fabric structure [30]. However, warp yarns are introduced to the fabric at selected places depending upon the end-use requirements.

Braided Fabric
2D braided fabric is the most widely used material in industrial textiles, especially in composite industry. It has one yarn set i.e. braiders (oriented in +θ and −θ directions) which are intertwined with each other to form the braided fabric surface,

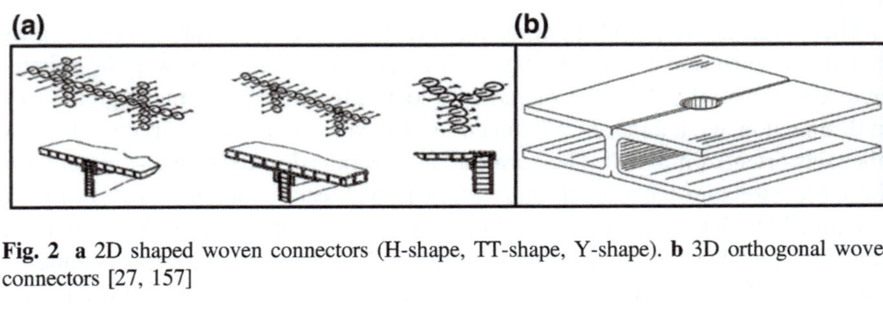

Fig. 2 **a** 2D shaped woven connectors (H-shape, TT-shape, Y-shape). **b** 3D orthogonal woven connectors [27, 157]

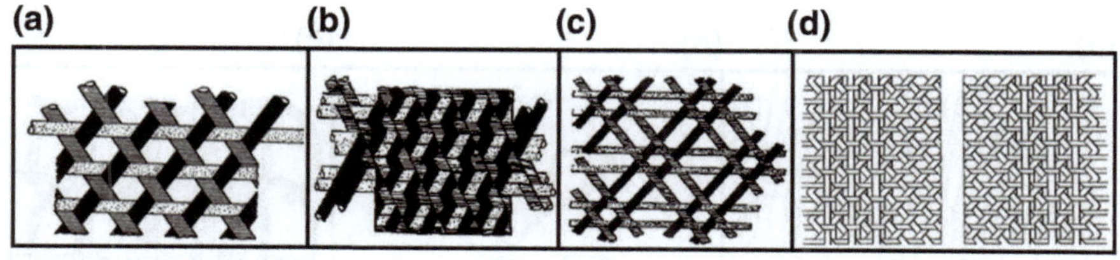

Fig. 3 Triaxial woven fabrics. **a** Loose fabric. **b** Tight fabric. **c** One variant of triaxial woven fabric. **d** Quart-axial woven fabric [28, 30]

as shown schematically in Fig. 4. Basically, there are three different braid patters namely diamond, regular and hercules braid which are all produced by traditional braiding techniques [31, 32]. 2D braided fabric can be layered according to the required thickness and consolidated to form a rigid composite. However, the resulting composite suffers from poor impact resistance because of crimp. It also has low delamination strength due to the lack of binder fibers (z-fibers) in the thickness direction [4]. Although, 2D layered stitched braided preform eliminates the delamination weakness, it can reduce the in-plane properties.

Triaxial Braided Fabric

Triaxial braided fabric has basically three sets of yarns: ±braid (±bias) and warp (axial). Braided yarns intertwine with each other around the axial yarns at about 45° angle whereas axial yarns lie throughout the structure. Therefore, the triaxial braided fabric is formed as shown in Fig. 4. The intertwining process is similar to that of traditional braided fabric, which means –braided yarns pass over and under the +braided yarns and this process is repeated across the fabric width and length. This type of braided fabric generally has large open areas between the axial yarns also known as intertwining regions. Dense fabrics can also be produced. However, it is not possible to braid a very dense structure like a traditional biaxial braided fabric. It was concluded that the axial directional properties of triaxial braided fabric is enhanced in comparison to biaxial braided fabric [33]. Another tubular triaxial

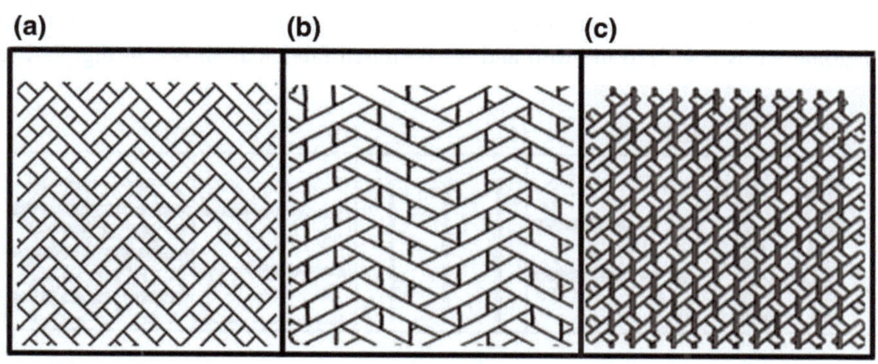

Fig. 4 **a** 2D traditional biaxial braided fabric. **b** Triaxial braided fabric [33]. **c** Non-intertwine triaxial braided fabric [34]

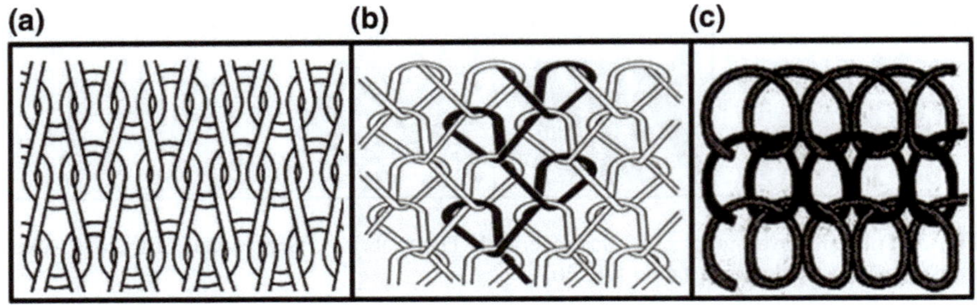

Fig. 5 **a** 2D weft knitted fabric. **b** Warp knitted fabric. **c** Spiral knitted fabric [35]

braided fabric which consists of ±bias yarns and warp (axial) yarns was developed [34]. ±Bias yarns are placed next to each other, and they are locked by warp (axial) yarns as shown in Fig. 4.

Knitted Fabric
Knitted fabric structure is formed by interlooping one yarn system into continuously connecting vertical columns (wale) and horizontal rows (course) of loops as shown in Fig. 5. Knitted fabrics are often specified according to their 'wale density' and 'course density'. The wale density is defined as the number of wales per unit length in the course direction. The course density is the number of courses per unit length in the wale direction. Both the wale and course densities give the stitch density [35, 36].

Uniaxial Knitted Fabric
Poor mechanical properties of knitted fabrics have led to structural modifications with inlay yarns in fabric length or width directions to develop fabrics suitable for composite applications as shown in Fig. 6. In weft or warp knitting, inlaid yarns are trapped inside the structure by weft or warp-type knitted loops. The tensile strength of uniaxial knitted fabric composites can be enhanced substantially in the inlaid directions [37].

Biaxial Knitted Fabric
Biaxial knitted structures are produced via insertion of warp (0°), weft (90°) or diagonal yarns (±45°) to the weft and warp knitted fabrics as shown in Fig. 7. The

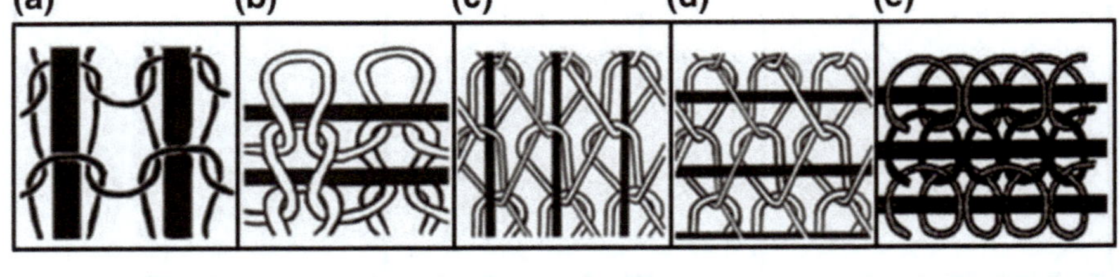

Fig. 6 **a** 2D warp in-laid weft knitted fabric. **b** 2D weft in-laid weft knitted fabric. **c** 2D warp in-laid warp knitted fabric. **d** 2D weft in-laid warp knitted fabric. **e** 2D weft in-laid spiral knitted fabric [37]

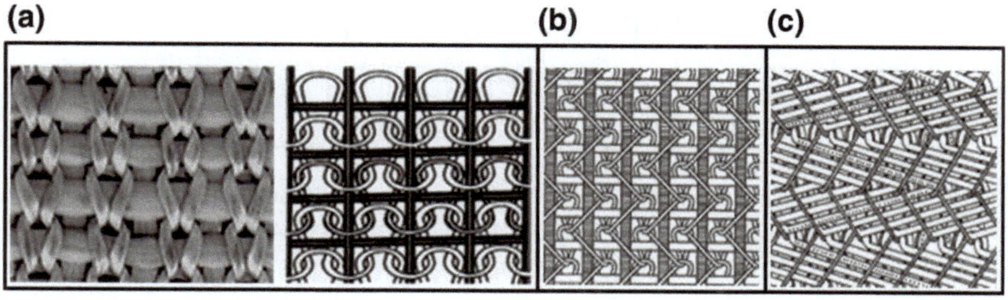

Fig. 7 **a** 2D weft in-laid 0°/90° knitted fabric and *schematic view*. **b** Warp in-laid 0°/90° knitted fabric. **c** Warp in-laid ± 45° knitted fabric [37, 102, 158, 159]

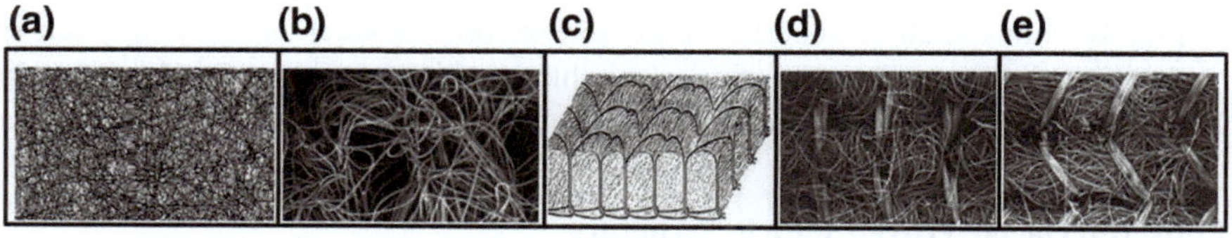

Fig. 8 **a** *Schematic view* of 2D nonwoven fabric by mechanical needling. **b** Hydroentanglement. **c** *Schematic view* of stitched nonwoven structure. **d** Knitting loop surface. **e** Knitting loop reverse surface [39, 160]

in-laid yarns in warp or weft knitted fabrics improve the directional mechanical properties.

Nonwoven Fabric

Nonwoven is a type of fabric made out of short fibers which are held together by various methods including needling, knitting, stitching and other entanglement techniques as thermal and chemical method. Thermal bonding requires fibers or powders with thermoplastic characteristics. In the chemical process, polymer dispersions are used as binders such that the consolidated fiber batt can be considered as a bi-component composite. Figure 8 shows the 2D nonwoven fabric by mechanical needling, hydroentanglement and stitching, respectively [38, 39].

2.2 Three Dimensional (3D) Architectures

Non-interlaced Fabric Structures

A uniaxial non-interlaced fabric preform has one fiber set oriented at 0°, whereas biaxial preform has two fiber sets oriented at 0/90°. A multiaxis non-interlaced fabric preform, on the other hand, has four fiber sets oriented at 0/90/±45° directions. The fiber sets are all wound around each other to form the preform structure as shown in Fig. 9. The yarns are all positioned in the in-plane directions of the preform structure without any interlacement with each other [25].

Fig. 9 **a** Unidirectional non-interlaced fabric schematic and actual fabric. **b** Biaxial non-interlaced fabric schematic and actual fabric. **c** Multiaxis non-interlaced fabric schematic and actual fabric [25]

Multistitched Fabric Structures

A multistitched fabric preform is fabricated by stitching 2D fabric layers together in the thickness direction. Fabrics can be stitched only in the warp (0°) direction; warp (0°) and weft (90°) directions; warp (0°), weft (90°) and ±bias directions as shown in Fig. 10. Lockstitch is generally used for preform production. Stitching can be made manually as well as by using a suitable stitching machine. Stitching can be applied to all fabric types such as braided fabrics, knitted fabrics or nonwoven fabrics [40].

Fully Interlaced Woven Fabric Structure

3D flat fully interlaced woven fabric structure has three sets of yarns as warp, weft and z-yarn. In order to form the structure, warp yarns are interlaced with weft yarns at each layer based on the weave pattern in the in-plane principal directions, whereas z-yarns are interlaced with warp yarns at each layer based on weave pattern in the out-of-plane principal directions. Figure 11 shows the 3D fully plain, 3D fully twill and 3D fully satin preform structures. If the warp and weft yarn sets are interlaced based on any weave pattern but the z-yarns are not interlaced but only laid-in orthogonally between each warp layers, these 3D woven structures are called semi-interlaced woven structures as shown in Fig. 12. 3D circular fully interlaced structure has three sets of yarns such as axial (warp), circumferencial (weft) and radial (z-yarn) yarns. In order to form the 3D circular fully interlaced structure, circumferential yarns are interlaced with axial yarns at each circular layer based on the weave pattern in the circumferential direction, whereas radial yarns are interlaced with axial yarns at each layer based on the weave pattern in the radial directions. Figure 13 shows the 3D fully plain, 3D fully twill and 3D fully satin circular woven preform structures [41, 42].

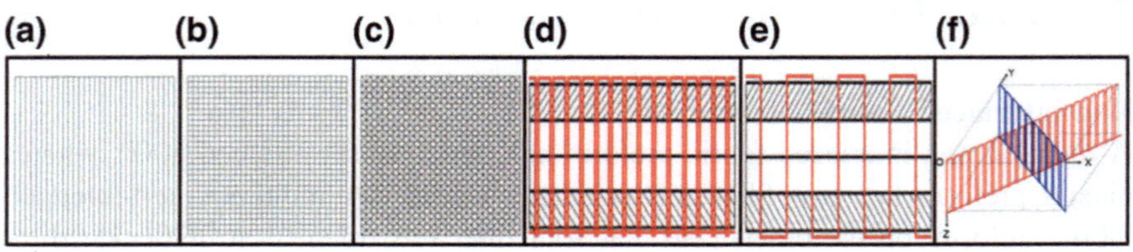

Fig. 10 *Schematic views* of multistitched 2D woven fabric. Stitching directions. **a** One direction. **b** Two direction. **c** Four direction; *cross-sectional view* of four directionally machine and hand stitched structures on **d** 0° **e** 90° **f** +45° and −45° [40]

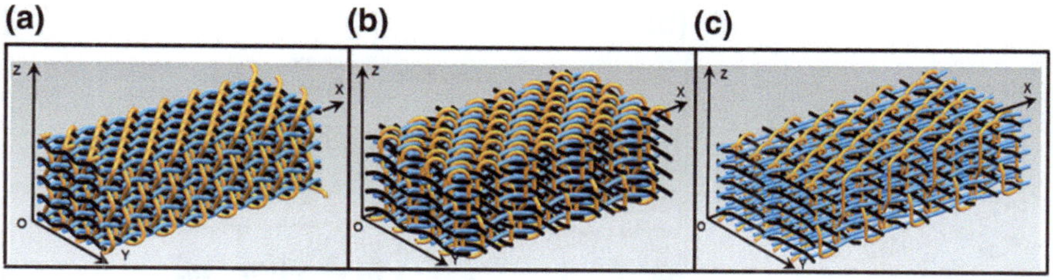

Fig. 11 3D fully-interlaced woven preform structures. General view of 5 layer computer aided drawing of **a** 3D plain. **b** 3D twill. **c** 3D satin woven preform structures [41]

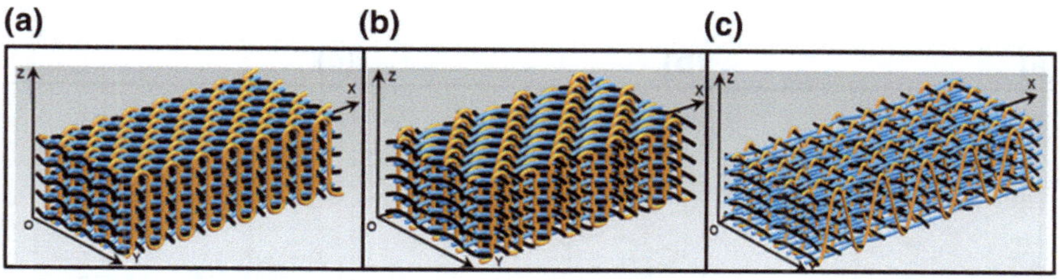

Fig. 12 3D semi-interlaced woven preform structures. General view of 5 layer computer aided drawing of **a** 3D plain-z yarn orthogonal. **b** 3D twill-z yarn orthogonal. **c** 3D satin-z yarn orthogonal woven preform structures [41]

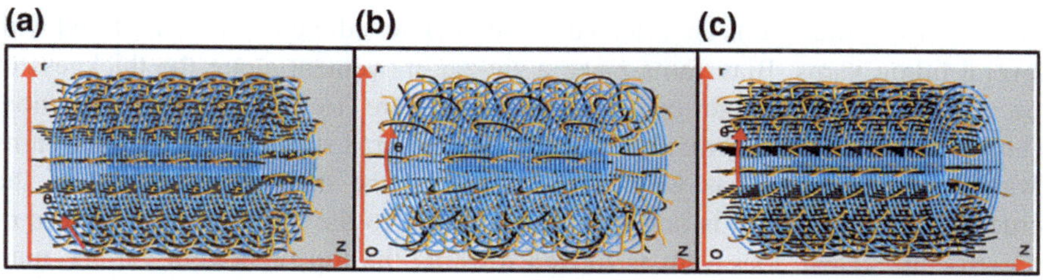

Fig. 13 3D fully-interlaced circular woven preform structures. General view of 5 layer computer aided drawing of **a** 3D plain. **b** 3D twill. **c** 3D satin circular woven preform structures [42]

Orthogonal Woven Fabric

3D orthogonal woven preform has three yarn sets: warp, filling, and z-yarns [41]. These sets of yarns are all interlaced to form the structure wherein warp yarns are longitudinal and the others are orthogonal. Filling yarns are inserted between the warp layers and double picks are formed. The z-yarns are used for binding the other yarn sets together to provide structural integrity. The unit cell of the structure is shown in Fig. 14 [41, 43].

3D angle interlock fabrics are fabricated by using a 3D weaving loom [44]. They are considered as layer-to-layer and through-the-thickness fabrics as shown in Fig. 15. Layer-to-layer fabric has four sets of yarns such as filling, ±bias and stuffer

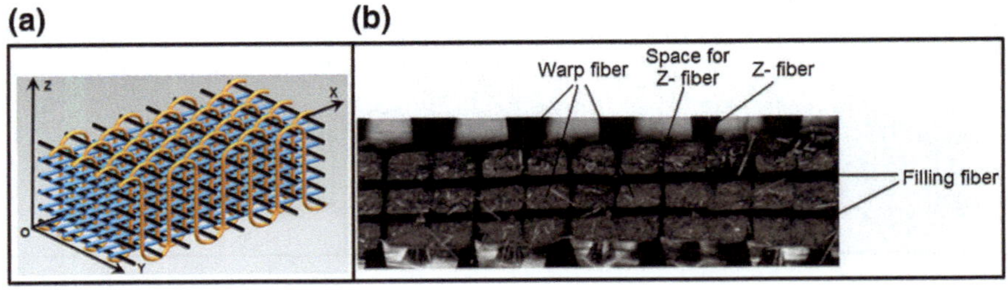

Fig. 14 **a** *Schematic view* of 3D orthogonal woven unit cell. **b** *Cross-sectional views* of 3D actual woven carbon fabric preform [41, 43]

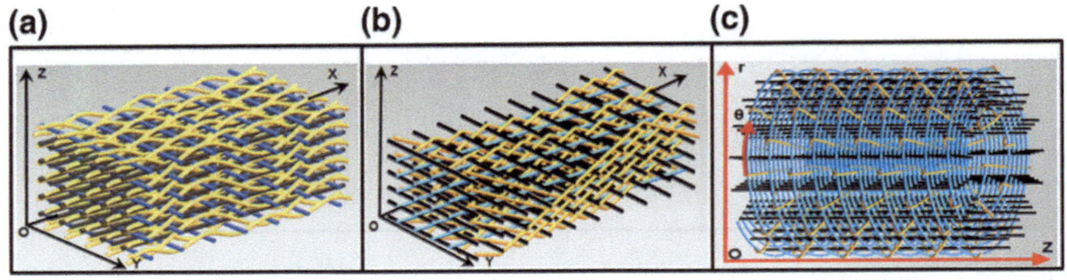

Fig. 15 General view of 5 layer computer aided drawing of traditional. **a** 3D angle interlock. **b** 3D through-the-thickness. **c** 3D circular orthogonal woven preform structures [41, 42]

yarns (warp). ±Bias yarns are oriented in the thickness direction and interlaced with several filling yarns. Bias yarns make a zig-zag movement along the thickness of the structure and change course in the structure to the machine direction. Through-the-thickness fabric also has four sets of yarns such as ±bias, stuffer yarns (warp) and fillings. ±Bias yarns are oriented in the thickness direction of the structure. Each bias yarn is oriented until reaching to the top or bottom surface of the structure. Then, bias yarn is moved towards the top or bottom surface until arriving to the edge. Bias yarns are locked by several filling yarns according to the number of layers [41].

3D circular weaving and fabric (or 3D polar weaving) was developed by Yasui et al. [45]. The preform has mainly three sets of yarns as axial, radial and circumferential that was formed to the cylindrical shape as shown in Fig. 15. In addition, central yarns were incorporated to form the rod. The circumferential yarns laid down between adjacent axial yarn layers, whereas radial yarns were inserted between adjacent axial yarn layers in the radial direction.

Multiaxis Woven Fabric

Multiaxis 3D woven fabric, method and machine based on lappet weaving principals were introduced by Ruzand and Guenot [46]. The fabric has four yarn sets such as ±bias, warp and filling. The bias yarns run across the full width of the fabric in two opposing layers on the top and bottom surfaces of the fabric, if required only one surface. They are held in position by selected weft yarns interlaced with warp

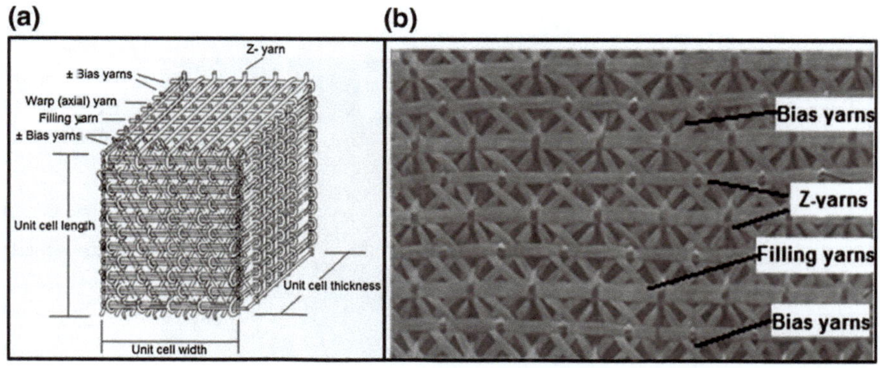

Fig. 16 **a** The unit cell of multiaxis fabric. **b** *Top surface* of multiaxis small tow size carbon fabric [48, 128]

binding yarns on the two surfaces of the structure. The intermediate layers between the two surfaces are composed of other warp and weft yarns which may be interlaced. Uchida et al. [47] developed a fabric called five-axis 3D woven which has five yarn sets such as ±bias, filling, warp and z-yarn. The fabric has four layers and sequences namely +bias, –bias, warp and filling from top to the bottom. All the layers are locked by the z-yarns.

Mohamed and Bilisik [48] developed a multiaxis 3D woven fabric, method and machine. The fabric has five yarn sets such as ±bias, warp, filling and z-yarn. Many of the warp layers are positioned in the middle of the structure. ±Bias yarns are positioned on the back and front faces of the preform and locked to the other yarn set by the z-yarns as shown in Fig. 16. This structure can enhance the in-plane properties of the resulting composites. Bilisik [49] developed a multiaxis 3D circular woven fabric, method and machine. The preform is basically composed of multiple axial and radial yarns along with multiple circumferential and ±bias layers as shown in Fig. 17. The axial yarns (warp) are arranged in radial rows and circumferential layers within the required cross-sectional shape. ±Bias yarns are placed at the outside and the inside ring of the cylinder surface. Filling (circumferential) yarns lay between each helical corridor of warp yarns. Radial yarns (z-yarns) are locked to the all yarn sets to form the cylindrical 3D preform. Cylindrical preform can be made with thin and thick wall sections depending upon end-use requirements.

3D Fully Braided Fabric

Florentine developed a 3D braided preform and a method [50]. The preform has a layered structure and yarns are intertwined with each other depending upon a predetermined yarn path. In this way, yarn passes through the thickness of the fabric and is biased such that the width of the fabric is at an angle between 10°–70°. The representative and the schematic views of the 3D braided preform with yarn paths are shown in Fig. 18. Brown developed a 3D circular braided fabric [51, 52]. The fabric has one set of fiber sets. They are intertwined with each other to make a

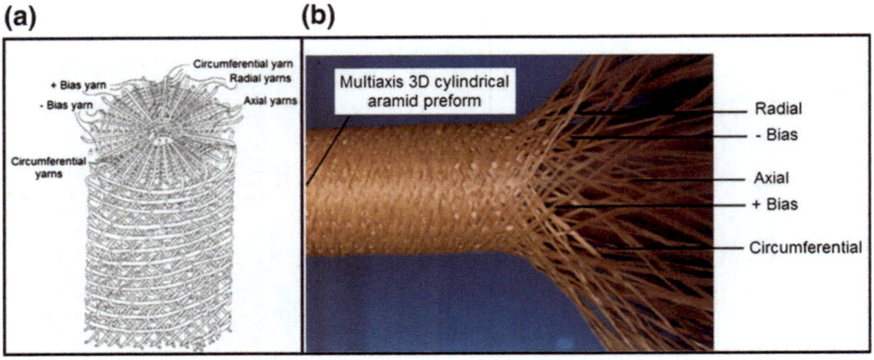

Fig. 17 **a** The unit cell of multiaxis 3D circular woven fabric. **b** Multiaxis 3D aramid circular woven fabric [49, 131]

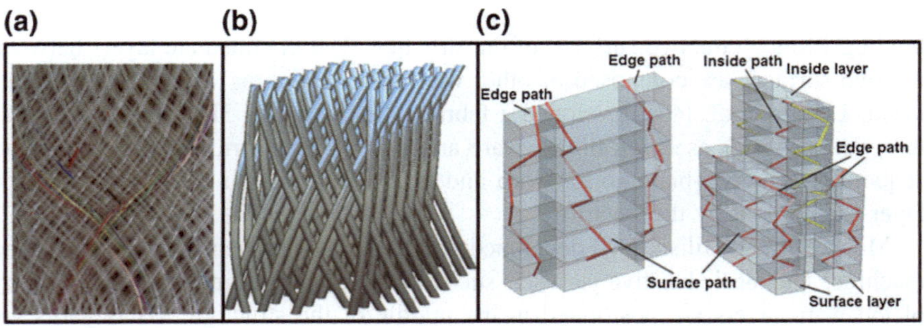

Fig. 18 **a** 3D braided representative preform. **b** Unit cell of 3D braided preform. **c** Braider yarn path on the edge and inside of the 3D representative braided preform with 4 layers (*left*) and 6 layers (*right*) [132]

circular fully braided structure. The fabric has a ±bias yarn orientation in the through-the-thickness of the cylinder wall and the cylinder surface at the helical path, as shown in Fig. 19. Tsuzuki [53] developed 3D various sectional braided preform in which four yarn carriers can surround a rotor and move in four diagonal directions. The addition and subtraction of braider yarns allow the making of various fabric geometries as I-beam, H-beam, TT-beam etc.

3D Axial Braided Fabric

A 3D circular axial braided structure can be formed by the maypole technique using two sets of yarn i.e. warp (axial) and braid. The braider yarns are intertwined with the fixed axial yarns by moving backwards and forwards radially around circumferential paths which allow more flexibility in the preform size, shape and microstructure. This type of braided structure is also called "solid braided fabric" as shown in Fig. 20 [54]. Brookstein developed a tubular fabric which consists of braiders (±bias yarns) and warp (axial) yarns. Braiders are intertwined around each axial yarn so that they lock each individual axial yarn in its place. This type of intertwining forms a helical structure as shown in Fig. 21. It is well-suited to

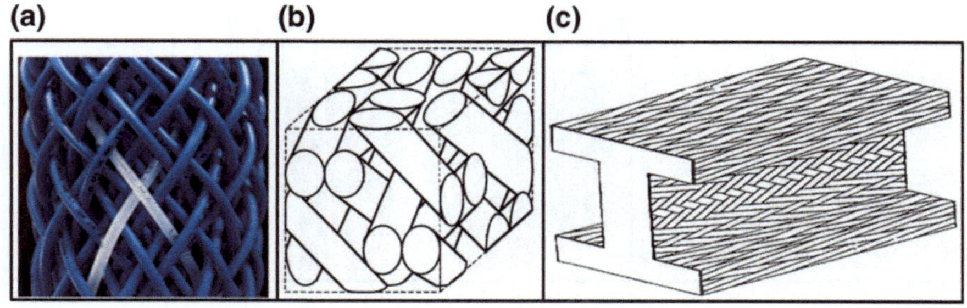

Fig. 19 **a** 3D circular representative braided preform [60]. **b** Unit cell of braided preform [161]. **c** *Schematic view* of 3D braided I-beam preform [53, 86]

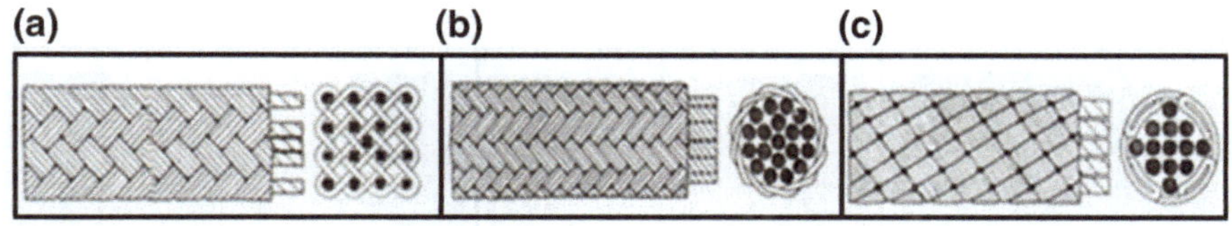

Fig. 20 Solid braid fabrics. **a** 4 × 4 axial braided fabric. **b** Axial round core braided fabric. **c** Axial spiral core braided fabric [54]

produce thick tubular structures and also has a potential for other geometries with a mandrel [55–57]. Another 3D braided preform in a 1 × 1 braid pattern was developed. The braider carrier and axial yarns are arranged in a matrix of rows and columns. The braider yarns are intertwined around each axial yarn row and column to the through-the-thickness direction as shown in Fig. 21. McConnell and Popper developed a 3D axial braided fabric [58]. The preform has a layered structure. Axial yarns are arranged according to a cross-sectional shape. Braided yarns pass through the opening of axial layers in the row and column directions of the arrangement. In this way, the braided yarns are intertwined to make a bias orientation in the through-the-thickness and on the surface of the structure.

Multiaxial 3D Braided Fabric

Multiaxial 3D braided structure has ±braider, warp (axial), filling and z-yarns. The braider yarns are intertwined with the orthogonal yarn sets to form the preform, as shown schematically in Fig. 22. The properties of the multiaxial 3D braided structure in the transverse direction are improved and the directional Poisson's ratios of the structure are identical [59]. Another multiaxial 3D braided structure has ±bias yarns placed in-plane, and warp (axial), radial (z-yarns) and ±braider yarns placed out-of-plane [60]. The braider yarns are intertwined with the axial yarns whereas ±bias yarns are oriented at the surface of the structure and locked by the radial yarns to the other yarn sets. Figure 22 shows the multiaxial cylindrical and conical para-aramid 3D braided structures. The properties of the multiaxial 3D braided structure in the transverse direction can be enhanced and the non-uniformity in the directional Poisson's ratios can be decreased [60].

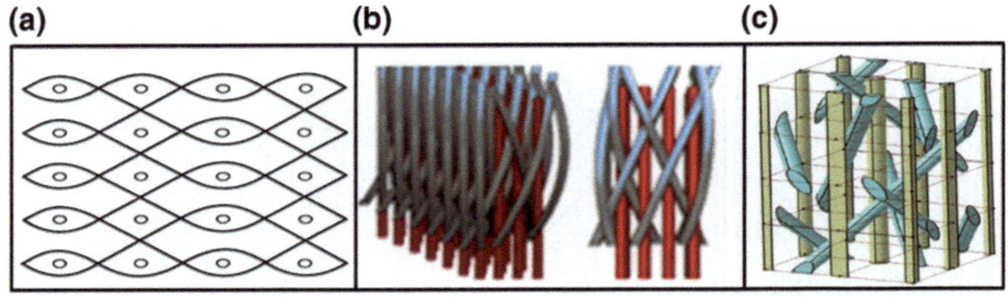

Fig. 21 **a** Unit cell of the 3D braided preform [57]. **b** 3D axial braided preform and unit cell [89]. **c** *Schematic view* of 3D axial braided preform [162]

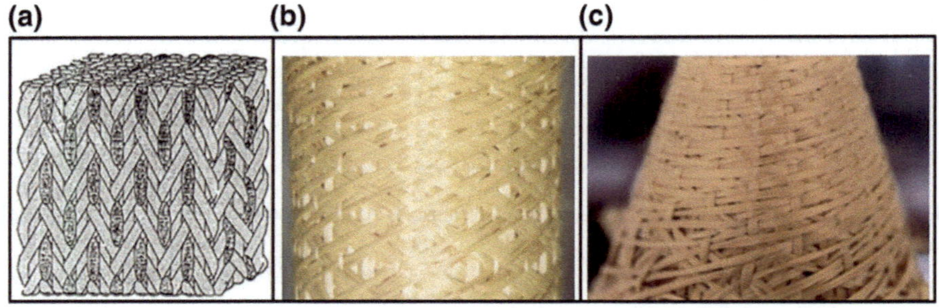

Fig. 22 **a** The unit cell of multiaxis 3D braided preform [59] and multiaxis 3D braided para-aramid preforms. **b** Cylindrical Kevlar® preform. **c** Conic Kevlar® preform [60]

3D Knitted Fabric

Wilkens [61] introduced multiaxis warp knit fabric for Karl Mayer Textilmaschinenfabric GmbH. The multiaxis warp knit machine developed by Naumann and Wilkens [62] uses warp (0° yarn), filling (90° yarn), ±bias yarns along with stitching yarn to produce the fabric as shown in Fig. 23. Wunner [63] developed a multiaxis warp knit machine for Liba GmbH. It uses four yarn sets such as ±bias, warp and filling (90° yarn) together with a stitching yarn. All layers are locked by the stitching yarn by using tricot pattern as shown in Fig. 23.

3D Knitted Spacer or Sandwiched Structure

A sandwich/spacer fabric is a 3D construction made of two separate fabrics connected by yarns or knitted layers [64]. The top and bottom fabrics can be weft or warp knitted fabrics or weft and warp in-lays that form the overall structure macro geometry. The fabric thickness is determined by the length of the connecting yarns/layers. Figure 24 shows various 3D knitted sandwich fabrics.

3D Nonwoven Fabric

3D nonwoven preform is produced using multiple 2D nonwoven webs connected by a stitching yarn in the thickness direction to provide structural integrity. Olry developed a 3D nonwoven preform using a method called Noveltex which relies on entanglement of fiber webs by needle punching [65]. A 3D nonwoven preform was

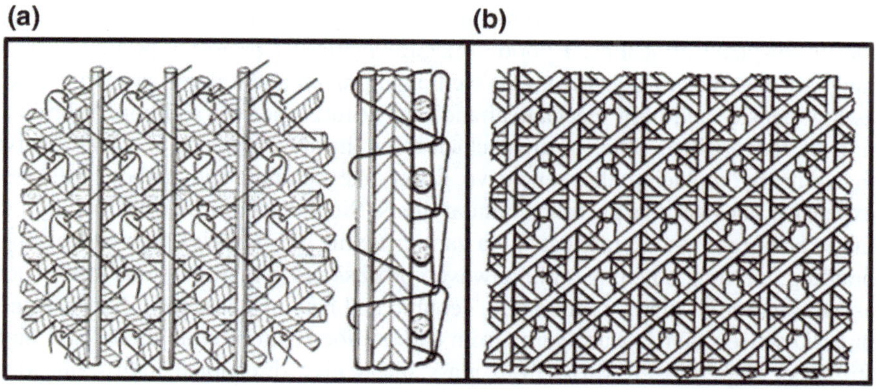

Fig. 23 **a** *Top* and *side view* of multiaxis warp knit fabric [62]. **b** Warp knit structure [63]

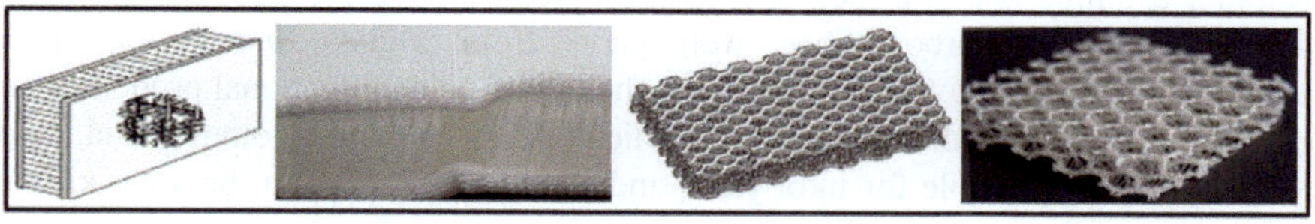

Fig. 24 Various developed actual and schematic 3D knitted sandwich or spacer fabrics [64]

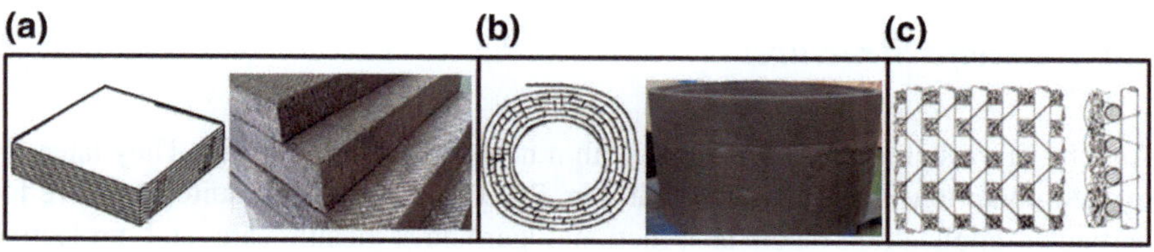

Fig. 25 3D nonwoven fabrics; **a** *schematic view* of flat 3D nonwoven preform (*left*) and 3D PAN based graphite felt composite (2600 °C) (*right*). **b** *Schematic view* of circular 3D nonwoven preform (*left*) and 3D PAN based graphite felt composite (2600 °C) (*right*) (c) *top* and *side views* of 3D biaxially reinforced nonwoven preform [65, 66, 163]

developed using hydroentanglement method to create the through-the-thickness fiber insertion. Another 3D nonwoven preform was developed using warp knitting technology. The preform has multiple warp and weft yarns along with a fiber web. They were all connected by stitching yarns to form an integrated structure as shown in Fig. 25 [66].

2.3 *Hybrid and Multi-layer Architectures*

2D and 3D multilayer fabric structures were exclusively explained in the previous sections. However, a brief discussion will be provided in this part. A hybrid fabric structure is defined as a fabric containing various types of fibers in the in-plane and the out-of-plane directions. Hybrid structural fabrics or preforms provide a wide range of materials to select from, as well as improved and tailored properties such as low density and high mechanical and impact resistance. In addition, hybridization of the preform offers a great chance for the development of cost-effective materials by replacing the expensive fibers like ceramic and carbon with less expensive ones such as glass and natural fibers. Single layer or multilayered hybrid fabric and preforms are fabricated via various textile technologies such as weaving, braiding, knitting, stitching and nonwoven process using high modulus fibers such as carbon, boron, glass, para-aramid, ceramic and natural fibers [67]. For instance, in a 2D woven fabric, warp yarns from carbon fibers can provide directional high stiffness, whereas weft yarns from glass fibers give directional high elongation properties depending upon the end-use requirements. Hence, the in-plane properties of the resulting 2D fabric structure for rigid composite application are improved. Another example is that a multilayer woven preform structure can be made using warp yarns from carbon fiber; weft yarns from S-glass and z-yarns from para-aramid. In this way, the in-plane and the out-of-plane directional properties are enhanced for particular thermo-mechanical loads. On the other hand, any hybridization is possible for intra-yarns and inter-yarns in woven, braided, knitted or stitched preform structures.

2.4 Auxetic Structures

Auxetic fabrics are textile structures with a negative Poisson's ratio. They laterally expand under tensile load as shown in Fig. 26. This makes them quite attractive for particular applications such as functional garments, health care and protective clothing due to their homogeneous porous structure under uniaxial tensile load, better formability into complex shapes and high energy absorption capability [68–70]. Auxetic fabrics are generally produced using knitting and braiding technologies due to fiber interlooping and interwinding type crossing during surface formation. It was reported that a type of warp knitted fabric structure show auxetic behavior under load as shown in Fig. 26 [71, 72]. These structures were constructed from wales of chain and inlay yarns. The wales were knitted from open loops using thicker, low-stiffness filaments, whereas a high stiffness filament is inlaid around the underlaping loops. There are examples of warp knitted auxetic fabrics with a double arrow head geometrical configuration where the auxetic effect is achieved by extending the fabric in the diagonal directions [73]. Weft knitted fabrics produced with flat knitting also showed auxetic properties. A weft knitted fabric based on an origami structure that was formed with connected parallelograms was fabricated [74]. Another type of weft knitted folded structures such as rotating squares and re-entrant hexagons were also developed [68]. When these materials are subjected to a bending force, they demonstrate a synclastic curvature property which makes them better fit with the curved surfaces.

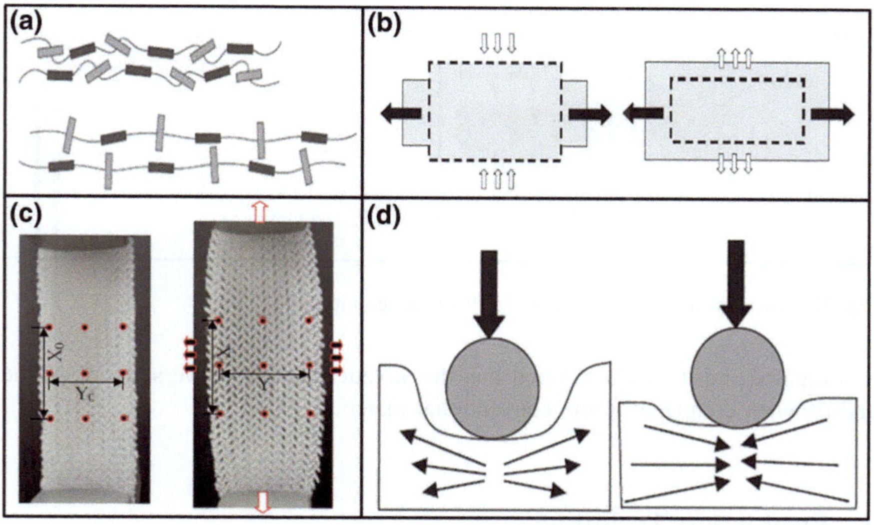

Fig. 26 **a** *Schematic view* of molecular model of auxetic material, before (*top*) and after (*bottom*) tensile load was applied **b** *Schematic view* of auxetic material under tensile load at conventional (*left*) and auxetic (*right*) materials. **c** *Actual view* of knitted fabric before (*left*) and after (*right*) tensile load was applied. **d** Indentation resistance of auxetic structure at conventional (*left*) and auxetic (*right*) structures [68–70]

A 3D auxetic fabric structure was developed by employing a combination of non-woven and stitching technologies for high impact-resistance composite applications [69]. Another type of 3D auxetic fabric structures that is produced using warp knitting technology is spacer fabric which can be used in sportswear, shoes, medical textiles (such as maternity dresses and smart bandages), functional garments (such as children's wear, protective pads for elbows and knees) and various sound absorption applications. It was shown that spacer fabrics display the in-plane auxetic effect in all directions. Generally, auxetic materials possess some enhanced properties, such as shear resistance, indentation resistance, synclastic curvature, crashworthiness and sound absorption. For instance, it was demonstrated that the auxetic foams exhibit a lower Young's modulus compared to conventional foams. However, the shear modulus of auxetic foams is higher than conventional foams. This means that the material becomes difficult to shear and easy to deform volumetrically. In addition, under the impact load, the conventional material flows away in the lateral directions which causes a reduction in density, whereas the auxetic material flows in the vicinity of the impact point and the material becomes denser at that point, resulting in an increase of the indentation resistance as shown in Fig. 26 [69, 70]. The auxetic materials have improved fracture toughness and crack resistance such that a crack will close up under loading due to the negative Poisson's effect. This makes them quite attractive for safety belts, blast curtains and bullet-proof vests. The variable permeability of the auxetic honeycomb structure was

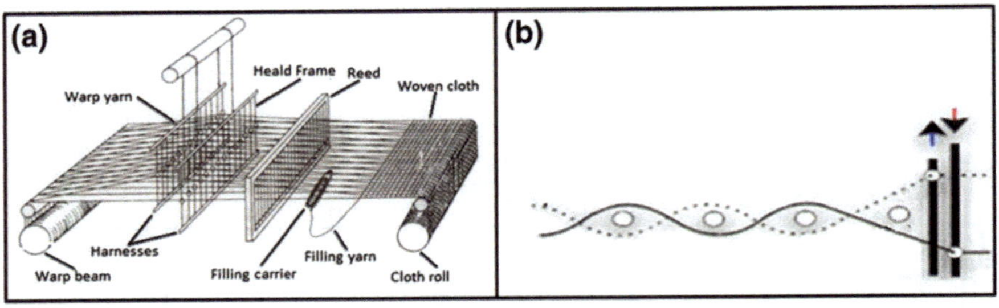

Fig. 27 *Schematic view* of **a** 2D weaving. **b** shedding unit [4, 76]

investigated and it was concluded that the auxetic materials have superior permeability when compared to the conventional materials [75].

3 Production Techniques

3.1 Weaving Technology

2D Weaving

2D woven fabric is the most widely used material in the composite industry with a share of about 70 %. It is made by using the traditional weaving machine as shown schematically in Fig. 27 [4, 76]. The weaving machine has warp let-off, fabric take-up, shedding, weft insertion and beat-up units. Recently, the weaving machine was modified to weave high modulus fibers such as carbon, E-glass, S-glass and para-aramid. Various weave patterns including plain, twill, satin and leno can be woven using such a machine, equipped with a shedding unit as shown in Fig. 27. Hybrid woven fabrics can also be manufactured by using various fiber types in warp and weft yarns [4].

Triaxial Weaving

The basic triaxial weave uses three sets of yarns such as ±bias (±warp) and filling [28]. They are interlaced to each other at about 60° angle to make the fabric. It is possible to add a warp yarn set in one of the principal directions to make a quart-axial fabric. Triaxial weaving machine consists of multiple ±warp beams, filling insertion, open beat-up, rotating heddle and take up unit as shown in Fig. 28. The ±warp yarns are taken from rotating warp beams located above the triaxial weaving machine. After leaving the warp beams, the warp ends are separated into two layers and brought vertically into the interlacing zone. The two yarn layers move in opposite directions i.e. the front layer to the right and the rear layer to the left. When the outmost warp end reaches the edge of the fabric, the motion of the warp layers is reversed so that the front layer moves to the left and the rear layer moves to the right. As a result, the warp makes the bias intersecting in the fabric. Shedding action is controlled by special hook heddles which are shifted after each

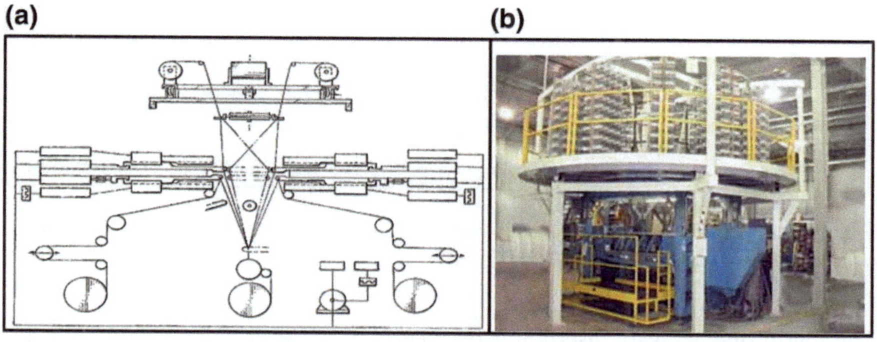

Fig. 28 **a** *Schematic view* of triaxial weaving. **b** Actual triaxial weaving [30, 164]

pick so that in principle they make a circular motion. The pick is beaten up by two comb-like reeds which are arranged opposite to each other in front of and behind the warp layers, penetrate into the yarn layer after each weft insertion and thus beat the pick against the fell of the cloth [29, 30]. In order to make quart-axial fabric, warp yarns are inserted to the triaxial woven fabric at selected places depending upon the end-use. After ±bias yarns rotate just one bobbin distance, heddles are shifted one heddle distance as well. Then, warp is fed to the weaving zone and heddles move to each other selectively to make the shed. Filling insertion takes place and open reed beats the filling to the fabric formation line. Take-up unit removes the fabric from the weaving zone [30].

3D Weaving

In order to make the representative 3D plain woven preform, the warp must be arranged in a matrix of rows and columns as shown in Fig. 29(a1). The first step is the one-step sequential movement of an even number of warp layers in the column direction (a2). This is carried out by using a 2D shedding unit (not shown). The second step is filling insertion between each warp layer in the row direction (a3). The third step is the one-step sequential movement of an even number of warp layers in the row direction (a4). This is again fulfilled by the 2D shedding unit.

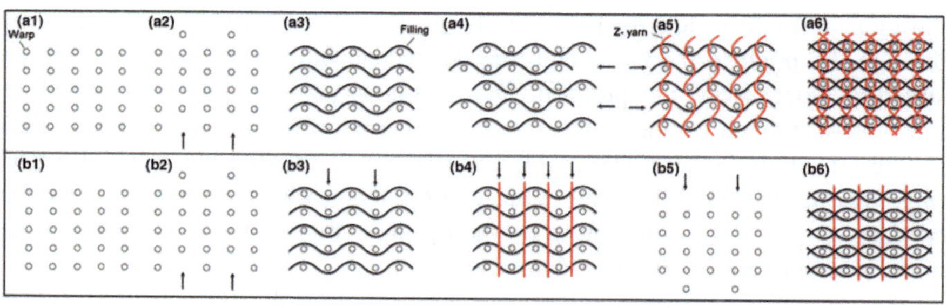

Fig. 29 3D weaving method to make representative fully-interlaced woven preforms; 3D plain woven preform (a1–a6), 3D plain-z yarn orthogonal woven preform (b1–b6) [41]

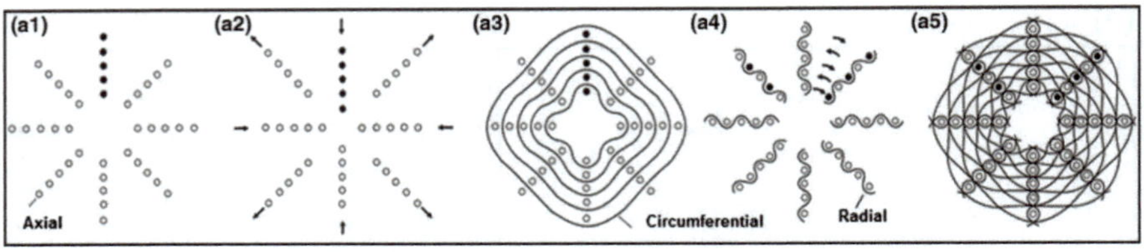

Fig. 30 3D weaving method to make representative fully-interlaced circular woven preform; 3D circular plain woven preform (a1–a5) [42]

The fourth step is z-yarn insertion between each warp layer in the column direction (a5). After the steps (a2–a5) are repeated, the 3D plain woven preform structure is achieved (a6). These steps are repeated depending on the preform length requirements. The 3D plain-z yarn orthogonal woven pattern is also shown in Fig. 29 (b1–b6) where the z-yarn insertion is fulfilled without any interlacement in the preform structure (b4–b6). The number of warp layers can be increased in the row and column directions depending upon preform dimensions [41].

In order to make a representative 3D circular plain woven preform, the axial yarns must be arranged in a matrix of circular rows and radial columns within the required circular cross section. It looks like a circular network in which the yarns are laid in between the adjacent axial yarns as shown in Fig. 30 (a1). The first step is the one-step sequential movement of an even and odd number of axial layers in the radial column direction (a2). This is carried out by using a 2D circular shedding unit (not shown). The second step is circumferential yarn insertion between each axial layer in the circular row direction (a3). The third step is the one-step sequential movement of an even and odd number of axial layers in the circular row direction (a4). This is again fulfilled by the 2D circular shedding. The fourth step is radial yarn insertion between each axial layer in the radial column direction (a4). After these steps (a2–a4) are repeated, the 3D circular plain woven preform structure is made (a5). The steps are repeated depending on preform length requirements. The 3D orthogonal circular woven perform unit cell has basically three yarn sets, namely, axial, circumferential and radial yarns. Axial yarns were arranged in a matrix of circular rows and radial columns. Circumferential yarns are laid down between each adjacent axial yarn row. They are single-end and are deposited through the preform length. Radial ends are positioned between each axial row through the preform thickness and they locked all other yarn sets to ensure the structural integrity of the preform. During the formation of the developed 3D fully-interlaced and orthogonal circular woven preform structures, the individual shuttle for circumferential yarn, which was mounted on each individually rotated ring, was used. In addition, the radial carriers reciprocated linearly to the radial corridor of the 2D shedding plane on the rig thus crossing the radial yarns in the preform structure (crossing shedding) [42].

The state-of-the-art weaving loom was modified to make 3D orthogonal woven fabrics [77]. For instance, one of the looms which has three rigid rapier insertions with dobby type shed control systems was converted to make 3D woven preform as

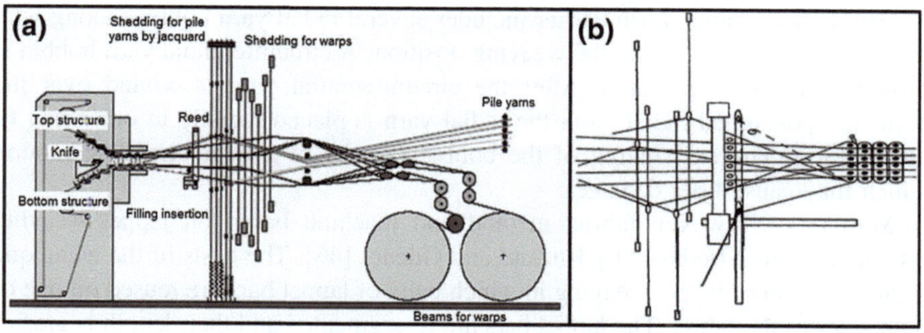

Fig. 31 **a** Traditional weaving loom. **b** New weaving loom making 3D orthogonal woven fabrics [77, 78]

shown in Fig. 31. The new weaving loom was also designed to make various sectional 3D woven preform fabrics [78]. On the other hand, specially designed weaving looms were developed to produce 3D orthogonal woven preform based structural parts such as billet and conical frustum. They are shown in Fig. 32. The first loom was developed based on the needle insertion principal [79], whereas the second loom relies on the rapier-tube insertion principal [80].

3D angle interlock (layer-to-layer) and the through-the-thickness fabrics were fabricated by 3D weaving loom [81]. In order to make the layer-to-layer fabric, ±bias yarns are oriented in thickness direction and interlaced with several filling yarns. Bias yarns make a zig-zag movement in the thickness direction of the structure and change course in the structure to the machine direction. In order to weave the through-the-thickness fabric, ±bias yarns are oriented in the thickness direction of the structure. Each bias yarn is oriented until reaching to the top or bottom surface of the structure. Then, bias yarn is moved towards the top or bottom surfaces until it arrives to the edge. Bias yarns are locked by several filling yarns depending on the number of layers [4]. 3D circular weaving method and fabric (or 3D polar weaving) was developed [44]. The device has a rotatable table holding the axial yarns together with a pair of carriers which extend vertically up and down to

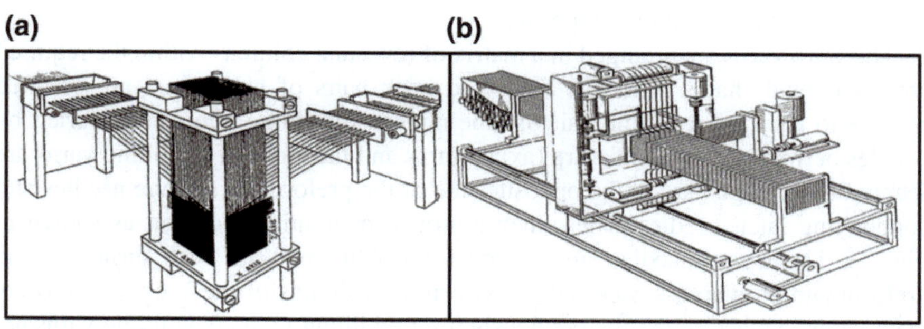

Fig. 32 **a** 3D weaving loom for thick part manufacturing based on needle. **b** Rapier-tube principals [79, 80]

insert the radial yarns. Each carrier includes several radial yarn bobbins along with a guide frame for regulating the weaving position. A circumferential yarn bobbin is placed radial to axial yarns. After the circumferential yarn is wound over the vertically positioned radial yarn, the radial yarn is placed radially to outer ring of the preform. The exchanging of the bobbins results in a large shedding motion which may cause fiber damage.

Multiaxis 3D woven fabric, method and machine based on lappet weaving principals were introduced by Ruzand and Guenot [46]. The basis of the technique is an extension of lappet weaving in which pairs of lappet bars are reused on one or both sides of the fabric. The lappet bars are re-segmented and their length is greater than the fabric width by one segment length. Each pair of lappet bars moves in opposite directions with no reversal in the motion of a segment until it fully extends past the opposite fabric selvedge. When the lappet passes the fabric width, the segment in the lappet bar is detached, its yarns are gripped between the selvedge and the guides, and then cut near the selvedge. The detached segment is then transferred to the opposite side of the fabric where it is reattached to the lappet bar and its yarns are subsequently connected to that fabric selvedge. Since, a rapier is used for weft insertion, the bias yarns can be consolidated into the selvedge by an appropriate selvedge-forming device employed for weaving. The bias warp supply for each lappet bar segment is independent and does not interfere with the yarns from other segments. Uchida et al. [47] developed a fabric called five-axis 3D woven which has five yarn sets such as ±bias, filling, warp and z-yarns. The process requires a bias rotating unit, filling and z-yarn insertion units, warp, ±bias and z-yarn feeding units and a take-up unit. Horizontally positioned bias chain rotates one bias yarn distance to orient the yarns while the filling is inserted to the fixed shed. Then, z-yarn rapier inserts the z-yarn to bind all yarns together and all z-yarn units are moved to the fabric fell line to carry out the beat-up action. Take-up unit removes the fabric from the weaving zone.

Mohamed and Bilisik [48] developed a multiaxis 3D woven fabric, method and machine where the fabric has five yarn sets as ±bias, warp, filling and z-yarns. Many warp layers are positioned in the middle of the structure. ±Bias yarns are positioned on the back and front faces of the preform and locked to the other set of yarns by the z-yarns as shown in Fig. 33. This structure can enhance the in-plane properties of the resulting composites.

The warp yarns are arranged in a matrix of rows and columns within the required cross-sectional shape. After the front and back pairs of bias layers are oriented relative to each other by the pair of tube rapiers, the filling yarns are inserted by needles between the rows of warp (axial) yarns and the loops of the filling yarns are secured by selvage yarn at the opposite side of the preform by selvage needles and cooperating latch needles. Then, they return to their initial position as shown in Fig. 33. The z-yarn needles are inserted from both front and back surface of the preform and pass across each other between the columns of the warp yarns to lay the z-yarns in place across the previously inserted filling yarns. The filling yarns are again inserted by filling insertion needles and secured by the selvage needle at the opposite side of the preform. Then, the filling insertion needles return to their

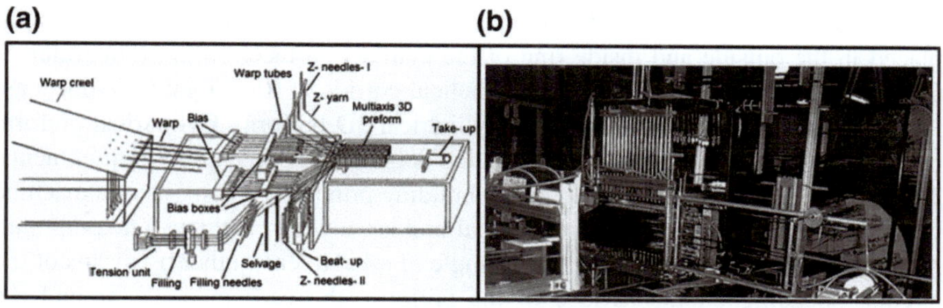

Fig. 33 **a** *Schematic view* of multiaxis weaving machine. **b** *Side view* of multiaxis weaving machine [48, 128]

starting position. Then, the z-yarns are returned to their starting position by the z-yarn insertion needles by passing between the columns of warp yarns once again and locking the bias yarn and filling yarns into place in the woven preform. The inserted filling, ±bias and z-yarns are beaten into place against the woven line, and the take-up system removes the woven preform. Bilisik [49] developed a multiaxis 3D circular woven fabric, method and machine. The preform is basically composed of multiple axial and radial yarns along with multiple circumferential and ±bias layers as shown in Fig. 34. The axial yarns (warp) are arranged in radial rows and

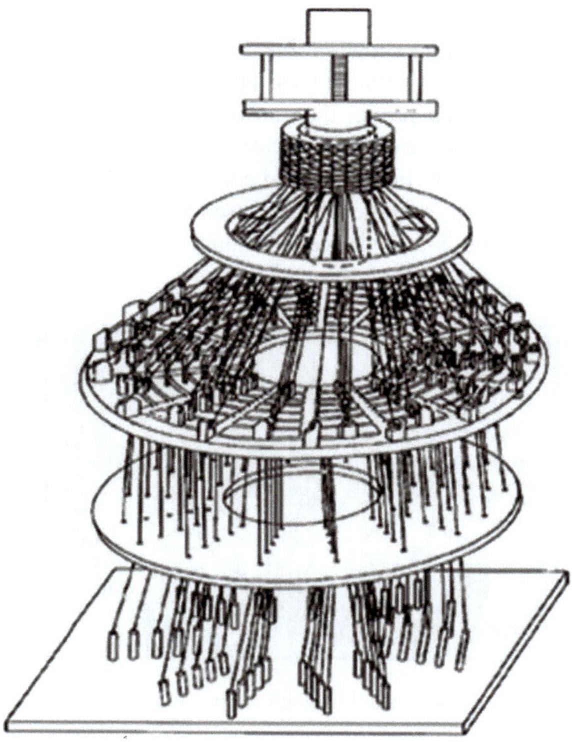

Fig. 34 *Schematic view* of multiaxis 3D circular weaving loom [49, 131]

circumferential layers within the required cross-sectional shape. ±Bias yarns are placed in the outside and inside ring of the cylinder surface. Filling (circumferential) yarns lay between each warp yarn helical corridors. Radial yarns (z-fiber) are locked to the all yarn sets to form the cylindrical 3D preform. Cylindrical preform can be made with a thin or thick wall section depending upon end-use requirements. The process is designed based on the 3D braiding principal. It consists of a machine bed, ±bias and filling ring carriers, a radial braider, a warp creel and a take-up unit. After the bias yarns start to orient at an angle of ±45° to each other by means of the circular shedding on the surface of the preform, the carriers rotate around the adjacent axial layers to wind the circumferential yarns. The radial yarns are inserted to each other by the special carrier units and lock the circumferential yarn layers with ±bias and axial layers together. A take-up system removes the structure from the weaving zone. This describes one cycle of the operation to weave a multiaxial 3D circular woven preform. It is expected that the torsional properties of the preform can be improved because of the bias yarn layers.

3.2 *Braiding Technology*

2D Braiding
2D braiding is a simple traditional textile based process to make bias fabrics. A typical braiding machine consists of a track plate, spool carrier, former, and a take-up unit as shown in Fig. 35. The track plate supports the carriers, which travel along the path of the tracks. The movement of the carriers can be provided by horn gears, which propel the carriers around in a maypole arrangement. The braider carriers are devices that carry the yarn packages around the tracks as shown schematically in Fig. 35 and control the tension of the braiding yarns. At the point of braiding, a former is often used to control the dimension and shape of the braid. The braid is then delivered through the take-up roll at a predetermined rate. If the

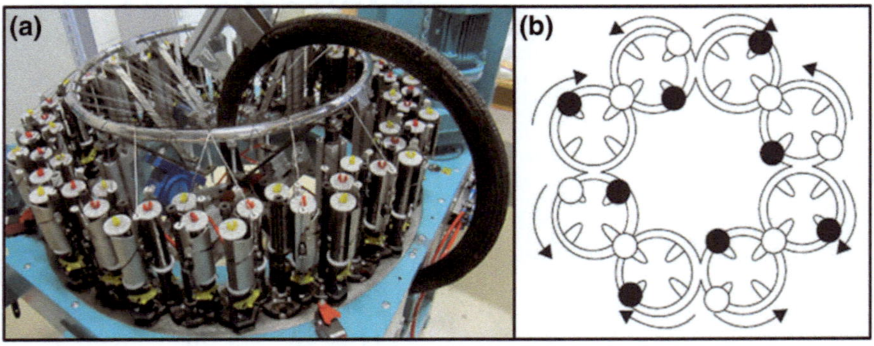

Fig. 35 **a** 2D circular braiding machine. **b** *Schematic views* of the predetermine carrier path [82, 165]

Fig. 36 2D triaxial braiding machine **a** By Boeing Inc. [83]. **b** By Fiber innovation Inc. [84]

number of carriers and the take-up speed are properly selected, the orientation of the yarn (braiding angle) and the diameter of the braid can be controlled. Braiding can take place in horizontal or vertical direction [82].

Triaxial Braiding

The Boeing Company developed a large scale 2D circular triaxial machine to make braided fabrics, as shown in Fig. 36. This tubular triaxial braided fabric has warp (axial) and ±bias yarns. The fabric was converted to various structural shapes with the aid of a mandrel [10, 83]. Fiber Innovation Inc. developed a large sized circular 2D triaxial braider machine. The machine has a circular bed, an axial guiding tube, a large braider carrier together with formation, mandrel and take-up units. The braider carrier moves around the axial fiber tubes according to a predetermined path to make ±bias orientation around the axial yarn. The fabric formation unit provides structural tightness at the fabric fell line, as shown in Fig. 36. Take-up unit removes the structure from the braiding zone. The structure thickness can be increased by over-braiding on the mandrel. It should be noted that braided fabrics can be cut and stitched to make complex-contoured shapes [84].

3D Fully Braided Fabric

By 4-Step Braiding Method

In the 4-step braiding process, there is one set of longitudinal yarns arranged in column and row directions in the cross-section. All these yarns are intertwined with each other by at least four distinct motions in each machine cycle. The braider carriers move simultaneously in predetermined paths relative to each other within the matrix to intertwine the braiding yarns to form the braided preform. Florentine developed a 3D braided preform and a method [50]. The preform has a layered structure and yarns are intertwined with each other depending upon a predetermined path. In this way, yarn passes through the thickness of the fabric and is biased such that the width of the fabric is at an angle between 10°–70°. The process has a rectangular array of individual row and column arrangements in the machine bed. Each individual row has a braider carrier to make four distinct cartesian motions, as shown in Fig. 37. Brown developed a 3D circular braided fabric [51, 52]. The fabric has one set of yarn sets. They are intertwined with each other to make a circular fully braided structure. The process requires concentric rings connected to a

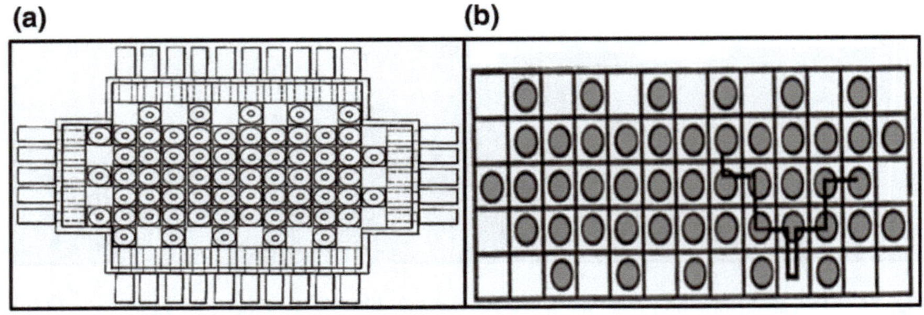

Fig. 37 **a** *Schematic view* of 3D braiding machine [50]. **b** Yarn carrier path [166]

common axis. Braid carriers are circumferentially mounted to the inside diameter of the ring. The ring is arranged side by side according to preform thickness. Rings rotate according to a predetermined path only one braid carrier distance. Then, the braid carriers are shifted in the axial direction. After that, the cycles are repeated in the above sequence. The fabric has ±bias yarn orientation through-the-thickness of the cylinder wall and the cylinder surface at the helical path, as shown in Fig. 38.

By Rotary Braiding
3D rotary braiding is an extension of maypole braiding that allows the braider carrier to move independently and arbitrarily over a base plate so that each braider yarn is placed and interlaced into a 3D preform [9, 12, 85]. Tsuzuki designed a 3D braider consisting of star shaped rotors arranged in a matrix of multiple rows and columns [86]. Four yarn carriers can surround a rotor and move in four diagonal directions which are determined by the rotation of the rotors, as shown in Fig. 39. The addition and subtraction of braider yarns allow making various fabric geometries such as I-beam, H-beam, TT-beam etc. It was noted that the speed of the machine was improved with the development of a rotary stepping actuator [53].

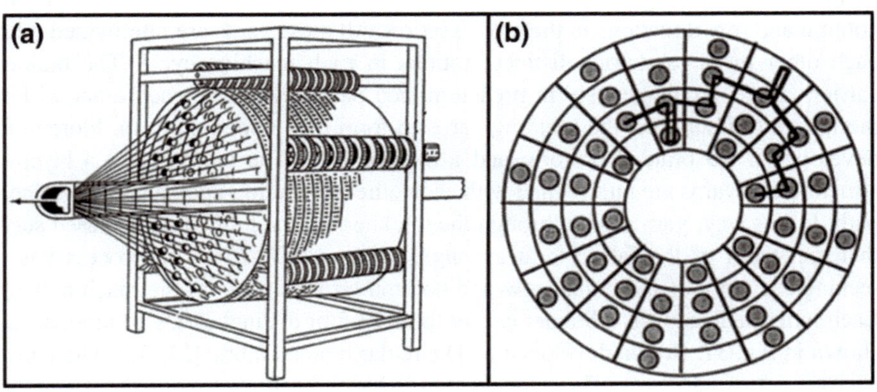

Fig. 38 *Schematic view* of **a** 3D circular braiding machine [51]. **b** Yarn carrier path [166]

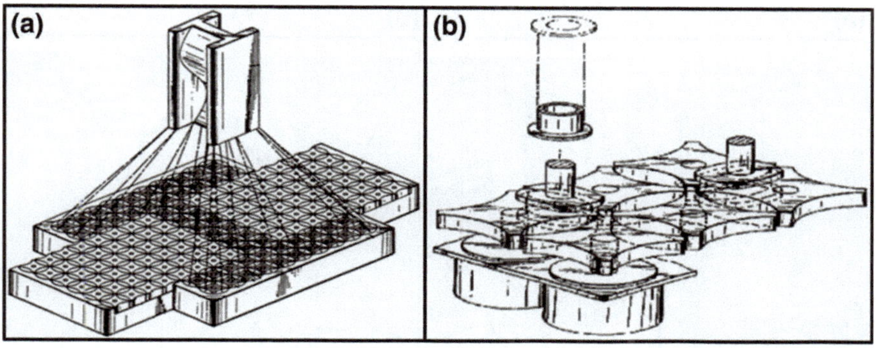

Fig. 39 *Schematic view* of **a** 3D rotary braiding machine. **b** Yarn carrier actuation unit [86]

3D Axial Braided Fabric

By Maypole Braiding Method

A 3D circular axial braided structure can be formed by the maypole technique which consists of two sets of yarn as warp (axial) and braid. The braider yarns are intertwined with the fixed axial yarns moving backwards and forwards radially around circumferential paths. Uozumi developed a 3D circular braided fabric which has ±bias (braider) and warp (axial) yarns [87]. Thick and various sectional fabrics, especially structural joint, end-fitting and flange tube were made by over-braiding [9]. The process, called multi-reciprocal braiding, is based on 2D circular triaxial braiding principals, as shown in Fig. 40. Brookstein developed a tubular fabric which consists of braiders (±bias yarns) and warp (axial) yarns. Braiders intertwined around each axial yarn so that they lock each individual axial yarn in its place. This intertwining forms a helix structure. In the process, a horn-gear type machine bed is arranged cylindrically so that the axial and braider carrier are positioned inside the diameter of the cylinder. In this way, adding layers and ensuring the structure compactness becomes easy. The process has warp (axial) yarns feeding; the braiders are actuated by a horn-gear mechanism to move in a

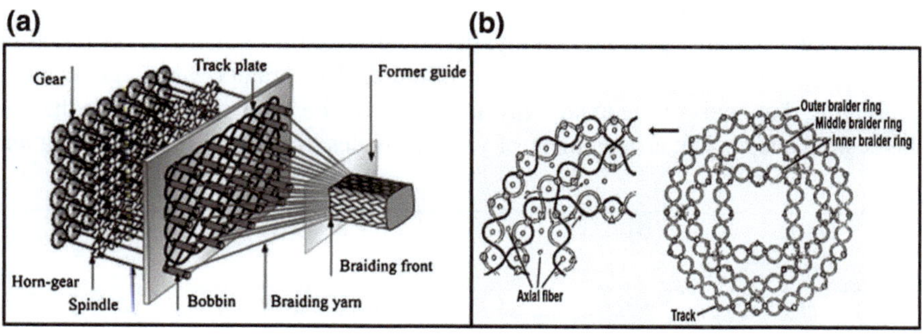

Fig. 40 **a** *Schematic view* of 3D circular axial braiding based on maypole method. **b** *Sectional schematic views* of machine bed and complete machine bed to show predetermined yarn path [9]

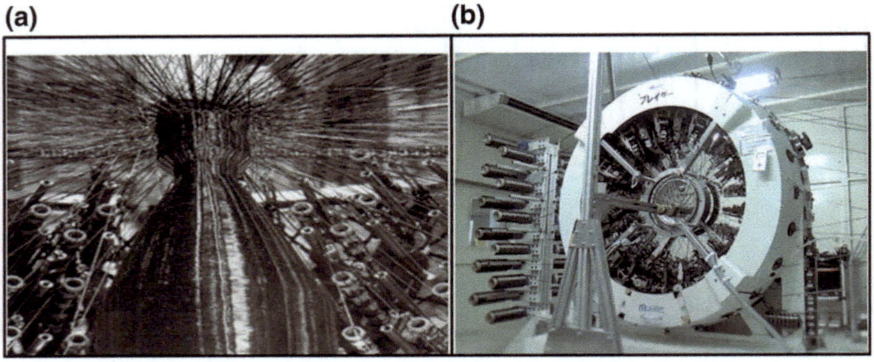

Fig. 41 **a** 3D circular braiding by maypole method [57]. **b** Another type of 3D axial braiding machine from Japan [88]

predetermined path around the axial yarn, as shown in Fig. 41. The fabric is removed by the take-up. It is well suited to produce thick tubular structures and also has the potential for other geometries with a mandrel [55–57]. Similar 3D axial braiding machine based on maypole method was also developed by Japan as shown in Fig. 41 [88].

By 4-step Braiding Method

In order to make a 3D braided preform in a 1 × 1 braid pattern, the braider carrier and axial yarns are arranged in a matrix of rows and columns as presented in Fig. 42. The first step is sequential and the reversal movement of the braider carriers in the column direction (b). The second step is sequential and the reversal movement of the braider carriers placed on the rapier in the row direction (c). The third step is again sequential and the reversal movement of the braider carriers in the column direction (d). The fourth step is again sequential and the reversal movement of the braider carriers placed the rapier in the row direction (e). These steps are repeated depending on preform length requirements. The number of braider carriers and axial yarn can be expanded in row and column directions depending upon preform dimensions [89].

By 2-step Braiding Method

In the 2-step braiding process, axial yarns are arranged in a matrix array based on the sectional geometry of the braided structure. The braider yarns move along alternating diagonals of the axial array and interlock the axial yarns holding them in the desired shape. The arrangement of yarns provides directional reinforcement and

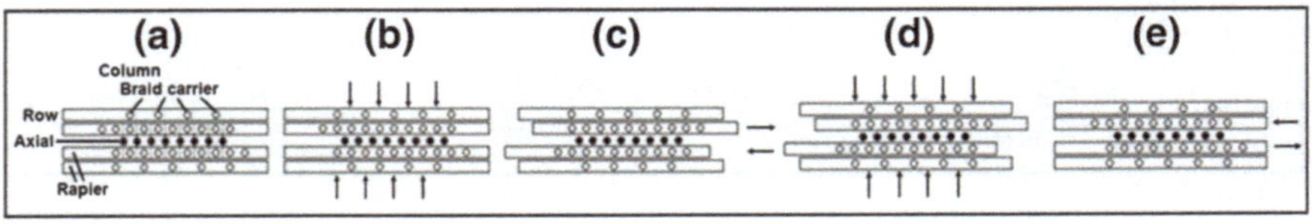

Fig. 42 3D axial braided preform fabrication principles (steps a–e) [89]

structural shape with a relatively small number of braider yarns. This requires fewer braider carriers in the process which eventually makes the process automation simple. The 2-step braiding process involves two distinct motions by each of the braider carriers [17, 90]. It was also shown that a variety of braided preforms including T-shape, H-shape, TT-shape and a braided bifurcation preform can be fabricated [17]. Mc Connell and Popper developed a 3D axial braided fabric [58]. The preform has a layered structure and axial yarns are arranged according to a cross-sectional shape. Braided yarns pass through the opening of axial layers to the row and column direction of the arrangement. In this way, the braided yarns are intertwined to make a bias orientation along the thickness and surface of the structure. The process requires a machine bed, axial unit, braid carrier and compaction unit. The braid carrier moves around the axial unit according to the predetermined path to make two distinct cartesian motions for creating braider type interlacements. The axial unit feeds the axial (0°) yarns in the machine direction. The compaction unit forms the preform, as shown in Fig. 43.

By Rotary Braiding Method
Schneider and Schneider developed a method and machine to make a 3D braided fabric which has multiple axial yarn networks and braider yarns [85, 91, 92]. The method is called 3D rotary braiding which is similar to Tsuzuki's rotor braiding. Figure 44 shows the flat and circular 3D axial braiding machine. The machine consists of horn gears which have a flat row-column array and each horn gear is

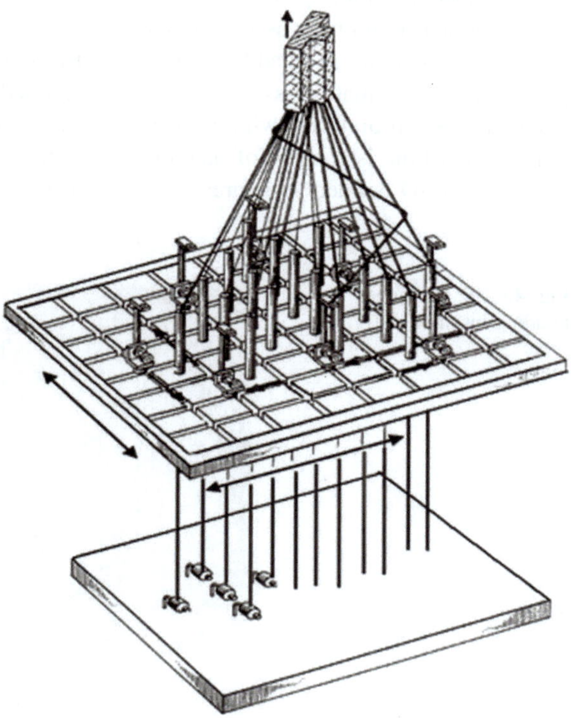

Fig. 43 *Schematic view* of the 3D axial braiding by 2-step braiding [58]

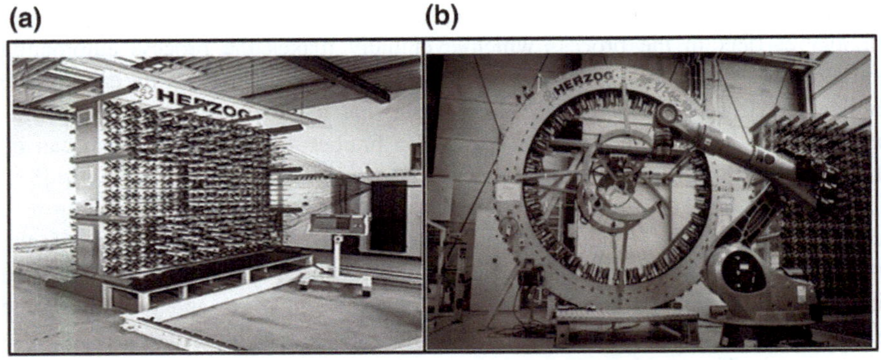

Fig. 44 **a** 3D flat. **b** 3D circular axial braiding machines [85, 91]

actuated by an individual servo control motor and equipped with a clutch-brake mechanism to control the step or rotation of each single horn gear, axial yarn guide and braider carrier. It is capable of making various sectional braided preforms with a computer aided design (CAD) tool. On the other hand, it was also reported that 3D axial braided preform by the rotary braiding method uses independently controlled rotary gripping forks which quickly transfer yarn carriers (in pairs) between the horn gears in accordance with the pre-programmed pattern [93, 94].

By Hexagonal Braiding Method

3D hexagonal braiding was developed to overcome the disproportional machine size/preform size ratios. In addition, most industrial machines are only able to braid preforms with a small cross section. It was claimed that hexagonal braiding can make large sized preforms without increasing the machine bed size. 3D hexagonal braiding machine is capable of handling micro filaments and manufacturing complex shaped 3D braided structures as shown in Fig. 45 [95].

Fig. 45 3D hexagonal braiding machine [95]

Multiaxis 3D Braided Fabric

By 6-step Braiding Method

Multiaxial 3D braided structure produced by the 6-step method has ±braider yarns, warp (axial), filling and z-yarns. The braider yarns are intertwined with the orthogonal yarn sets to form the multiaxis 3D braided preform. The properties of the multiaxial 3D braided structure in the transverse direction are enhanced and the directional Poisson's ratios of the structure are identical. In this process, there are six distinct steps in each cycle. Steps 1 and 2 are identical to those in the 4-step method. Step 3 inserts yarn in the transverse direction. Steps 4 and 5 are identical to the steps 1 and 2, and step 6 inserts yarn in the thickness direction. Another multiaxial 3D braided structure produced by the 6-step method has ±bias yarns placed in the in-plane direction of the structure, and warp (axial), radial (z-yarns) and ±braider yarns placed in the out-of-plane direction of the structure [60]. The braider yarns are intertwined with the axial yarns whereas ±bias yarns are oriented at the surface of the structure and locked by the radial yarns to the other yarn sets. In this process, there are six distinct steps in each cycle. In steps 1 and 2, ±braider yarns are intertwined around the axial yarns as in the 4-step method. In step 3, ±bias yarns are laid down on the surface of the structure. In step 4, the radial yarns move in the thickness direction of the structure and lock the ±bias yarns to the ±braider and axial yarns. In steps 5 and 6, the ±braider yarns are intertwined around the axial yarns as in the 4-step method.

By Multi-step Braiding Method

Kostar and Chou developed a multi-step braiding process which was based on a computer algorithm [96]. In this way, the yarns make a large interlacement angle at the thickness of the fabric which results in a large sized unit cell. The yarns may also change to more positions in the unit cell compared to the unit cell in the 4-step and 2-step processes. The algorithm can also calculate the steps needed for the production of unusual braids which include surrogate materials such as fasteners or structures containing additional yarns or voids [21].

3.3 Non-Woven Technology

2D Nonwoven Fabric

Fabrication of a 2D nonwoven fabric starts with the formation of a fiber web by employing various techniques such as dry-laying, wet-laying and spun-laying. After the web formation, the webs are consolidated by suitable techniques. These techniques include mechanical ones such as needling, stitching and water-jet entangling; chemical methods such as impregnating, coating and spraying or cohesion-based techniques like calendering, air blowing and ultrasonic impact [39].

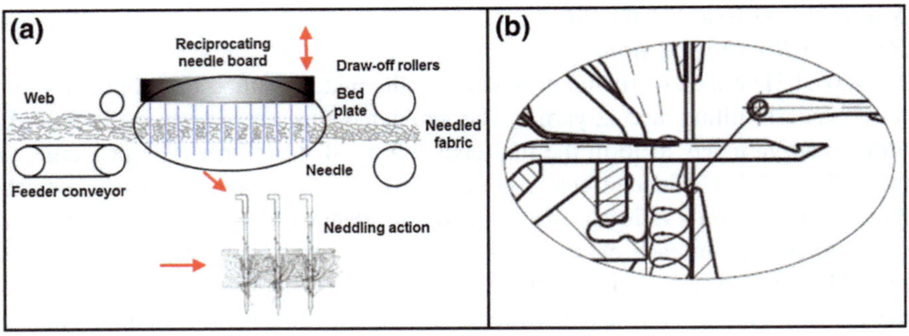

Fig. 46 **a** Principle of needling a fiber web [39, 97]. **b** *Schematic view* of stitching method in which loop formation cycle of a stitch bonding machine is shown [39]

By Needling Method

Short fiber based webs can be bonded by mechanical needling process to orient the fibers both in the in-plane and the out-of-plane directions as schematically shown in Fig. 46 [39, 97]. Pressure application during the process is critical since it increases the so called friction-lock among fibers and improves the degree of bonding in the felt. Important processing parameters during needling are needle design, needle density per fabric width, the stroke frequency, the delivery speeds and the working width. The main principle of the needling process is to reorient a portion of horizontal fibers to ensure that they are aligned in the fabric out-of-plane (thickness) direction. Barbed needles are used for this purpose. During the fiber reorientation, a fiber movement takes place within the felt as the fibers are caught by the needles and change their position. This fiber movement results in dimensional changes in the web which affects the areal mass at a particular location. It is possible to minimize the friction between fibers and needle to prevent fiber breakages by using fiber finishing agents [39]. Web feeding and take-up speeds are important process parameters. The stitch density which is the number of penetrations per square area of the felt is calculated by using Eq. 1:

$$Ed = \frac{n_h.N_D}{V_v.10^4} \tag{1}$$

where, E_d is the stitches per area (stroke per cm^2), n_h is the number of lifts (per min.), N_D is the number of needles by nonwoven fabric width (per m), V_v is the web take-up speed (m/min).

By Stitching Method

In stitch-bonding method, the fiber webs are strengthened by stitching loops in the thickness direction of the web. The loop formation technique is shown in Fig. 46 [38, 39]. The main elements for this process are the compound needle, closing wire, compound needle hook and guide. The nonwoven fabric process consists of a carding machine, a cross lapper and a stitch bonding machine. Compound needle and closing wire bar are connected to the driving cams and the knocking over sinker. Stitch bonding machines are equipped with one or two guide bars. Lapping is achieved by means of two movements namely swinging and shogging. The swinging action is conducted with the aid of a rotary cam and a crank drive. Shogging allows the use of a cam disc. The application of two guide bars makes it possible to use one for the pillar stitch and the other for tricot-stitch. The main process parameter that determines the degree of bonding is the number of loops per unit area which is a function of wale density (number of wales per unit length) and course density (number of courses per unit length). The density of stitched loops is determined by machine gauge and the stitch length. Twisted yarns, textured filaments and film yarns can be used for stitching.

By Hydroentanglement Method

In this method, the nonwoven fabric is consolidated by the striking fluid jets. During the jet action, some fibers are reoriented in the out-of-plane direction of the web to lock the remaining fibers similar to needling process. The web is soaked from the bottom side only after they have passed the jets which neutralize the part of the web densification [39, 98]. The fibers on the side facing, the arriving jets are influenced differently from the fibers at the bottom which lies on the permeable web support. The main process parameter that determines the bonding efficiency is the jet speed. The following relation can be used to find the jet speed.

$$v_w = a.\sqrt{\frac{2.\Delta p}{\rho}} \qquad (2)$$

where, v_w is the speed of the jet at the exit point (m/min), Δ_p is the pressure differential between the nozzle element and the surroundings (Pa), ρ is the density of the medium (g/cm^3) and a accounts for the friction.

By Thermal and Chemical Methods

Thermal bonding process comprises hot-air treatment, calendering and welding of nonwovens, whereas chemical bonding includes the application of binder dispersions, then curing and drying of the impregnated webs [39].

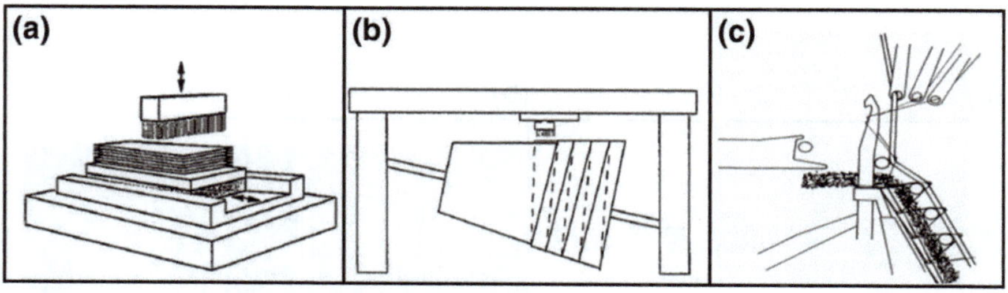

Fig. 47 **a** 3D flat nonwoven machine. **b** 3D circular nonwoven machine [65, 66]. **c** *Schematic view* of 3D biaxially reinforced nonwoven structure made by warp knitting machine [39]

3D Nonwoven Fabric

By Needling Method

P. Olry developed a 3D nonwoven process based on the needling method [65]. In this method, multiple nonwoven webs are laid layer by layer and multiple needles are used to entangle the fibers so that a portion of the fibers are aligned in the thickness direction. Therefore, the 3D nonwoven fabric is strengthened in the thickness direction. Figure 47 shows 3D flat and circular nonwoven machines schematically. Fukuta developed a 3D nonwoven process based on the hydroentanglement method where multiple fluid jets are used to make through the thickness fiber entanglement [66].

By Stitching Method

3D nonwoven structure was made by warp knitting principal. The web is fed to the warp knitting machine. Warp yarns are fed by one or more guide bars, whereas the weft yarns are inserted between the warp yarns and the nonwoven web. Bias yarns are laid over the warp layers. The web, warp yarns, weft yarns and bias yarns are all locked by the stitching yarns. The stitching yarns are inserted to the web by multiple compound needles as shown in Fig. 47 [38, 39].

3.4 Knitting Technology

2D Knitted Fabric

2D knitted fabric is produced by mainly two methods as weft knitting and warp knitting. In weft knitting, latch needles are arranged circumferentially in the axial and radial direction of the machine bed. Yarn guiding bars lay the yarns to the axial latch needles which are mounted on the cylinder. Both axial and radial latch needles interloop the yarns to make 2D circular weft knitted fabric for various composite applications. Figure 48 shows the interlooping action, 2D glass weft knitted fabric and cylinder section of a 2D circular weft knitting machine [37]. In warp knitting, the main components are the following: A yarn feeding unit, multiple yarn guiding bars, multiple axial latch needles, a sinker and a fabric take-up unit. The guide bars are at the front of the machine, completing their underlap shogging. The sinker bar

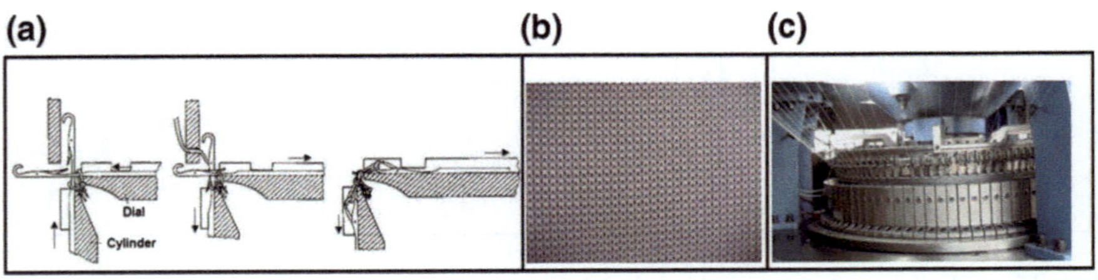

Fig. 48 **a** *Schematic views* of 2D weft knitted fabric during formation. **b** 2D weft knitted glass fabric. **c** 2D weft knitting machine [37, 64]

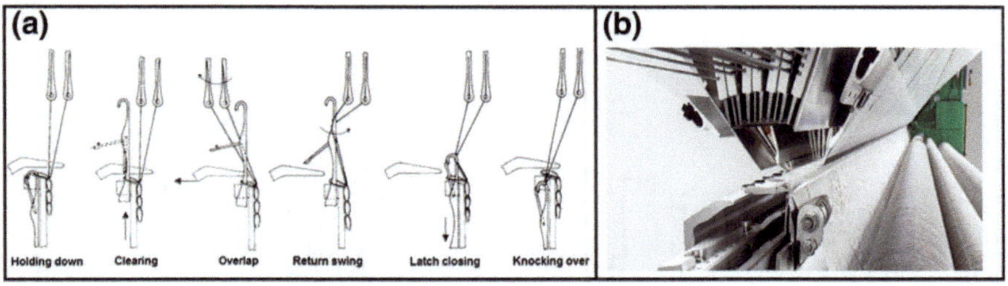

Fig. 49 **a** *Schematic views* of the warp knitting action to form the 2D warp knitted fabric structure using the latch. **b** Actual 2D warp knitting machine [35, 37, 167]

moves forward to hold the fabric down whilst the needle bar starts to rise from knocks-over (holding down action). As the needle bar rises to its full height, the old overlaps slip down on to the stems after opening the latches. The sinker bar then starts to withdraw to allow the guide bars to overlap (clearing action). The guide bars swing to the back of the machine and then make a shogging for the overlap (overlap action). As the guide bars swing to the front, the yarns wrap into the needle hooks (return swing action). The needle bar descends so that the old overlaps contact and close the latches, trapping the new overlaps inside. The sinker bar now starts to move forward (latch closing action). As the needle bar continues to descend, its head passes below the surface of the trick-plate, drawing the new overlap through the old overlap and as the sinkers advance over the trick-plate, the underlap shogging of the guide bar is commenced (knocking-over and underlap action). These knitting actions and the machine are shown in Fig. 49 [37, 99].

3D Knitted Fabric

By Weft Knitting Method

3D weft knitting method was developed by Offermann [100, 101]. The weft knitting machine consists of warp and weft feeding, warp yarn guide track, weft yarn carrier, stitch yarn carrier, yarn feeding unit and fabric take-up unit as shown

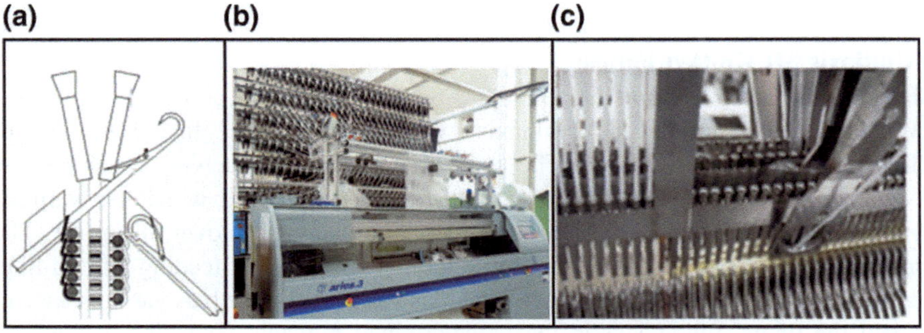

Fig. 50 **a** *Schematic views* of 3D weft knitting methods. **b** 3D knitting machine. **c** Weft yarn carrier during knitting [100, 103]

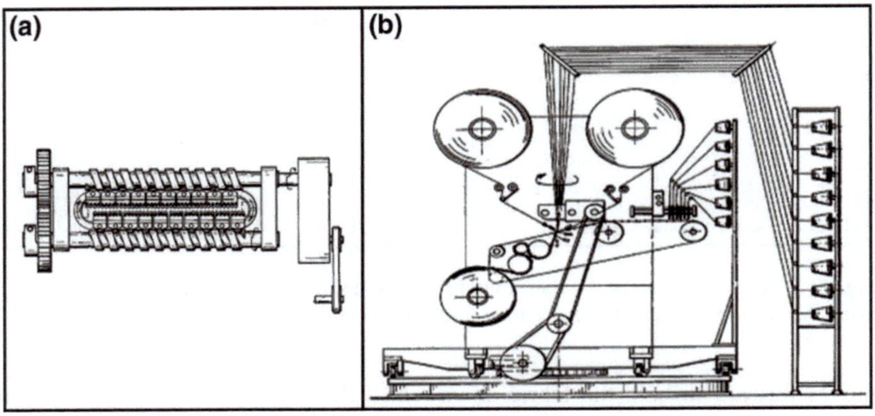

Fig. 51 **a** Bias indexing mechanism. **b** Warp knitting machine [62]

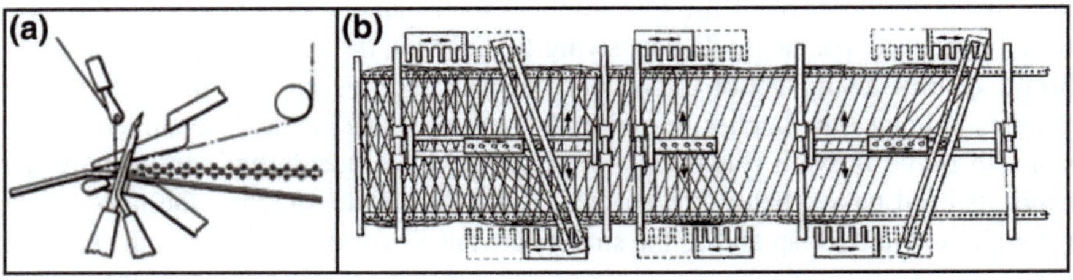

Fig. 52 **a** Stitching unit. **b** Warp knitting machine [63]

in Fig. 50. Two layers of warp yarns are laid by warp yarn guide track. Two layers of weft yarns are laid over the warp layers by the weft yarn carriers. The stitching yarn locks the warp and weft yarn sets using multiple latch needles in which stitched yarns were structured as weft loops. Simple and complex sectional knitted preforms were fabricated by the special take-up device. The critical process parameters are warp and weft densities, stitching density, yarn feeding and fabric take-up ratios [102, 103].

Multiaxis 3D Knitted Fabric

By Warp Knitting Method

Wilkens [61] introduced multiaxis warp knit fabric for Karl Mayer Textilmaschinenfabric GmbH. The multiaxis warp knit machine developed by Naumann and Wilkens [62] produces multiaxis warp knit fabric which has warp (0° yarn), filling (90° yarn), ±bias yarns and stitching yarns as shown in Fig. 51. The machine has ±bias beam, ±bias shifting unit, warp beam feeding unit, filling laying-in unit and stitching unit. After the bias yarn rotates one bias yarn distance to orient the fibers, filling lays-in the predetermined movable magazine to feed the filling in the knitting zone. Then, the warp ends are fed to the knitting zone and the stitching needle locks all yarn sets to form the fabric. To eliminate bias yarn inclination in the feeding system, whole machine bed rotates around the fabric. Stitching pattern i.e. tricot or chain, can be arranged depending upon end-use requirement. Wunner [63] developed a multiaxis warp knit machine for Liba GmbH. It uses four yarn sets such as ±bias, warp and filling (90° yarn) along with stitching yarn. All layers are locked by stitching yarn using tricot pattern as shown in Fig. 52. The process requires a pinned conveyor bed, a fiber carrier for each yarn set, a stitching unit, yarn creels and a take-up unit.

4 Properties of Advanced Architectures

4.1 2D Architectures

Woven Fabric Structure

2D biaxial woven composite has high in-plane properties compared to the 3D woven composites due to the absence of z-yarns and high directional volume fraction. It shows consistent dry fabric properties and good drapability, and is produced with a highly automated process. But, it has low out-of-plane properties due to the lack of z-yarns. In addition, the biaxial woven fabric is the most economical structure in the composite industry.

Scardino and Ko [104] reported that triaxial fabric has better properties in the bias directions compared to the biaxial fabric. Comparisons have revealed a 4-fold tearing strength and 5-fold abrasion resistance compared with the biaxial fabric with the same setting. Elongation and strength properties were roughly the same. Schwartz [105] analyzed triaxial fabrics and compared them with the leno and the biaxial fabrics. He defined the triaxial unit cell and proposed the fabric moduli at crimp removal stage. It was concluded that equivalency in all fabrics must be carefully defined to explore usefulness of the triaxial fabric. Schwartz [105] suggested that when the equivalence is determined, triaxial fabric shows better isotropy compared to the leno and the plain fabric. Isotropy plays an important role in fabric bursting and tearing strength as well as shearing and bending properties. Skelton [106] proposed a bending rigidity relation depending upon the angle of orientation. Triaxial fabric is highly isotropic in bending which means that it is not depended on the orientation angle. The stability of the triaxial fabric is much greater than that of an orthogonal fabric with the same percentage of open area. The triaxial fabric exhibits greater isotropy in its bending behavior and a greater shear resistance than a comparable orthogonal fabric.

Braided Fabric Structure

Process-property relations of 2D circular biaxial braided fabric were examined. A step response model was developed to obtain the temporal change in braiding angle under unsteady-state conditions [107]. The formation and flow patterns of 2D circular braided fabric structure were examined during the consolidation process.

It was reported that inter-fiber spacing in the low braid angle appeared to dominate the permeability. Permeability and porosity may result in a non-uniform flow pattern during liquid molding [108]. In addition, the permeability of the 2D circular braided perform was measured through a radial flow experiment and compared with the permeability predicted by a 3D finite volume method and Darcy's law [109]. The effect of braiding angle on the mechanical properties of 2D biaxial braided composite was analyzed. The bending modulus and bending strength of the 2D braided composite decreased with the increasing in the braiding angle [110]. On the other hand, it was also demonstrated that the highest static bending property of braided composites was achieved with the smallest braiding angle [111] which was around 15°. Also, the elastic modulus and strength properties of 2D biaxial 2 × 2 pattern braided fabric composites were studied by a 3D finite element micromechanics model. They were compared with equivalent 2 × 2 twill fabrics to analyze their fracture modes under various loading requirements [112].

The biaxial compressive strength properties of 2D triaxial braided cylinders were investigated. It was shown that the axial compression strength was sensitive to waviness in the fiber path. Compression and tension strength in the braid direction of 2D triaxial braided fabric composite were significantly lower than those in the axial direction [113]. Another study was performed on the uniaxial and biaxial compression properties of 2D triaxial braided fabric composite. It was reported that the laminated plate theory provided good stiffness predictions for low braid angle, while a fiber inclination model worked well for preforms with various braid angles [114]. The burst strengths of 2D biaxial and triaxial braided cylindrical fiber perform composites were studied. In a biaxial braided cylinder, the initial cracks are aligned in the direction of the fiber tows, while in the triaxial braided cylinder, the initial cracks appear first in the longitudinal direction [115].

An analytical model based on the unit cell geometry and the averaging technique was developed for the prediction of the geometric characteristics and the elastic modulus of 2D triaxial braided fabric composites. It was reported that the braid angle and the fiber volume fraction were obtained from the geometric model. The analytical model utilizes coordinate transformation and the averaging of stiffness and compliance constants on the basis of the volume fraction of each reinforcement and matrix material. It was found that the averaging method was more accurate when the braid angle was small or when the bundle size of axial yarns was much larger than that of the braider yarns [116]. An energy based macrostructure model was also developed to predict the effective stiffness of the repeated unit cell of 2D triaxial braided composite [117].

Nonwoven Fabric Structure

Properties of 2D nonwoven fabric structure depend on fiber type and size, packing density (fiber volume fraction), pore size and distribution in the web volume, fiber orientation in the web [38]. The packing density (α) of a web is defined as the ratio of the volume occupied by the fibers to the whole volume of the web as defined by Eqs. 3 and 4.

$$\alpha = \text{total fiber volume/total web volume} = \frac{V_f}{V_{web}} = \frac{W_f/\rho_f}{tA} = \frac{Basis\ weight}{t\rho_f} \quad (3)$$

where, V_f is the volume of fibers (%); V_{web} is the volume of the web (%); W_f is the weight of fibers = weight of the web (gram); ρ_f is the fiber or polymer density (g/cm^3); t is the thickness of the web (mm) and A is the area of the web (mm^2).

Porosity (ε) is the fraction of the void volume to the volume of the web i.e.:

$$\varepsilon = 1 - \alpha \quad (4)$$

Mechanical properties of the nonwoven fabric are of primary importance for all composite structural applications. Nonwovens are composed of fibers and their strength is obtained from the fiber strength and the bonding strength among the fibers. Nonwovens are anisotropic materials whose directional strength is determined mainly by the fiber modulus, fiber orientation distribution in the web and the packing density.

Knitted Fabric Structure

2D knitted preforms are coarse structure due to 3D loop formation. These preforms distribute the stress throughout the fabric structure. They are highly extensible, and have lower flexural rigidity [118]. It was reported that the knit preform properties were greatly influenced by the fiber strength and modulus, knitted structure, stitch density, pre-stretch parameters and incorporation of in-lays [35]. Deformation behavior of knitted preforms can be predicted by initial load-elongation properties of knitted fabrics. The knitting process parameters influence the knitted preform during fabrication. Knittability of high performance yarns depends on frictional properties, bending strength, stiffness, and yarn strength [37]. Brittleness, high stiffness, and high coefficient of friction of such yarns, require low tension during yarn input, fabric take-up tension setting, and loop length control [103]. Also, knittability of high performance yarns mainly depends on yarn-to-metal friction characteristics. Positive yarn feeding control and tension compensator improve the dimensional stability of the knitted preform. It was demonstrated that yarn bending rigidity and inter-yarn coefficient of friction were important determinants for loop shape while the loop length of high performance yarns, glass yarns in particular, was found to vary with needle diameter, stitching cam setting and machine setting [37, 64, 103].

4.2 3D Architectures

Non-interlaced Fabric Structures

Non-interlaced/non-z single layer and multilayered uniaxial, biaxial and multiaxis composite structures were analyzed. It was found that the tensile and flexural strength of E-glass/polyester structures depended on the yarn orientation and the number of layers. When the preform packing density increased, the tensile and flexural strength of the E-glass/polyester structures increased. This also indicated that the tensile and flexural properties of the composite structures were proportional to their total fiber volume fraction. On the other hand, when all the E-glass/polyester structures were produced with the same preform packing density, the structures produced from coarse fibers showed higher tensile and flexural strength compared to the structures produced from fine fibers due to high fiber volume fraction. All structures displayed a mode-I delamination failure which occurred in the out-of-plane direction as a form of interlayer splitting, parallel to the direction of the applied tensile load. Local interlayer failures were observed between warp/warp in uniaxial, warp/filling in biaxial, bias/bias and bias/filling in multiaxis structures due to the non-z yarn reinforcement [25, 119]. Another study showed that the specific energy absorption based on the structure density was slightly better in the non-interlaced/non-z E-glass/polyester composite plate with para-aramid soft layered dense woven fabric structure compared to that of the 3D woven carbon/epoxy and non-interlaced/non-z E-glass/polyester composite plates with para-aramid soft layered loose woven fabric structure. Damage propagation in the 3D woven structure was smaller than that of the non-interlaced/non-z multiaxis structure, and the impact damage was restricted by the z-fiber. Carbon fiber showed the brittle behavior during energy absorption, but E-glass fiber showed high extension and distributed the energy around the impacted zone [120].

Multistitched Fabric Structures

Warp and weft directional specific tensile strength and modulus of unstitched structure were higher than those of the four- and two-directional lightly and densely multistitched structures as stitching caused minor filament breakages during stitching process. On the other hand, the warp and weft directional specific tensile strains of unstitched structure were slightly lower than all multistitched structures. It was found that when the number of stitching directions, and stitching density in structures increased, their warp and weft directional tensile strength and modulus decreased. It was also generally found that the warp directional specific tensile strength and modulus of the composite structures were slightly higher than those of the weft direction. These results indicated that stitching yarn type, stitching directions and stitching density generally influenced the warp and weft directional tensile properties of multistitched E-glass/polyester woven composites. The failure of warp and weft directional 2D unstitched woven E-glass/polyester composite structures was observed as a form of matrix breakages, partial fiber breakages in their surfaces, and fiber splitting in their boundary regions. They had a complete delamination in their cross-sections. The failures of warp and weft directional 2D multistitched woven E-glass/polyester composite structures were observed as a form of matrix breakages, partial and complete filaments and yarn (tow) breakages in their surfaces. They had a local delamination in their cross-sections and the delamination did not propagate to the large areas due to multiple stitching. The failure was confined at a narrow area due to multistitching and resulted in catastrophic fiber breakages. In addition, it was found that the warp and weft directional specific damaged areas of unstitched structure were higher than those of the multistitched structures. It was considered that the damage tolerance-performance of the multistitched structures was enhanced due to stitching (in particular, four-directional stitching) [121]. In addition, it was found that stitching yarn type, stitching directions, stitching density, and amount of nano materials generally influenced the bending properties of multistitched E-glass/polyester woven composites [122].

The specific short beam strengths of unstitched structures were lower than those of the multistitched/nano structures. When the nano silica material in the unstitched E-glass/polyester composite structure increased, the specific short beam strengths of the unstitched/nano structures increased. When the stitching directions increased from two to four, the short beam strengths of all hand-stitched structures slightly increased [123]. It was found that the material type, material particle size, amount of materials, stitching density and stitching type influenced the damaged areas of the developed composite structures. Generally, the unstitched/nano and unstitched/micro E-glass/polyester woven composite structure had small damaged areas in the front face but they had large damaged areas in the back face. The machine-multistitched structure had also a small damaged area in the front face, but it had a very small damaged area in the back face. This indicated that adding the silica based nano or micro material to the composite structures caused a stiff front face that resisted the impact load on this face. In contrast, the machine multistitching confined the damaged zone to the surrounding area; in particular, it contained the fiber breaking propagation in the damaged area. Thus, the multistitching suppressed the impact energy relatively at a small area of the composite structure and adding the nano material to the multistitched structure can further enhance the damage resistance performance [40].

Fully Interlaced Woven Fabric Structure

Geometrical properties of the representative 3D fully interlaced woven preforms were analyzed and the results are shown in Fig. 53. Crimps in the 3D fully-interlaced and semi-interlaced representative woven preform structure were calculated based on the structure dimensions and the uncrimped representative yarn lengths [41]. The following relations can be used:

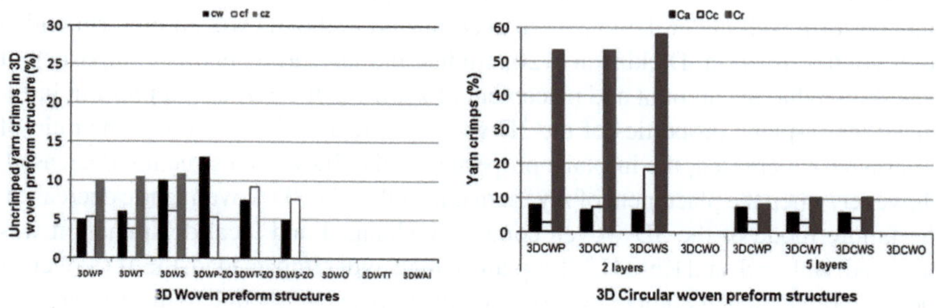

Fig. 53 Relationship between the yarn crimps and the various five layer 3D fully interlaced flat and circular woven preform structures, respectively [41, 42]

$$cw(\%) = (lw - Sl) \times 100/Sl \tag{5}$$

$$cf(\%) = (lf - Sw) \times 100/Sw \tag{6}$$

$$cz(\%) = (lzt - St \times 100)/St \tag{7}$$

where, *cw* is the warp crimp (%), *lw* is the uncrimped warp length (cm), *Sl* is the structure length (cm), *cf* is the filling crimp (%), *lf* is the uncrimped filling length (cm), *Sw* is the structure width (cm), *cz* is the z-yarn crimp (%), *lzt* is the uncrimped total z-yarn length (cm) and *St* is the structure thickness. In addition, crimps in the 3D fully-interlaced representative circular woven preform structure were calculated based on the structure dimensions and the uncrimped representative yarn lengths [42]. The following relations can be used:

$$Ca(\%) = (la - Sl) \times 100/Sl \tag{8}$$

$$Cc(\%) = (lc - Ssl) \times 100/Ssl \tag{9}$$

$$Cr(\%) = (lrt - St \times 100)/St \tag{10}$$

where, *Ca* is the axial crimp (%), *la* is the uncrimped axial length (cm), *Sl* is the structures length (cm), *Cc* is the circumferential crimp (%), *lc* is the uncrimped circumferential length (cm), *Ssl* is the structures outside surface length (cm), *Cr* is the radial crimp (%), *lrt* is the uncrimped total radial length (cm) and *St* is the structures wall thickness.

Orthogonal Woven Fabric Structure

Gowayed and Pastore [124] reviewed computation methods for the 3D woven fabric. The developed analytical methods were stiffness averaging, fabric geometry and inclination models. They were based on classical lamination theory, and a micromechanical approach. Gu [125] reported that the take-up rate in 3D weaving affects the directional and total volume fraction of the 3D woven fabrics. High packing density can be achieved if the beat-up acts twice to the fabric formation line. Friction between brittle fibers such as carbon, and machine parts must be kept low to prevent filament breakages. Cox et al. [5] stated that the 3D woven preform with a low volume fraction may perform well under the impact load compared to the 3D woven preform with a high volume fraction. Dickinson [126] studied the 3D carbon/epoxy composites. It was shown that the amount and placement of z-yarn in the 3D woven preform influenced the in-plane properties of the 3D woven structure. When the volume ratio of z-yarns was increased, the in-plane properties of the 3D woven structure decreased. However, when the placement of z-yarn in unit cell of the 3D woven fabric decreased, the failure mode of the 3D woven composite changed and local delamination was occurred. Bobcock and Rose [127] explained that, under impact load, the 3D woven or the 2D fabric/stitched composites confined the impact energy due to the z-yarn.

Multiaxis Woven Fabric Structure

Five-axis 3D woven fabric composite was characterized by Uchida et al. [20]. Tensile and compression results from the multiaxis woven composite and the stitched 2D composite were comparable. Open hole tensile and compression results of the multiaxis 3D woven structure were high compared to that of the stitched 2D composite. Compression After Impact (CAI) test showed that the multiaxis 3D woven composite was better than that of the stitched 2D structure. In addition, damaged area under the CAI load was small for the 5-axis 3D woven composite compared to that of the stitched 2D composite.

Bilisik [128] identified the important process parameters for the multiaxis 3D flat woven fabric. These were bias angle, width ratio, packing, tension and fiber waviness. The bias angle is the angle between bias fiber and warp fiber to the machine direction. Bias fiber is oriented by discrete tube-block movement. One tube-block movement is about 15°–22° based on processing parameters. If the structure requires any bias angle between 15° and 75°, the tube-block unit must be moved by one, two, or three tube distance as shown in Fig. 54a. It is also concluded that a small bias angle variations was occured in the loom state and the out-of-loom state an average 4°. The multiaxis weaving width was not equal to that of the preform. This difference was defined as the width ratio (preform width/weaving width). This was not currently the case in traditional 2D or 3D orthogonal weaving. The width ratio was almost 1/3 for the multiaxis weaving. This was caused by an excessive filling length during insertion. It was reported that fiber density and pick variations were observed. Some of the warp yarns accumulated at the edges were similar to those of the middle section of the preform. When the preform cross-section was examined, it was found that a uniform yarn distribution could not be achieved for the whole preform volume. These indicated that the light beat-up did not apply enough pressure to the preform, and the layered warp yarns were redistributed under the initial tension. In part, the crossing of bias yarn prevented the z-yarn from sliding the filling yarns towards the fabric line where the filling was curved. Probably, this problem is unique to the multiaxis weaving. Hence, it can be concluded that rigid beat-up was necessary. This unique problem could be solved by a special type of open reed, if the width ratio is considered the main design

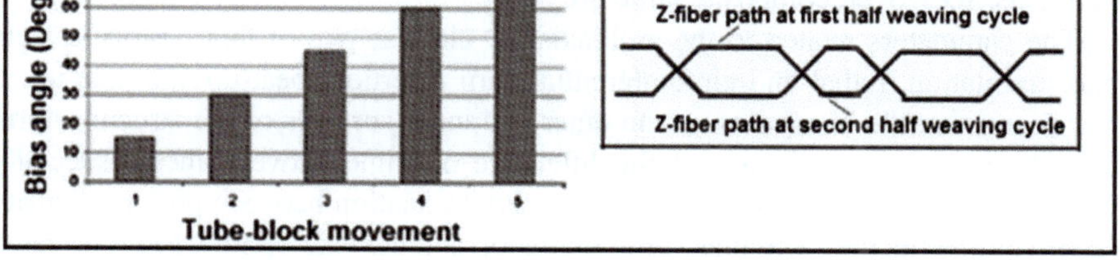

Fig. 54 **a** Relations between tube-block movement and bias angle. **b** z-fiber path in the multiaxis 3D preforms [128]

Table 1 Multiaxis 3D woven preform elastic constants from multiaxis 3D weaving [130]

			Carbon fiber	Epoxy matrix
			Thornel™ T-300 PAN	(Tactix™ 123)[3]
Material properties	Tensile strength (MPa)		3450	76.50
	Tensile modulus (GPa)		230	3.45
	Modulus of rigidity (GPa)		88.50	1.30
	Elongation (%)		1.62	5.70
	Poisson's ratio (ν)		0.27	0.31
	Density (g/cm^3)		1.76	1.16
			Preform 1	Preform 2
Bias angle (°), (measured)			30°	40°
Fractional volume (%), (measured at preform)	+ Bias		9.43	11.7
	−Bias		9.43	11.7
	Warp		10.5	13.7
	Filling		5.42	4.77
	z-yarn		3.67	5.61
	Total volume (%)		38.4	47.5
Elastic constants (Calculated)	Modulus of elasticity (GPa)	E_{11}	48.33	48.00
		E_{22}	19.87	23.85
		E_{33}	9.86	14.24
	Modulus of rigidity (GPa)	G_{12}	10.42	15.65
		G_{23}	2.78	3.47
		G_{31}	2.80	3.47
	Poisson's ratio	ν_{12}	0.446	0.530

parameter [129]. Dry volume fraction in the fabricated preform showed that increasing the fiber content in the warp or the bias and the filling fiber sets resulted in a high total preform volume fraction and a reduced porosity in the fiber set crossing points [129]. The stiffness averaging method was applied on the multiaxis 3D carbon/epoxy composites [130]. Table 1 shows the directional tensile and shear elastic constants of multiaxis carbon/epoxy composite structure. It was demonstrated that yarn orientation in the preform influenced the shearing properties of the multiaxis 3D woven composite structure.

The parameters related to the multiaxis 3D circular woven fabric-process were bias orientation, radial and circumferential yarn insertion, beat-up and take-up. It was found that the bias yarns on the outer and inner surfaces of the structure form helical paths and there was a slight difference of angle between them especially making the thick wall preforms. There is a certain relation between preform density (fiber volume fraction) and bias yarn orientation and the take-up rate. More research may be required to understand relations between those processing parameters and the preform structural parameters. The excessive yarn length during circumferential

Table 2 Specifications of the multiaxis 3D circular woven aramid (Kevlar) preforms [131]

Preform unit cells	Multiaxis 3D woven	
	Preform-I	Preform-II
Material		
Axial fiber	1100 dtex-Kevlar 29[a]	1600 dtex-Kevlar™ 49[b]
Circumferential fiber	1100 dtex-Kevlar 29	1600 dtex-Kevlar™ 49
Radial fiber	1100 dtex-Kevlar 129[c]	1600 dtex-Kevlar™ 49
± Bias fiber	1100 dtex-Kevlar 29	1600 dtex-Kevlar™ 49
Preform structure		
Axial	7 layers × 36 rows	7 layers × 36 rows
Circumferential	6 layers	6 layers
Radial fiber	36 ends (single end)	36 ends (single ends)
	(one radial fiber for every axial row)	(one radial fiber for every axial row)
+ Bias fiber	2 layers × 36 rows	2 layers × 18 rows
−Bias fiber	2 layers × 36 rows	2 layers × 18 rows
Cross section	Cylindrical section	Conical section
Dimensions (Outside diameter × wall thickness, mm.)	100 × 5	150(bottom) × 70 (top) × 5
Preform tightness	Very high	Low
Diameter ratio	7	4.7–10

[a]Kevlar™ 29, [b]Kevlar™ 49, [c]Kevlar™ 129: It is an para-aramid based fibers and Kevlar 29, Kevlar 49 and Kevlar 129 are a trade name produced by DuPont

yarn insertion is because diameter ratio (preform outer diameter/outermost ring diameter) is not equal to 1. The diameter ratio depends on the number of the rings. When the excessive circumferential yarn is not retracted, it causes waviness in the structure. However, there must be an adequate tension on the circumferential yarns to get proper packing during beat-up. Circumferential yarn ends in each layer, which is equivalent to the filling in the flat weaving, are six during insertion as seen in Table 2 [131]. This resulted in a high insertion rate. It is realized that there is a relation between number of layers and radial yarn retraction. If the number of layers in the preform increases, yarn retraction in the radial carrier also increases. The retraction must be kept within the capacity of the radial carrier. It is also observed that tension level in the radial yarn is kept high compared to that of the circumferential yarns because of easy packing and applying tensioning force to the bias crossing points which resist the radial yarn movement during structure formation at the weaving zone.

3D Fully Braided Fabric Structure

Geometric relations on the representative 3D fully braided structure were investigated in terms of the unit cell angle, the unit cell yarn length and the unit cell yarn path. It was demonstrated that braid patterns influence the 3D braided and 3D axial braided unit cell structures produced by the 4-step method. Patterns on odd numbered rows resulted in fully interconnected integral unit cell structures, whereas patterns on even numbered rows resulted in layer-to-layer interconnection on the edge of the unit cell structure where there was an empty pocket between each braided layer. The unit cell structure has a fine intertwine in the 1 × 1 pattern, whereas it has a coarse intertwine for other braid patterns. On the other hand, the number of layers affects the 3D braided and 3D axial braided unit cell structures: when the number of layers increases, the thickness of the unit cell structure increases for all braid patterns. In addition, for the same layer number, the thickness of the unit cell structure in the 1 × 1 pattern is less than that of other patterns. This indicated that all braid patterns except 1 × 1 resulted in a coarse form of the unit cell structure [132]. Jamming conditions considerably affect 3D braided and 3D axial braided unit cell structures for all braid patterns. Minimum jamming decreases the width of the unit cell structures, whereas maximum jamming increases their width. Width reduction of the unit cell structure in the 1 × 1 pattern was high compared to that of the 2 × 1, 3 × 1 or 4 × 1 patterns. However, the width increment of the unit cell structure in the 1 × 1 pattern was slightly higher than that of other patterns. Also, minimum jamming increased the densities of the 3D braided and 3D axial braided unit cell structures, whereas maximum jamming decreased their densities [89, 132]. It was shown that braid pattern slightly influences the yarn angles in 3D braided and 3D axial braid unit cell structures produced by the 4-step method. Jamming conditions affects the yarn angles in 3D braided and 3D axial braided unit cell structures. Minimum jamming decreased the surface angle of the 3D braided and 3D axial braided unit cell structures whereas maximum jamming increased their surface angle. Increasing the layers caused increase in braider and surface yarn lengths, and multilayer yarn length in the 3D braided and 3D axial braided unit cell structures. However, increasing the number of layers also decreased the surface arc length and corner yarn length, as well as edge yarn lengths. It was found that jamming conditions did not affect yarn length in 3D braided and 3D axial braided unit cell structures [89, 132]. The study showed that increasing the layer number created an additional yarn path, the "multilayer yarn path", on the edge of the 3D braided and 3D axial braided unit cell structures, and this could affect the mechanical behavior of the 3D braided and 3D axial braided composites. This was considered especially important for the manufacturing of near-net-shape thick 3D braided and 3D axial preforms and composites [89, 132].

Modeling studies on 3D braided fabric composite generally include the geometric model of unit cells, identification of key process parameters such as the pattern and the take-up rate, the limiting geometries of braided fabric jamming, the microstructural characteristics such as braid yarn orientation and fiber volume fraction, and the properties of the fiber and matrix [133]. The mathematical model based on the unit cell approach predicted the structural features of 3D braided composites such as fiber orientation, fiber volume fraction, and inter yarn voids from the key process variables of braiding pattern, the take-up rate, and the yarn geometry. The limiting geometry was computed by considering yarn jamming in the structure. Using the yarn jamming factor, it is possible to identify the complete range of allowable geometric arrangements for 3D braided preform [134]. The microstructure of 3D braided 1 × 1 pattern preforms was analyzed and the mathematical relationships among the structural parameters, such as the yarn packing factor, the fiber orientation, the fiber volume fraction and the braiding pitch, were derived. It was noted that the unit cell size and shape changed during the consolidation of braided preform. Therefore, this affects the properties of the 3D braided preforms [135].

The fabric geometry model (FGM) was developed to characterize the 3D braided preform composite with regard to fiber and matrix, and the processing parameters. 3D braided unit cell geometry in FGM requires two basic components: the fabric geometry and the determination of the fiber volume fraction. Fabric geometry is a function of the take-up rate during fabric formation, while yarn displacement values, in terms of the number of yarns, depend on row and column motions. The orientation of the yarns in a 3D preform depends on the fabric construction, the fabric shape, and the dimensions of the braiding loom [136]. The yarn orientation angle tends to decrease as the number of yarns in the fabric increases. For the same number of yarns in the fabric, the yarn orientation angle decreases as the linear density of the fabric decreases. The effective Young's modulus and Poisson's ratio of 3D braided composites with internal crack were characterized by using the homogenization theory and the modified finite element method [137, 138]. The effects of cut edges, filament bundle size and braid pattern were examined through the tensile, compressive, flexural and shear tests. It was found that the specimens were sensitive to cut edges where the tensile strength of the cut and shaped graphite-epoxy preform composite was reduced [139]. Braid pattern also had an important effect on the tensile strength. For instance, changing the yarn orientation from a 1 × 1 to a 1 × 3 braid pattern reduced the yarn orientation angle. As a result, the preform's tensile strength increased [139]. In general, the tensile strength and the modulus of 3D braided composites tend to increase as filament bundle size increases. Although the strength and modulus of braided composites were significantly higher than those of the 0°/90° woven composite, the Poisson's ratio of the braided composites was very large, leading to instability in the transverse direction; by adding transverse yarns, the Poisson's ratio of the braided composite was reduced, but at the cost of a reduction in strength and modulus [139].

3D Axial Braided Fabric Structure

Braid topology on 3D axial braided preform produced by the 2-step method can be used to analyze the effects of yarn size and spacing, and the pitch length on the resulting braided fabric geometry [140]. Another study on 3D axial braided preform demonstrated that once the yarn sizes, preform contour sizes, pitch length and number of axial and braider yarns are measured, the braider yarn orientation and yarn volume fraction can be estimated [141]. The axial yarns carried most of the load in the axial direction, and the braider yarns were the main load carriers in the transverse direction. Therefore, it was desirable for the orientation angle of the braiders to be large [142]. 3D flat axial braided composites were analyzed by a 3D finite element model based on a representative volume element under periodical displacement boundary conditions, which simulates the spatial configuration of the braiding yarns and the axial yarns. The software ABAQUS was adapted to study the mechanical properties and the meso-scale mechanical response of the 3D axial braided composites [143]. A fiber-inclination model was developed to predict the strength of the 3D axial braided 1 × 1 pattern perform composite produced by the 4-step method. The analysis was based upon the transverse isotropy of unidirectional laminae and the Tsai-Wu polynomial failure criterion. The results showed that braider angle has a significant influence on the tensile modulus and the strength. The transverse angle has an obvious influence upon Poisson's ratio and axial yarns can improve the tensile properties of 3D braided flat composites [144, 145].

3D Nonwoven Fabric Structure

3D nonwovens are anisotropic materials in which the fiber strength and modulus, fiber length and thickness, fiber volume fraction and fiber angle in the in-plane and the out-of-plane directions are important preform properties [38]. The general stress–strain relationship is given by:

$$\sigma = C : \varepsilon \tag{11}$$

where σ is the stress, a second order tensor, ε is the strain, also a second-order tensor, and C is the stiffness constant, a fourth-order tensor [38]. For two-dimensional nonwovens, the in-plane directional stiffness of the nonwoven is calculated with Eqs. 12 and 13 based on fiber web theory [38].

$$C_{11} = E_f \int_0^{\pi} cos^4 f(\theta) d\theta \tag{12}$$

$$C_{12} = E_f \int_0^{\pi} cos^2\theta sin^2\theta f(\theta) d\theta \tag{13}$$

where C_{11} is the stiffness constant in the x-direction on the plane perpendicular to the x-axis, C_{12} is the stiffness constant in the y-direction on the plane perpendicular to the x-axis, E_f is the fiber modulus and $f(\theta)$ is the distribution of fiber orientation.

Directional stiffness constants of a web can be obtained, given the fiber modulus and the fiber orientation distribution in the web. The fiber web theory can be applied to the 3D nonwoven preform in which some amount of z-fiber is oriented in the thickness direction of the web [98, 146]. There is a good agreement between the fiber web theory and the experimental measurement of needle punched web where the fibers between two bonded points are straight and that fibers are rigidly bonded. It is tedious and time consuming to determine the fiber orientation distribution in a web. However, several practical methods have been developed to measure the fiber orientation such as the X-ray diffraction, the laser light diffraction, the light reflection and the refraction intensity [147].

3D Knitted Fabric Structure

The properties of 3D knitted structures including multilayered weft or warp knitted fabrics and 3D multiaxis warp knitted preforms were studied by various researchers. The fiber volume fraction of the 3D multilayered weft knitted structure can be calculated with Eq. 14.

$$V_f = \frac{n_k D_y L_s C W}{9 \rho_f A t} \times 10^{-5} \tag{14}$$

where n_k is the number of plies of the fabric in the composite, D_y is the yarn linear density (tex), L_s is the length of yarn in one loop of the unit cell (cm), C is the course density (end/cm), W is the wale density (end/cm), ρ_f is the density of fiber (g/cm^3) and A is the planar area over which W and C are measured (cm^2) and t is the thickness of the composites (cm).

It was concluded that the fiber content of the weft knitted fabric composites can be increased by increasing D_y by using coarser yarns [35]. In general, the coarser yarns are difficult to knit and the coarsest yarn knittable is dependent on the yarn type and knitting needle size. In addition, the maximum V_f is limited by the knitting needles used in the knitting machine based on the relation $N = C/W$, where N is the stitching density. Hence, V_f is proportional to the structure parameters L_s and N. The maximum V_f can be achieved with increasing stitch density or tightness of the knitted fabric. It was claimed that the maximum attainable volume fraction of knitted fabric composite can be 40 % [35, 64]. It was stated that the failure mechanisms for weft knitted structures were depended on both the wale and the course directional crack propagations, and demonstrated better interlaminar fracture toughness properties due to the 3D loop structure [37]. The failure process of the weft knitted structures under tensile load includes crack branching, loop to loop friction, yarn bridging and fiber breakages. It was also shown that an increase in loop length or stitch density has opposite effects on the tensile strength and impact performance of the weft-knitted composites. The plain weft knitted structure exhibited good energy absorption capacity. Matrix cracking, matrix/fiber debonding, and fiber breakage were the major damage mechanisms [35, 37, 103].

3D loop structure was studied and it was proposed that the loop structure was constituted by sets of arcs. Adoption of the arc shape loop geometry into micromechanical technique considers the influence of knitting parameters and estimation of elastic properties of knitted composites [148]. On the other hand, the loopy nature of the yarn in the preform and 3D architecture of the loop renders the preform heterogeneous, thus complicating the analytical procedure for predicting mechanical properties. It was explained that multiaxis 3D knitted fabrics suffer from limitation in fiber architecture, through-thickness reinforcement due to the thermoplastic stitching yarn and three dimensional shaping during molding. For this reasons, the multiaxis 3D knitted fabric was layered and stitched to increase damage resistance and reduce production cost [1].

5 Examples of Application

Traditional as well as contemporary fabric structures are increasingly gaining acceptance due to their attractive specific performance and low cost for usage in technical textiles [149], defense and civilian areas such as transportation, automobile, energy and marine industries [150]. Biaxial, triaxial and more sophisticated multiaxis 3D fabric structures based on weaving, braiding, knitting, stitching and nonwoven technologies, are used as structural elements in medical, space and rocket industry [151]. Examples of these elements are plate, stiffened panel and beams and spars, shell or skin structures [152], and hip and medical device and prosthesis [153, 154]. Recently, Atkinson et al. [155] explored that using the nano-based high modulus fibers in the 3D fabrics could improve their mechanical properties.

5.1 *Structural Components*

2D and 3D woven, braided, knitted and nonwoven composites as structural components for various industrial applications fulfilled the general requirements such as low cost, manufacturability, good mechanical performance and energy absorption, corrosion resistance, repairability and recyclability, fuel economy and low noise level [156]. Typical structural components in various industrial applications are knot elements for space frame-like structures, beams, shells, exhaust, seats and chassis. For instance, the use of woven and braided composites in structural applications allows a significant reduction in component number and provides a substantial weight reduction compared with metal [82]. In addition, 2D and 3D woven, braided and knitted composites as T-joints and T-shape connectors, cones, pipes, and I-beams are attractive applications in general engineering fields. 2D and 3D nonwoven composites can be used in construction industry as a roofing and tile underlay, thermal and noise insulation, house wrap. Some geotextile applications of 2D or multilayered 3D nonwoven composites are asphalt overlay, soil stabilization, drainage, sedimentation and erosion control. Industrial applications of 2D or 3D nonwoven structures are cable insulation, battery separators, satellite dishes and coating [156].

5.2 *Ballistic Applications*

2D and 3D woven fabric and rigid ballistic plate are used extensively to protect the human and goods from various threats such as projectile, blast, fragment and high energy explosives. In addition, 2D and 3D braided, knitted and nonwoven fabrics and rigid composites can be utilized as protective products for vehicular crash guards, composite helmet, interlinings, insulation and protective industrial workwear and fire fighter suits [37, 156]. 2D and 3D woven, braided, knitted and nonwoven structures for soft and rigid ballistic applications are made by using high modulus and high strength fibers like para-aramid, polyethylene and glass fibers. Ballistic structures are manufactured as multilayer to obtain required structural thickness by using the above mentioned 2D fabrics. In some cases, to enhance the out-of-plane properties, those 2D fabrics are formed by stitching or quilting. Especially, 2D woven fabric for soft ballistic applications is very effective due to the in-plane crimp between yarn sets since the crimps act as a secondary energy absorbing mechanism due to local intra-yarn frictional forces generated during impact loads.

5.3 Space and Aerospace Applications

2D and 3D woven and braided fabrics are used in aerospace applications as soft space suits for astronauts, space shuttle components and aircraft seat cushions. They are currently employed in critical structures of both civil and military aircrafts such as the fuselage, the wings and the skin of the aircraft. Other areas of usage are top and side tail units, fuselage panelling, leading edges on side rudders, and engine panelling. It was reported that multiaxis 3D warp knitted composites are also being evaluated for rotor blades, outer skin and ballistic protection for helicopters. 3D weft or warp knitted ceramic composite was also developed for use as a structural parts for jet engine vanes, radomes and rudder tip fairing [37].

5.4 Automotive Applications

Fiber based dry and soft textile fabrics or structures are the main interior materials for automobiles as well as trains, aircrafts and ships. This provides the users well-being and comfort. They also withstand daylight and ultraviolet radiation. The maintenance cost is low and they are easy to care. 2D and 3D woven and knitted fabrics are used as airbag, and car seat parts which has better air permeability and moisture removal properties and car seat cover for the aesthetic and durability requirements. 2D and 3D textile preform composite structures are widely used as parts of the suspension, gears, drive belts, tires, heater hoses, battery separators, brake and clutch linings, air filters, gaskets and crash helmets. 2D and 3D warp knitted fabrics and nonwovens are widely used as preassembled interior components such as boot liners, seatbacks, door panels, oil and cabin air filters, molded bonnet liners, heat shields, wheelhouse covers, parcel shelves and shelf trim. 2D and 3D braided preform and composites have been used in racing car bodies, structural members such as beams which are made up of foam cores over braided with a carbon preform structure, aprons and spoilers, connecting rods, drive shafting and flexible couplings. Car noses, monocoques and bumpers are also made from braided carbon structures. They reduce the weight and improve the crash behavior [37, 156].

5.5 *Medical Applications*

2D and 3D fiber based structures are used in protective medical apparel such as baby diapers, feminine hygiene products, adult incontinence items, dry and wet pads, nursing pads or nasal strips, operation drapes, gowns and packs, face masks, surgical dressings. 2D and 3D woven, braid, warp knitted and nonwoven structures find also more functional applications as in vascular prosthesis due to good mechanical properties and better ingrowth of tissue to seal the prosthesis walls, grafts for inborn vessel anomaly or arteriosclerotic damage, soft tissue such as skin and cartilage, artificial tendons and ligaments, wound dressing, absorbable and non-absorbable sutures, stents, tissue engineering scaffolds as to repair or regenerate tissues through combinations of implanted cells-biomaterial scaffolds-biologically active molecules, blood filters, plasters, compression bandages, surgical hosiery and hospital bedding. It was also demonstrated that 2D and 3D fabrics are dimensionally stable, similar mechanical properties with the human organs and biocompatible [39, 82, 156].

5.6 *Sports Applications*

2D and 3D woven and braided composite structures are employed in various sports activities especially golf, baseball and tennis. The specific applications are roller blades, bike frames, golf stick, tennis rackets, baseball stick, ski and surf equipment and footwear. 3D warp knitted spacer fabrics are also extensively used in both sports shoes and garments due to its lightweight, springiness, washability and air permeability properties [156].

6 Conclusions

Two dimensional and three dimensional fabric architectures for composite application and their fabrication techniques have been reviewed in this chapter. Two dimensional woven, braided, knitted and nonwoven fabrics have been widely used for making various structural composite parts in civilian and defense related areas. However, composite structures from biaxial layered fabrics are prone to delamination between layers due to the lack of z-fibers and have crimp which lowers the properties. Biaxial fabric method and techniques are well developed. Triaxial fabrics have an open structure and low fabric volume fraction. But, in-plane properties of triaxial fabric are homogeneous due to bias yarns. On the other hand, biaxial and triaxial braided fabrics have size and thickness limitations. Similar to biaxial fabbrics, triaxial fabric method and techniques are also well developed.

3D woven fabrics have multiple layers and suffer no delamination due to the z-fibers. But, 3D woven fabrics have low in-plane properties. 3D braided fabrics have multiple layers and they exhibit no delamination due to intertwine type out-of-plane interlacements. However, 3D braided fabrics have low transverse properties due to lack of yarns equivalent to the filling yarns in 3D woven fabrics. They also have size and thickness limitations. Similar to 2D fabrics, various 3D woven and braided fabrication method and techniques are commercially available. Additionally, various unit cell based models on 3D woven, braided and knitted structures have been to define the geometrical and mechanical properties of these structures. Most of the unit cell based models include micromechanics and numerical techniques.

Multiaxis 3D knitted fabrics which have four layers consolidated with stitching, possess no delamination problems and in-plane properties are enhanced due to the ±bias yarn layers. But they have limitations for multiple layering and layer sequences. Multiaxis 3D knitting method and techniques have been well developed. Similarly, multiaxis 3D woven fabrics have multiple layers and no delamination problems due to the z-fibers and in-plane properties are enhanced due to ±bias yarn layers. Multiaxis 3D braided fabrics also have multiple layers and no delamination, and their some properties are improved for the placement of the bias layer in the outer surface of the structure. But multiaxis 3D technique is still being its early development stages and needs to be commercialized for the future. 3D woven, braided, knitted and nonwoven, and multiaxis 3D textile preforms are potential candidate in use space and aerospace industries for various structural parts as stiffened panels, combustion chamber and connectors, tubes, exit cones. These end-uses could be expanding towards civilian areas as automotive, civil engineering and medical field.

References

1. Dow MB, Dexter HB (1997) Development of stitched, braided and woven composite structures in the ACT Program and at Langley Research Center (1985 to 1997). NASA/TP-97-206234
2. Kamiya R, Cheeseman BA, Popper P et al (2000) Some recent advances in the fabrication and design of three dimensional textile preforms: a review. Compos Sci Technol 60:33–47
3. Ko FK, Chou TW (1989) Textile structural composites. Elsevier, New York
4. Chou TW (1992) Microstructural design of fiber composites. Cambridge University Press, Cambridge
5. Cox BN, Dadkhah MS, Morris WL et al (1993) Failure mechanisms of 3D woven composites in tension, compression and bending. Acta Metall Mater 42:3967–3984
6. Brandt J, Drechsler K, Filsinger J (2001) Advanced textile technologies for the cost effective manufacturing of high performance composites. Paper presented at the RTO AVT specialist meeting on low cost composite structures, Loen, Norway, 7–11 May 2001

7. Mohamed MH (1990) Three dimensional textiles. Am Sci 78:530–541
8. Bilisik A, Mohamed MH (1994) Multiaxis 3D weaving machine and properties of multiaxial 3D woven carbon/epoxy composites. Paper presented at the 39th international SAMPE symposium, Anaheim, USA, 11–14 Apr 1994
9. Uozumi T, Iwahori Y, Iwasawa S et al (2001) Braiding technologies for airplane applications using RTM process. Paper presented at the 7th Japan international SAMPE symposium, Tokyo, Japan
10. Furrow KW (1996) Material property evaluation of braided and braided/woven wing skin blade stiffeners, NASA contractor report: 198303
11. Ko FK (1987) Braiding. In: Reinhart TJ (ed) Engineered materials handbook. ASM International, Ohio, pp 519–528
12. Bogdanovich A, Mungalov D (2002) Recent advancements in manufacturing 3D braided preforms and composites. In: Bandyopadhyay S (ed) Proc. ACUN-4 composite systems-macro composites, micro composites, nanocomposites. University of New South Wales, Sydney
13. Bluck BM (1969) High speed bias weaving and braiding. US Patent 3426804, 11 Feb 1969
14. Maistre MA (1974) Process and apparatus for producing wire braids. GB Patent 1356524, 12 June 1974
15. Florentine RA (1983) Magnaweave process-from fundamentals to applications. Text Res J 53:620–623
16. Weller RD (1985) AYPEX: A new method of composite reinforcement braiding, 3D composite materials. In: NASA conference publication no 2420, 1985
17. Popper P, McConnell R (1987) A new 3D braid for integrated parts manufacture and improved delamination resistance-the 2-step process. In: Proceedings of 32nd international SAMPE symposium and exhibition, Anaheim, CA, USA
18. Dexter HB, Hasko GH (1996) Mechanical properties and damage tolerance of multiaxial warp-knit composites. Compos Sci Technol 51:367–380
19. Mohamed MH, Bilisik AK (1995) Multilayered 3D fabric and method for producing. US Patent 5465760, 14 Nov 1995
20. Uchida H, Yamamoto T, Takashima H (2000) Development of low cost damage resistant composites. http://www.muratec.net/jp. Accessed 14 May 2008
21. Kostar TD, Chou TW (1994) Microstructural design of advanced multistep three dimensional braided performs. J Compos Mater 28:1180–1201
22. Bilisik K (2011) Multiaxis three dimensional (3D) woven fabric. In: Vassiliadis SG (ed) Advances in modern woven fabrics technology. InTech-Open Access, Rijeka, pp 79–106
23. Kevra Advanced Composite Technology (2015) http://www.kevra.fi/en/Products/Reinforcements/. Accessed 20 Mar 2015
24. Bhatnagar A, Parrish ES (2006) Bidirectional and multiaxial fabric and fabric composites. US Patent 7073538, 11 July 2006
25. Bilisik K, Yilmaz B (2012) Multiaxis multilayered non-interlaced/non-Z e-glass/polyester preform and analysis of tensile properties of composite structures by statistical model. Text Res J 82:336–351
26. Cox BN, Flanagan G (1997) Handbook of analytical methods for textile composites. NASA contractor report 4750
27. Abildskow D (1996) Three dimensional woven fabric connector. US Patent 5533693, 9 July 1996
28. Dow NF (1969) Triaxial fabric. US Patent 3446251, 27 May 1969
29. Dow NF, Tranfield G (1970) Preliminary investigations of feasibility of weaving triaxial fabrics (doweave). Text Res J 40:986–998
30. Lida S, Ohmori C, Ito T (1995) Multiaxial fabric with triaxial and quartaxial portions. US Patent 5472020, 5 Dec 1995
31. Brunnschweiler D (1953) Braids and braiding. J Text I 44:666–686
32. Brunnschweiler D (1954) The structure and tensile properties of braids. J Text I 45:55–77

33. Rogers CW, Crist SR (1997) Braided preform for composite bodies. US Patent 5619903, 15 Apr 1997
34. Klein JT, Broughton RM, Beale DG (1999) Braided fabric and method of forming. US Patent 5899134, 4 May 1999
35. Hamada H, Ramakrishna S, Huang ZM (1999) Knitted fabric composites, 3-D textile reinforcements in composite materials. In: Miravete A (ed) 3-D textile reinforcements in composite materials. Woodhead Publishing Ltd, Cambridge, pp 180–216
36. Ramakrishna S, Hamada H, Kotaki M et al (1993) Future of knitted fabric reinforced polymer composites. In: Proceedings of 3rd Japan international SAMPE symposium, Chiba, 7–10 Dec 1993
37. Padaki NV, Alagirusamy R, Sugun BS (2006) Knitted preforms for composite applications. J Ind Text 35:295–321
38. Chapman RA (ed) (2010) Applications of nonwovens in technical textiles. Woodhead Publishing Limited, Oxford
39. Albrecht W, Fuchs H, Kittelmann W (eds) (2003) Nonwoven fabrics: raw materials, manufacture, applications, characteristics, testing processes. Wiley, Weinheim
40. Bilisik K, Yolacan G (2014) Experimental characterization of multistitched two dimensional (2D) woven e-glass/polyester composites under low velocity impact load. J Compos Mater 48:2145–2162
41. Bilisik K, Sahbaz N, Bilisik NE, Bilisik HE (2013) Three dimensional (3D) fully interlaced woven preforms for composites. Text Res J 83:2060–2084
42. Bilisik K, Sahbaz N, Bilisik NE, Bilisik HE (2014) Three-dimensional circular various weave patterns in woven preform structures. Text Res J 84:638–654
43. Bilisik K (2010) Multiaxis 3D woven preform and properties of multiaxis 3D woven and 3D orthogonal woven carbon/epoxy composites. J Reinf Plast Comp 29:1173–1186
44. Crawford JA (1985) Recent developments in multidirectional weaving. NASA publication no. 2420
45. Yasui Y, Anahara M, Omori H (1992) Three dimensional fabric and method for making the same. US Patent 5091246, 25 Feb 1992
46. Ruzand JM, Guenot G (1994) Multiaxial three-dimensional fabric and process for its manufacture. International Patent WO 94/20658, 15 Sept 1994
47. Uchida H, Yamamoto T, Takashima H et al (1999) Three dimensional weaving machine. US Patent 6003563, 21 Dec 1999
48. Mohamed MH, Bilisik AK (1995) Multilayered 3D fabric and method for producing. US Patent 5465760 A, 14 Nov 1995
49. Bilisik K (2000) Multiaxial three dimensional (3D) circular woven fabric. US Patent 6129122, 10 Oct 2000
50. Florentine RA (1982) Apparatus for weaving a three dimensional article. US Patent 4312261, 26 Jan 1982
51. Brown RT (1988) Braiding apparatus. UK Patent 2205861 A, 31 May 1988
52. Brown RT, Ratliff ED (1986) Method of sequenced braider motion for multi ply braiding apparatus. US Patent 4621560, 11 Nov 1986
53. Tsuzuki M (1994) Three dimensional woven fabric with varied thread orientations. US Patent 5348056, 20 Sept 1994
54. Kostar TD, Chou T-W (1999) Braided structures, 3D textile reinforcements in composite materials. In: Miravete A (ed) 3-D textile reinforcements in composite materials. Woodhead Publishing Ltd, Cambridge, pp 217–240
55. Brookstein DS, Rose D, Dent R et al (1996) Apparatus for making a braid structure. US Patent 5501133, 26 Mar 1996
56. Brookstein DS, Skelton J, Dent JR et al (1994) Solid braid structure. US Patent 5357839, 25 Oct 1994
57. Brookstein DS (1991) A comparison of multilayer interlocked braided composites with other 3D braided composites. Paper presented at the 36th international SAMPE symposium, Anaheim, 15–18 Apr 1991

58. McConnell RF, Popper P (1988) Complex shaped braided structures. US Patent 4719837, 19 Jan 1988
59. Chen JL, El-Shiekh A (1994) Construction and geometry of 6 step braided preforms for composites. Paper presented at the 39th international SAMPE symposium, Anaheim, CA, 11–14 Apr 1994
60. Bilisik AK (1998) Multiaxial and multilayered 8-step circular braided preform for composite application. Paper presented at the 8. international machine design and production conference, Middle East Technical University, Ankara, 9–11 Sep 1998
61. Bilisik K (2012) Multiaxis three dimensional weaving for composites: a review. Text Res J 82:725–743
62. Naumann R, Wilkens C (1987) Warp knitting machine. US Patent 4703631, 3 Nov 1987
63. Wunner R (1989) Apparatus for laying transverse weft threads for a warp knitting machine. US Patent 4872323, 10 Oct 1989
64. Ciobanu L (2011) Development of 3D knitted fabrics for advanced composite materials. In: Attaf B (ed) Advances in composite materials—ecodesign and analysis. InTech Open Access, Rijeka, pp 161–192
65. Olry P (1988) Process for manufacturing homogeneously needled a three-dimensional structures of fibrous material. US Patent 4,790,052, 13 Dec 1988
66. Ko FK (1999) 3-D textile reinforcements in composite materials. In: Miravete A (ed) 3-D textile reinforcements in composite materials. Woodhead Publishing Ltd, Cambridge, pp 9–41
67. Vandeurzen P, Ivens J, Verpoest I (1999) Mechanical modelling of solid woven fabric composites. In: Miravete A (ed) 3-D textile reinforcements in composite materials. Woodhead Publishing Ltd, Cambridge, pp 67–99
68. Ge Z, Hu H (2013) Innovative three-dimensional fabric structure with negative Poisson's ratio for composite reinforcement. Text Res J 83:543–550
69. Wang Z, Hu H (2014) Auxetic materials and their potential applications in textiles. Text Res J 84:1600–1611
70. Wang Z, Hu H, Xiao X (2014) Deformation behaviors of three-dimensional auxetic spacer fabrics. Text Res J 84:1361–1372
71. Ugboluea SC, Kim YK, Warner SB et al (2010) The formation and performance of auxetic textiles. Part I: theoretical and technical considerations. J Text I 101:660–667
72. Ugboluea SC, Kim YK, Warner SB et al (2011) The formation and performance of auxetic textiles. Part II: geometry and structural properties. J Text I 102:424–433
73. Subramani P, Rana S, Oliveira DV et al (2014) Development of novel auxetic structures based on braided composites. Mater Des 61:286–295
74. Liu Y, Hu H, Lam JKC et al (2010) Negative Poisson's ratio weft-knitted fabrics. Text Res J 80:856–863
75. Sanami M, Ravirala N, Alderson K et al (2014) Auxetic materials for sports applications. Proc Eng 72:453–458
76. Textile Learner Blog (2015). http://textilelearner.blogspot.com.tr/2011/06/weaving-weaving-mechanism_643.html. Accessed 20 Mar 2015
77. Deemey S (2002) The new generation of carpet weaving machines combines flexibility and productivity, Technical notes, Van de Wiele Incorporations
78. Mohamed MH, Zhang ZH (1992) Method of forming variable cross-sectional shaped three dimensional fabrics. US Patent 5085252, 4 Feb 1992
79. King RW (1977) Three dimensional fabric material. US Patent 4038440, 26 July 1977
80. Fukuta K, Nagatsuka Y, Tsuburaya S et al (1974) Three dimensional fabric, and method and loom construction for the production thereof. US Patent 3834424, 10 Sept 1974
81. Crawford JA (1985) Recent developments in multidirectional weaving. NASA publication no. 2420, pp 259–269

82. Bilisik K, Karaduman NS, Bilisik NE (2015) Applications of braided structures in transportation. In: Fangueiro R, Rana S (eds) Braided structures and composites: production properties mechanics and technical applications. Taylor and Francis, Boca Raton (Accepted for Publication)
83. Wilden KS, Harris CG, Flynn BW et al (1997) Advanced technology composite fuselage-manufacturing, The Boeing Company, NASA contractor report 4735
84. Fiber innovations Inc. (2002) Technical documents, 8 Jan 2002
85. Schneider M, Pickett AK, Wulfhorst B (2000) A new rotary braiding machine and CAE procedures to produce efficient 3D braided textiles for composites. Paper presented at the 45th international SAMPE symposium, Long Beach, CA, 21–25 May 2000
86. Tsuzuki M, Kimbara M, Fukuta K et al (1991) Three dimensional fabric woven by interlacing threads with rotor driven carriers. US Patent 5067525, 26 Nov 1991
87. Uozumi T (1995) Braid structure body. US Patent 5438904, 8 Aug 1995
88. Core77 web site (2015) http://www.core77.com/blog/materials/video_of_lexus_360-degree_carbon_fiber_loom_19146.asp. Accessed 20 Mar 2015
89. Bilisik K (2011) Three dimensional (3D) axial braided preforms: experimental determination of effects of structure-process parameters on unit cell. Text Res J 81:2095–2116
90. Spain RG (1990) Method for making 3D fiber reinforced metal/glass matrix composite article. US Patent 4916997, 17 Apr 1990
91. Langer H, Pickett A, Obolenski B et al (2000) Computer controlled automated manufacture of 3D braids for composite. Paper presented at the Euromat symposium, Munich, Germany
92. Schneider M, Pickett AK, Langer H (2000) Exemplary CAE design tools for textile reinforced composites by means of FE-analysis. Paper presented at the Euromat symposium, Munich, Germany
93. Mungalov D, Duke P, Bogdanovich A (2007) High performance 3-D braided fiber preforms: design and manufacturing advancements for complex composite structures. SAMPE J 43:53–60
94. Mungalov D, Bogdanovich A (2004) Complex shape 3-D braided composite preforms: structural shapes for marine and aerospace. SAMPE J 40:7–21
95. Schreiber F, Theelen K, Schulte S et al (2011) 3D-hexagonal braiding: possibilities in near-net shape preform production for lightweight and medical applications. Paper presented at the 18th international conference on composite materials, Jeju Island, Korea, 21–26 Aug 2011
96. Kostar TD, Tchou TW (1994) Process simulation and fabrication of advanced multistep 3-dimensional braided performs. J Mater Sci 29:2159–2167
97. National Programme on Technology Enhanced Learning (2015) http://nptel.ac.in/courses/116102005/3. Accessed 20 Mar 2015
98. Tausif M, Russell SJ (2012) Characterisation of the z-directional tensile strength of composite hydroentangled nonwovens. Polym Testing 31:944–952
99. Dewalt PL, Reichard RP (1994) Just how good are knitted fabrics? J Reinf Plast Comp 13:908–917
100. Offermann P, Hoffmann G, Engelmann U (1996) Mehrlagengestricke und verfahren zu seiner herstellung. DE Patent 4419985C2, 4 Apr 1996
101. Offermann P, Hoffmann G, Engelmann U (2001) Multilayer knitted structure and method of producing the same. US Patent 6244077, 12 June 2001
102. Cebulla H, Diestel O, Offermann P (2002) Fully fashioned biaxial weft knitted fabrics. AUTEX Res J 2:8–13
103. Cherif C, Krzywinski S, Diestel O et al (2012) Development of a process chain for the realization of multilayer weft knitted fabrics showing complex 2D/3D geometries for composite applications. Text Res J 82:1195–1210
104. Scardino FL, Ko FK (1981) Triaxial woven fabrics: Part I, Behavior under tensile, shear and burst deformations. Text Res J 51:80–89
105. Schwartz P (1981) The mechanical behavior of fabrics having three, non-orthogonal thread directions (triaxial) and the equivalence of conventional fabrics. Dissertation, NCSU

106. Skelton J (1971) Triaxial woven fabrics: their structure and properties. Text Res J 41:637–647
107. Nishimoto H, Ohtani A, Nakai A et al (2010) Prediction method for temporal change in fiber orientation on cylindrical braided performs. Text Res J 80:814–821
108. Long AC (2001) Process modelling for liquid moulding of braided performs. Compos A Appl Sci 32:941–953
109. Song YS, Chung K, Kang TJ et al (2004) Prediction of permeability tensor for three dimensional circular braided preform by applying a finite volume method to a unit cell. Compos Sci Technol 64:1629–1636
110. Nasu S, Ohtani A, Nakai A et al (2010) Deformation behavior and mechanical properties of braided rectangular pipes. Compos Struct 92:752–756
111. Fujihara K, Yoshida E, Nakai A et al (2007) Influence of micro-structures on bending properties of braided laminated composites. Compos Sci Technol 67:2191–2198
112. Goyal D, Tang XD, Whitcomb JD et al (2005) Effect of various parameters on effective engineering properties of 2 × 2 braided composites. Mech Adv Mater Struct 12:113–128
113. Smith LV, Swanson SR (1996) Effect of architecture on the strength of braided tubes under biaxial tension and compression. J Eng Mater Technol 118:478–484
114. Smith LV, Swanson SR (1993) Response of braided composites under compressive loading. Compos Eng 3:1165–1184
115. Tsai JS, Li SJ, Lee LJ (1998) Microstructural analysis of composite tubes made from braided preform and resin transfer molding. J Compos Mater 32:829–850
116. Byun JH (2000) The analytical characterization of 2-D braided textile composites. Compos Sci Technol 60:705–716
117. Yan Y, Hoa SV (2002) Energy approach for prediction of mechanical behavior of 2-D triaxially braided composites. Part II: Parameter analysis. J Compos Mater 36:1233–1253
118. Ramakrishna S, Hamada H, Rydin R et al (1995) Impact damage resistance of knitted glass fiber fabric reinforced polypropylene composite laminates. Sci Eng Comp Mater 4:61–72
119. Bilisik K, Yolacan G (2011) Multiaxis multilayered non-interlaced/non-Z e-glass/polyester preform composites and determination of flexural properties by statistical model. J Reinf Plast Comp 30:1065–1083
120. Bilisik K (2011) Experimental determination of ballistic performance of newly developed multiaxis non-interlaced/non-Z e-glass/polyester and 3D woven carbon/epoxy composites with soft backing aramid fabric structures. Text Res J 81:520–537
121. Bilisik K, Yolacan G (2014) Warp and weft directional tensile properties of multistitched biaxial woven e-glass/polyester composites. J Text I 105:1014–1028
122. Bilisik K, Yolacan G (2014) Warp and weft directional bending properties of multistitched biaxial woven e-glass/polyester nano composites. J Ind Text. doi:10.1177/1528083714523163
123. Bilisik K (2010) Multiaxis 3D weaving: comparison of developed two weaving methods-tube-rapier weaving versus tube-carrier weaving and effects of bias yarn path to the preform properties. Fibers Polym 11:104–114
124. Gowayed YA, Pastore CM (1992) An integrated approach to the mechanical and geometrical modeling of textile structural composites. Presented at the sixth conference on advanced engineering fibers and textile structures for composites, FIBER-TEX'92. North Carolina State University, Raleigh
125. Gu P (1994) Analysis of 3D woven preforms and their composite properties. Dissertation, NCSU
126. Dickinson LC (1990) Evaluation of 3D woven carbon/epoxy composites. Dissertation, NCSU
127. Babcock W, Rose D (2001) Composite preforms. AMPTIAC Newsl 5:7–11
128. Bilisik K (2010) Dimensional stability of multiaxis 3D woven carbon preform. J Text I 101:380–388

129. Bilisik K (2010) Multiaxis 3D weaving: comparison of developed two weaving methods—tube-rapier weaving versus tube-carrier weaving and effects of bias yarn path to the preform properties. Fiber Polym 11:104–114
130. Bilisik K, Mohamed MH (2010) Multiaxis three dimensional (3D) flat woven preform-tube carrier weaving. Text Res J 80:696–711
131. Bilisik K (2010) Multiaxis three dimensional (3D) circular woven preforms—"Radial crossing weaving" and "Radial in-out weaving": preliminary investigation of feasibility of weaving and methods. J Text I 101:967–987
132. Bilisik K, Sahbaz N (2012) Structure-unit cell base approach on three dimensional (3D) representative braided preforms from 4-step braiding: experimental determination of effect of structure-process parameters on predetermined yarn path. Text Res J 82:220–241
133. Byun JH, Chou TW (1996) Process-microstructure relationships of 2-step and 4-step braided composites. Compos Sci Technol 56:235–251
134. Du GW, Chou TW, Popper P (1991) Analysis of 3-dimensional textile preforms for multidirectional reinforcement of composites. J Mater Sci 26:3438–3448
135. Chen L, Tao XM, Choy CL (1999) On the microstructure of three-dimensional braided performs. Compos Sci Technol 59:391–404
136. Ko F (1985) Development of high damage tolerant, net shape composites through textile structural design. In: Proceedings of 5th international conference on composite materials ICCM-V, San Diego, CA
137. Byun JH, Chou TW et al (1989) Modeling and characterization of textile structural composites-A review. J Strain Anal Eng 24:253–262
138. Zeng T, Wu LZ, Guo LC et al (2005) A mechanical model of 3D braided composites with internal transverse crack. J Compos Mater 39:301–321
139. Macander AB, Crane RM, Camponaschi ET Jr (1984) The fabrication, processing and characterization of multidimensionally braided graphite/epoxy composite materials. David Taylor Naval Ship Research and Development Centre, DTNSRDC/SME-84–66, July 1984
140. Kuo WS (1997) Topology of three-dimensionally braided fabrics using pultruded rods as axial reinforcements. Text Res J 67:623–634
141. Li W, Hammad M, El-Shiekh A (1990) Structural analysis of 3D braided preforms for composites. Part II: Two step performs. J Text I 81:515–537
142. Li W (1990) On the structural mechanics of 3D braided preforms for composites. Dissertation, North Carolina State University
143. Xu K, Xu XW (2008) Finite element analysis of mechanical properties of 3D five-directional braided composites. Mat Sci Eng A Struct 487:499–509
144. Sun HY, Qiao X (1997) Prediction of the mechanical properties of three-dimensionally braided composites. Compos Sci Technol 57:623–629
145. Yang JM, Ma CL, Chou TW (1986) Fiber inclination model of three dimensional textile structural composites. J Compos Mater 20:472–484
146. Cox HL (1952) The elasticity and strength of papers and other fibrous materials. Br J Appl 3:72–74
147. Tsai PP, Bresee R (1991) Fiber orientation distribution from electrical measurements, Part I: Theory. Inda Jnr 3-36–40
148. Ramakrishna S, Fujita A, Cuong NK et al (1994) Tensile failure mechanisms of knitted glass fiber fabric reinforced epoxy composites. In: Proceedings of 4th Japan international SAMPE symposium & exhibition, Tokyo, 24–28 Sep 1995
149. Hearle JWS (1994) Textile for composites. Text Horiz 11:11–15
150. Mouritz AP, Bannister MK, Falzon PJ et al (1999) Review of applications for advanced three dimensional fiber textile composites. Compos A Appl Sci 30:1445–1461
151. Beyer S, Schmidth S, Maidi F et al (2006) Advanced composite materials for current and future propulsion and industrial applications. Adv Sci Tech 50:178–181
152. Yamamoto Y, Hirokawa T (1990) Advanced joint of 3D composite materials for space structure. Presented at the 35th international SAMPE symposium, Anaheim, CA, 2–5 Apr 1990

153. Donnet JB, Bansal RC (1990) Carbon fibers. Marcel Dekker Inc., New York
154. Bilisik K (2009) Multiaxis three-dimensional (3-D) woven and braided preform unit cells and implementation of possible functional characterization for biomedical applications. Artif Organs 33:A101
155. Atkinson KR, Skourtis C, Hutton SR (2008) Properties and applications of dry-spun carbon nanotube yarns. Adv Sci Tech 60:11–20
156. Jinlian HU (2008) 3-D fibrous assemblies: properties, applications and modeling of three dimensional textile structures. Woodhead Publishing Limited, Cambridge
157. Pu G, Mohamed MH (2001) Shaped three dimensional engineered fiber preforms with insertion holes and rigid composite structures incorporating same and method thereof. US Patent 6,283,168, 4 Sep 2001
158. Technische Universitat Dresden (2015) http://tu-dresden.de/die_tu_dresden/fakultaeten/fakultaet_maschinenwesen/itm/forschung/forschungsthemen/gl_gestrick. Accessed 20 Mar 2015
159. Gehring GG, Reisfeld A Jr (2001) Pointed thrust weapons protective fabric system, US Patent 6233978B1, 22 May 2001
160. Stanford University (2015) https://micromechanics.stanford.edu/advanced-homogenization-techniques-soft-matter-materials. Accessed 20 Mar 2015
161. Bilisik K (2013) Three dimensional braiding for composites: a review. Text Res J 83:1414–1436
162. Li D, Lu Z, Chen L et al (2009) Microstructure and mechanical properties of three-dimensional five-directional braided composites. Int J Solids Struct 46:3422–3432
163. CFC Carbon Ltd. (2015) http://www.cfccarbon.com/graphite-felt/pan-rigid-graphite-felt.html . Accessed 20 Mar 2015
164. Triaxial (2015) http://www.triaxial.us/Triaxial%20Fabric%20History%204.php. Accessed 20 Mar 2015
165. Technische Universitat München (2015) http://www.lcc.mw.tum.de/en/research-groups/process-technology-for-fibers-and-textiles/braiding-technology/. Accessed 20 Mar 2015
166. Pastore CM (1988) A processing science model for three dimensional braiding. Dissertation, Drexel University
167. Textile Learner Blog (2015) http://textilelearner.blogspot.com/2011/06/knitting-action-of-single-needle-bar_204.html#ixzz3OXswDfAO. Accessed 20 Mar 2015

Synthetic Fibres for Composite Applications

Davide Pico and Wilhelm Steinmann

Abstract This chapter gives the details of various synthetic fibres (both organic and inorganic such as glass, carbon, aramides, polyolefins, ceramic fibres, etc.) used to reinforce composite materials for conventional as well as very high-tech applications. Production and properties of these fibres and also the most common applications in fibre reinforced composites are included in this chapter.

1 Overview on Synthetic Fibers

Synthetic fibres are nowadays the most important material class for fibre reinforced composites, especially for application fields with high demands on mechanical, thermal and chemical stability of the material, such as aerospace, automotive or energy conversion. In this chapter, the most relevant synthetic fibres for composite applications are described. For each material class, the production process, the resulting fibre structure and fibre properties as well as the applications fields of the composite materials are summarized. The materials described in this chapter are:

- Glass fibres
- Basalt fibres
- Carbon fibres
- Ceramic fibres (oxide and non-oxide ceramics)
- Polymeric fibres (aramides, UHMWPE and other)

Although all of the fibre classes are important for industrial applications, the amount of fibres produced per year differs over several orders of magnitude for the

D. Pico (✉) · W. Steinmann
Institut für Texiltechnik (ITA) der RWTH Aachen University,
Otto-Blumenthal-Straße 1, 52074 Aachen, Germany
e-mail: davide.pico@ita.rwth-aachen.de

W. Steinmann
e-mail: wilhelm.steinmann@ita.rwth-aachen.de

© Springer Science+Business Media Singapore 2016

S. Rana and R. Fangueiro (eds.), *Fibrous and Textile Materials for Composite Applications*, Textile Science and Clothing Technology,
DOI 10.1007/978-981-10-0234-2_4

Table 1 Market data for different synthetic reinforcement fibres [1–3]

Fiber type	Production [t/y]	Price [€/kg]
Glass fibers	4,750,000	2–8
Basalt fibers	13,500	2–3
Carbon fibers (PAN)	150,000	15–50
Carbon fibers (pitch)	1,680	95–3,000
Oxide ceramics	15–20	200–800
Non-oxide ceramics	15–20	1,000–20,000
Aramide fibres	63,000	26–35

different materials. Furthermore, there are large differences in the fibre price, which is mostly driven by the complex production processes for certain fibre types. In Table 1, an overview on the produced fibres and their price is given.

For most of the applications, glass fibres (or basalt fibres as an alternative) provide sufficient properties, making them the preferable materials because of the lower price. This results in the highest market share among all synthetic reinforced fibres. If special properties are needed, other types of fibres are necessary:

- PAN-based Carbon fibres: Carbon fibres are used for applicatopns with special demands on mechanical properties because of the highest stiffness among all fibre classes and a high specific strength.
- Pitch-based Carbon fibres: Pitch-based CF are applied if a higher stiffness is required compared to PAN-based carbon fibres.
- Ceramic fibres: Ceramic fibres are used for applications with the highest demand on temperature stability. Fibres with the highest price combine good mechanical properties with excellent temperature stability.
- Aramide fibres: A special property of aramide fibres is their high impact strength, making them a preferable material for ballistic applications.
- UHMWPE fibres: Ultra high molecular weight polyethylene has a specific strength comparable to carbon fibres, with a lower stiffness and high elongation. Therefore, the energy to break the fibres is the highest among all fibre types.

The mechanical properties (tensile strength and Young's modulus) of the most important fibre types are summarized in Fig. 1. Within the figure, typical ranges of the properties are indicated. The fibre types are described within the following sections in this chapter.

2 Carbon Fibres

Carbon fibres are defined as fibrous materials with a carbon content of more than 90 % [4]. Carbon fibres where firstly discovered by Edison [5], who carbonized a cellulosic fibre and used the filament in a light bulb. In the 1950s and 1960s, first carbon fibres based on polyacrylonitrile were developed, which is still the most important material for carbon fibre production [6–10].

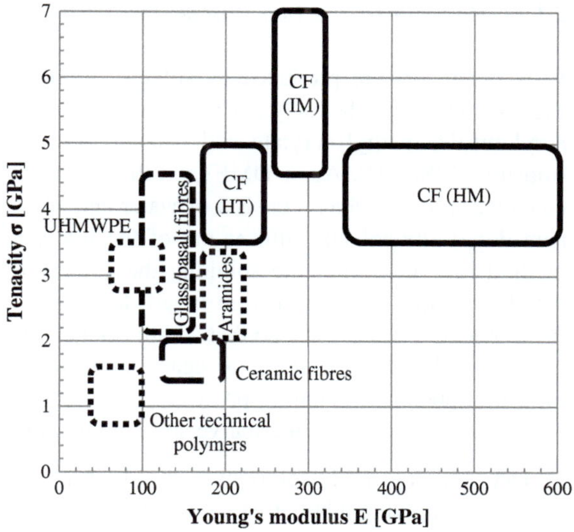

Fig. 1 Mechanical properties of the most important synthetic fibres for composite applications

2.1 *Fibre Manufacturing*

In this section, different methods of producing carbon fibres are described. Since pure carbon cannot be processed directly to fibres within a reasonable temperature and pressure range, carbon fibres are converted from so-called precursor materials. The most common precursor material with a market share of more than 90 % is polyacrylonitrile (PAN), for which the carbon fibre manufacturing is described in detail. The material and process chain for PAN-based carbon fibres is displayed in Fig. 2. Furthermore, pitch-based carbon fibres and alternative precursor systems are described.

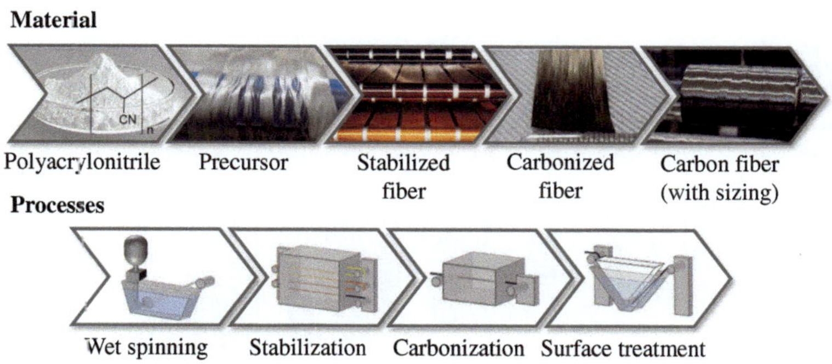

Fig. 2 Material and process chain for PAN-based carbon fibres

2.1.1 PAN-Based Carbon Fibres: Precursor

PAN precursors are produced by solvent spinning, mostly by wet spinning (See Fig. 3). Thereby, the polymer (normally a linear copolymer which consists of acrylonitrile, methyl acrylate and itaconic acid) is solved in an organic solvent (mostly DMSO, DMAc or DMF) and then spun into in a coagulation bath. In the coagulation bath, which consists of water and solvent (less concentration as in the spin dope), the solvent diffuses out of the filaments, which are solidified by that method [11, 12]. After the spinning the solvent is washed out in further bathes which contain water and solvent in still lower concentrations. During that process the filaments are stretched, so that a high orientation is achieved [13]. For achieving higher levels of orientation, a steam stretching process is added in a typical industrial precursor spinning process.

Beside wet spinning, alternative spinning methods exist to produced PAN fibres:

1. Dry spinning: PAN solutions with a high polymer concentration are spun into air. By applying a hot air stream with temperature above the boiling point of the solvent, the solvent evaporates and the fibres solidify. However, the air stream leads to a collapsing of the fibre cross-section, which results in non-round shapes and makes dry spinning a non-preferable method for precursor production [11, 12].
2. Air-gap spinning: Compared to wet spinning, the dope is first spun into an air gap before entering the coagulation bath. Due to lower friction in the air gap compared to the liquid, a higher stretching can be applied in the first part of the process. This has two positive effects: Firstly, a higher total stretch rate can be applied. Therefore, the throughput and winding speed can be increased, which leads to a higher productivity in precursor production. Secondly, the fibre diameter decreases in the air gap and thinner fibres enter the coagulation bath, which leads to a more homogeneous fibre structure due to shorter diffusion length during coagulation. Furthermore, thinner filaments do not collapse during

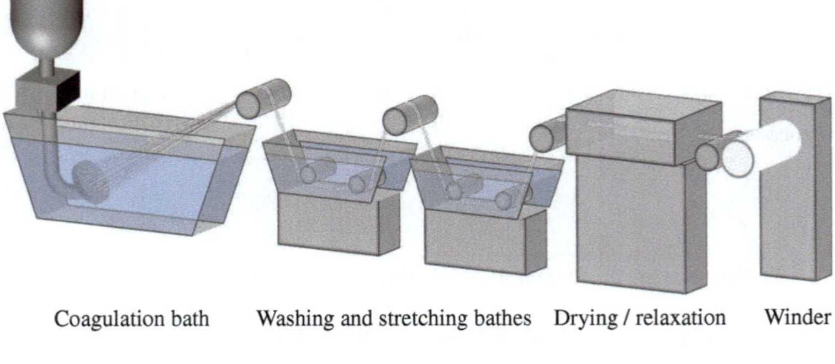

Fig. 3 Illustration of wet spinning process for PAN precursors

coagulation so that a completely round cross-section can be achieved. Air-gap spinning is therefore industrially applied for making carbon fibre precursors [11].

3. Melt spinning: Even though PAN cannot be directly molten (as described above), there are two approaches to plastify the polymer. Firstly, different co-monomers can be used to lower the melting point below the decomposition temperature. However, alternative stabilization techniques (e.g. by UV light) have to be found, since the polymer would otherwise melt in the stabilization furnace [14]. Secondly, water or carbon dioxide can be added under high pressure during extrusion of the material, which act as plasticizer. Since both materials would evaporate from the filaments after extrusion spontaneously and leave larger pores in the precursor material, a controlled evaporation has to be achieved by applying a pressure chamber after extrusion [15]. Even though melt spinning is more efficient because of higher winding speeds and environmentally friendly because of no solvents necessary for the process, an economic advantage of the process could still not be achieved due to high investment costs into the specialized machinery.

2.1.2 PAN-Based Carbon Fibres: Thermal Conversion

The thermal conversion (by convective heating of the fibres) is composed of the process steps stabilization and carbonisation. During the stabilization the fibers are prepared for the carbonation, so that they become incombustible. Furthermore, the first ring structures (ladder polymer consisting of two linear main chains) are formed [16]. The stabilization is composed of the exothermic part reactions cyclization, dehydration and oxidation [17]. The carbonisation (pyrolysis) takes place in inert atmosphere, mostly under nitrogen [18]. The inert gas avoids an oxidation at higher temperatures. The purpose is the elimination of the non-carbon atoms and the formation of turbostratic arranged graphene layers. For high modulus fibers, another process step (graphitization) follows among argon atmosphere at temperatures up to 3,000 °C [11] (Fig. 4).

Beside convective heating, alternative heating methods are suitable for carbon fibre conversion. Currently, research is carried out on the following conversion methods:

1. *Heating by direct (contact) heat transfer*: A method for lowering the energy consumption in the stabilization process is using contact heat instead of convective heating, e.g. by using hot godets. It was shown that energy savings of around 25 % are possible. Beside positive effects on the efficiency of the process, the fibre temperature can be controlled more efficiently, since heat from the chemical reactions is transferred efficiently to the surface of the godets. Furthermore, smaller temperature gradients close to linear heating profiles can be realized by applying a temperature gradient on the godet surface [19].

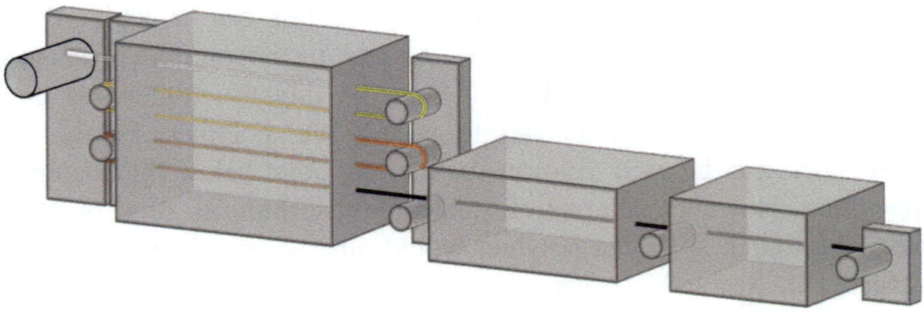

Fig. 4 Illustration of thermal conversion processes of PAN precursors to carbon fibres (without additional graphitization step for high modulus fibres)

2. *Microwave heating*: Precursor, stabilized and carbonized fibres can be heated by microwave radiation due to dipoles in their chemical structure or their electrical conductivity respectively. Beside possible energy savings, a more homogeneous heating across the fibre diameter can be achieved if microwave radiation is not directly absorbed on the surface. Another method of applying microwave technology in carbonization, which is already industrially available, is heating the carbonization furnace itself by microwave radiation instead of conductive heating [20].
3. *Microwave assisted plasma heating*: Microwave radiation can be used to ignite a plasma on the fibres' surface. At atmospheric pressure, high plasma temperatures can be achieved, which are suitable for carbonization. With the help of plasma, a homogeneous heating is also possible. However, inert gases have to be used to prevent the combustion of the fibre [21].

2.1.3 PAN-Based Carbon Fibres: Surface Treatment

The fiber surface is almost inert after the thermal conversion. Therefore, further process steps (surface activation and application of sizing) are necessary to enable a bonding (chemical and mechanical) for load transfer between the filament surface and the matrix material in fiber reinforced composites. An activation of the carbon fiber surface has the purpose of implementing functional groups on the surface and, depending on the method, reaching a increase of surface roughness for enhanced mechanical interlock.

In industrial processes, the anodic oxidation is currently the state-of-the-art technology for surface activation. For this process, carbon fibers are conducted through a bath which contains an electrolytic solution (salt solution, mostly with

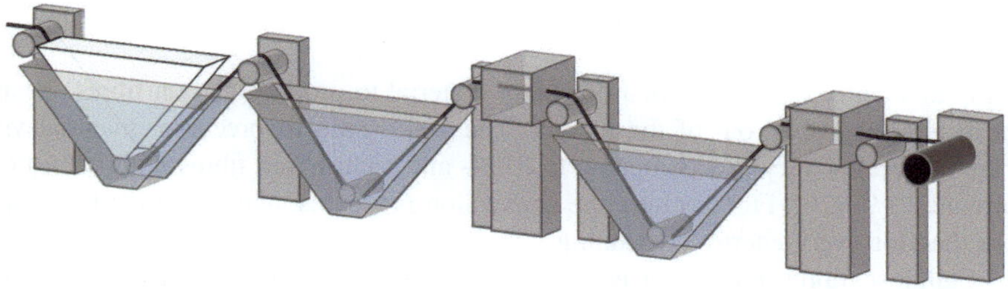

Fig. 5 Illustration of post-treatment of PAN-based carbon fibres (surface activation and sizing)

ammonium bicarbonate (NH_4HCO_3)) [11]. Between carbon fiber (served as anode) and a cathode, which is located in the bath, a voltage is applied driving the electrolysis process. By the use of the anodic oxidation is the carbon fiber surface functionalized by oxygenic groups like carboxyl, carbonyl and hydroxyl groups. The degree of the functionalization can be controlled by the type and concentration of salt in the electrolysis bath, the voltage as well as the dwell time of the carbon fibers in the bath [22] (Fig. 5).

Beside electrochemical activation of carbon fibres, the activation with plasma processes is currently researched intensively. With the help of atmospheric or low pressure plasma and using different reaction gases, the variety of functional groups, which can be implemented on the fiber surface, increases. Therefore, plasma activation is a promising method for enabling carbon fibers for alternative matrix materials [23, 24].

2.1.4 Pitch-Based Carbon Fibres

Pitch-based precursors are the second most important materials for industrial carbon fibre production, although their market share is only in the range of a few per cent. In 2012, only 1,580 tons (equal to 1.6 % of the total PAN-based carbon fibres) were produced. The prices of pitch-based carbon fibres range from 95 up to 3,000 €/kg [25]. However, pitch-based carbon fibres have unique properties, like the highest Young's modulus achievable among all materials, high thermal conductivity and a negative thermal expansion coefficient.

The main steps in the manufacturing process of pitch-based carbon fibres are the synthesis of the base material suitable for melt spinning, the melt-spinning process itself, stabilisation, carbonisation and graphitisation, followed by a surface treatment and sizing application.

2.1.5 Cellulose-Based Carbon Fibres

Although cellulose was the first precursor material to produce carbon fibres, it has been almost pushed out of the market. The yield (ratio of precursor material vs. carbon fibre) is very low compared to PAN- and picht-based fibres with values of around 30 % [26]. Furthermore, the conversion processes take very long time and are therefore very energy consuming.

Cellulose fibres are produced with solvent spinning. Two different methods (derivatizing and non-derivatizing) are suitable for cellulose spinning. In a derivatisation process, cellulose is modified with suitable chemicals forming covalent bonds with the hydroxyl groups, so it can be solved and spun. After spinning, a regeneration processes is needed. In non-derivatizing processes cellulose is directly solved, spun and therefore requires no regeneration process [27]. First carbon fibres were produced from viscose (rayon), whereas better mechanical properties in the final carbon fibre can be achieved from lyocell fibres [28].

Cellulose fibres are converted to carbon fibres in a stabilisation as well as in a carbonisation process. Stabilisation can be carried out continuously or discontinuously. Fibres are stabilised by a thermal-chemical decomposition of organic bonds. This process takes up to 10 days (compared to a few hours for the stabilisation of PAN) [11]. Carbonisation is divided into low temperature (LT) carbonisation and high temperature (HT) carbonisation.

2.1.6 Alternative Precursors

Beside the precursor systems mentioned above, which are industrially used to produce carbon fibres, alternative precursor materials are in focus of current research. The following materials are of great importance:

1. *Polyolefine-based precursors*: Polyolefins are a promising alternative as a precursor material for low-cost carbon fibres due to their high availability and low market price (1.00–1.50 €/kg) [29–32]. Furthermore the fibre production by melt spinning for polyolefin precursors, in contrast to the solvent spinning for PAN precursors, offers an additional reduction of production costs. Melt spinning is faster (up to 6,000 m/min), and therefore more energy efficient and more eco-friendly, because no complex and expensive solvent disposal is required. However, a chemical stabilization process (mostly by sulfuric acid treatment) is necessary.
2. *Lignin-based precursors*: First approaches for the production of lignin-based carbon fibres were performed during the 1960s and 1970s by Nippon Kayaku Company (Tokyo, Japan) [33]. The application of lignin as a carbon fibre precursor is mostly motivated by cost reduction efforts. Lignin can be either melt- or solution-spun and then converted in conventional (thermal) conversion processes [34].

2.2 Fibre Structure

The final structure of carbon fibres is formed during the conversion proces. In precursors primary crystallites in the orthorhombic structure of PAN [35] can be detected. These are orientated uniaxial along the fiber axis [36], in which the backbone of the polymer chain is oriented in a regular conformation in one direction of the unit cell and determines the lattice parameter which is orientated in the fiber direction. Distance and arrangement of the polymer chains in the other directions determine the symmetry and the two other lattice parameters perpendicular to the fiber axis. On the contrary to typical melt-spun fibers, small crystallites in the dimensions of a few nanometers are formed and polymer chains in the amorphous parts are also oriented along the fibre axis due to high stretching [12].

During the stabilization process a decline of the crystalline order and the crystallinity can be detected [37, 38]. Especially, the crystal size decreases. The stabilization temperature and duration is correlated with the structural characteristics [39].

After the carbonization the fiber structure consists of parallely ordered graphene sheets consisting of hexagonal carbon rings (mostly in sp^2 hybridization). The overall crystallinity increases in the carbonization process compared to stabilized fibres. The graphene layers are oriented turbostratic, but with are preferred orientation along the fiber axis in PAN-based carbon fibres. For pitch-based fibres, a three-dimensional graphite structure with a fixed correlation legth between the layers can be detected [40]. It was shown that the surface treatment of carbon fibers has an influence on the near-surface crystal structure [41]. Furthermore, structural changes in the volume of carbon fibres can also be induced by plasma surface activation [42]. In Fig. 6, the qualitative change of structural parameters along the process chain is displayed.

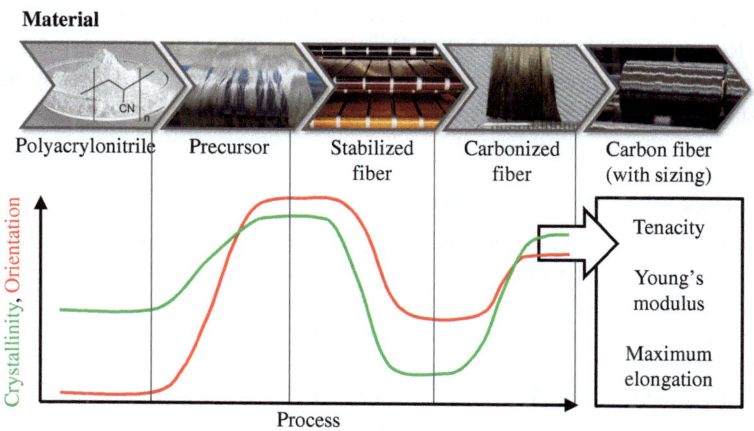

Fig. 6 Structural changes (crystallinity and orientation) along the process chain for PAN-based carbon fibre

In all types of fibres, stacking faults and nano-sized pores can be found in the crystalline structure. The arrangement of these structures in a filament (forming the nano structure) mainly depends on the precursor type. Qualitative models for the nano structure of PAN- and pitch-based fibres have been developed mainly from small-angle X-ray scattering data.

2.3 *Fibre Properties*

2.3.1 Mechanical Properties

For the tensile strength, a theoretical value of about 20 GPa could be achieved in a perfectly orientated graphite fiber [43]. This value is not achieved in reality because of the imperfect orientation of the structures in the fiber, structural defect (in the micro dimension: stacking faults, imperfections; in the macro dimension: pores, cracks) and non-carbon atoms. The tenacity of commercially available carbon fibers reaches values of up to 7 GPa.

Depending on the processing conditions, several carbon fibre properties can be achieved. They are classified into different carbon fibre types. High tension fibres (HT) have a high tensile strength, a Young's modulus of about 200–250 GPa and a maximum elongation between 1 and 2 %. Intermediate modulus-fibres (IM) even show a higher tenacity, in combination with a higher Young's modulus at the same maximum elongation. High modulus fibres (HM) have a higher Young's modulus, but with a considerably lower strength (<4.5 GPa) and a maximum elongation below 1 %. Pitch-based carbon fibres have a considerably higher Young's modulus compared to PAN-based carbon fibres, why they are called ultra high modulus fibres (UHM). The fibre properties for the different classes are summarized in Table 2 [44].

Furthermore, carbon fibres can be classified according to their number of filaments. Rovings with less than 24,000 filaments are called small tows, rovings with more than 24,000 filaments are called heavy tows. However, the filament diameter is constant with 7 μm for most of the fibre types.

Table 2 Properties of different carbon fibre classes [17, 44]

Fiber type	Tensile strength [GPa]	Young's modulus [GPa]	Maximum elongation [GPa] (%)
High tension (HT)	3–5	200–250	1–2
Intermediate modulus (IM)	4–7	250–350	1–2
High modulus (HM)	2–4.5	350–450	<1
Ultra high modulus (UHM)	≈3	>700	<0.5

2.3.2 Electrical Properties

Carbon fibres have anisotropic electrical properties, since the graphite structure in the materials provides electrical conductivity in the direction of graphene layers, whereas it is semiconducting in between the layers [45].

The specific resistivity is usually in the range of 10^{-3} $\Omega \cdot cm$ for PAN-based carbon fibres and 10^{-4} $\Omega \cdot cm$ for pitch-based carbon fibres. The electrical conductivity of carbon fibres can be used to control the carbonization process [46]. Due to the increase of conductivity by an increasing amount of carbon atoms in the structure, the degree of carbonization can be deducted from the electrical properties. Furthermore, electrical properties are also correlated to the mechanical properties of the fibres. Especially, a correlation to Young's modulus can be found, since it depends mainly on the orientation of graphite layers in the fibre. If a high orientation is achieved, both conductivity and modulus increase [47].

2.4 Carbon Fibre Producers

In the past, only a few producers (Toray Group, Toho Tenax Group, Mitsubishi Rayon, Formosa Plastics, Hexcel) supplied carbon fibres to the market. The production capacity of these companies is still growing. However, new producers entered the market in the recent years, especially supplying carbon fibres for industrial applications. Especially DowAksa (a joint venture between the American chemical company Dow and the Turkish acrylic fibre producer Aksa) and SGL Group (especially with their joint venture SGL ACF with BMW AG) have installed larger production plants. Furthermore, Chinese suppliers, only producing low quality fibres in the last decades, have installed lines for producing high quality carbon fibres. An overview on the installed and planned name plate capacities of the most important carbon fibre producers worldwide is given in Table 3 [1, 48].

Fibres with high strength (beyond 5 GPa) and high modulus (beyond are still exclusively produced by the established carbon fibre companies like Toho Tenax and Toray. New producers mainly focus on HT fibres with properties suitable for mass applications [1, 44, 48]. An overview on the mechanical properties of the fibres produced by different companies is given in Fig. 7.

2.5 Application in Composites

In this section, the application of carbon fibres in different industrial sectors is described. For these applications, the most relevant fibre types, textile structures and composite manufacturing techniques are explained.

Table 3 Installed (and planned) name plate capacities of carbon fibre producers for 2015 [44, 48]

Type	Manufacturer	Country	Name plate capacity [t/a]
Small tow (≤24 k)	Toray group	Japan	27,100
	Toho Tenax group	Japan	13,900
	Mitsubishi Rayon	Japan	10,800
	Formosa plastics group	Taiwan	8,750
	Chinese producers (accumulated)	China	18,150
	Hexcel	USA	11,200
	Cytec	USA	4,000
	DowAksa	Turkey	3,600
	Kemrock	India	4,000
	Hyosung	South Korea	6,500
	Taekwang industrial	South Korea	2,000
	Composite holding company	Russia	1,500
	Total		111,500
Heavy tow (>24 k)	Zoltek group (owned by Toray group)	USA	26,000
	SGL group (including SGL ACF)	Germany	9,000
	Bluestar	China	3,700
	Kemrock	India	1,500
	Mitsubishi Rayon	Japan	2,700
	Toray group	Japan	300
	Total		43,200

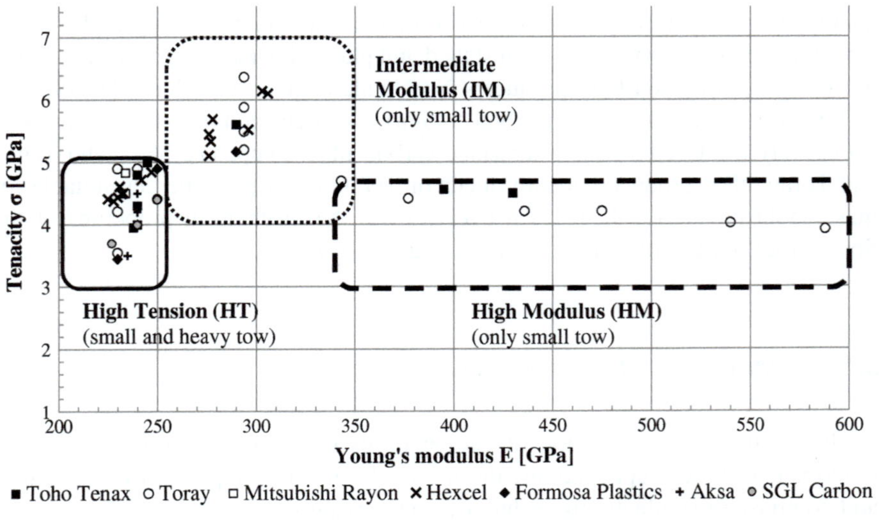

Fig. 7 Mechanical properties of products of the different carbon fibre producers assigned to the classification of carbon fibres

2.5.1 Aerospace Applications

The application of carbon fibres in civil aviation, military aviation and space vehicles was the first large scale application of carbon fibres materials and is still the most important field, where carbon fibres are applied. In the recent years, starting in the 1970s, where only a few hundred kg of carbon fibres were used in one airplane, the amount of carbon fibres and CFRP in airplanes has been constantly growing. In the Boeing 787,30 tons of CFRP are used for one airplane, representing over 50 % of the total weight. The total demand of carbon fibres for aerospace applications was about 7,500 tons in 2011 and is supposed to grow to 12,500 tons in 2020 [1, 48, 49].

For aerospace applications, only small tows of intermediate modulus or high modulus fibres have been certified and are applied for the manufacturing of aircrafts. These fibres are usually processed to a prepreg with epoxy resins in the form of fabrics or tapes. Consolidation is mostly performed in autoclave processes, but techniques like automated tape laying or automated fibre placement have become more important in the last years [1, 50].

2.5.2 Automotive Applications

In the recent years, the application of carbon fibre reinforced plastics in automotive has been massively increasing. The first cars built from CFRP were mainly racing cars, where lightweight construction was used to increase the performance. The application of CFRP was then transferred to luxury sports cars, also with the aim of creating high performance vehicles. Since only small amounts of parts need to be produced for these applications, manufacturing techniques like hand lay-up or autoclave processes are applied. Since these cars are in the upper price segment, fibres suitable for aerospace applications (small tows with enhanced mechanical properties) can be used even though their price is high [1, 51].

Within the last years, first examples for CFRP in mass production of vehicles were established, especially the i3 and i8 series of BMW AG, Munich, Germany. Within the field of electromobility, more lightweight materials are needed to save the extra weight by large capacity batteries, which cannot be achieved by other fibre types. For these mass products, usually resin transfer moulding (RTM) is applied to achieve both low cycle times and good performance. Furthermore, large tows (50 k) are usually applied because of their low price. However, the price of CFRP produced with these techniques is still too high to compete with standard materials like aluminium [1, 52].

2.5.3 Application in Wind Energy

Most of rotor blades for wind energy plants are made out of glass fibres. However, the size of the rotor blades is limited, because the material cannot withstand the mechanical load due to the high speed of the ends of the rotor blade. Therefore, carbon fibres are applied in wind energy for high capacity (>2 MW) plants, which are usually installed in offshore wind parks. Hybrid materials (glass and carbon) are applied, whereas carbon fibres are used at the ends and edges of the rotor blades. Like pure fiberglass-based rotor blades, these hybrid blades are industrially produced by large scale vacuum infusion [1, 53].

2.5.4 Other Applications

Beside the main applications mentioned above, carbon fibres are applied in the following fields:

- Pressure vessels: Pressure vessels can be produced by filament winding or braiding of carbon fibres. CFRP pressure vessels have a high stability at low weight, making them preferable for the transportation of gaseous media over long distances, e.g. in regions with no piping systems [54].
- Power cables: By applying carbon fibres (beside aluminium or copper) in power cables, a lower weight at the same performance level can be achieved. Because of the lower weight, longer distances between pylons for power transmission lines can be achieved [1].
- Industrial applications/machinery parts: CFRP can be applied in industrial applications. It is a preferable material if a high stiffness or a low weight (especially for moving or rotating parts, leading to lower energy consumption) is needed [1, 55].
- Sporting goods: Typical sporting goods made from CFRP are bicycle parts. However, also tennis rackets or golf clubs can be made out of CFRP. Because of the higher price, carbon fibre based sporting goods are mostly used in the professional area, so that the market volume for carbon fibres is very small [1, 44].
- Reinforcement of concrete: Another large scale application is the replacement of steel reinforcement by carbon fibres in concrete. Because of their good corrosion resistance, mechanical properties and resistance against the concrete itself, carbon fibres are a preferable material for textile reinforced concrete (TRC). Beside advantages of the fibre material itself compared to steel, TRC provides more design flexibility for complex shaped parts because of the good drapability of textiles compared to steel [56].

2.5.5 Carbon Fibre Demand

Based on the demands of the different industries described above, the actual and projected carbon fibre demand can be indicated. An overview with a conservative (confirmed) and aggressive forecast scenario is given in Fig. 8. Especially for wind energy and automotive applications, huge growth rates can be expected depending

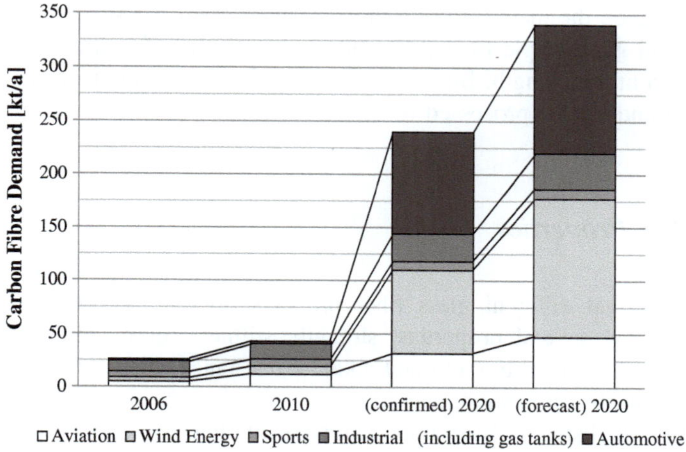

Fig. 8 Market demand (past values and forecast) on carbon fibres of different industries (according to [1, 57])

on the market penetration of CFRP in these areas. The demand for aerospace applications is also supposed to grow, depending on the application of larger volume of CFRP in single aisle airplanes [1, 48, 49].

3 Glass and Basalt Fibers

Glass fiber development has a very long history. There are some indications that glass fibers were developed in ancient Egypt around 1600 B.C. More evidence of glass fibers were discovered in the Middle Ages in Venice, Italy and Murano, Italy. Initially the technique of production was based on drawing heated rods into fine filaments. In 1713 Ferchault de Reaumur developed fabrics with glass fiber from heated rods. At the beginning of 1900 W. V. PACZINSKY proved the feasibility of endless glass fibers. The industrial production of continuous glass fiber and the use of modern technology started in 1930 by Owens-Illinois Glass Co. of Newark, Ohio, U.S.A. Glass fibers for reinforcement are produced nowadays based on the process developed by Owens-Illinois Glass Co., U.S.A. [58].

Glass fibers are today defined by the International Bureau for the Standardization of Man-Made Fibers (BISFA) as "fibers in textile form, obtained by drawing molten glass" [59]. Since the 1980s glass fibers have seen a continuous growth due to the successful use as reinforcement principally in fiber-reinforced plastics.

Glass fibers are made of inorganic oxides and present an amorphous structure. They present a constant diameter along the fiber. Their morphology and production process distinguishes the glass fibers for reinforcement applications from other products like glass wool and micro fiber for isolation.

Depending on the manufacturing process, glass fibers are subdivided into glass filaments and glass staple fibers. The first is practically of unlimited length. The second has a limited length. Both have a well-defined constant diameter between 8 and 24 μm and can be processed to textile products.

3.1 Fiber Properties

There are several different glass fiber products on the market. The fibers are characterized by several properties: strength, stiffness, elastic modulus, thermal stability, chemical stability (acid and alkaline resistance) dielectric, price, density, light scattering, IR absorption, and so on.

In Table 4 are listed the most important types of fibers for reinforcement function and their most significant properties.

Fibers used as reinforcement in composites are almost exclusively in the categories of E-glass, ECR-glass, AR-glass and S-glass.

E-Glass is the most common glass fiber used in composites. Due to its low fiberizing temperature and the absence of expensive raw materials, E-Glass fibers are the most important low-cost fibrous reinforcement in relation to their mechanical properties.

Table 4 Properties and chemical composition of textile glass (nominal values, wt %) [58, 60, 61]

Types of glass					
Property	Multi-purpose	Acid resistant	Alkali resistant	High strength	High temperature resistant
	E	ECR	AR	S	Quartz
SiO_2, %	52–56	60.1	62	65	100
Al_2O_3, %	12–15	13.2	0.8	25	
CaO, %	21–23	22.1	5.6		
MgO, %	0.4–4	3.1		10	
B_2O_3, %	4–6				
F_2, %	0.2–0.7	0.1			
Na_2O, %	≤1	0.6	14.8		
ZrO_2, %			16.7		
K_2O, %	≤1	0.2			
Fe_2O_3, %	0.2–0.4	0.2			
TiO_2, %	0.2–0.5	0.5			
Density, g/cm^3	2.55	2.62	2.68	2.49	2.15
Young's modulus, GPa	72	80	72	88–91	69
Tensile strength, GPa	3.1–3.8	3.1	1.7	4.4–4.6	3.4
Process parameters					
Fiber forming, temp, °C	1160–1196	1260		1565	2300
Liquidus temp, °C	1065–1077	1200		1500	1670

ECR-Glass is a special boron-free E-Glass. It is used when higher acid resistance (CR = "chemical resistant") and strength are required.

AR-Glass fibers are almost exclusively used in cement and in strong alkaline mediums (AR = "alkali resistant"). Due to the massive use of ZrO_2 the price of this fiber is around 6–8 times higher than the one of the E-Glass.

S-glass fibers are fibers with a very high percentage of SiO_2. S-Glass fibers are a special product used exclusively when high strength (S = "strength"), high stiffness and high temperature resistance are required.

3.2 *Fiber Manufacturing*

3.2.1 Raw Material

Glass chemical composition for glass fibers are obtained by mixing different natural rocks.

The most important oxides for glass fiber production are listed in Table 4. They are SiO_2, Al_2O_3, CaO, MgO, B_2O_3, Na_2O, ZrO_2, F_2. The most important mineral rocks for the production of glass fibers are Glass-making sand SiO_2 for silica, Kaolinite $Al_2Si_2O_5(OH)_4$ and Bauxite $Al(OH)_3$ for alumina, Colemanite $CaB_3O_4(OH)_3{\cdot}H_2O$ for boron oxide, Dolomite $CaMg(CO_3)_2$ for calcium and magnesium oxides, Limestone $CaCO_3$ for calcium, Fluorspar CaF_2 for fluorine.

By comminution, each raw material is reduced to small particles and collected in different silos.

Before being charged in a melting oven, the minerals are weighted and mixed into a batch.

3.2.2 Melting Process

The fiberization process requires a homogenous and crystal free melt. That can be obtained at temperatures over the T_l (Liquidus). In commercial glasses Tl is reached between 1300 and 1600 °C.

There are two commercial melting processes for the manufacture of glass filaments: the indirect marble melting process and the direct melt process.

In the indirect marble melting process, the production of glass and the glass spinning process are separated. Glass is initially produced in the form of glass marbles or beads (diameter of 15–20 mm), which are then re-melted in a second phase in the glass forming process (two-stage process) (Fig. 9).

The direct melt process (Fig. 10) consists in a single melting process. The raw materials are mixed, melted and directly fiberized. In an initial phase the raw materials are fused into the furnace at a high temperature in order to reduce the viscosity. That improves the material homogenization and eliminates the bubbles of

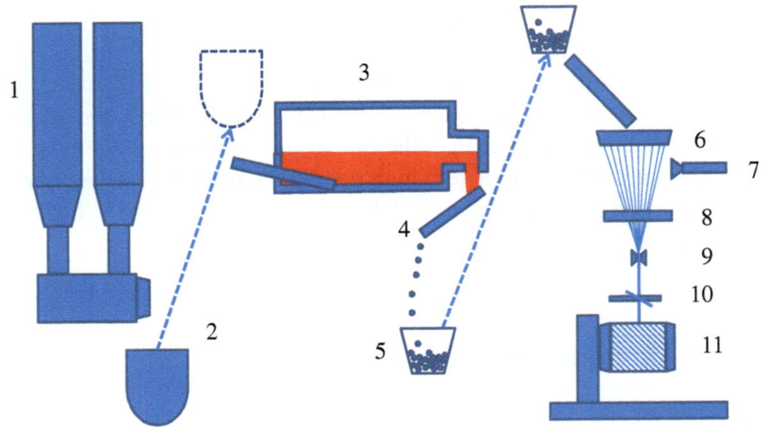

Fig. 9 Marble melting (redrawn from [62]) (1) Mixing silos; (2) Furnace feed; (3) Furnace; (4) Marble forming; (5) Transport; (6) Marble bushing; (7) Waterspray; (8) Sizing applicator; (9) Strand formation; (10) Traversing; (11) Cake winder

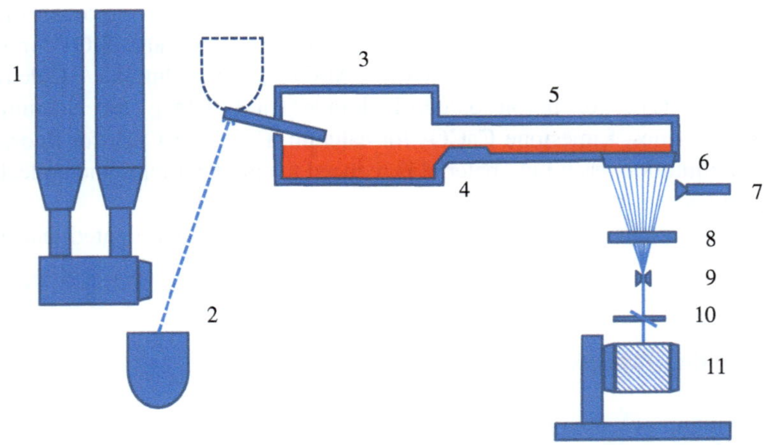

Fig. 10 Direct melting (redrawn from [62]) (1) Mixing silos; (2) Furnace feed; (3) Furnace; (4) Refiner; (5) Forehearth; (6) Bushing; (7) Water spray; (8) Sizing applicator; (9) Strand formation; (10) Traversing; (11) Cake winder

gas and air bolted into the melt. The E-Glass melting process requires a furnace temperature of about 1400 °C.

The melt flows from the first part of the furnace to the refiner. The temperature is lowered to around 1370 °C and at the end the melt is uniform and without bubbles. The third part of the furnace is the forehearth where the temperature is lowered to 1225–1260 °C. From the forehearth the glass melt flows to the bushings where it is fiberized. The forehearth is connected to several bushings [58, 63].

3.2.3 Fiber Forming

The glass fiberization process is based on a temperature quenching of the melt due mainly to the fast drawing.

The melt reaches the bushing made of platinum and rhodium with a temperature above the liquid phase. The glass melt flows spontaneously trough the nozzles (Fig. 11) due to its own hydrostatic pressure and the force of gravity. The number of nozzles per bushing in E-Glass fiberization is up to 5000. The melt viscosity and the surface tension oppose a resistance to its spontaneous flow. Those two parameters are regulated by the temperature. The temperature in the bushing is regulated by a last direct electrical resistance trough the platinum rhodium alloy. In order to be drawn, the viscosity of the melt must be usually between 30 and 100 Pa s. The fibers are then formed by the application of a drawing force up to 240 km/h to the melt.

3.2.4 Sizing and Surface Treatment

The sizing applicator has the function to homogenously wet the filaments with sizing right after the bushing.

During the glass fiber production the sizing reduces the friction between the fibers helping to protect them from damages. Moreover, the Sizing improves the fiber quality in terms of workability (textile process) and end-use application (composites).

Sizing is an aqueous dispersion and consists of different elements. The most important elements of a sizing for glass fiber reinforcements are coupling agents and film former. Coupling agents are used to improve the adhesion between the glass fiber and the specific polymeric matrix. The most important coupling agents for glass fibers are silanes. Film former improve the wettability, the strength and the cohesion of the roving. Other elements like antistatic and lubricant agents are used principally for a better manufacture of glass fibers in fabrics.

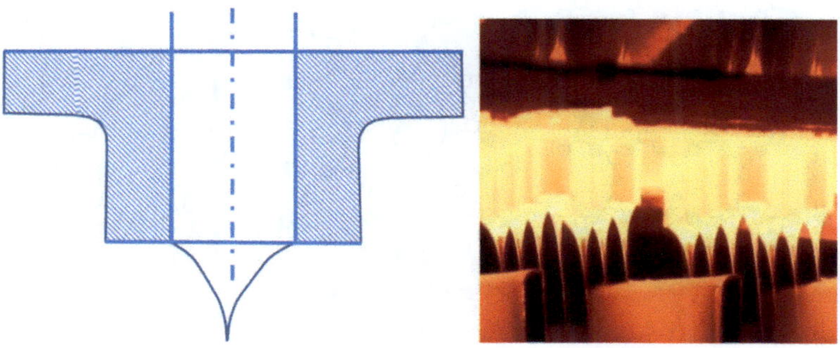

Fig. 11 (*left*) Bushing nozzle and (*right*) photo of bushing during fiberization

3.3 Fiber Structure

In order to get a glass structure the melt must be frozen from the liquid (T > Tl). The cooling rate must be superior to its crystallization. Glass and crystal differ in the organization of the atoms during the solidification phase. This deviation can be detected by measuring the specific volume. In Fig. 12 the solidification behavior of crystal and amorphous glass in terms of specific volume is shown. Starting from the liquid phase ($T > T_l$) the melt is cooled to T_l where it starts to solidify. At this point, depending on the cooling rate the solidification will happen at constant temperature T_l to a crystal lattice and a lower specific volume. A high cooling rate will not permit an organization of the atoms in a regular lattice and the resulting structure will be amorphous [64].

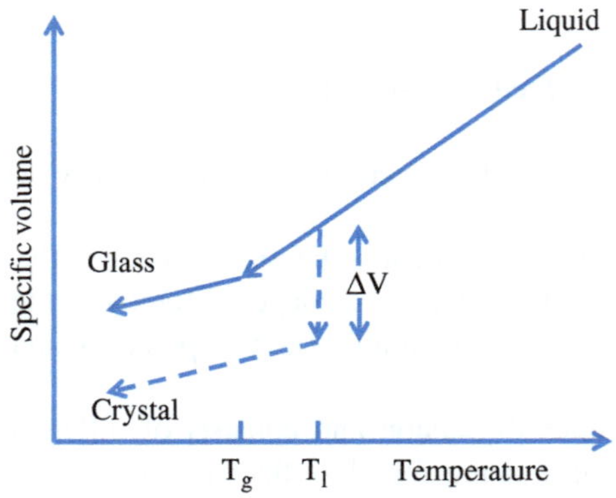

Fig. 12 Variation of the specific volume with temperature from liquid to glass and crystal structure [64]

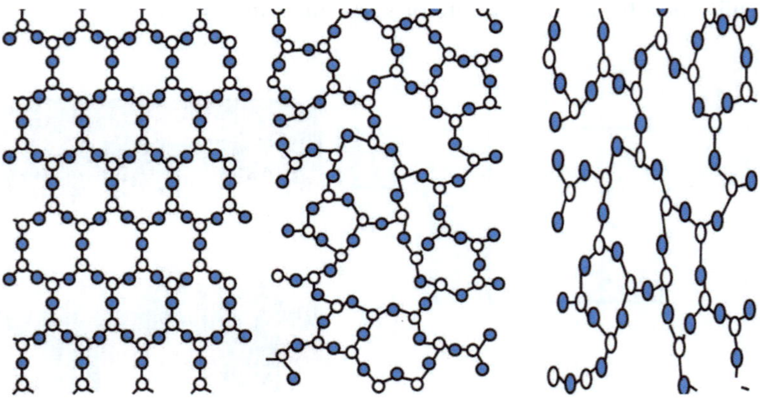

Fig. 13 Draw of crystalline (*left*), amorphous bulk (*middle*) and anisotropic amorphous (*right*) structure in glass fiber

Glass fibers are characterized by an amorphous structure. The non-periodic atomic arrangement is achieved by quenching the melt during the fiber forming. The velocity of melt solidification is much higher compared to the velocity of its crystallization.

In terms of material structure, glasses have no periodic atomic arrangement and have a time dependent glass transformation behavior.

Depending on the solidification process, glass present different properties and structure characteristics. During the fiber drawing, the melt is quenched and stretched. Due to the freezing of the isotropic liquid at high draw velocity, the structure result stretched along the fiber orientation. The stretching of the melt during the drawing process results in an anisotropic amorphous structure [65]. In Fig. 13 three examples are drawn of structure arrangement related to the melt solidification rate and process.

3.4 Application in Composites

The glass fibers for reinforcement applications are on the market in different forms. There are three main glass fiber products on the market: simple fibers (staple fibers, roving and yarn), 2D fabrics (nonwovens, woven fabrics, knitted fabrics, braided fabrics and noncrimp fabrics) and 3D fabrics (3D Nonwoven, multilayer fabrics, woven spacer fabrics and 3D braided fabrics).

It is already evident from the foregoing how varied the application of the fiber reinforcement can be. The most important application is in plastic matrices as GFRP (glass fiber reinforced plastics). They combine good mechanical properties with chemical resistance capability, have a great potential for lightweight construction and provide extensive freedom in shaping. Products made of glass fiber reinforced plastics have not only industrial applications, but are integrated in everyday life. The most important applications of GFRP are drawn in Fig. 14, for a better overview.

The most important sector is construction (36 %) followed by transport (34 %). The other two major glass fiber application sectors are electric/electronic (14 %) and sports and leisure (14 %) [66] (Fig. 14).

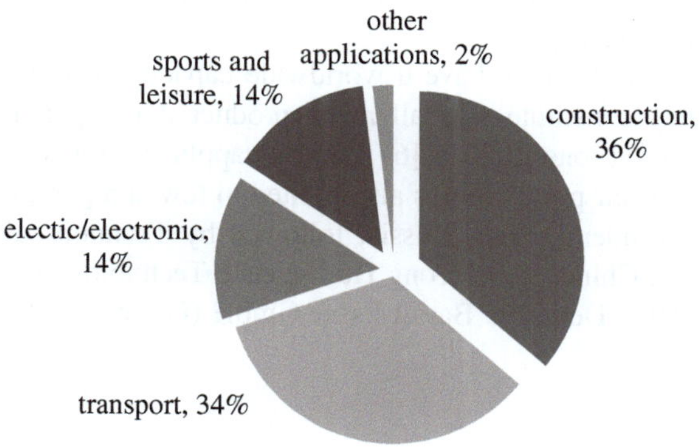

Fig. 14 Distribution of GRP production in Europe per application sector [66]

Other emerging applications are in the energy sector (wind energy). A special application of glass fiber (AR-Glass fiber) is in cement matrix as fiber-reinforced concrete (FRC) and textile reinforced concrete (TRC) in substitution to the asbestos fibers.

3.5 Special Fibers: Basalt Fibers

Basalt fibers are principally produced from natural basalt rocks. Basalts are volcanic rocks that came out from the mantel during a volcanic activity (eruption) as magma on the surface or on the see plate. The chemical composition of basalt fiber depends mostly on the geographical provenience.

Basalt is an inorganic material, a silicate composed by a mix of different oxides (SiO_2 45–52 %, Al_2O_3 12–16 %, Fe_xO_y 6–18 %, alkaline earth oxides 10–20 % and alkaline oxides 2–8 %).

The basalt fibers were developed in the Soviet Union at the beginning of the 60s but their commercialization started only after 1991.

These fibers are produced by the melt spinning process also used for commercial glass fibers (Fig. 10). In some case, fibers with a chemical composition similar to the one of basalt are obtained from a mixing process of different raw materials. In this case a two-step process is used (Fig. 9).

The basalt fiber production is very difficult to control principally because of the natural variation in chemical composition of rocks.

The small fiberization temperature range of basalt does not allow the use of big bushings with a high number of nozzles like the ones used for E-Glass. Basalt is more difficult to fiberize and therefore the bushing is limited to a maximum of 1000 nozzles compared to 5000 nozzles in E-Glass bushings.

Basalt fibers, if compared to E-Glass fibers, have a very high thermal and chemical stability; very high strength (up to 4.84 GPa) and stiffness (up to 110 GPa). Optically, basalt fibers are colored dark grey to gold brown, depending on the quantity of iron oxides (%) and the oxidation state (Fe_2O_3/FeO).

Basalt fibers, like glass-fibers, find application in different textile products. The main applications are in fiber reinforced concrete (mostly as rebars) and fiber reinforced plastics.

Basalt fibers have a worldwide capacity of production of around 15,000 tons, they constitute a small niche product if compared to glass fibers (more than 4 million tons in 2011 [67]) and its application is still limited.

Their production is also limited to few companies. The most important producer is Kamenny Vek, Russia, followed by Technobasalt, Ukraine, GBF Basalt Fiber Co., China, Hebei Tong Hui Science Technology Co., China, Isomatex (Belgium), DBF—Deutsche Basalt Faser GmbH (Germany), Mafic (Ireland) [68].

4 Ceramic Fibers

Ceramics are inorganic, not metal, compounds that are difficult to solve in water with at least 30 % crystallinity [69]. They usually exist in a polycrystalline form.

Technical ceramics are in addition built for technological use and have therefore specific properties as technical material.

Those materials are characterized by high hardness, high thermal resistance, high stiffness, low friction and high electrical insulation. On the other hand they are brittle, sensible to thermal shocks and flexural strength.

Ceramic can be produced directly as technical components or as ceramic fibers in a spinning process. Due to their morphology, technical ceramic fibers are eligible for special applications, in particular in combination with other materials as fibrous reinforcement. Ceramic fibers are used as reinforcement in metallic and ceramic matrices. The ceramic fibers in a ceramic or metallic matrix improve the basic materials in terms of mechanical properties, especially in the field of strength and modulus of elasticity along the fiber orientation [70].

4.1 Fiber Properties

Ceramic fibers can be divided into oxide and non-oxide ones. Those two type present different bond behavior and different chemical and physical characteristics. Oxide ceramics have a strong ionic bond due to the large difference of electronegativity between oxygen and the bonding partner. This bond is stronger than metallic bonds, but lower than covalent, which is reflected in the technical characteristics. Oxide ceramics are characterized by high chemical and thermal stability and hardness. The most important commercial oxide ceramic fiber is aluminum oxide (Al_2O_3).

Non-oxide ceramics have no oxygen bond. The bonds are predominantly covalent. For this reason, they have a higher chemical inertness, a higher elasticity modulus and can be used in higher temperature ranges. In applications with high temperatures, they are not prone to oxidation or corrosion. The most important commercial non-oxide ceramic fiber is silicon carbide (SiC).

The mechanical properties of commercial oxide ceramic fibers at room temperature are reported in Fig. 15.

Non oxide ceramic fibers have mechanical properties comparable with carbon fibers. Oxide ceramic fibers have lower strength (around 2 GPa) and E-modulus.

Ceramic fibers present a very high E-Modulus between 200 and 400 GPa that is comparable to carbon fibers. Oxide ceramic fibers compared with non-oxide ceramic fibers have lower strength (around 2 GPa against 3–6 GPa).

However, their mechanical properties at room temperature are only secondary if compared to their mechanical properties at high temperatures. The strength of aluminum and silicon carbide fibers are drawn in Fig. 16 as a function of temperature

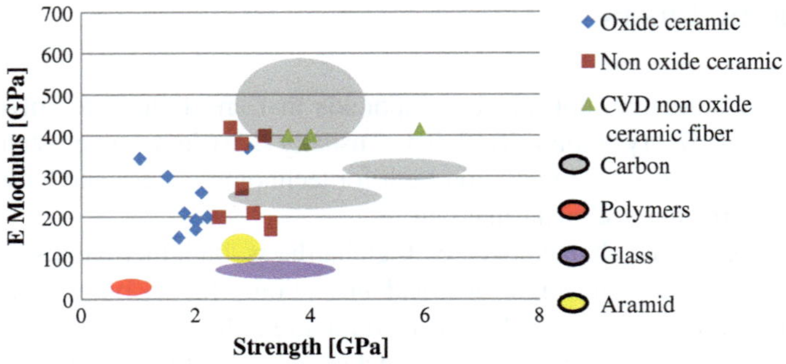

Fig. 15 Strength and E-modulus of commercial ceramic fibers compared to other fibrous reinforcements

Fig. 16 Fiber strength as a function of temperature after 1 h of exposure to air [71]

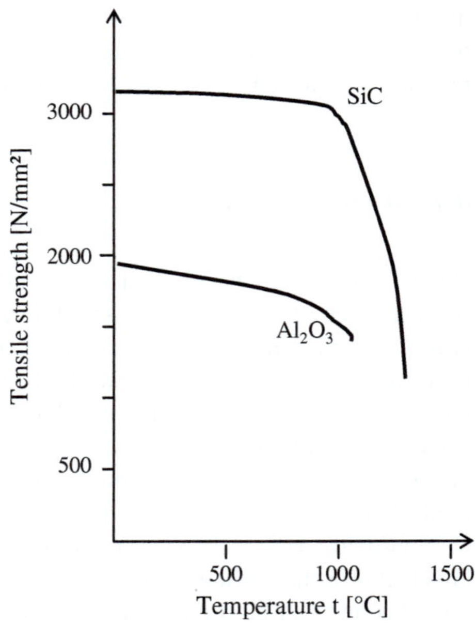

after 1 h of heat exposure in air. In spite of the presence of carbon, SiC fibers are not prone to oxidation or corrosion with high temperature.

Ceramic fibers are commercially produced in three different ways: direct processes based on Sol-Gel and polymeric precursor and indirect process based on chemical vapor deposition (CVD).

4.2 Fiber Manufacturing

The most important processes for the production of ceramic fibers are sol gel process, polymeric precursor and CVD (Fig. 17).

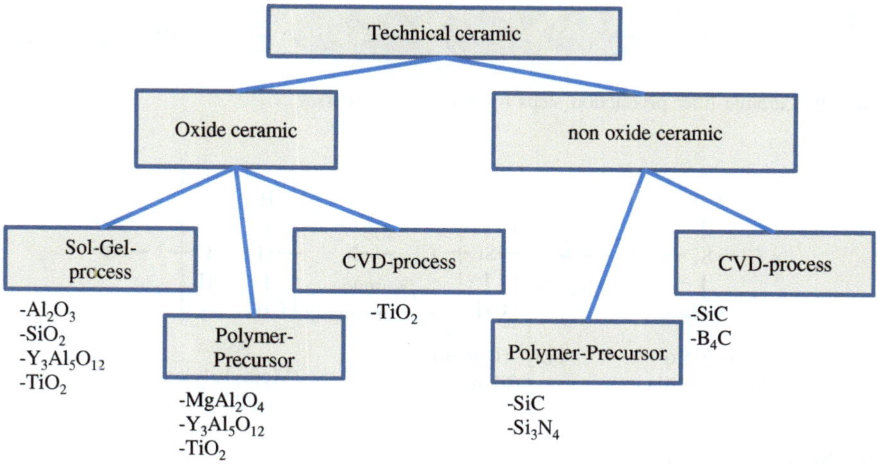

Fig. 17 Ceramic fibers and their production process [72, 73]

Considering the commercial ceramic fiber, those are principally produced by sol-gel (aluminum oxide), polymer precursor or CVD-process (silicon carbide).

4.2.1 Sol Gel Process

The sol-gel method is a wet-chemical production that uses soluble salts as spinning material that are transformed in oxides by using a calcination process [72]. The solution is continuously changed in its consistency towards a gel. Such a network contains liquid and solid phases. Reactions of hydrolysis and polycondensation form the colloid (liquid phase with dispersed solids). The salts are molecular or colloidal dispersion in a solvent, mostly water [74]. In addition, organic polymers like polyethylene oxide or polyvinyl alcohol are used as solvent in order to optimize the rheological behavior of the spinning process. A summary of the manufacturing steps are shown in Fig. 18 [75].

4.2.2 Polymers Precursor Process

In the polymer precursor process, ceramic fibers are produced by using polymer-based precursors. A general overview of the process steps is shown in Fig. 19.

Fig. 18 Ceramic fiber production steps by sol gel process

Fig. 19 Ceramic fiber production steps by polymeric precursor

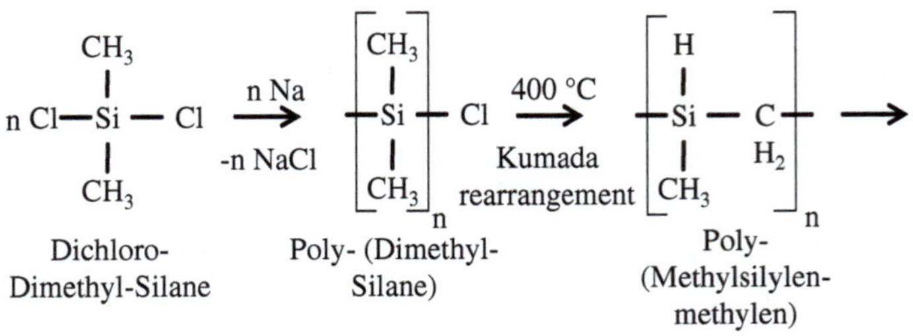

Fig. 20 Yajima polymerization process [77]

At first the polymers are synthetized. The most important polymers for the production of SiC are the polycarbosilanes PCS. Those polymers contain Si-C bonds. Depending on the polymerization process, different PCS can be obtained.

In Fig. 20 the reactions occurring during the Yajima polymerization are shown. Firstly Dimethyldichlorosilane is converted into polydimethylsilane (PDMS) during a catalytic dehalocoupling reaction. After that PDMS is converted into PCS by the Kumada rearrangement (polycondensation) at 400 °C and is ready to be spun [76].

After the synthesis of PCS the polymer is fiberized into a green fiber. Green fibers are obtained by melt spinning of PCSs under inert gas. The next step, hardening of the green fibers consists in a thermal treatment (200 °C 1 h) in order to obtain a cross-linked material similar to thermoset. The organic solvents are than removed and the precursor is converted in the pyrolysis furnace under a nitrogen atmosphere at 1200 °C from a green fiber into SiC ceramic fibers [76].

4.2.3 CVD Chemical Vapor Deposition Process

In the gas phase deposition, a material is transferred into the gas phase. The concept of vapor deposition is drawn in Fig. 21.

The vapor is partially transferred from the carrier gas through the reaction chamber. The gas molecules react on the surface of the substrate and remain there as a layer. Decomposition products are than removed. Both carbon and wolfram filaments are used as substrate in industrial production.

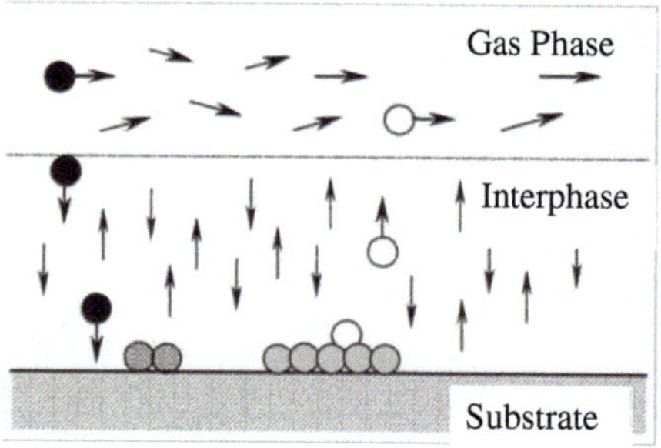

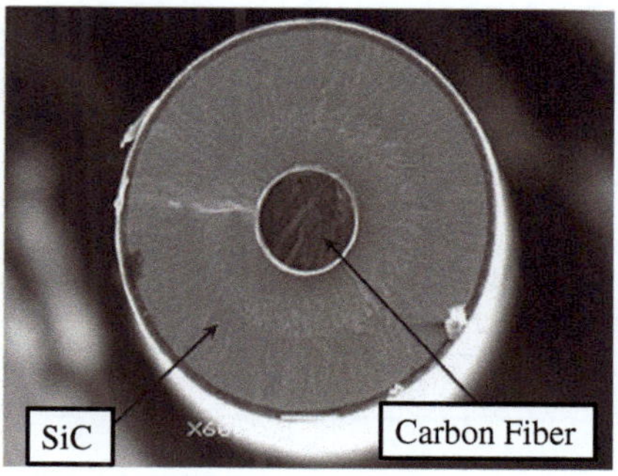

Fig. 21 CVD process (*left*) and SiC ceramic fiber (*right*). *Source* Specialty Materials, Inc

4.3 Application in Composites

The most important application of ceramics is related to their thermal stability. Their main application is at very high temperature (mostly over 800 °C) where metals and other conventional materials already reach their limits.

Metals and ceramics can be reinforced by using ceramic fibers. In order to absolve their function as reinforcement, the fibers must have high stiffness and maintain their mechanical properties during the composite manufacture and during the operating time.

Ceramic matrix composites CMC are nowadays the most important reinforced material in high temperature applications. The addition of ceramic fiber modifies the properties of the ceramic matrix by improving its thermo-shock resistance and its brittle behavior. In Fig. 22. the operating temperature of different materials and their mechanical properties (specific strength) is drawn.

In Fig. 23 are presented the actual manufacturing processes for CMCs. Depending on the type of matrix (oxide or non-oxide), its porosity and the requirements of the final product a defined process will be used.

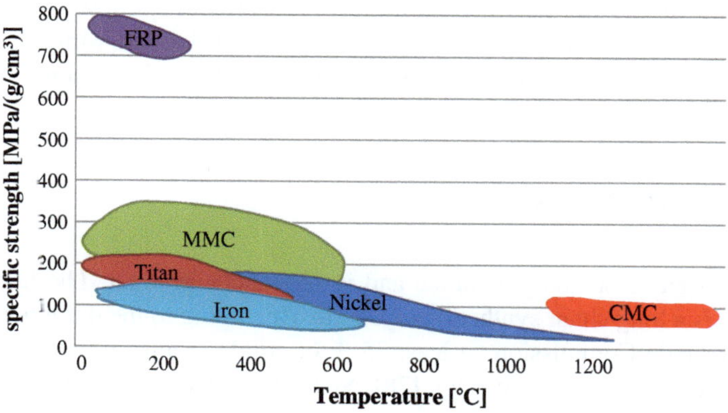

Fig. 22 Dependence of specific strength to the operating temperature of different materials

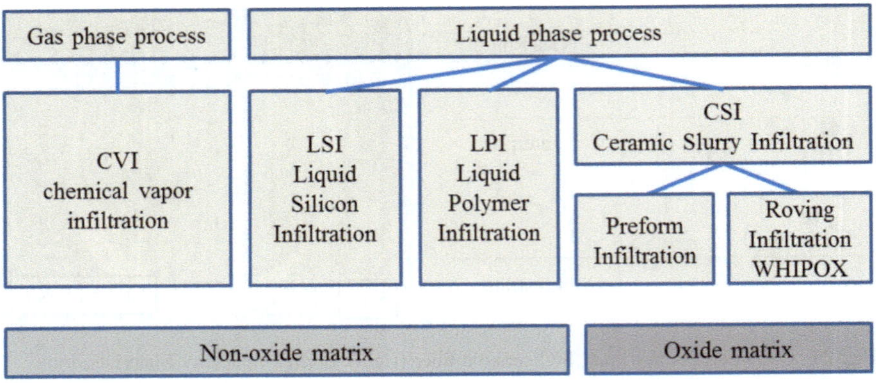

Fig. 23 CMC manufacturing processes

CMC find applications in:

- aerospace heat shield systems (manufactured with CVI) use SiC at an operating temperature of 1500 °C,
- gas turbine components manufacturing system LPI, SiC fibers in SiC matrix.
- burner uses a CSI manufacturing process and the product made of aluminum oxide has an operating temperature above 1000 °C in oxygen.
- Brake disk are produced by using the LSI process; operate at a temperature above 500 °C. In spite of the other applications where thermal stability is the most important parameter, the use of CMC in brakes is related to other advantages compared to common metal brakes like the chemical stability (no salt corrosion), the lightness and no braking system fade.

5 Polymeric Fibres

Different classes of polymeric fibres are of great importance for fibre reinforced composites. In general, polymeric fibre materials provide the lowest density compared to other material classes and a ductile failure behavior, while their thermal stability is limited. The material classes are described in this section.

5.1 Aramide Fibres

Aramide means aromatic polyamide and is a synthetic product. The fiber forming substance is a long chain synthetic polyamide. According to the definition of the U. S Federal Trade Commission, at least 85 % of amide groups in the polymer chain are attached to two aromatic rings [78]. Stephanie Kwolk has developed them at DuPont in the early 1960s [79].

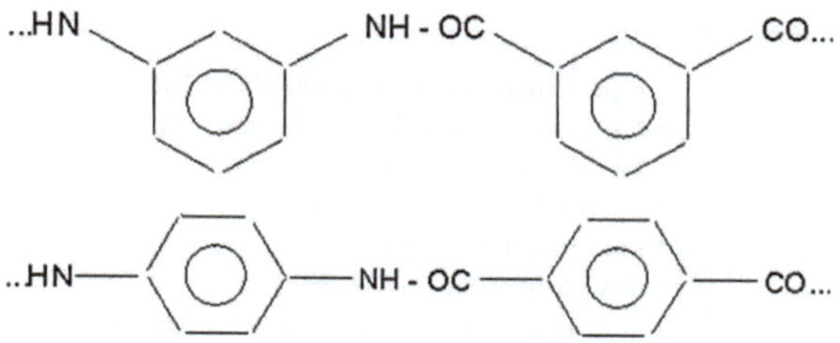

meta aramide

para aramide

Fig. 24 Molecular structure of aramides

5.1.1 Fibre Material and Manufacturing

Two types of aramide fibres exist, on the one hand meta aramide and on the other hand para aramide (See Fig. 24). The meta aramide fibres have significantly worse mechanical properties than para aramide, but instead they are flame-retarding. The reason for these differences in properties are the different connections between the aromatic rings. The backbone of the para aramide is stretched, while the backbone of meta aramide is arranged with an angle of 120° [80].

Aramide fibres are manufactured by solvent spinning methods, either wet or dry spinning, because they already degrade below their melting point. Mostly, sulphuric acid is used as a solvent [81].

During the wet spinning the spinning pump presses the dissolved substance through a pipe and a filter till the spinneret. The spinning solvent diffuses out of the filaments in the coagulation bath, so that solid filaments are formed. The last part is the stretching and washing of the filaments. The typical winding speed is 300 m/min [81].

In the dry spinning process, the spin dope is also pressed through the spinneret and then extruded in air. The solvent can then evaporate in hot air, so that the filaments are formed. Typical winding speeds are in the range of 200 m/min [81].

5.1.2 Fibre Properties

The characteristics of the aramid fibre are summarized in the following Table 5.

5.1.3 Aramide Fibre Market

The maufacturers of aramide fibres are mostly Tejin (Twaron®, Technora®) and DuPont (Kevlar®, Nomex®). They dominate the market of aramide fibre and have a market share of 85 % worldwide. DuPont has an estimated production capacity of 27,600 t/a and Tejin one of 26,400 t/a. The whole market represents probably about 63,000 t/a, whereas 2,500 t are used for applications in fibre-reinforced composites [83].

Table 5 Properties of aramide fibres [82]

Mechanical properties	• high tensile strength (1500–2500 MPa) • high tensile modulus of elasticity (60–150 GPa) • low density (1.38–1.44 g/cm^3) • good vibration absorption • low material fatigue • low elongation (2.7–4 %)
Thermal properties	• high temperature resistance (370–500 °C) • low thermal expansion • low thermal conductivity
Yarn properties	• number of filaments: 65–6000 • diameter of filaments: 12–14 μm

Based on expected growth rates the aramide fibre market has such an important relevance. That is shown by the fact that the global demand for aramid fibres is projected to reach about 3 billion $ by 2019. The increasing fields are security and protection, optical fiber reinforcement, and aerospace applications. Currently, security and protection applications are the largest market with a share of 31 % (2013) [84].

5.1.4 Application in Composites

Aramides are applied un various industries like safety and protection, lightweight and civil construction, transportation or electrical industry. An important application of the safety and protection industry is the ballistic protection. Aramide fibres are used for bulletproof vests or protective helmets because of the high energy absorption of the fibres. But also heat-protective clothing and helmets are an application of aramide fibre in this industry [82].

The low density of the fibres is relevant for the lightweight construction. Moreover, they are used for the reinforcement of civil constructions. In these applications, the high tensile strength and the high energy absorption of the aramide fibre are the reasons for their application [82].

5.2 Ultra High Molecular Weight Polyethylene (UHMWPE)

UHMWPE is a special type of polyethylene with the highest molecular weight. The most important commercial product is Dyneema by DSM [85].

5.2.1 Fibre Material and Manufacturing

UHMWPE has extremely long chains with molecular weight numbering between 2 and 6 million. Due to the high molecular weight, the material cannot be processed by melt spinning. Therefore, a special sol-gel spinning process is used to form the fibres. In this process, nearly all polymer chains are orientated in the fibre direction and they are linked by van der Waal bonds. Because of the long molecule chains the total bond energy between the chains is very high, forming a high strength fibre. Furthermore, a high crystallinity of 80 % and more can be reached, resulting in a high Young's modulus [85, 86].

5.2.2 Fibre Properties

The fibre proberties are shown in the Table 6 below. Because of the low density, the specific strength of UHMWPE fibres exceeds the specific strength of the most carbon fibre types [85].

5.2.3 Application of UHMWPE

UHMWPE is used for protective clothing, fishing lines, nets and ropes. It is also used in the automotive industry for headlights and wipers. Other industry applications are Electronic Stability Programme (ESP) control modules, gear housing, steering-angle sensors, door control devices or airbag connectors [85].

5.3 *Other Types of Polymeric Fibres*

Beside the most important fibre types explained in the sections above, other types of synthetic fibres can be used for composite applications. Some examples are explained here:

Table 6 Properties of UHMWPE fibres

Mechanical properties	• high tensile strength (2.8–3.2 GPa) • lowest density 0.93 g/cm^3 • high Young's modulus (670–725 MPa) • high wear resistance
Thermal properties	• melting point (130–136 °C) • low thermal conductivity
Yarn properties	• number of filaments: 65–6000 • yarn count: 3–250 tex • diameter of filaments: 21–23 μm

- *Rayon for tire cord applications*: Viscose (rayon) can be processed to technical yarns by wet spinning. The most important application is the reinforcement of tires with so called tire cords. Furthermore, rayon staple fibres can be used for reinforcing molding grades of technical thermoplastic materials [87].
- *Thermoplastic staple fibres for concrete reinforcement*: Staple fibres made of polyproplyene (also polyamides or polyesters) are added to concrete during the construction to modify the failure behavior. Fibres with small diameter can help to modify the propagation of cracks at their interface to create a more ductile failure [88].
- *Liquid crystal polymers (LCP)*: Liquid crystal polymers (LCP are processed to high performance fibres in the melt spinning process. The crystallites are oriented in the molten state in the capillaries during spinning. Due to their molecular structure, LCP fibres offer high strength and especially good cut and impact resistance. They are applied in aerospace applications or as reinforcement for cables and hoses [89].
- *Thermoplastic fibres for hybrid yarns/textiles*: Thermoplastic fibres are added to common reinforcement fibres (usually glass or carbon). A mixing of the two fibre types can be performed at filament level during spinning (e.g. Twintex [xxx]) or an additional mixing process (e.g. co-mingling) to form hybrid yarns or during textile processing for making hybrid textiles. These pre-impregnated textile structures made of hybrid yarns or hybrid textiles can then be consolidated to thermoplastic composites with the thermoplastic yarn component acting as matrix material. Compared to state-of-the-art organo sheets, better drapability can be achieved. Especially for hybrid yarns mixed at filament level, shorter cycle times with lower pressure during hot pressing can be achieved due to shorter flow pathes for the thermoplastic material [90, 91].
- *All-thermoplastic composites*: Thermoplastic materials with different melting points can be mixed and then consolidated to all-thermoplastic composites [92]. Beside conventional manufacturing techniques, these materials can mixed in form of polymer blends or bicomponent fibres in the spinning process (e.g. in core-sheath or islands-in-the-sea geometry). The polymer with the higher melting point should form fibrils in the fibre, acting as reinforcement after consolidation of the material with the lower melting point (forming the matrix material) [93].
- *Self-reinforced composites*: Among the all-thermoplastic composites, especially self-reinforced composites made of the same polymer material with different melting points are of great interest because of good recycling behavior. Possible material combinations are atactic or syndiotactic polypropylene reinforced with isotactic polypropylene or standard poly(lactide acid) reinforced with high-melting point stereocomplex poly(lactide acid) [94].

References

1. Roberts T (2011) The carbon fibre industry worldwide 2011–2020: an evaluation of current markets and future supply and demand. Materials Technology Publications, Watford
2. Jec Group (2011) Global glass-fibre production: changes across the board. http://www.jeccomposites.com/news/composites-news/global-glass-fibre-production-changes-across-board. Accessed 24 Jun 2015
3. Clauß B (2008) Ceramic matrix composites. In: Krenke W (ed) Fibers for ceramic matrix composites. Wiley, Weinheim, pp 1–20
4. Wlochowicz A (1984) Kohlenstoffasern aus Pech, ihre Herstellung und Eigenschaften Textiltechnik 34(11):595
5. Edison TA (1879) U. S. Pat 223:898
6. Houtz RC (1950) Orlon acrylic fibre: chemistry and properties. J Text Res 20:786–801
7. Shindo A (1959) Japanisches Patent 28287
8. Shindo A (1962) Japanisches Patent 29270
9. Johnson W, Phillips LN, Watt W (1964) The production of carbon fibres Britische Patentmeldung GB 1,110,791
10. Johnson W, Watt W, Phillips LN, Moreton R (1965) Improvements in or relating to carbonisable fibre and carbon fibre and their production. British Patent GB 1,166,251
11. Morgan P (2005) Carbon fibres and their composites. Taylor & Francis, Boca Raton
12. Masson J (1995) Acrylic fiber technology and applications. Marcel Dekker, New York
13. Gries T, Rixe S, Steffens M, Cremer C (2002) Faserstoff-Tabellen nach P. A. Koch: Polyacrylfasern, 6. Ausgabe Eigenverlag, Aachen
14. Huang J, Baird DG, McGrath JE (2013) Melt-spinning of polyacrylonitrile fibers as carbon fiber precursors. Paper presented at the 245th ACS national meeting and exposition, New Orleans, Louisiana, 7–11 April 2013
15. Kilic S, Michalik S, Wang Y, Johnson JK, Enick RM, Beckman EJ (2007) Phase behavior of oxygen-containing polymers in CO_2. Macromolecules 40:1332–1341
16. Beyer H (1998) Lehrbuch der Organischen Chemie. 23.Aufl.. S. Hirzel Verlag, Stuttgart
17. Foley A, Frohs W, Hauke T, Heine M, Jäger H, Sitter S (2008) Carbon fibers. In: Ullmann's encyclopedia of industrial chemistry, fibers. Chap. 5 Synthetic inorganic. Wiley, Weinheim, p 291ff
18. Fitzer E, Manocha LM (1998) Carbon reinforcements and carbon/carbon composites. Springer, Berlin
19. De Palmenaer A, Langner C, Linke O, Lüpfert L, Seide G, Gries T, Fourné R (2014) Stabilization of PAN fibers by contact heat transfer. Chem Fibers Int 65(1):45–46
20. Menendez JA, Arenillas A, Fidalgo B, Fernandez Y, Zubizarreta L, Calvo EG, Bermudez JM (2010) Microwave heating processes involving carbon materials. Fuel Process Technol 91 (1):1–8
21. Kim S-Y, Kim SY, Lee S, Jo S, Im Y-H, Lee H-S (2015) Microwave plasma carbonization for the fabrication of polyacrylonitrile-based carbon fiber. Polymer 56(15):590–595
22. Gulyas J, Földes E, Lazar A, Pukanszky B (2001) Electrochemical oxidation of carbon fibers: surface chemistry and adhesion. Compos A 32:353–360
23. Erden S, Kingslei KCH, Lamoriniere S, Lee A, Yildiz H, Bismarck A (2010) Continuous atmospheric plasma oxidation of carbon fibres: influence on the fibre surface and bulk properties and adhesion to polyamide 12. Plasma Chem Plasma Process 40:471–487
24. Santos AL, Botelho EC, Kostov KG, Nascente PAP, da Silva LLG (2013) Atmospheric plasma treatment of carbon fibers for enhancement of their adhesion properties. IEEE Trans Plasma Sci 41(2):319–324
25. Schürmann H (2005) Konstruieren mit Faser-Kunststoff-Verbunden. Springer, Berlin
26. Donnet JB, Wang TK, Peng JCM (eds) (1998) Carbon fibers, 3rd edn. Dekker, New York
27. Fink HP, Fischer S (2005) Celluloseverarbeitung - umweltfreundliche Technologien auf dem Vormarsch. Praxis der Naturwissenschaften - Chemie in der Schule 54(7):18–25

28. Wu Q, Pan D (2002) A new cellulose based carbon fiber from a lyocell precursor. Text Res J 72:405–410
29. Otani S (1995) On the carbon fiber from the molten pyrolysis products. Carbon 3(1):31–34
30. Paiva MC, Lin C, Haynie T, Kotasthane P, Ogale AA, Kennedy JM, Edie DD (2001) Carbon fibers from alternative precursors. Paper presented at the international conference on carbon, Lexington, 14–19 July 2001
31. Morales J (2013) Polyethylene. Global overview SPI flexible film & bag. Paper presented at the SPI flexible film and bag conference, Nashville
32. Sagel E (2012) Polyethylene global overview IHS (Hrsg.): Expo Foro Pemex, Mexiko-Stadt
33. Kadla JF, Kubo S, Venditti RA, Gilbert RD, Compere AL, Griffith W (2002) Lignin-based carbon fibers for composite fiber applications. Carbon 40(15):2913–2920
34. Kubo S, Kadla JF (2005) Kraft lignin/poly(ethylene oxide) blends: effect of lignin structure on miscibility and hydrogen bonding. J Appl Polym Sci 98:1437–1444
35. Colvin BG, Storr P (1974) The crystal structure of polyacrylonitrile. Eur Polymer J 10:337–340
36. Anghelina VF, Popescu IV, Gaba A, Popescu IN, Despa V, Ungureanu D (2010) Structural analysis of PAN fiber by X-ray diffraction. J Sci Art 10:89–94
37. Yu M, Wang C, Bai Y, Wang Y, Xu Y (2006) Influence of precursor properties on the thermal stabilization of polyacrylonitrile fibers. Polym Bull 57:757–763
38. Mukhopadhyay SK, Zhu Y (1995) Structure-property relationships of PAN precursor fibers during thermo-oxidative stabilization. Text Res J 65:25–31
39. Lee S, Kim J, Ku BC, Kim J, Joh HI (2012) Structural evolution of polyacrylonitrile fibers in stabilization and carbonization. Adv Chem Eng Sci 2:275–282
40. Anderson DP (1991) Carbon fiber morphology, II: expanded wide angle X-ray diffraction studies of carbon fibers. Wright Research & Development Center, US Air Force
41. Nohara LB, Filho GP, Nohara EL, Kleinke MU, Rezende MC (2005) Evaluation of carbon fiber surface treated by chemical and cold plasma processes. Mater Res 8:281–286
42. Saelhoff AK, Jäger M, Steinmann W, Gries T (2014) Surface treatment of carbon fibres—increasing the interlaminar shear strength in CFRP. In: Dörfel A (ed) Proceedings of the 8th Aachen-Dresden international textile conference, Dresden
43. Chand S (2000) J Mater Sci 5:1303–1313
44. Warnecke M, de Palmenaer A, Veit D, Seide G, Gries T (2013) Fibre-table carbon fibres. Shaker, Aachen
45. Eiswirth M, Schwankner M (1982) Graphit und seine Verbindungen Praxis der Naturwissenschaften. Chemie 31:137–143
46. Frohs W (1989) Untersuchungen zum thermischen Abbau von Polyacrylnitril (PAN) – Precursorfasern zu Carbonfasern im Temperaturbereich von 500 bis 2800 °C. Dissertation, Eigenverlag, Universität Karlsruhe
47. Peebles LH (1995) Carbon fibres—formation, structure, and properties. CRC Press Inc, Florida
48. CompositesWorld (2015) Supply and demand: advanced fibers. http://www.compositesworld.com/articles/supply-and-demand-advanced-fibers-2015. Accessed 04 Aug 2015
49. Walker A (2014) Next generation carbon fibre. Paper presented at the GOCarbonFibre Conference, Cologne, 9–10 October 2014
50. Jäger H (2010) Carbonfasern und ihre Verbundwerkstoffe: Herstellungsprozesse, Anwendungen und Marktentwicklung. Süddeutscher Verlag onpact
51. Aucken A (2014) Cytec. Paper presented at the GOCarbonFibre conference, Cologne, 9–10 October 2014
52. Verdenhalven J (2014) CFRP, the steel of the 21st Century … or the story of fishes. Paper presented at the GOCarbonFibre Conference, Cologne, 9–10 Oct 2014
53. Monk C (2014) Carbon fibre—challenges and benefits for use in wind turbine blade design. Paper presented at the GOCarbonFibre conference, Cologne, 9–10 Oct 2014
54. Mafeld A (2014) The global market for composite pressure vessels—drivers, challenges and trends. Paper presented at the GOCarbonFibre conference, Cologne, 9–10 Oct 2014

55. Regan B (2014) Carbon fibre for energy storage applications. Paper presented at the GOCarbonFibre conference, Cologne, 9–10 Oct 2014
56. Chen PW, Chung DDL (1995) Carbon-fibre-reinforced concrete as an intrinsically smart concrete for damage assessment during dynamic loading. J Am Ceramic Soc 78(3):816–818
57. Witten E, Kraus T, Kühnel M (2014) Composite market report 2014: market developments, trends, challenges and opportunities
58. Loewenstein KL (1993) The manufacturing technology of continuous glass fibers. Elsevier, Amsterdam
59. BISFA (2000) Terminology of man-made fibres. BISFA, Brussels
60. Wallenberger FT, Watson JC, Li H (2001) Glass Fibers. In: ASM Handbook 21. ASM International, Materials Park (OH)
61. Chawla K, Tekwani B (2013) Studies of glass fiber reinforced concrete composites. Int J Struct Civil Eng Res 2(3)
62. Pico D, Wilms C, Seide G, Gries T, Kleinholz R, Tiesler H (2010) "Fibers, 12. Glass Fibers" Ullmann's encyclopedia of industrial chemistry 7, Wiley, Weinheim [u.a.], 2012. doi:10.1002/14356007
63. Gardiner G (2009) The making of glass fiber, composites technology 15(2), Gardner Publications Incorporated
64. Zarzycki J (1991) Glasses and the vitreous state. Cambridge University Press, Cambridge (Cambridge solid state science series)
65. Ya M, Deubener J, Yue Y (2008) Enthalpy and anisotropy relaxation of glass fibers. J Am Ceram Soc 91:745–752. doi:10.1111/j.1551-2916.2007.02100.x
66. Witten E, Schuster A (2010) Composites-Marktbericht: Marktentwicklungen, Herausforderungen und Chancen AVK – Industrievereinigung verstärkte Kunststoffe
67. N.N. (2011) Global glass-fibre production: tailoring better for needs, JEC Compos Mag 66, 16–18
68. Pico D, Wilms C, Seide G, Gries T (2011) Natural volcanic rock fibers. Man-Made Fiber Yearb 2011, pp 45–46
69. Hennicke HW (1967) Zum Begriff Keramik und zur Einteilung keramischer Werkstoffe. Berichte d. Deutsch. Ker. Gesellschaft 44:209–211
70. Flemming M, Ziegmann G, Roth S (1995) Faserverbundbauweisen. Springer, Berlin
71. Kochendörfer R, Krenkel W (2003) Möglichkeiten und Grenzen faserverstärkter Keramiken. In: Krenkel W (ed) Keramische Verbundwerkstoffe. Wiley, Weinheim, pp 1–22
72. Krenkel W (2003) Keramische Verbundwerkstoffe. Wiley, Weinheim
73. Kroschel M (2001) Amorphe B/Si/C/N-Hochleistungskeramiken aus Einkomponentenvorläu-fern. Universität Bonn Dissertation, Bonn
74. Hench LL, West JK (1990) The sol-gel process. Chem Rev 90(1):33–72
75. Chawla KK (2001) Composite materials—science and engineering. Springer, Berlin
76. Wallenberger FT, Bingham PA (2010) Fiberglass and glass technology. Springer, New York
77. Brachtel G (2004) Keine Keramik ohne Organik - Festschrift 125 Jahre keramische Ausbildung an der FH Koblenz. Koblenz
78. Fibermax composites. http://www.aramid.eu/. Accessed 24 Jun 2015
79. Fibermax composites. http://www.aramid.eu/history.html. Accessed 24 Jun 2015
80. Brandrup J, Immergut E, Grulke E, Abe A, Bloch D (eds) (1999) Polymer handbook, 4th edn. Wiley, New York
81. Blumberg, Hillermeier, Krüger (1982) Aramid-Prozess. Melliand Textilberichte
82. Wulfhorst B, Büsgen A (1989) Faserstofftabelle nach P.-A. Koch: Aramidfasern. Chemiefasern/Textilindustrie 39:1263–1270
83. Wulfhorst B, Gries T, Veit D (2006) Textile technology. Hanser, Munich
84. MarketsandMarkets (2014). http://www.marketsandmarkets.com/Market-Reports/aramid-fibers-market-112849061.html. Accessed 24 Jun 2015
85. Dyneema (2015). http://www.dsm.com/products/dyneema/en_GB/home.html. Accessed 04 Aug 2015

86. Ticona (2001). http://www.hipolymers.com.ar/pdfs/gur/diseno/GUR%20%28PE-UHMW%29.pdf. Accessed 24 Jun 2015
87. Kauffman GB (1993) Rayon: the first semi-synthetic fiber product. J Chem Educ 70(11):887
88. Ahmed S, Bukhari IA, Siddiqui JI, Quereshi SA (2006) A study on properties of polypropylene fiber reinforced concrete. Paper presented at the 31st conference on our world in concrete & structures, Singapore, 16–17 Aug 2006
89. Geary JM, Goodby JW, Kmetz AR, Patel JS (1987) The mechanism of polymer alignment of liquid-crystal materials. J Appl Phys 62:4100
90. Kravaev P, Stolyarov O, Seide G, Gries T (2013) A method for investigating blending quality of commingled yarns. Text Res J 83:122–129
91. Choi BD, Diestel O, Offermann P (1999) Commingled carbon/PEEK hybrid yarns for use in textile reinforced high performance rotors. Paper presented at the 12th international conference on composite materials (ICCM), Paris, 5–9 July 1999
92. Biron M (2007) Thermoplastics and thermoplastic composites. Elsevier, Amsterdam
93. Tavanaie MA, Shoushtari AM, Goharpey F, Motjahedi MR (2013) Matrix-fibril morphology development of polypropylene/poly(butylenes terephthalate) blend fibers at different zones of melt spinning process and its relation to mechanical properties. Fibers Polym 14(3):396–404
94. Alcock B, Cabrera NO, Barkoula N-M, Loos J, Peijs T (2006) The mechanical properties of unidirectional all-polypropylene composites. Comp Part A 37:716–726

Natural Fibers for Composite Applications

Malgorzata Zimniewska and Maria Wladyka-Przybylak

Abstract The chapter presents selected vegetable fibers such as flax, hemp, jute, kenaf, abaca, sisal, coir in terms of their properties and short description of commonly applied processes of fiber extraction. Processes of retting and preliminary processing influence fiber quality and related with it appropriate fibre distribution into the matrix, what is responsible for mechanical performance and quality of composites reinforced with vegetable fibres. Vegetable fibers can be used to reinforce composites in the form of loose mass of short fibers, roving, yarn, fabric or nonwoven depending on needs, methods and the type of composite formation. Natural fiber containing composites can be manufactured by almost all production techniques. Chemical and physical modifications of natural fibers are used for increasing adhesion and compatibility between fibers and the polymer matrix. Methods of surface modifications of natural fibres as well as methods of the composites fabrication are discussed along with their applications.

1 Introduction

In recent decades, traditional natural fibers have achieved a strong position as reinforcement materials for composites. The definition of "natural fibers" is discussed by researchers representing different textile sectors and it is possible to come across an opinion that only fibers made from natural resources should be classified as natural fibers. The most correct and representative definition of natural fibers has been proposed by the FAO for the needs of International Year of Natural Fibers IYNF 2009: "(…) Natural fibers are greatly elongated substances produced by plants and animals that can be spun into filaments, thread or rope (…)" [1].

M. Zimniewska (✉) · M. Wladyka-Przybylak
Institute of Natural Fibres and Medicinal Plants, Poznań, Poland
e-mail: gosiaz@iwnirz.pl

M. Wladyka-Przybylak
e-mail: maria.przybylak@iwnirz.pl

© Springer Science+Business Media Singapore 2016
S. Rana and R. Fangueiro (eds.), *Fibrous and Textile Materials for Composite Applications*, Textile Science and Clothing Technology,
DOI 10.1007/978-981-10-0234-2_5

General system of natural fiber classification, based on the above mentioned definition, covers two main fiber groups:

- Plant or vegetable fibers
- Animal fibers

From the composite view point, vegetable fibers mainly bast, leaf and fruit fibers show the most attractive properties for application to composites.

The classification system for vegetable fiber related to its anatomical origin contains the following main types of fibers [2]:

- seed fibers—represented by cotton, kapok
- bast fibers—such as flax, hemp, kenaf, ramie, jute, nettle
- leaf fibers represented by agaves like sisal and henequen, pineapple, curaua, abaca (manila), cabuya, chambira, pita
- fruit fibers represented by coir,
- wood such as hardwood and softwood,
- grass and reed such as bamboo, wheat, rice, oat, barley, elephant grass and others.

Fibrous plants can be classified in terms of their intended use. In this classification system, bast fibers are classified into the group of fibrous plants cultivated only for fiber production. The second group contains plants grown for other goods, for example fruits—coconut production and fibers constitute only a by-product of the plant—coir [3].

The world production of vegetable fibers has stabilized since 2009, what is presented in the Fig. 1, however decreasing of the all types of natural fiber share in total fiber production is observed due to the growth of chemical fibers production in 2012. The share of natural fibers decreased from 41.9 % in 2008 to 37.4 %, while production of manmade fibers increased respectively from 58.1 to 62.6 % during the same period, as it is visible in Fig. 2.

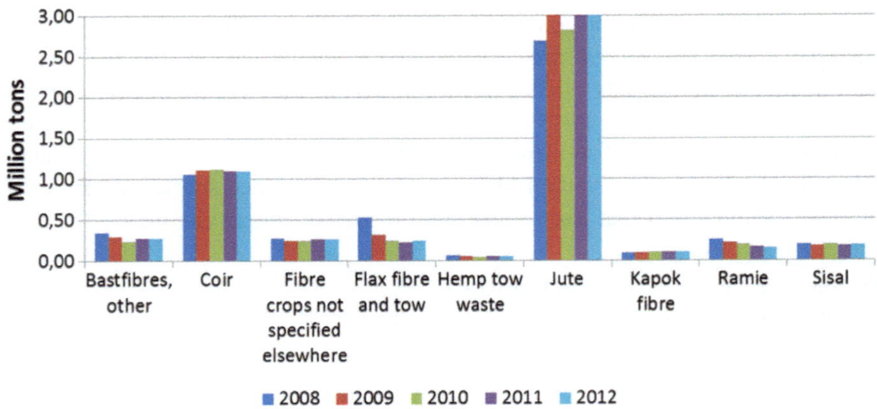

Fig. 1 World production of vegetable fiber excluding cotton 2008 to 2012 [4]

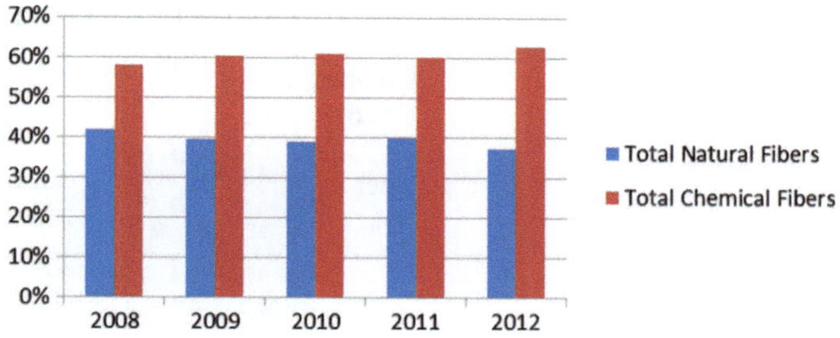

Fig. 2 Fiber share in total fiber production, 2008–2012 [4]

Maintaining natural fibers production on stable level results from rapid development of the sector of composites reinforced with natural fibers.

Big diversity of fibrous plants (more than 2000 types are known) gives researchers possibility to look for new wide applications of vegetable fibers, in which the advantages of the fibers can bring benefits by improvements in properties of final products. However, the only plants with the highest fiber contents are used in industrial scale for composite manufacture due to easy and effective technologies of their extraction and processes. The most important fibers applied for composite reinforcement are: flax, hemp, jute, kenaf, sisal, coir and abaca. The volume of their production is significant in the fibers total production.

According to the European Confederation of Linen and Hemp (CELC), European production of flax fibers was 160 000 metric tons in 2013 (115 000 tons of long fiber and 45 000 tons of short fibers), accounting for approximately 80 % of the world production, estimated around 200 000 tons [5]. Increasing of flax importance in textile and technical sector is reflected in the growth of flax production in Europe in 2014, Table 1.

In 2014 10 % of total hemp fiber production and 6 % of flax fiber production was used in the composite sector, Figs. 3 and 4.

The share of flax and hemp fibers in different applications is presented in Figs. 3 and 4.

Great interest in replacement of glass fibers with bast fibers in composites results from favorable properties of natural fibers for composite like low density and good mechanical properties [6]. Natural fibers are significantly lighter than glass fibers, with a density of 1.15–1.50 g/cm^3 [7] versus 2.4 g/cm^3 for E-glass [8].

Biodegradability and environmentally friendly character of plant fibers decide about their application for green composite production. As a renewable resource,

Table 1 Flax production in Europe [5]

Year	2008	2009	2010	2011	2012	2013	2014
Flax plantation area (* 1000 ha) France, Belgium, The Netherlands	84.5	70.1	70.1	76.8	83.3	73.2	81.8
Flax long fibers production (* 1000 t)	95.3	109.3	119.1	85.7	67.9	114.8	132.9
Flax short fibers production (* 1000 t)	94.0	59.4	51.7	51.3	45.1	45.0	49.8

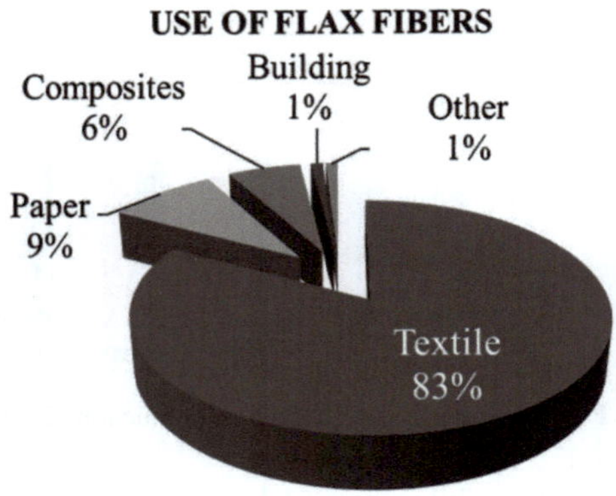

Fig. 3 Use of flax fibers in different branches of industry [5]

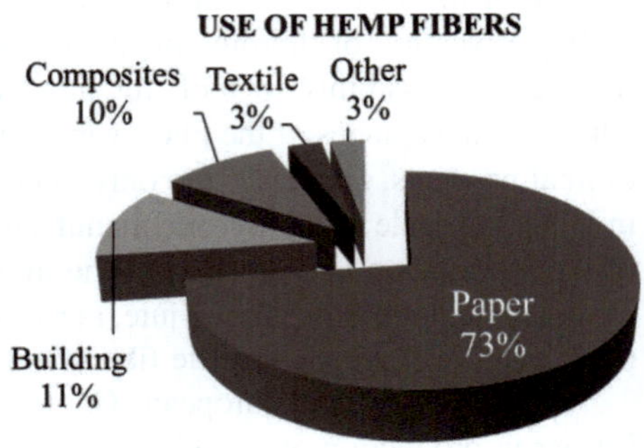

Fig. 4 Use of hemp fibers in different branches of industry [5]

natural fibers are used for manufacturing of recyclable "green" products in industrial processes leading reduction of carbon emission and minimizing waste. A good example of action contributing to CO_2 reduction is cultivation of fibrous plans. Cultivation of hemp (*Cannabis sativa L.*) on 1 ha area causes absorption of approximately 2.5 tons of atmospheric CO_2 during one vegetation season, whereas jute absorbs as much as 2.4 tons of carbon per ton of dry fiber [9]. In contrast, producing one ton of polypropylene emits into the atmosphere more than 3 tons of carbon dioxide [2, 10, 11].

Advantages of natural fibers in terms of their application as reinforcement of composites are clear, but the materials have several bottlenecks such as: non-homogeneity resulting from fiber nature, poor wettability, incompatibility with some polymeric matrices and high ability of fibers to moisture absorption and related with it swelling [2, 6]. In case of flax fibers transverse swelling area is 47 %, the swelling of jute is a little bit lower and reaches 40 %. High ability of bast fibers for moisture sorption from surrounding area is presented in Table 2. The amount of bonded water is strongly related to air humidity.

To reduce the fiber disadvantages, different types of fiber modification processes are conducted. The processes are described in the further part of the chapter.

To achieve high mechanical performance of composites reinforced with vegetable fibers, it is necessary to pay attention to fiber distribution into the matrix, because this aspect strongly influences the quality of the processing and of final

Table 2 Effect of relative air humidity on hygroscopicity of selected bast fibers

Fiber	Moisture content in condition of different relative humidity of air [%]					
	30	40	50	60	70	100
Flax	7.5	8.3	9.1	9.9	10.7	23
Hemp	8.0	8.7	9.4	10.1	10.	24
Jute	6.7	7.9	8.6	9.4	10.1	24

composite [12]. For this reason, processes of retting, fiber extraction and preliminary processing should be focused on maximizing separation of fiber bundles from woody non-cellulosic materials, cleaning and dividing to smaller fiber complexes with the size closer to elementary fibers. Selection of appropriate technology of fiber extraction is a key element in production of vegetable fibers characterizing by desired quality, uniformity and purity to ensure proper fiber distribution into the matrix. Presentations of selected vegetable fibers presented below contain short description of commonly applied processes of fiber extraction.

2 Characteristics of Selected Vegetable Fiber

Vegetable fibers can be used to reinforce composites in the form of loose mass of short fibers, roving, yarn, fabric or nonwoven depending on needs, methods and type of composite formation. The quality and properties of the fibers are important aspects in composite manufacturing. Fibers extracted from the plants occur in the form of bundles glued by pectin, which has to be separated to elementary fibers and cleaned from the woody particles and other impurities. The effective dividing of fiber bundles and separation from woody parts results in more uniform fiber distribution in polymer matrices. Depending on the type of fibrous plants, different technologies of fiber isolation and processes are applied. There is diversity of retting methods employed to bast fiber extraction, e.g. water retting, dew retting, enzymatic, chemical, osmotic degumming as well as mechanical decortication. Water retting produces fibers with the best quality and uniformity, but limitation in use of this method is related to severe water pollution and negative environmental impact. At the other end of the bast fibers extraction methods is mechanical process of decortication, which does not need water use but produces fibers with low quality, which can be applied only for special application, where the quality is of no importance [13].

2.1 *Flax fiber*

Flax (*Linum usitatissimum*) is one of the oldest fibrous plant used by our ancestors for textile manufacturing, but the fibers are still very attractive for traditional and new advanced application [14].

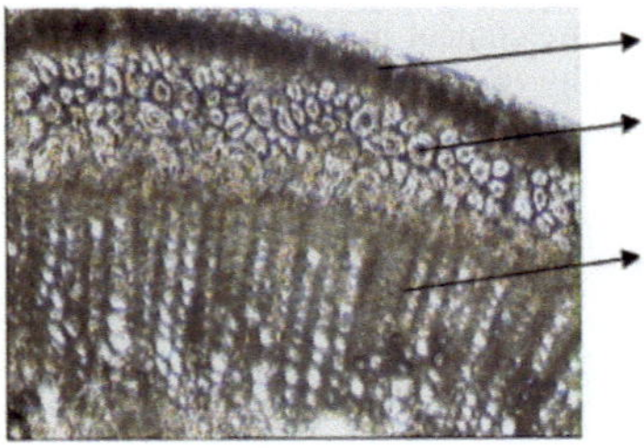

Fig. 5 Image of cross section of fibrous plant (flax) stem *(own database of INFMP)*

Flax fibers must be extracted from phloem surrounding the stem of the fibrous plants, mainly dicotyledonous ones. Figure 5 presents cross section of fibrous plant stem. Bark or skin protects the plant against moisture evaporation, sudden temperature changes as well as gives partly mechanical reinforcement to the stem. Fibers are located in the phloem and occur usually in bundles under the skin, they support the conductive cells of the phloem and provide strength to the stem. The xylem material—woody core is in the middle part of the plant.

There are several methods to isolate the fibers from the stem. Some of them are based on retting but it is possible to extract fibers with the use of mechanical processes only e.g. with the decortication technology. The aim of the processes is to degrade the pectin and other cementing compounds that bind the bast fibers and fiber bundles to other tissues and thereby to separate fibers from non-fiber materials [15].

Traditionally, two main retting methods have been used for flax i.e. water retting and dew retting and both are primarily microbial processes. In water retting flax stems are pulled and submerged into water for 5–7 days and then dried and sun bleached in the field. Anaerobic bacteria, primarily pectin degraders, are the main organisms in water retting. Application of water retting is strongly limited due to harmful environmental impact and for this reason the dew retting is the most often used method for fibers extraction. Dew retting is carried out by pulling flax stems and laying them in even layers of rows on field for the moisture to encourage indigenous fungi to colonize and grow on the stem.

As the dew retting is long and weather dependent process, it is very difficult to obtain repeatable properties of extracted fibers. For this reason other methods of fiber retting or degumming have been developed e.g. chemical or enzymatic retting, osmotic degumming, steam explosion or ultrasonic treatment. All the processes are significantly shorter than dew retting and their conditions are controlled [16, 17]. The key objective of retting processes is quality of the fibers suitable to final application in terms of fiber fineness, elementarization of fiber bundles, purity, tenacity and uniformity of all properties. In case of the enzymatic retting, good results are reached with the use of pectinase-rich commercial enzyme products e.g. multi-enzyme complex, which is prepared from a selected strain of Aspergillus, shows activity against branched pectin-like compounds and also contains a wide range of carbohydrases including arabanase, cellulase, glucanase, hemicellulase, and xylanase [15, 18, 19].

Table 3 Flax fibers characteristics [7]

<table>
<tr><th colspan="3">Flax (Linum usitatissimum)</th></tr>
<tr><th>Plant</th><th>Fiber</th><th>Information</th></tr>
<tr><td></td><td></td><td rowspan="3">Country of origin: Europe, Asia
Applications: textiles, production of particleboards, thermo insulating materials, paper, composites, food (oil), used in cosmetics industries
Linear density [tex]: 0.2–2.0
Length [mm]: 13–40
Density [g/cm³]: 1.50
Breaking tenacity [cN/tex]: 40–80
Diameter [µm]: 17–20
Thermal stability [°C]: 150
degradation:175–205
Young's Modulus [GPa]: 100
Chemical composition [%]:
<table><tr><td>Cellu lose</td><td>Hemi cellulo ses</td><td>Pectin</td><td>Lignin</td><td>Fat/ wax</td></tr><tr><td>62-71</td><td>16-18</td><td>1.8-2.0</td><td>2.0-2.5</td><td>1.5</td></tr></table></td></tr>
<tr><th>Longitudinal view</th><th>Cross- section</th></tr>
<tr><td>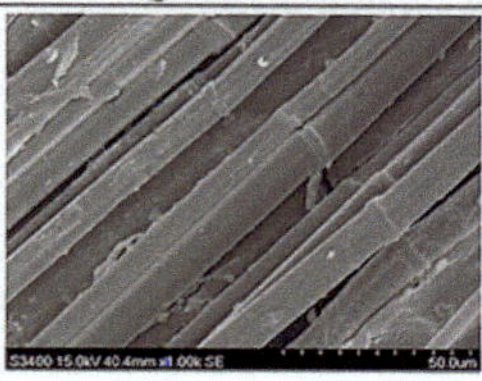</td><td>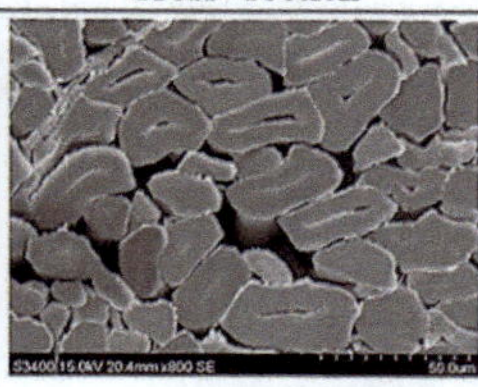</td></tr>
</table>

Decortication process is conducted with green fiber without retting. The quality of bast fiber obtained by this method is lower in comparison to fiber after retting. Decorticated fibers are thick, strong, non-divisible, heavily contaminated by remnants of plant tissues (wood, epidermis). This makes decorticated fiber a raw material suitable mainly for non-textile applications [20].

After retting, the mechanical processes called scutching are conducted for fiber separation, removing of impurities, woody parts and other non-fibrous materials. Other processes are conducted depending on the type of fibers and assumed application. Long flax is hackled to clean, align and remove short fibers to prepare for a wet spinning system. Short fibers are processed with a carding machine. The key flax fiber properties for composite application are listed in the Table 3.

2.2 *Hemp Fibers*

Hemp (*Cannabis sativa*) belongs to the Mulberry family. It is an annual plant, which can be monoecious and dioecious, its growing season lasts from mid-April to mid-September [21]. After harvesting hemp plants, the retting process is conducted in a similar way to flax retting, e.g. water and dew retting is applied. Decortication process of green stalks is an alternative to retting methods used for fiber extraction. The technical and mechanical processes applied during primary processing after retting cover scutching and then hackling for long fibers and carding for short hemp fibers. The technologies of processing and spinning hemp fibers are similar to those used for flax. The chemical composition of hemp fibers is very similar to flax,

Table 4 Hemp fiber characteristics [7]

<table>
<tr><th colspan="3">Hemp (Cannabis sativa)</th></tr>
<tr><th>Plant</th><th>Fiber</th><th>Information</th></tr>
<tr><td></td><td></td><td rowspan="3">Country of origin: Europe, Asia
Applications: rope production, textiles, composites, food (hempseed oil), cosmetics industries
Linear density [tex]: 0.3–2.2
Length [mm]: 15–25
Density [g/cm³]: 1.48–1.49
Breaking tenacity [cN/tex]: 47–80
Diameter [μm]: 15–30
Thermal stability [°C]: 150
degradation:175–205
Young's Modulus [GPa]: 96
Chemical composition [%]:
<table>
<tr><td>Cellu lose</td><td>Hemi cellulo ses</td><td>Pectin</td><td>Lignin</td><td>Fat /wax</td></tr>
<tr><td>67-75</td><td>16-18</td><td>0.8</td><td>2.9-3.3</td><td>0.7</td></tr>
</table></td></tr>
<tr><th>Longitudinal view</th><th>Cross- section</th></tr>
<tr><td></td><td>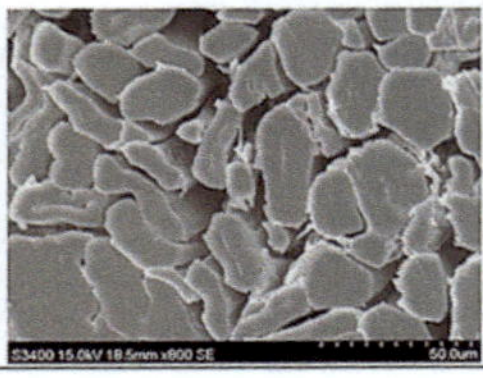</td></tr>
</table>

however hemp fiber is more coarse, thicker, longer and stiffer and for this reason, hemp is usually used for technical applications.

The key properties of hemp fibers from the composite view point are presented in Table 4.

2.3 *Jute*

Jute plays the most important role in total world vegetable fiber production, excluding cotton, Fig. 1 The white jute (*Corchorus capsularis*) is the best known among over thirty *Corchorus* spieces [22]. Jute fibers occur as bundle fibers, which are composed of several fibers glued together by non-cellulosic substances mainly pectin and lignin, similarly to flax, hemp and other bast fibers. The fibers must be extracted from stem by retting, stripping and/or decortication process. Traditional water retting is usually used for jute plants, where whole stems are soaked in water for 10–20 days to enable bacteria to operate in order to remove pectin and other non-cellulosic substances and destroying woody parts of stem. Stripping, i.e. the process of removing the fibers from the stalk, is conducted after the completion of retting. Separation of retted fibers from woody parts of stem is conducted manually. Extracted fibers are washed and dried. An alternative method for jute water retting is ribbon retting. This method allows for reduction of water consumption, because only the ribbons of green bark extracted from the stem are immersed in water, the quantity of water required for ribbon retting is much lower in comparison to traditional water retting.

Table 5 Jute fibers characteristics [7]

<table>
<tr><th colspan="3">Jute (Corchorus capsularis)</th></tr>
<tr><th>Plant</th><th>Fiber</th><th>Information</th></tr>
<tr><td></td><td></td><td rowspan="4">Country of origin: Asia, South America, Europe, Brazil
Applications: sacks and all kind of dressing materials, packing, conveyor belts, upholstery, decorative fabrics, floor covering materials, composites, geotextiles and paper pulp
Linear density [tex]: 1.4–3.0
Length [mm]: 2–3
Density [g/cm³]: 1.44–1.49
Breaking tenacity [cN/tex]: 23.9–27.6
Diameter [μm]: 14–20
Thermal stability [°C]: 150 degradation:175–205
Young's Modulus [GPa]: 64
Chemical composition [%]:
<table>
<tr><td>Cellu lose</td><td>Hemi cellulo ses</td><td>Pectin</td><td>Lignin</td><td>Fat /wax</td></tr>
<tr><td>59-71</td><td>12-13</td><td>0.2-4.4</td><td>11.8-12.9</td><td>0.5</td></tr>
</table></td></tr>
<tr><th>Longitudinal view</th><th>Cross- section</th></tr>
<tr><td>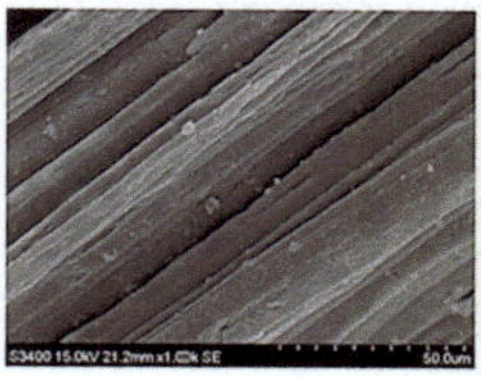</td><td>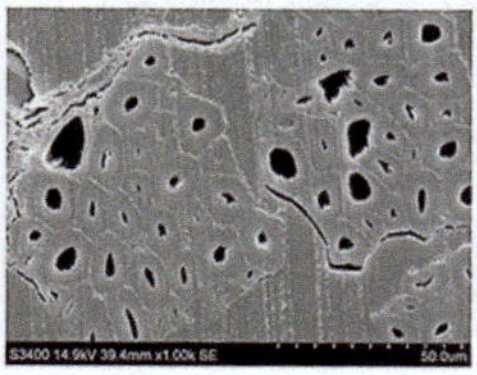</td></tr>
</table>

Another method of jute fiber extraction and scraping out barky material is mechanical decortication. The time of decortication of jute plant is determined precisely. The best stage in the growth of the plant for decortication is 115 ± 10 days after germination. Decortication should be done immediately after harvesting. Optimum diameter of stems for jute decortication is around 12 mm [20, 22]. Jute fiber is much more coarse and stiff in comparison to flax and hemp because it contains more lignin, about 12 % in their chemical composition. Jute is characterized by worse mechanical properties. The high lignin content causes technological limitations in preliminary processes and spinning of jute fibers, what influences the final jute application, mainly to non-textile products.

The key parameters of jute fibers are presented in Table 5.

2.4 Kenaf

Kenaf (*Hibiscus cannabinus L.*), known as Mesta or Ambari, is considered as a variety of jute, what is strongly related with similarity of kenaf and jute fiber properties. Recently, kenaf has become more attractive due to development of natural fiber reinforced composites and possibility of growing kenaf in Europe. Even though, it is a plant native to tropical climate regions, kenaf has been cultivated in the Southern Europe from the beginning of the 20th century. Conventional water retting in natural water reservoirs e.g. rivers is the common method applied for kenaf, but due to generation of harmful water pollution, many studies have been undertaken to develop alternative, more environmental friendly retting processes [23, 24].

Table 6 Kenaf fiber characteristics [7]

Kenaf *(Hibiscus cannabinus)*		
Plant	**Fiber**	**Information**
		**Country of origin:** Asia, Africa, America, Europe **Applications:** rope production, paper production, thick wound dressing fabric **Linear density [tex]:** 1.9–2.2 **Length [mm]:** 1.5–11 **Density [g/cm^3]:** 1.2 **Breaking tenacity [cN/tex]:** 25.4 **Diameter [µm]:** 14–33 **Thermal stability [°C]:** 150 degradation:175–205 **Young's Modulus [GPa]:** 53
Longitudinal view	**Cross- section**	**Chemical composition [%]:**
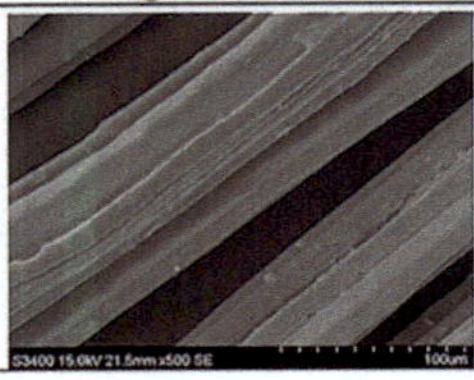	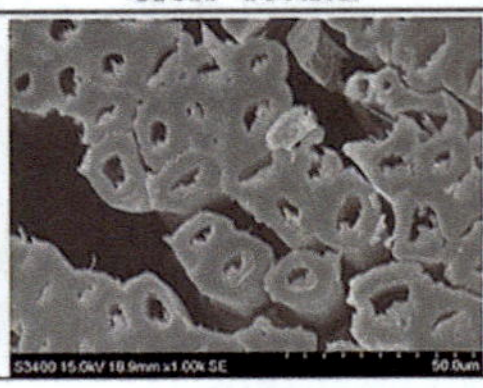	

Cellu lose	Hemi cellulo ses	Pectin	Lignin	Fat /wax
44-72	15-19	-	9-19	2.2

Kenaf plant is grows high and is resistant to pests and diseases but there is still a low level in technical advancement in mechanization of its cultivation and harvesting. Kenaf fibers after water retting are manually extracted and then is washed and dried. Mechanical cleaning and separation from woody parts, e.g. scutching and processing is conducted at the flax processing line.

A more important advantage of kenaf fibers is their the lowest density among other bast fibers, e.g. 1.2 g/cm^3. This aspect makes kenaf very attractive in terms of using it to reinforce composites.

The most important parameters of kenaf fibers are presented in Table 6.

2.5 Sisal

Sisal (*Agave sisalana*) is a monocotyledonous plant that belongs to the *Agavaceae* family. Sisal fiber, contrary to bast fibers, is derived from the leaves of the plant. It is an attribute of leaf fibers that they are harder than fibers extracted from the bast of plants, like flax, hemp, jute and kenaf.

The fibers in sisal plant occur longitudinally in the leaf near its surface. The first stage of fiber extraction is harvesting, e.g. cutting off sisal leaves, which is usually conducted once a year. Extracting process is carried out manually or with the use of mechanical decortication to scrap away and remove pulpy material. After decortication the fibers are washed in water and then sun dried [25, 26].

A coarse and strong fiber, sisal with low density is being increasingly used in composite materials for cars, furniture and construction as well as in plastics and paper products.

The most important parameters of sisal fibers are presented in Table 7.

Table 7 Sisal fiber characteristics [7]

<table>
<tr><th colspan="3">Sisal (Agave sisalana)</th></tr>
<tr><th>Plant</th><th>Fiber</th><th>Information</th></tr>
<tr><td></td><td></td><td rowspan="3">Country of origin: South and Central America, Eastern Africa
Applications: ropes, strings, bags fabrics, plaiting, mats, dart targets, fishnets
Linear density [tex]: 28.6–48.6
Length [mm]: 0.8–8
Density [g/cm³]: 1.2
Breaking tenacity [cN/tex]: 57.2
Diameter[μm]:7–47
Thermal stability [°C]: 150 degradation:175–205
Young's Modulus [GPa]: 9.4–22
Chemical composition [%]:
<table>
<tr><th>Cellu lose</th><th>Hemi cellulo ses</th><th>Pectin</th><th>Lignin</th><th>Fat /wax</th></tr>
<tr><td>66-73</td><td>12-13</td><td>0.8</td><td>9.9</td><td>0.3</td></tr>
</table></td></tr>
<tr><th>Longitudinal view</th><th>Cross- section</th></tr>
<tr><td>
</td><td>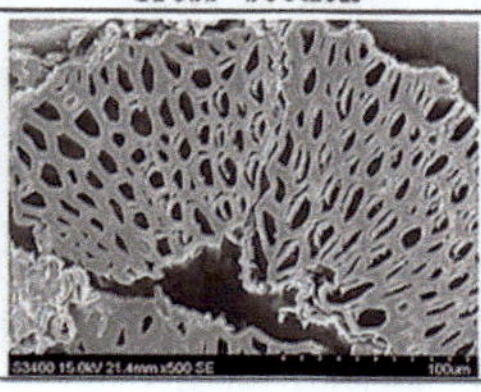
</td></tr>
</table>

2.6 *Abaca*

Abaca (*Musa textiles nee*) belongs to banana family of plants. Abaca is a leaf fiber, composed of long slim cells that form part of the leaf supporting structure. The matured abaca plant consists of about 12 to 30 stalks radiating from a central root system [27]. Abaca fiber is extracted from the leaf sheath in a traditional way by manual stripping or mechanical stripping e.g. spindle stripping. When either of the process is used, tuxying is employed. Tuxying is the process of separating primary fibers from the outer sheath and secondary fibers from the inner leaf sheath. The separated outer leaf sheath is called tuxy. Alternatively, abaca fibers can be extracted with the use of the decortication process, where the fibers are recovered from the whole leaf sheath, both the primary and secondary fibers. Decortication process ensures obtaining much more fibers from the leaves, e.g. 3–3.5 % in comparison to manual extraction, where it is possible to recover about 1 % of fibers. After stripping or decortication the ribbons of fibers, which can be up to 3 m long, are hung up to dry. Abaca fibers are processed in a similar way to other hard fibers like sisal [28].

The main parameters of abaca fibers are presented in Table 8.

2.7 *Coir*

Coir is a fiber extracted from the outer fibrous covering of the fruit of coconut palm, botanically known as (*Cocos nucifera*) [29]. After manual separation of the nut

Table 8 Abaca fiber characteristics [7]

<table>
<tr><th colspan="3">Abaca (Musa textilis nee)</th></tr>
<tr><th>Plant</th><th>Fiber</th><th>Information</th></tr>
<tr><td></td><td></td><td rowspan="3">Country of origin: Philippines, Java, Sumatra, Borneo, Central and South America
Applications: plaiting, thick fabrics, fishnets, sails, ship ropes, paper, boards used in construction
Linear density [tex]: 4.2–44.4
Length [mm]: 6
Density [g/cm³]: 1.5
Breaking tenacity [cN/tex]: 54–72
Diameter [μm]: 10–30
Thermal stability [°C]: 150 degradation:175–205
Young's Modulus [GPa]: 12
Chemical composition [%]:<table><tr><th>Cellu lose</th><th>Hemi cellulo ses</th><th>Pectin</th><th>Lignin</th><th>Fat /wax</th></tr><tr><td>63-68</td><td>19-20</td><td>0.5</td><td>5.1-5.5</td><td>0.2</td></tr></table></td></tr>
<tr><th>Longitudinal view</th><th>Cross- section</th></tr>
<tr><td></td><td>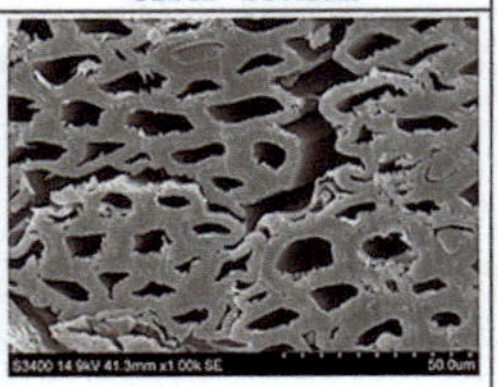</td></tr>
</table>

from the fibrous coconut husk, the husks are processed by various retting techniques, most commonly in ponds of brackish waters or in salt backwaters or lagoons. By retting the fibers are softened and can be decorticated and extracted by beating, which is usually done by hand. After hackling, washing and drying (in the shade) the fibers are loosened manually and cleaned. Alternatively, mechanical decortication process can be used for fiber extraction from the husks after only 5 days of immersion in water tanks. Crushing the husks in a breaker opens the fibers and the coarse long fibers are separated from the short woody parts and the pith. The stronger fibers are washed, cleaned, dried, hackled and combed. The quality of the fiber is greatly affected by these procedures. There are two types of coir: the more commonly used brown fiber, which is obtained from mature coconuts, and finer white fiber, which is extracted from immature green coconuts after soaking for up to 10 months. Mature coir fibers contain more lignin, a complex woody chemical, and less cellulose than bast fibers. Among vegetable fibers, coir has one of the highest concentrations of lignin, making it stronger but less flexible than cotton and unsuitable for dyeing. The tensile strength of coir is low compared to abaca, but it has good resistance to microbial action and salt water damage and needs no chemical treatment [30–34].

The main parameters of coir fibers are presented in Table 9.

Table 9 Coir fibers characteristics [7]

Coir *(Cocos nucifera)*		
Plant	**Fiber**	**Information**
		**Country of origin:** Western, Central and Southern Africa, India, The Ivory Coast **Applications:** brushes, mattresses, bags, ropes, upholstery, automotive industry **Linear density [tex]:** 50 **Length [mm]:** 0.3–3 **Density [g/cm³]:** 1.2 **Breaking tenacity [cN/tex]:** 397 **Diameter [μm]:** 7–30
Longitudinal view	**Cross- section**	**Thermal stability [°C]:** 150 degradation:175–205 **Young's Modulus [GPa]:** 3.7
	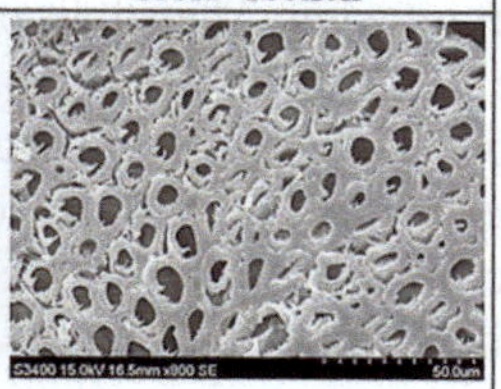	**Chemical composition [%]:**

Cellu lose	Hemi cellulo ses	Pectin	Lignin	Fat /wax
36-43	0.2	3-4	41-45	-

3 Vegetable Fiber Textiles as Reinforcement of Structural Composites

Most developments in the area of natural fiber reinforced composites have focused on randomly distributed short fiber composite systems. Natural vegetable fibers, mainly flax, hemp and kenaf can be used for composites not only in the form of loose mass of short fibers, but also in continuous forms like roving, yarn or textiles in 2D or 3D structures, woven fabrics, knitted fabrics, and nonwovens. The reinforcement of composites can be prepared from pure natural fibers or from blends of natural fibers and polymer filament or staple fibers [35].

A study on long bast fibers application in form of roving to reinforce composites was conducted by Shah et al. [36]. They utilized wet-spun flax roving to flax/polyester composite for manufacturing and mechanical testing of rotor blades suitable for wind turbines. This blade made of flax base composite was compared to E-glass/polyester composite blade. Prior to spinning, the dew-retted flax slivers were soaked in a hot dilute solution of caustic soda to improve roving regularity, purity and promote better fiber/matrix adhesion. The E-glass fibers were surface treated with an epoxy size that is suited to both polyester and epoxy resin systems. Stitched aligned fabrics were prepared from both raw materials: flax roving and E-glass. The authors found that the blade prepared with flax composite is 10 % lighter (fiber mass saving of 45 %) than the identical construction E-glass blade. It means that flax is a potential structural replacement to E-glass for similar composite small wind turbine blade applications, although flax blade cannot compete against an E-glass blade in terms of stiffness.

The structure of vegetable fiber textiles dedicated to reinforce composites has strong effect on conditions of composite formation as well as on composite properties, what was confirmed by many researchers [37, 38].

Utilization of different bast fibers: flax, hemp and jute in form of roving and yarn with high and low twist to reinforce of composite was investigated by Shah [38]. The high twist yarns made of jute and hemp as well as low twist yarn from flax commingled with polyester and flax roving were used for preparation of unidirectional fabric. Four layers of the mat were used to produce composite in two standard thermosetting resin systems e.g. polyester and low viscosity epoxy. Resin injection was achieved by vacuum infusion. The study proved that low twist plant fibers reinforcement can produce high quality composites that can compete and even outperform glass fiber composites in terms of specific stiffness, without any fiber surface modification, however, the plant fiber composites showed poor strength performance (tensile, flexural and impact). The composites reinforced with vegetable fibers yarn with low twist require significantly longer fill times and those composites are more homogenous in comparison to composites reinforced with high twist natural fibers yarn.

The properties of flax and hemp fabric in relationship to mechanical and structural parameters of yarns and the fabric prepared from these yarns in terms of composite reinforcement application were investigated in [37]. The flax yarns were prepared from flax noils, cottonized flax fibers, flax long fibers short hemp fibers with the use of different spinning systems, respectively: dry flax spinning system, cotton ring spinning system, wet long flax spinning system and wet spinning system for short fibers. The fabrics prepared from these yarns were used as reinforcement for laminates by film stacking layers of fabric with polypropylene sheets. This study proved that structural parameters of yarn as well as the fabric influence the properties of laminates reinforced with textiles. Lowering the density of yarn or fabric causes reduction of its mechanical properties, which might be expected to result in worsening of the quality of the composite reinforced by the fabric. In fact, a lower density in the textiles means better conditions of adhesion between fibers and matrices and hence improved mechanical properties of composites. The type of fibers used for yarn spinning, as well as the spinning system has an effect on the number of twists and hence yarn tenacity, which affects the mechanical properties of the fabric. In the case of yarn made of short fibers it is necessary to apply a higher number of twists per meter in the spinning process in comparison to long fibers, what means worsening condition for composite formation.

Goutianos et al. [39] conducted research on development of high-performance natural fiber composite systems for structural applications using continuous textile reinforcements like unidirectional tapes or woven fabrics based on flax fibers. The authors focused on optimization of the yarn to be used for manufacturing the textile reinforcement by balancing requirements related to spinning and textile processing with mechanical properties. Different types of fabrics, e.g. biaxial plain weaves, unidirectional fabrics and non-crimp fabrics were produced and evaluated as reinforcement in composites based on unsaturated polyester, vinyl ester, vinyl ester hybrid thermosetting resin and as reference—epoxy resin. The composites were manufactured with the use hand lay-up, vacuum infusion, pultrusion and resin transfer moulding (RTM). The developed flax based composites cannot directly compete in terms of strength with glass fiber composites but they are able to compete with these materials in terms of stiffness, especially if the low density of flax is taken into account. Flax fiber based materials are characterized by favourable properties in comparison to non-woven glass composites [39].

Investigation in the reinforcement of composites has been carried out not only with the use of 100 % vegetable fibers, but also with the use of blends with manmade fibers and as a hybrid mix with other polymers. This topic was studied in [40]. Fabrics dedicated to the reinforcement of laminated composites were made from two groups of raw materials: flax noils and thermoplastic polymer fibers: poly-propylene, polyester and polylactic acid (PLA).

Mixing of natural fibers and manmade polymer to obtain hybrid yarns dedicated to woven reinforcements was conducted with the application of the following methods:

- Mixing of both types of short fibers in mass at the carding stage,
- Twisting of two yarns—100 %flax and polymer filament,
- Core yarn formation with the use of flax fibers as a core and polymer filament as a sheath.

Investigation of the textile reinforcement made of mix of short fibers was conducted in the following sequence: fibers → yarn → fabric → composite, it is illustrated in Fig. 6.

Application of different spinning methods allowed for obtaining diversity in the distribution of flax and polymer fibers in cross sections of the yarns. Fiber distribution in the yarn plays a significant role in the creation of proper conditions for fiber wettability and adhesion between fibers and matrices, which in consequence has an effect on the composite formation and its properties [41–43].

Blended yarns were spun using different spinning systems, e.g. cotton ring spinning system, woollen spinning system, twisting two yarns: flax and man-made yarn and hollow spindle spinning system, where polymer filament was applied as a sheath covering the flax fiber core. Fiber distribution in the cross section of hollow spindle yarn was presented in Fig. 7, where flax fibers were concentrated as a core in the central part of the yarn cross section and were surrounded by man-made polymer filaments, what ensured very good mechanical properties of the yarn.

The yarns were used for preparation of woven fabric characterized by similar structures. For the formation of laminates the following thermoplastic matrices were

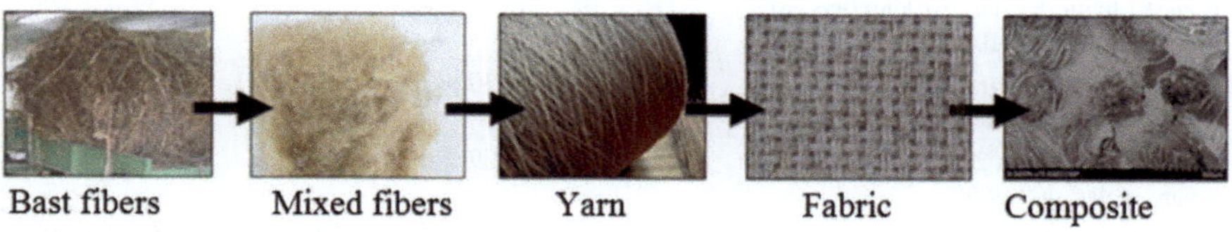

Fig. 6 Scheme of technological chain of bast fibers and fabric application to composite

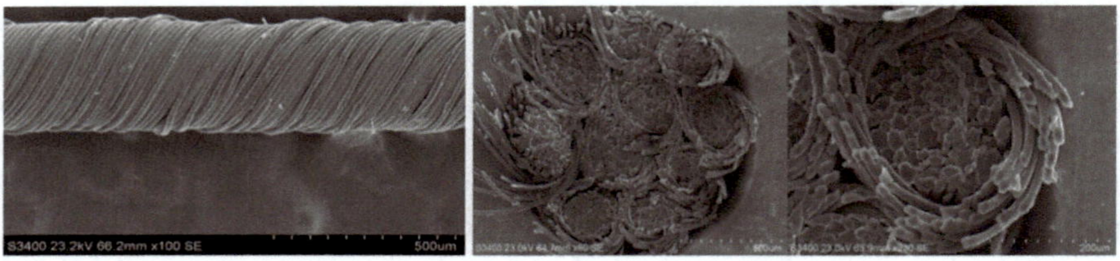

Fig. 7 SEM images of hollow spindle yarn FLAX/PLA (Core-Flax) 100tex

applied: PLA, polypropylene (PP) and polyethylene terephthalate (PET). The laminates were formatted with the use of the vacuum consolidation technique for thermoplastics. The results of the study proved that the structure of natural fiber reinforcement of the composite is responsible for ensuring proper conditions of adhesion between the fibers and matrices, e.g. resin penetration and fiber bonding, and finally for the composite mechanical properties. The factors influencing composite quality are the homogeneity of fiber distribution in the matrices, the yarn and fabric structure, mainly their density and uniformity, and applied matrices. The mechanical properties of yarns and fabrics used in the study as reinforcements did not influence the tensile strength of the composites [40]. The fabric woven from a blended yarn made using the hollow spindle spinning system showed the best structure suitable for composite formation of all tested samples, e.g. parallel position and uniformity of flax fiber distribution in the yarn core, effective PLA filament contact with flax fibers and low textile density. Mechanical properties of the laminate based on the hybrid fabric made of hollow spindle yarn showed the highest value.

In other study [35] composite reinforcement 100 % flax fabric was compared with fabrics made of flax yarns and KEVLAR, glass fibers and polyimide P84. All the fabrics were prepared in plain weave and were used for five-layer composites with the use of ASHLAND G105 vinylester resin as a matrix by vacuum infusion method. Mechanical properties of the composites are presented in Table 10.

Table 10 Mechanical properties of composites reinforced by flax based fabrics [40]

Fabric description	Tensile strength [MPa]	Tensile modulus [MPa]	Flexural strength [MPa]	Flexural modulus [MPa]
Blended fabric Kevlar 1610dtex/flax, Mass per square meter: 230 g/m^2	49 ± 2	4835 ± 15	100 ± 9	3650 ± 542
Blended fabric Kevlar 270dex/flax, Mass per square meter: 290 g/m^2	59 ± 11	4649 ± 385	67 ± 6	2399 ± 356
Blended fabric Kevlar 1610dtex/flax, Mass per square meter: 310 g/m^2	110 ± 19	6899 ± 826	98 ± 20	3974 ± 966
Blended fabric polyimide 29,5x3tex/flax, Mass per square meter: 290 g/m^2	48 ± 6	4330 ± 706	79 ± 11	2878 ± 394
Blended fabric glass 200dtex/flax, Mass per square meter: 400 g/m^2	80 ± 5	6539 ± 64	97 ± 7	4070 ± 497
Homophase flax Mass per square meter: 220 g/m^2	61.57 ± 7	4890 ± 624	70.5 ± 6	5897 ± 497

In case of blended fabrics, hydrophilic flax yarn occurs together with hydrophobic Kevlar, glass fibers or polyimide P84. Such a raw material composition improves adhesion between the fabric and the resin. The composite reinforced by flax/p-aramid KEVLAR1610 dtex fabric (310 g/m2) is characterized by the best mechanical properties. The blended fabric prepared with use of glass fibers shows also high tensile strength and the highest bending module. This means that the glass fiber and Kevlar 1610dtex (310 g/m2) improve mechanical properties of composites. The composite reinforced by pure flax fabrics have better mechanical properties than the composite reinforced by fabrics made of flax/Kevlar (290 and 230 g/m2) and flax/P84. The study confirmed suitability of applying flax fabrics as composite reinforcement. Due to highly hydrophilic properties of pure flax, composites reinforced by the blended fabrics with Kevlar or glass fibers showed better mechanical properties. Chemically modified flax fabric, as a result of improvement in adhesion between fibers and matrix, can serve as valuable reinforcement for composites, ensuring high mechanical properties comparable to glass fibers.

Salman et al. [44] investigated composites reinforced with kenaf plain woven fabric. Three types of thermoset resins were used for the composite samples fabricated using the vacuum infusion technique, namely, epoxy, unsaturated polyester, and vinyl ester resin and fiber weight content for each type of resin was 35 %.The authors found that the tensile, flexural, and impact strengths of the woven kenaf/epoxy composite were superior to the others tested composites but the variation of the mechanical properties, for the same type of the thermoset, due to the variation of the fiber orientation, was significant. The results obtained from this study suggested that woven kenaf had better properties for use as a reinforcement in the thermoset composites sector.

4 Fabrication of Natural Fiber Composites

The composites reinforced with lignocellulosic materials, both wood and natural fibers, are known as *Wood Plastic Composites* (WPC) or *Natural Fiber Reinforced Composites* (NFRC) and are found more and more often in new applications in various sectors of economy. They differ from other wood materials such as MDF and plywood, particle board and OSB [45].

The composites are materials that contain at least two different components of different properties, clearly separated one from the other. Its components fill the total volume in such a way that it has better properties than those of the separate components. Natural fibers prove suitable for a wide range of composites materials for various applications. Physical and mechanical properties of selected natural fibers differ from those of glass fibers

Figure 8 shows the growing trend in the publication and citing numbers regarding composites reinforced with natural fibers (*according to the Web of Science*). The keywords used for the search were: *natural fiber, reinforced polymer, composite*

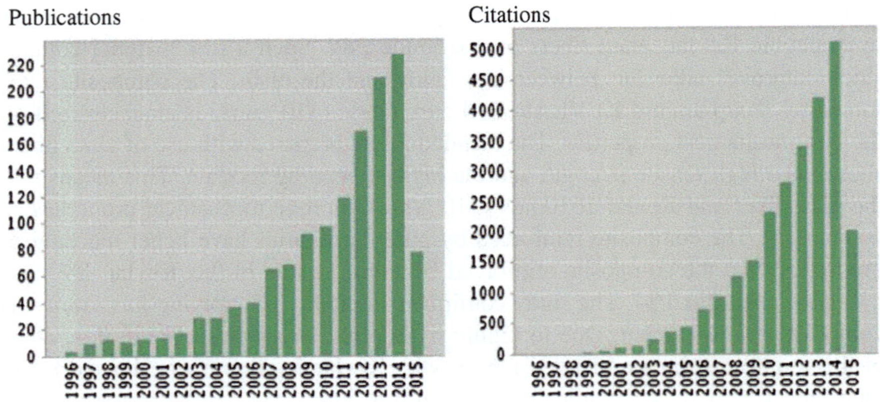

Fig. 8 The number of publications and citations regarding NFRCs within the last 20 years

As Fig. 8 shows in recent years we have experienced a rapid growth of interest in NFRCs. Lignocellulosic materials are suitable for production of a wide range of composite materials for various applications and are an excellent alternative to glass fibers and inorganic fillers [46]. The trend is mainly caused by pressure from decision makers from many European countries on the industry to reuse and recycle materials. NFRCs are used in the transport industry for production of car elements, railway carriages, in military applications, in construction, also as ceiling and floor panels, for packaging and other consumer products. In Table 11 selected vegetable fibers applied as reinforcement with various thermoplastic and thermoset polymeric matrices are presented.

Table 11 Thermoplastic and thermoset polymeric composites reinforced with different natural fibers [76]

Thermoplastic	**Vegetable fibers**
Polypropylene (PP)	flax, hemp, jute, sisal, green coir
Polyethylene (PE)	green coir, sisal,
High density polyethylene (HDPE)	sisal,
High Impact Polystyrene (HIPS)	green coir, sisal,
Thermoset	**Vegetable fibers**
Polyester	flax, hemp, jute, sisal, coir
Polyurethane (PU)	sisal, coir,
Epoxy	flax, hemp, jute, sisal, coir,
Phenolic	flax, jute, sisal
Vinylester	hemp, jute, sisal, coir
Starch/EVOH	coir
Soy protein	sisal

4.1 Fabrication Techniques

Natural fiber-containing composites can be manufactured by almost all production techniques. Resins used for the manufacture of composites can be hardened without using elevated temperatures and products can be obtained by laying layers of mats or fabrics made of reinforcing fibers on a simple one-sided mold, followed by saturation with a resin composition, performed e.g. by brushing. Such a simple, manual method of product formation, which consist of contacting starting materials with uncomplicated one-sided mold, without applying high temperature and pressure, is called the contact method or hand lay-up method. This is the most popular method of manufacturing products made of polymer-containing composites. The simplicity of this method enables fast start of the production of new products at low cost. It is enough to make the model of a product designed. The advantage of the method is the absence of limitations concerning dimensions of a product and the possibility of manufacturing products having quite a complicated shape. The contact method can be used for the manufacture of a variety of products, e.g. cases for laboratory and electronic equipment, machine housing, elements of car body and whole body of a car, sail boats and fish cutters, swimming pools, chemical apparatus, tanks and silos.

The contact method imposes, however, some limitations on the quality and construction of products. Only one surface of a product manufactured by this method is smooth. Thickness of product walls, formed by hand and consisting of many layers is not uniform. It should be kept in mind that the quality of a product manufactured by the contact method depends, to a considerable extent, on well-designed production technology, complying with procedures and operations contained therein as well as on self-observation, craftsman's skills and reliability of a worker involved in the production.

A variant of the contact method is the spray method. In the latter method filament fibers are used, instead of reinforcing fibers in the form of mats and fabrics. The filament fibers are cut with the help of special devices and sprayed on a mold together with a resin composition covering the mold with a kind of a loose scum. Then the latter is pressed down towards the mold by means of brushes and rollers, i.e. in the way similar to that used in the contact method. The spray technique enables to obtain fiber-reinforced composites in a cheaper way and to mechanize some stages of manufacturing process as well as it forces to set the manufacturing process in order.

The manufacture of goods made of chemosetting resins is often vacuum-aided and this requires an air-tight mold composed of two parts. By forming composites in the vacuum-aided way it is possible to gain such advantages as better resin penetration and better product de-aeration, higher content of reinforcing fibers, smoother internal surface of a product, air-tight sealing of the process.

For the manufacture of closed-shape products the method of press molding with flexible punch is applied as well. A protective helmet or headpiece can be molded by using a punch made of e.g. polyurethane, flexibility of which enables to press the composite to the walls of a matrix.

Another, widely used manufacturing method of goods made of polymer composites, is so-called RTM (Resin Transfer Molding). In this method a resin composition is forced under a slight overpressure into a tight mold which has been loaded with reinforcing fibers. Sometimes, in addition to the pressure necessary to force resin into mold, a vacuum is employed in order to close the mold and to increase penetration of the reinforcement. The two-part molds used in the RTM method have to be more rigid and made with greater accuracy than those used when applying vacuum only and this fact enables to obtain better quality and repeatability of products. The RTM method is more effective and it makes possible to manufacture composites of higher quality and better repeatability than the contact method. In the former method, the compositions of resin can be modified by adding fillers, to a considerably larger extent.

Of particular importance to the manufacture of composites is the pultrusion method [47].

Pultrusion is a manufacturing process for producing continuous lengths of reinforced plastic structural shapes with constant cross-sections. Raw materials are a liquid resin mixture (containing resin, fillers and specialized additives) and flexible textile reinforcing fibers. The process involves pulling these raw materials (rather than pushing, as is the case in extrusion) through a heated steel forming die using a continuous pulling device. The reinforcement materials are in continuous forms such as rolls of fiber mat and natural fiber. As the reinforcements are saturated with the resin mixture ("wet-out") in the resin bath and pulled through the die, the gelation, or hardening of the resin is initiated by the heat from the die and a rigid, cured profile is formed that corresponds to the shape of the die [48].

Products manufactured by the above method are characterized by a high content of reinforcing fibers (of order of 70 % and in some special products even as much as 90 %) and the fibers are orderly laid out. The strength of the products is very high, oriented and programmable. The pultrusion method enables to manufacture bars and sections as well. Fibers are pulled through a tank filled with a resin composition or melted polymer and then they pass through forming dies which serve also for resin excess removal as well as through a heating zone in which hardening occurs (in the case of thermoplastics a cooling zone is applied).

In the method of filament winding a bundle of filament fibers is pulled through binder-containing tank and then it is wound round a rotating core (the role of which is to give shape to a product). By appropriate programming of the movements of the tank and the core on a special device (a winder), a desired arrangement of reinforcing fibers in the product can be achieved. This technique makes it possible to manufacture pipes, tanks and other products in the form of solids of revolution. By using special fibers and resins, elements of rocket launchers and ballistic missiles can be produced by the filament winding method. The last two of the discussed methods are characterized by very high effectiveness due to high mechanization of the processes of saturating and forming.

The above review points to a rich variety of manufacturing techniques of polymer composites containing chemosetting resins.

The choice of the thermoplastics that can make matrices for natural fiber-reinforced composites is limited, because above 230 °C a fast deterioration of fiber strength occurs. For this reason the thermoplastics that can be processed below that temperature are applied. Composites based upon a thermoplastic matrix are manufactured, among others, by a press molding method. Moreover, more complex methods are used to make products known as SMC (Sheet Molding Compound) and BMC (Bulk Molding Compound). While in the former case the molding compound is in the form of sheets, in the latter it makes a shapeless mass or loose granulated product. Properties of SMC and BMC products can be modified by using different kinds and combinations of resins and reinforcing fibers, hardeners, fillers and upgrading additives. These composites have found application in automotive industry, where prototype bumpers, elements of truck driver's cabins, etc., are produced by using flax fibers and polyester resin.

Bulk molding compounds make also suitable materials for the manufacture of products of smaller dimensions and lower strength requirements, such as structural elements and housings of office machines, elements of tape recorders and video recorders, circuit-breakers, controllers, household equipment, street light housings, etc.

Sheet molding compounds can be applied to the production of bumpers and body elements, particularly for tractors and vehicles manufactured in limited series (trucks, delivery vans, special cars). They have found application also in railway and air transportation (window frames, partition walls, passenger compartment ceilings, luggage racks, etc. Their other applications include park benches, seats for amphitheaters and stadiums, bill-posts, traffic limiting elements such as barriers, road guards.

4.2 Natural Fiber Surface Modification

The main disadvantages of natural fibers in respective composites are the poor compatibility between fiber and matrix and their relatively high moisture absorption. Therefore, natural fiber modification is considered leading to a change of the fiber surface properties to improve their adhesion with different matrices.

Reinforcing fibers can be modified by physical and chemical methods. Chemical and physical modifications of natural fibers are usually performed to correct for the deficiencies (described above) of these materials, especially to impart bonding and adhesion, dimensional stability and thermoplasticity. Surface modification of natural fibers can be used to optimize properties of the interface.

Physical methods involve surface fibrillation, electric discharge (corona, cold plasma), etc. Cold plasma is already a very effective method to modify the surface of natural polymers without changing their bulk properties. This method has been used for increasing adhesion and compatibility between two polymers [49–51].

Physical treatments change structural and surface properties of the fiber and thereby influence the mechanical bonding with the matrix.

There are some treatment methods of partially physical and chemical in nature, among which mercerization [47, 52, 53] and liquid ammonia treatment [47, 54] should be mentioned.

Mercerization leads to the increase in the amount of amorphous cellulose at the expense of crystalline cellulose. The important modification expected here is the removal of hydrogen bonding in the network structure. As a result of sodium hydroxide penetration into crystalline regions of parent cellulose (cellulose I), alkali cellulose is formed. Then, after washing out unreacted NaOH, the formation of regenerated cellulose (cellulose II) takes place.

The treatment with liquid ammonia has been used mainly for cotton. Its development occurred since the late 1960s as an alternative to mercerization. Liquid ammonia, due to its low viscosity and surface tension, penetrates quickly the interior of cellulose fibers, forming a complex compound after the rupture of hydrogen bonds. The molecule of ammonia is relatively small and is able to increase distances between cellulose chains and penetrate crystalline regions.

The original crystal structure of cellulose I changes to cellulose III after liquid ammonia treatment, and at the next stage, cellulose III changes to cellulose I again after hot water treatment. Liquid ammonia treatment of natural fibers results in their deconvolution and smoothing their surfaces. At the same time, fiber cross-section becomes round and lumens decrease.

Conventional chemical modification is usually carried out through typical esterification and etherification reactions of lignocellulose hydroxyl groups.

Esterification usually involves the reactions with organic acids or anhydrides. Many esters are possible depending on the nature of organic acid (anhydride) used in the reaction. Esters containing from 1 to 4 carbon atoms are formate, acetate, propionate and butyrate; laurate has 12 carbon atoms and stearate 18 carbon atoms. Maleate and fumarate are esters of dicarboxylic acids containing double bonds in the carbon chain. Double bond-containing esters with longer chains confer thermoplasticity on the lignocellulosic materials [47].

The most popular esterification method is acetylation, which has already been developed in commercial scale, first in the United States [55], then in Russia [56]. Within the UK, BP chemicals performed an economic appraisal of the process and built a small pilot plant to modify fibers, but have now discontinued this work [57]. Within Europe several groups are working towards a scale up of the process [58–60].

Substitution of hydrophilic hydroxyl groups in lignocellulose by hydrophobic radicals alters properties of fibers, especially equilibrium moisture content (EMC), depending of degree of substitution. Modification alters polarization of fibers and make them more compatible to non-polar matrix.

Organosilanes are the main group of coupling agents for glass fiber-reinforced polymers. Silanization of natural fibers minimizes disadvantageous effect of moisture on properties of composites and at the same time increases adhesion between fibers and polymer matrix, which results in upgrading composite strength. The effectiveness of the modification depends, among others, on the type of silane used, its concentration in solution, temperature and time of fiber silanization, moisture content and volume contribution of fibers to composite [61–63].

The surface of natural fibers can be modified by grafting copolymerization. Grafting efficiency, grafting proportion and grafting frequency determine the degree of compatibility of cellulose fibers with a polymer matrix. The grafting parameters are influenced by the type and concentration of initiator, by the monomer to be grafted and the reaction conditions [64]. The polymerization reaction is initiated at the surface of the fibers by incorporation of peroxides or oxidation–reduction agents, or treatment with gamma radiation or cold plasmas.

The cellulose is treated with an aqueous solution containing selected ions and is exposed to a high-energy radiation. Then, cellulose molecules crack and radicals are formed. Afterwards, the radical sites of the cellulose are treated with a suitable solution (compatible with the polymer matrix), for example vinyl monomer [62], acrylonitrile [65], methyl methacrylate [66], polystyrene [67]. The resulting copolymer possesses properties characteristic of both fibrous cellulose and grafted polymer.

The lignin content in the fiber is a governing factor on the extent of acrylonitrile grafting. The accessibility of the monomer molecules to the active centers of cellulose is easier in water than in an organic solvent. It was found that without the presence of lignin, the grafting reaction could favorably proceed between cellulose and acrylonitrile.

The treatment of cellulose fibers with hot polypropylene-maleic anhydride (MAPP) co-polymers provides covalent bonds across the interface [68].

There are two ways of obtaining biocomposites from natural fibers and polymer. In the first one, pretreated fibers with maleated polymer are reinforced with desired polymer matrix. In the second method, fibers, polymer and maleic anhydride with addition of peroxide initiator in one step processing are reactively extruded and then processed with molding or injection to obtain a final composite.

Expanding the application range for the composites to some branches of industry such as construction and transport make reduced flammability a crucial factor and requirement. The knowledge on flammability of the composites reinforced with lignocellulosic materials and methods used for flame retardancy of these materials are necessary to allow for the application of these materials in the aforementioned sectors of the economy.

Reduction of composite flammability can be achieved by reducing the flammability of the polymer matrix and/or of fiber reinforcement or of the composite as a whole.

The latest review on flammability of composites reinforced with natural fibers and methods used for their FR treatment was done by Kandola and Chapple&Anandijiwala. [69, 70]. Numerous flame retardancy treatments were presented by Mngomezulu et al. when reviewing the literature on biofiber and biocomposite flammability [71]. Many of the treatments were adopted from the fire retardancy of thermoplastic and thermosetting polymers and textile materials.

The effect of fiber surface modification on the mechanical performance of oil palm fiber is shown in Table 12.

Table 12 Mechanical performance of parent and modified oil palm fibers [72]

Fiber	Tensile strength (MPa)	Young's modulus (MPa)	Elongation at break (%)
Untreated	248	6700	14
Mercerized	224	5000	16
Acetylated	143	2000	28
Peroxide-treated	133	1100	24
Permanganate-treated	207	4000	23
γ-irradiated	88	1600	25
TDIC-treated	160	2000	T2
Silane-treated	273	5250	14
Acrylated	275	11,100	26
Acrylonitrile-grafted	95	1700	24
Latex-coated	98	1850	23

5 Properties of Natural Fiber Composites

The properties of natural fiber reinforced composites depend on a number of parameters such as volume fraction of the fibers, fiber aspect ratio, fiber–matrix adhesion, stress transfer at the interface and orientation.

Both the matrix and fiber properties are important in improving mechanical properties of the composites. According to Nabi Saheb and Jog [73] the tensile strength is more sensitive to the matrix properties, whereas the modulus is dependent on the fiber properties. To improve the tensile strength, a strong interface, low stress concentration, fiber orientation is required, whereas fiber concentration, fiber wetting in the matrix phase, and high fiber aspect ratio determine tensile modulus. They also say [68] that the aspect ratio is very important for determining the fracture properties. In short-fiber reinforced composites, a critical fiber length exists that is required to develop its full stressed condition in the polymer matrix. Fiber lengths shorter than this critical length are required to develop its full stressed condition in the polymer matrix. Fiber lengths shorter than this critical length lead to failure due to deboning at the interface at a lower load. On the other hand, for fiber lengths higher than the critical length, the fiber is stressed under applied load and thus results in higher strength of the composite [68].

As said by Bledzki and Gassan [74] the mechanical properties of composites are influenced mainly by the adhesion between matrix and fibers. The adhesion properties can be changed by pretreating the fibers. So special processing, such as chemical and physical modification methods were developed. Moisture repellency, resistance to environmental effects, and not last, the mechanical properties are improved by the treatments.

The role of matrix in a fiber reinforced composite is to transfer stress between the fibers, to provide a barrier against an adverse environment and protect the surface of the fibers from mechanical abrasion. Taj et al. [75] say that the matrix plays a major role in the tensile load carrying capacity of a composite structure. The matrix in the composite is of critical importance. Most of the composites used in the industry today are based on polymer matrices. Polymer resins have been divided broadly into two categories: Thermosetting (e.g. epoxy, unsaturated polyester, vinyl ester, phenolic epoxy resin, etc.) and Thermoplastics (e.g. high density polyetene HDPE, low density polyethene LDPE, Polypropylene PP, Poly-Vinyl chloride PVC, Normal polystyrene PS etc.

6 Vegetable Fibers Based Composites Application

An expansion in industrial usage of vegetable fiber composites as engineering materials and trying to replace traditional fiberglass with natural fibers as composite reinforcement is observed in many areas, like automobiles, aircrafts components, building plates, construction, packaging, sport equipment, electrical parts, medical prosthesis and other.

The current trends in application of composites and their market share given in percent are the following [2]:

- Construction—30 %,
- Automotive industry—25 %,
- Industrial equipment—10 %,
- Electronics—9 %,
- Sport—8 %,
- Steel industry and shipbuilding—6 %,
- Electrical engineering—6 %,
- Aviation and space industries—3 %,
- Medicine—1 %,
- Railways—1 %,
- Wind power plants—1 %.

The vehicles manufacturing sector, in particular, is adopting vegetable fibers composites at a fast rate in both interior and exterior components [77].

The trend watching in industry allows to assume that automotive and aviation industries are most likely to develop dynamically in terms of using composites reinforced with natural plant materials [78]. Strong development of production of various parts for cars and aircrafts with the use of such materials is visible. The elements containing natural fibers have better electrostatic properties, absorption of vibrations, thermal insulation and sound attenuation as compared with traditional polymers. Lower weight of the composite affects the total weight of a vehicle, thus reducing the fuel consumption and consequently the costs of driving and finally it will have beneficial effect on the environment. Polymeric materials reinforced with natural fibers can be applied in the production of car parts like bumpers, brake lining and clutches. Replacing chemical fibers (glass, carbon) with the natural ones enables to produce reinforcement significantly less harmful for the environment and easier to recycle [79].

The German and Austrian car industry alone employed 8.5–9 kton flax fibers in the years 2000 and 2001 [80]. The introduction of every new car model increases the demand—depending on the model—by 0.5–3 kton per year. In cars in which natural fibers are employed, presently 5 to 10 kilo natural fibers (flax, hemp, jute, etc.) per car are used. If the total European car industry would employ natural fibers, this would mean a market potential for about 80 to 160 kton per year of natural fibers for compression moulded parts in cars [81].

Many automotive components are produced in natural composites, mainly based on polyester or PP and fibers like flax, hemp or sisal. The adoption of natural fiber composites in this industry is led by motives of (a) price (b) weight reduction and (c) marketing ('processing renewable resources') rather than technical demands. The range of products is restricted to interior and non-structural components like door panels, seat backs, headliners, package trays, rear shelves, dashboards trunk liners and interior parts. Today nearly 50 % of vehicle internals are made of polymeric materials; in the developed countries, the average use of plastic in a vehicle is 120 kg, and the global average is around 105 kg which accounts for 10–12 % of the total vehicle weight [82].

Figure 9 presented one of the sample of natural fibers composites application for automotive needs in Mercedes-Benz.

Research on development of partially eco-friendly injection molded hybrid long fiber reinforced thermoplastics dedicated to automotive bumper was conducted by Jeyanthi et al. [84]. The composite with kenaf fiber was produced by impregnation process to improve the desired mechanical and thermal properties and recycling of the automotive components. The authors found that kenaf long fiber reinforced composites could be utilized in automotive structural components such as bumper beams, front end modules and also in interiors of automobiles. Moreover, recycling properties of kenaf composite are immense compared to its mechanical and thermal properties. The study proved that kenaf fiber composites can undeniably replace the commercial glass fibers based material for automotive components.

Fig. 9 Flax, hemp, sisal, wool, and other natural fibers are used to make 50 Mercedes Benz E-Class components [83]

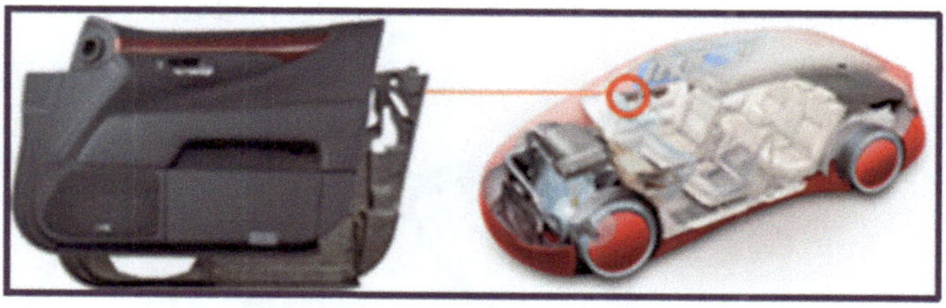

Fig. 10 Door trim using material made from kenaf and PLA [85]

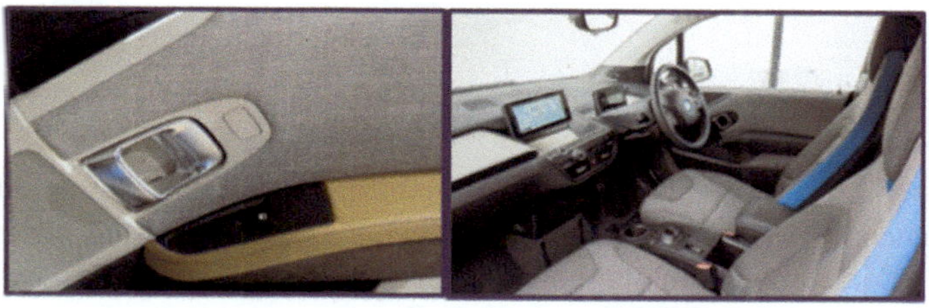

Fig. 11 Elements of BMW i3 made of composites reinforced with kenaf [86]

Kenaf fiber low density, high strength and insulation characteristics help reduce car weight while maintaining the same level of durability as conventional products as well as a high level of heat resistance and shock resistance. The advantages decided that composites reinforced with kenaf fibers have been utilized by Toyota for door trim, spare tyre cover, deck board, Fig. 10 and by BMW for black surrounds, door panels—interior innovations, Fig. 11. Kenaf based composites are used as well by Ford, Fig. 12.

Apart from wide application of vegetable fibers based composite for automotive purposes, it is worth to know about other interesting direction of natural fibers reinforced materials Fig. 13.

Museeuw Bikes, a Flemish company who designs and produces frames for race bicycles, has implemented a new material for the bike frame. Instead of using the

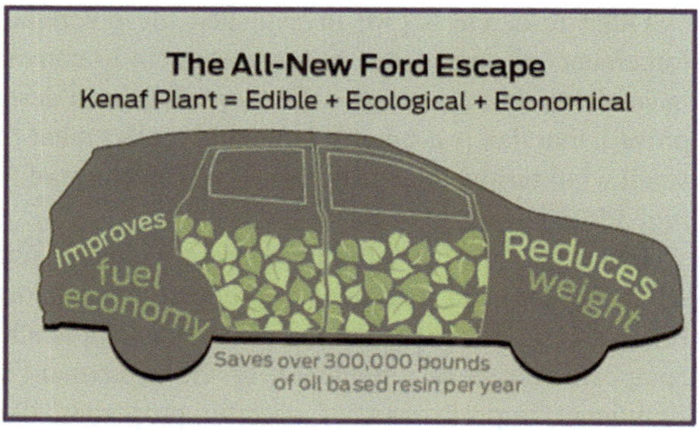

Fig. 12 Ford Escape with elements made of composite reinforced with kenaf [87]

Fig. 13 Bicycle with frame made of flax-carbon composite [89]

classic carbon fiber, a combination of flax and carbon fiber is used as reinforcement material in composites. Two bicycle types are built from carbon-flax composite: the MF1 has a flax-carbon ratio of 50-50, this means that the quantity flax and carbon is equally distributed and IMF5 frame, 80 % flax is present in the frame [91].

Flax is chosen over other natural fibers because flax has the best mechanical strength properties. It has lower density in comparison to carbon, good vibration absorbing properties due to the links between cellulose and hemicellulose chains—low-energy bonds that are known as the Van der Waals force. During vibration, these contact bonds break easily, allowing the molecules to move slightly. Thanks to this potential for movement, the fiber is able to absorb the vibrations.

There are many similarities between flax and carbon fibers (such as a coefficient of expansion that is practically zero and an elongation at break of about 1:5 %), so it is possible to make high-performance hybrid composite parts that are reinforced with both flax and carbon. The resulting bicycle has definite advantages: the carbon gives stiffness and the flax absorbs the vibration from flaws in the road. There is also the environmental aspect, since these fibers are renewable and non-polluting, and have a very small carbon footprint. The bicycle flax frames absorb micro-shocks up to 20 % better than a full carbon bike. Carbon flax and absorption of micro-shocks carbon fiber composite materials have major advantages for building high quality racing frames [89].

Other researchers [36] investigated the potential of sustainable flax fiber reinforcements of composites as a replacement to conventional E-glass reinforcements in small wind turbine blades. The topic has been described in 5.2 section. The study proved, that flax is a suitable structural replacement to E-glass for similar composite small wind turbine blade applications. The photo of flax based blade is presented on Fig. 14.

Investigation of possibility to use composites reinforced with vegetable fibers: sisal, banana and Roselle to produce helmet outer shell was conducted by Prasannasrinivas and Chandramohan [90] Application of natural fibers reinforced composites leads to a reduction of the component's weight and furthermore to a significant improvement of specific properties like impact strength and crash

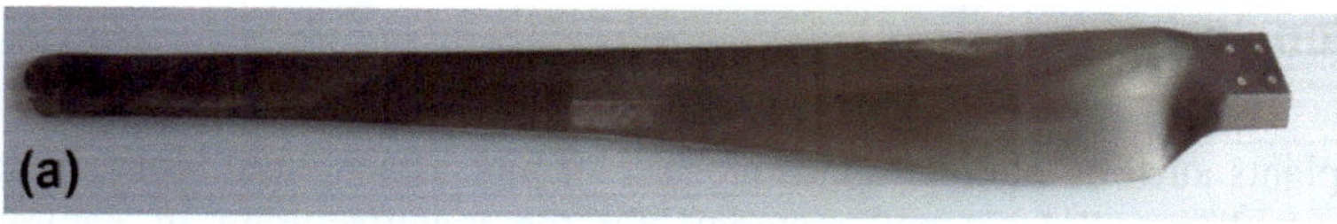

Fig. 14 Images of the (*a*) flax/polyester blades [94]

behavior. The authors concluded that hybrid combination of natural fiber composite material can be a replacement for plastic in helmet outer shell manufacturing.

Designed and built by Antithesis Design Works, this first generation kayak and paddle offers the weight savings of composites with the toughness of traditional plastic boats. Half the weight of a standard whitewater boat, two thirds of the composite reinforcement is hemp [91]. Designed and built 100 % sustainable with hemp composite blades and bamboo shaft is presented on Fig. 15.

A chair from composite reinforced with hemp Fig. 16 has been designed for a lightweight manufacturing process stemming from the car industry: the renewable raw materials hemp and kenaf are compressed with a water-based thermoset binder to form an eco-friendly, lightweight and yet strong composite [92].

Fig. 15 Hemp composite blades [91]

Fig. 16 "Hemp chair" D [92]

7 Conclusions

Fiber plants such as flax, hemp, ramie, kenaf, jute, abaca, sisal, coir are seen as promising lignocellulosic raw materials for the manufacturing of natural fiber reinforced composites for different applications.

Vegetable fiber usability to composite application and their commercial utilization is strongly related to advantages and disadvantages of natural fibers:

Advantages of natural fibers [93]:

- Low specific weight, which results in higher specific strength and stiffness than glass.
- This is a benefit especially in parts designed for bending stiffness.
- It is a renewable resource, the production requires little energy, CO_2 is used while oxygen is given back to the environment.
- Producible with low investment at low cost, which makes the material an interesting product for low-wage countries.
- Friendly processing, no wear of tooling, no skin irritation
- Thermal recycling is possible, where glass causes problems in combustion furnaces.
- Good thermal and acoustic insulating properties

Disadvantages of natural fibers:

- Lower strength properties, particularly its impact strength
- Variable quality, depending on unpredictable influences such as weather.
- Moisture absorption, which causes swelling of the fibers
- Restricted maximum processing temperature.
- Lower durability, fiber treatments can improve this considerably.
- Poor fire resistance
- Price can fluctuate owing to harvest results or agricultural politics

Their great advantage is biodegradability and the fact that when combined with polymers or natural resins they are as strong as steel yet of lower density. Such composites may be used for vehicles, building elements, furniture, machine constructions, insulating materials, gardening and agriculture equipment, tropical housing and even grape holding structures.

Recycling of natural fiber-reinforced composites is relatively easy and convenient. This fact makes one of the most important factors in forecasting the future growth of production and consumption of these materials.

References

1. http://www.naturalfibers2009.org
2. Zimniewska M, Władyka-Przybylak M, Mańkowski J (2011) Cellulosic bast fibers, their structure and properties suitable for composite applications, chapter of book: Cellulose fibers, bio-, and nano- polymer composites. Springer, Germany, pp 97–119

3. Pickering LK (2008) Properties and performance of natural-fiber composites. Woodhead Publ Ltd 2008:3–66
4. Townsend T World production of natural and manmade fibers. http://dnfi.org/
5. CELC & CIPALIN, based on CELC data www.mastersoflinen.com
6. Wambua P, Ivens J, Verpoest I (2003) Natural fibers: can they replace glass in fiber reinforced plastics? Compos Sci Technol 63:1259–1264
7. Zimniewska M, Kicinska-Jakubowska A Fact sheet—plant fibers. www.dnfi.org
8. Mohantya AK, Misraa M, Hinrichsen G (2000) Biofibers, biodegradable polymers and biocomposites: an overview, macromolecules and material engineering, 2000, 276/277, pp 1–24
9. Kołodziej J, Mańkowski J, Kubacki A (2007) Właściwości energetyczne odpadów z przerobu lnu i konopi w porównaniu z innymi surowcami roślinnymi. Biuletyn Informacyjny PILiK Len i Konopie nr 6 Poznan: 35–42
10. Plackett D, Sodergard A (2005) In: Mohanty AK, Misra M, Drzal LT, Selke SE, Harte BR, Hinrichsen G (eds) Natural fibers, biopolymers and biocomposites. CRC Press, Boca Raton, p 569
11. Sahari J, Sapuan SM (2011) Natural fiber reinforced biodegradable polymer composites. Rev Adv Mater Sci 30:166–174
12. Coroller G, Lefeuvre A, Le Duigou A, Bourmaud A, Ausias G, Gaudry T, Baley C (2013) Effect of flax fibers individualisation on tensile failure of flax/epoxy unidirectional composite, 2013 Elsevier Ltd. Compos A Appl Sci Manuf 51:62–70
13. Paridah MT, Basher AB, SaifulAzry S, Ahmed Z (2011) Retting process of some bast plant fibers and its effect on fiber quality: a review. BioResources 6(4):5260–5281
14. Czerniak L, Kirkowski R, Kozlowski R, Zimniewska M (1998) The earliest traces of flax textiles in central Europe, Kujawy, Poland, Natural Fibers, Special edn, 1998/1:18–19
15. Akin DE (2013) Linen most useful: perspectives on structure, chemistry, and enzymes for retting flax, vol 2013. Hindawi Publishing Corporation, ISRN Biotechnology. http://dx.doi.org/10.5402/2013/186534
16. Evans JD, Akin DE, Foulk JA (2002) Flax-retting by polygalacturonase-containing enzyme mixtures and effects on fiber properties. J Biotechnol 97:223–231
17. Konczewicz W (2015) Physical phenomena occurring in the process of physical-mechanical degumming of fiber from flax straw. Text Res J 85(4):380–390
18. Akin DE, Foulk JA, Dodd RB, McAlister III DD (2001) Enzyme-retting of flax and characterization of processed fibers. J Biotechnol 89:193–203
19. Mooneya C, Stolle-Smits T, Schols H, de Jong E (2001) Analysis of retted and non-retted flax fibers by chemical and enzymatic means. J Biotechnol 89:205–216
20. Zimniewska M, Frydrych I, Mankowski J, Trywianska W (2013) Process control in natural fiber production. In: Process control in textile manufacturing, woodhead publishing series in textiles, chapter 2: process control in fiber production and yarn manufacture, no 131, pp 81–108
21. Sponner J, Toth L, Cziger S, Franck RR (2005) Hemp. In: Franck RR (ed) Bast and other plants fibers. Woodhead Publishing in Textiles
22. Krishnan KB, Doraiswamy I, Chellamani KP (2005) Jute, bast and other plant fibers. In: Franck RR (ed) Woodhead Publishing in Textiles, pp 24–92
23. Hongqin Yu, Chongwen Yu (2010) Influence of various retting methods on properties of kenaf fiber. J Text Inst 101(5):452–456
24. Akubueze EU, Ezeanyanaso CS, Orekoya EO, Akinboade DA, Oni F, Muniru SO, Igwe CC (2014) Kenaf fiber (Hibiscus cannabinus L.): a viable alternative to jute fiber (Corchorus genus) for agro-sack production in Nigeria. World J Agric Sci 10(6):308–331
25. Yu C (2005) Sizal, bast and other plants fibers. In: Franck RR (ed) Woodhead Publishing in Textiles, pp 228–273
26. Naik RK, Dash RC, Pradhan SC (2013) Sisal fiber extraction: methods and machine development agricultural engineering today, vol 37, Issue 4, pp 27–30

27. Vijayalakshmi K, Neeraja CYK, Kavitha A, Hayavadana J (2014) Abaca fiber, transactions on engineering and sciences, vol. 2, Issue 9, pp 16–19
28. Franck RR (2005) Bast and other plants fibers. In: Franck RR (ed) Woodhead Publishing in Textiles, 315–320
29. Mathai PM (2005) Coir, bast and other plants fibers. In: Franck RR (ed) Woodhead Publishing in Textiles, pp 274–312
30. FAO: Rolf W (2000) Boehnke, improvement in drying, softening, bleaching, dyeing coir fiber/yarn and in printing coir floor coverings. http://www.fao.org
31. Umayorubhagan V, Albert GMI, Ray CIS (1995) Physico-chemical analysis of the water of Pottakulam Lake at Thengapattanam in Kanyakumari district (Tamil Nadu). Asian J Chem Rev 6:7–12
32. Nandan SB, Abdul Azis PK (1995) Benthic polychaetes in anoxic sulfide biomes of the retting zones in the Kadinamkulam Kayal. Int J Environ Stud 47:257–267
33. Abbasi SA, Nipaney PC (1993) Environmental impact of retting of coconut husk and directions for the development of alternative retting technology. Pollut Res 12:117–118
34. Ravindranath AD, US Sarma (1995) Bioinoculants for coir retting, CORD 11:34–38
35. Zimniewska M, Myalski J, Koziol M, Mankowski J, Bogacz E (2012) Natural fibers textile structures suitable for composite materials. J Nat Fibers 9(4):229–239
36. Shah DU, Schubel PJ, Clifford MJ (2013) Can flax replace E-glass in structural composites? A small wind turbine blade case study. Compos Part B 52:172–181
37. Zimniewska M, Bogacz E (2009) Preliminary study on flax yarn suitable for composite application. In: Monograph: natural fibers—their attractiveness in multi-directional applications, edited by Gdynia Cotton Association, Gdynia, pp 178–183
38. Shah DU, Schubel PJ, Clifford MJ, Licence P, Warrior NA (2011) Yarn optimisation and plant fiber surface treatment using hydroxyethylcellulose for the development of structural bio-based composites. In: 18th international conference on composite materials, 2011, Jeju Island, Korea
39. Goutianos S, Peijs T, Nystrom B, Skrifvars M (2006) Development of flax fiber based textile reinforcements for composite applications. Appl Compos Mater 13(4):199–215
40. Zimniewska M, Stevenson A, Sapieja A, Kicińska-Jakubowska A (2014) Linen fibers based reinforcements for laminated composites. Fibers Text East Eur 22 3(105):103–108
41. Krucińska I, Klata E, Ankudowicz W, Dopierała H (2001) Influence of the structure of hybrid yarns on the mechanical proper-ties of thermoplastic composites. Fibers Text East Eur 9 (2):38–41
42. Klata E, Borysiak S, Van de Velde K, Garbarczyk J, Krucińska I (2004) Crystallinity of polyamide-6 matrix in glass fiber/polyamide-6 composites manufactured from hybrid yarns. Fibers Text East Eur 12(3):64–69
43. Krucińska I, Gliścińska E, Mäder E, Häßler R (2009) Evaluation of the influence of glass fiber distribution in polyamide matrix during the consolidation process on the mechanical properties of GF/PA6 composites. Fibers Text East Eur 17(1):81–86
44. Salman SD, Sharba MJ, Leman Z, Sultan MTH, Ishak MR, Cardona F (2015) Physical, mechanical, and morphological properties of woven kenaf/polymer composites produced using a vacuum infusion technique. Int J Polym Sci Article ID 894565. http://www.hindawi.com/journals/ijps/aa/894565/
45. Kozłowski R, Władyka-Przybylak M, Helwig M, Kurzydłowski K (2004) Composites based on lignocellulosic raw materials. Mol Cryst Liq Cryst 415–418:301–321. ISSN 0888-5885
46. Faruk O, Bledzki A, Fink H, Sain H (2014) Progress Report on Natural Fiber Reinforced Composites. Macromol Mater Eng 299:9–26
47. Young RA (1996) Utilization of natural fibers: characterization, modification and applications. In: Lea AL et al (eds) Lignocellulosic-plastics composites, UNESP, Sao Paulo, Brazil
48. www.strongwell.com
49. Iorio I, Leone C, Nele L, Tagliaferri V (1997) Plasma treatments of polymeric materials and Al alloy for adhesive bonding. J Mater Process Technol 68:179–183

50. Petash W, Räuchle E, Walker M, Elsner P (1995) Improvement of the adhesion of low energy polymers by a short time plasma treatment. Surf Coat Technol 74–75:682–688
51. Tu X, Young RA, Denes F (1994) Improvement of bonding between cellulose and polypropylene by plasma treatment. Cellulose 1:87–106
52. Bledzki A, Gassan J, Lucka M (2000) (in Polish) Renesans tworzyw sztucznych wzmocnionych wloknami naturalnymi (Natural Fiber—Reinforced Polymers Come Back). Polimery 45(2):98–108
53. Paukszta D (2000) The structure of modified natural fibers used for the preparation the composites with polypropylene. SPIE Int Soc Opt Eng 4240:38–41
54. Yanai Y Non-Resin Shrink-Proof Process, Celtopia. Nisshinbo Industries Inc. Miai Plant, Aichi, Japan (unpublished)
55. Koppers' Acetylated Wood (1961) New Materials Technical Information No. (RDW-400), E-106
56. Otlesnov Y, Nikitina N (1977) Latvijas Lauksaimniecibas Akademijas Raksti 130:50
57. Sheen AD (1992) The preparation of acetylated wood fiber on a commercial scale. Pacific rim bio-based composites symposium; chemical modification of lignocellulosics. FRI Bulletin 176:1–8
58. Beckers EPJ, Militz H (1994) Acetylation of solid wood. In: Second pacific rim bio-based composites symposium, Vancouver Canada, pp 125–134 (1994)
59. Militz H, Beckers EPJ, Homan WJ (1997) Int. Res. Group Wood Pres., Doc. No. IRG/WP 97-40098
60. Rowell RM Chemical Modification of Natural Fibers to Improve Performance. This proceedings
61. Abdalla A, Pickering K (2002) The use of silane as a coupling agent for wood fiber composites. In: Proceedings of the 3rd Asian-Australasian conference on composite materials (ACCM-3), Auckland, New Zealand
62. Abdalla A, Pickering K, MacDonald AG (2002) Mechanical propierties of thermoplastic matrix composites with silane-treated wood fiber. In: Proceedings of the 6rd international conference on flow processes in composite materials. Auckland, New Zealand
63. Bledzki AK, Gassan J (1997) Natural fiber reinforced plastics. In: Cheremisinoff NP (ed) Handbook of engineering polymeric materials. Marcel Dekker, Inc., New York
64. Kroschwitz JI (1990) Polymers: fibers and textiles. Wiley, New York
65. Mohanty AK, Patmaik S, Singh BC (1989) J Appl Polym Sci 37:1171
66. Escamila G, Trugillo GR, Franco PJH, Mendizabal E, Puig JE (1970) J Polym Sci 66:339
67. Maldas D, Kokta BV, Daneaulf C (1989) J Appl Polym Sci 37:751
68. Han GS, Saka S, Shiraisi N (1991) Composites of wood and polypropylene. Morphological study of composites by TEM-EDXA, Mokuzai Gakkaishi 3:241
69. Kandola BK (2012) Chapter 5: flame retardant characteristics of natural fiber composites. In: Natural polymers: volume 1: composites: eds Maya J John, Sabu Thomas, the royal society of chemistry, vol 1, pp 86–117. ISBN: 978-1-84973-402-8
70. Chapple S, Anandjiwala R (2010) Flammability of natural fiber-reinforced composites and strategies for fire retardancy: a review. J Thermoplast Compos Mater 23:871–893
71. Mngomezulu ME et al (2014) Carbohydrate poymers. Rev Flammabl biofibers biocompos 111:149–182
72. Sreekala MS, Kumaran MG, Thomas S (2000) Effect of chemical modificatons on the mechanical performance of oil palm fiber reinforced Phenol Foraldehyde Composites. In: Capparelli Mattoso LH et al (eds) Natural polymers and composites. Embrapa Instrumentacao Agropecuaria, Sao Carlos
73. Nabi Saheb D, Jog JP (1999) Natural fiber polymer composites: a review. Adv Polym Technol 18(4):351–363
74. Bledzki AK, Gassan J (1999) Composites reinforced with cellulose based fibers. Prog Polym Sci 24:221–274
75. Taj S, Munawar MA, Khan S (2007) Natural fiber reinforced polymer composites. Rev Proc Pak Acad Sci 44(2):129–144

76. Monteiro SN, Calado V, Margem FM, Rodriguez RJS 'Thermogravimetric stability behavior of less common lignocellulosic fibers—a review'. Materials Science Department, Military Institute of Engineering (IME), Rio de Janeiro, Brazil
77. Pereira PHF, Rosa MF, Cioffi MOH, Benini KCCC, Milanese AC, Voorwald HJC, Mulinari DR (2015) Vegetal fibers in polymeric composites: a review. Polímeros 25(1):9–22
78. Czaplicka-Kolarz K (2008) Foresight technologiczny materiałów polimerowych w Polsce—analiza stanu zagadnienia. Poznan
79. Kozłowski R (1997) The potential of natural fibers in Europe. In: Industrial applications lignocellulosic—plastics composites, Sao Paulo, Brasil
80. Kiziltas A, Gardner DJ Utilization of carpet waste as a matrix in natural filler filled engineering thermoplastic composites for automotive applications. Advanced Engineered Wood Composite (AEWC) Center, University of Maine, Orono, USA
81. Bos HL (2004) The potential of flax fibers as reinforcement for composite materials. Technische Universiteit Eindhoven, Eindhoven
82. Ozen E, Kiziltas A, Kiziltas EE, Gardner DJ Natural fiber blends—filled engineering thermoplastic composites for the automobile industry. Advanced Engineered Wood Composite (AEWC) Center, University of Maine, Orono, ME 04469, USA
83. Sue Elliott-Sink (2005) Special report: cars made of plants. www.edmunds.com/advice/fueleconomy/articles/105341/article.html. Accessed 12 April 2005 (downloaded 28 August 2006)
84. Jeyanthi S, Janci Rani J (2012) Influence of natural long fiber in mechanical, thermal and recycling properties of thermoplastic composites in automotive components. Int J Phys Sci 7 (43):5765–5771
85. http://www.toyota-boshoku.com/
86. http://www.bmw.com
87. http://www.at.ford.com/news/
88. Vanwalleghem J (2010) Study of the damping and vibration behavior of flax-carbon composite bicycle racing frames. Master in de ingenieurswetenschappen: Werktuigkunde-Elektrotechniek, Faculteit Ingenieurswetenschappen, Universiteit Gent
89. http://www.museeuw-bicycles.com/technology/
90. Prasannasrinivas R, Chandramohan D (2012) Analysis of natural fiber reinforced composite material for the helmet outershell—a review. Int J Curr Res 4(03):137–141
91. http://www.envirotextile.com/pl/natural-fiber-composites
92. http://thinkgreen.typepad.com/blog/2011/04/chair-natural-fiber.html
93. Brouwer WD (Rik), Natural fiber composites in structural components: alternative applications for sisal? http://www.fao.org/docrep/004/y1873e/y1873e0a.htm#fn30
94. www.flaxcomposites.com

Metallic Fibers for Composite Applications

K. Shabaridharan and Amitava Bhattacharyya

Abstract Chapter 6 talks about the different types of metallic fibers used in composites materials. Properties of metallic fibers, fabrication of composites, and the properties and application of these composites are discussed in detail.

1 Introduction

Composites are grossly defined as a material consists of at least two distinctly distinguishable phases or constituents. The continuous phase is termed as matrix while the discontinuous phase is often denoted as reinforcements. The reinforcements, usually stronger and stiffer, are distributed and dispersed in a comparatively less strong and stiff matrix material. The reinforcements share the load and in some cases, especially when a composite consists of fiber reinforcements dispersed in a weak matrix (e.g., carbon/epoxy composite), the fibers carry almost all the load. The strength and stiffness of such composites are, therefore, controlled by the strength and stiffness of constituent fibers. Hence, the physical properties of matrix materials are improved by introduction of reinforcing materials, while the reinforcements are protected (from chemical and physical attacks) by the matrix phase in a composite structure. The matrix also shares the load when there is not much difference between the strength and stiffness properties of reinforcements and matrices (e.g., SiC/Titanium composite). However, the primary task of a matrix is to act as a medium for transferring the load to reinforcements. Being a continuous phase, it distributes and transfers the external forces uniformly to all the reinforcing materials preventing the localized stress build up on the material. It also holds the reinforcements together [18].

K. Shabaridharan
Department of Textile Technology, Bannari Amman Institute of Technology, Sathyamangalam, Erode 638401, India

A. Bhattacharyya (✉)
PSG Institute of Advanced Studies, Coimbatore 641004, India
e-mail: amitbha1912@gmail.com; abh@psgias.ac.in

© Springer Science+Business Media Singapore 2016

S. Rana and R. Fangueiro (eds.), *Fibrous and Textile Materials for Composite Applications*, Textile Science and Clothing Technology,
DOI 10.1007/978-981-10-0234-2_6

Besides, the matrix usually influences the functional properties such as hygral, thermal, electrical, magnetic, etc. of a composite. For example, an aluminum matrix is more preferred to obtain a good thermally conducting composite with SiC fibers than a titanium matrix as both SiC fibers and titanium matrix possess poor thermal transmission property.

The classifications of composites are commonly based on either the forms of reinforcements or the matrices used. There are two broad classes of composites; fiber reinforced composites and particle reinforced composites (or particulate composites) according to reinforcement type. Based on matrices, there are three groups such as polymers, metals (and their alloys) and ceramics. Composites are also grouped in several other ways. One important class of composites is laminated composites or laminates which consists of two or more layers of planar composites in which each layer (also called lamina or ply) may be of the same or different materials. Laminated and sandwich laminate composite structures are very strong and stiff, and are commonly used for lightweight structural applications. Most of the reinforcements used for making structural composites are fibers. Short fibers are discontinuous fibers and may also be treated as particles with cylindrical shapes. Flakes or platelets are also commonly used. They are less expensive than short fibers, and can be aligned to obtain improved in plane directional properties compared to those of short fibers. Metal flakes (say, aluminum) are used to improve the thermal and electrical conductivity of the composite, whereas ceramic flakes can be added to the matrix to increase the resistivity. Aluminum alloys reinforced with silicon carbide particles are found to exhibit higher strength and stiffness. The aluminum alloy AA2124-T6 matrix reinforced with the silicon carbide particles (volume content by 40 %) has shown improved tensile strength (690 MPa) and Young`s modulus (150 GPa). Ceramic grains of borides, carbides, oxides, nitrides, silicides, etc. can produce a class of particulate composites known as cermets, having low density with good thermo-mechanical properties when properly dispersed in metal matrix [25].

The major advantage of composites over mono phase materials is the weight reduction. A composite structure with reduced payload can deliver better performance than a pure or alloy structure. Among all classes of composites, fiber reinforced composites are the most important in terms of tensile properties enhancements at commercial scale. The strength of a continuous or staple fiber is utilized in most of the commercial composite materials. For this, the reinforcing fibers should have very good tensile properties with as light weight as possible. Most of the natural and synthetic fibers were researched as reinforcements for their different application potentials. Glass fibers are the earliest known fibers used to reinforce materials. Ceramic and metal fibers were subsequently found out and put to extensive use, to render composites stiffer and more resistant to heat [33].

2 Types of Metallic Fibers

Metallic fibers are manufactured fibers composed of metal, plastic-coated metal, metal-coated plastic, or a core completely covered by metal. Metallic fibers are an attractive class of fibers for composite applications considering their exceptional mechanical properties; however, the specific strength or strength to weight ratio (strength divided by density) of metallic fibers is inferior to high performance carbon, polymer or ceramic fibers due to their higher density. They also have disadvantages like poor corrosion resistance and inferior bonding strength with matrix. But, the surface of metallic fibers can be coated with ceramics to address these problems. Till today, they are the popular choice for applications like infrastructure building, electromagnetic interference shielding (EMI shielding), etc. [1].

A number of metals are available in various fibrous forms. However, a number of practical factors limit the number of fiber types for composite applications. The most important influencing factors are availability, abundance, processing difficulties and cost. Cost of metal fibers is determined by a number of factors such as fiber type, abundance, production procedure, diameter, form and so on. The use of the metal fiber is driven by the application, which in turn is also affected by the cost of the materials. Most metal fibers are extremely fine wire produced by bundle drawing process. The process involves stepwise reduction of wire diameter using a pair of hot rollers and subsequent annealing to stabilize the structure (Fig. 1).

A drawn wire of 50 microns diameter or less is defined as a fiber. 300 and 400 series stainless steel, nichrome, inconel, hastelloy x, carpenter 20cb3, nickel, 80/20 nickel chromium, titanium and tantalum are widely used to produce metal fibers. The typical fiber diameter ranges from 2 to 50 microns. They are available in the form of chopped fiber, sliver, continuous tow, air laid web, continuous filament yarn and broken fiber (Fig. 2).

Stainless steel fibers are the most potential metal fibers in terms of availability, application and material properties. Chromium present in steel produces a self-healing oxide film on these metals which results in high corrosion resistance. Further improvement is achieved by adding noble metals such as nickel and molybdenum. A wide range of mechanical properties obtainable from these alloys, along with their

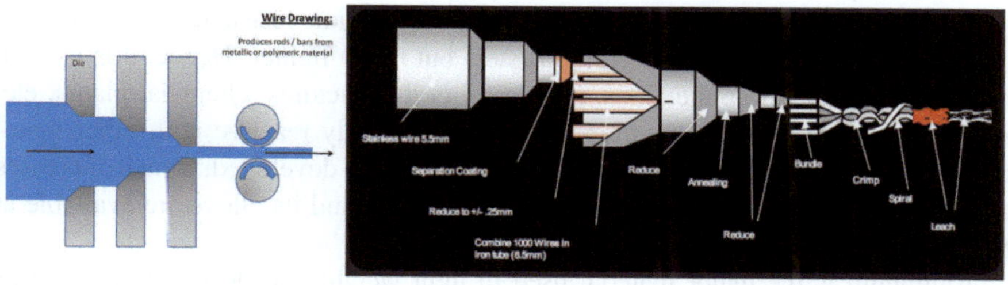

Fig. 1 Schematic representation of bundle drawing process (Ref. http://aerotechgolfshafts.com/steelfiber/)

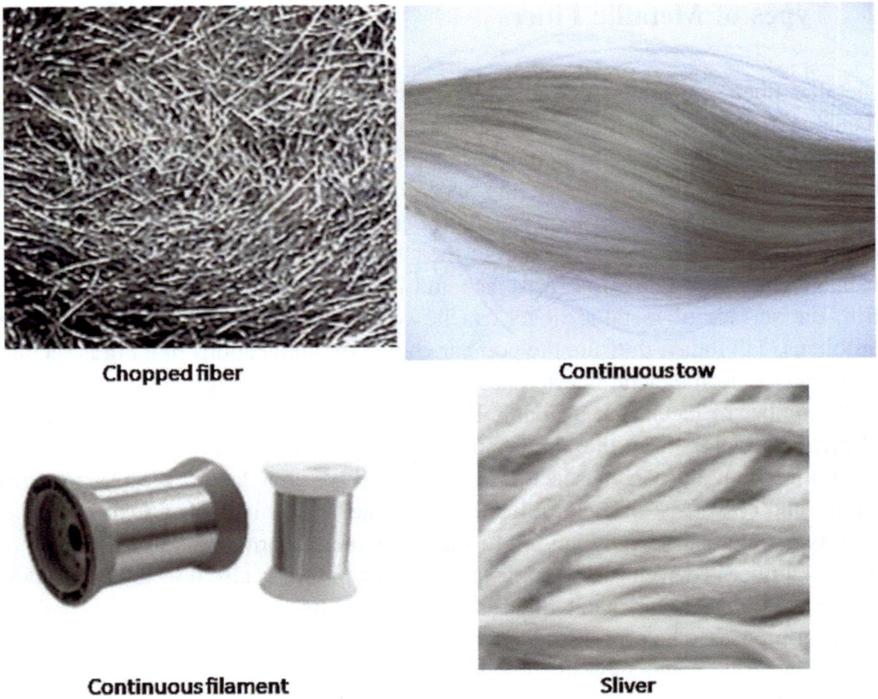

Fig. 2 Different forms of metal fibers [1]

corrosion resistance, and in some cases, resistance to oxidation at elevated temperatures, makes them extremely versatile in their application [7]. Among four classes of stainless steel (austenitic, martensitic, ferritic and precipitation hardened), austenitic steels (AISI300 series) is available in the form of fine wire and fiber for its ease of production and use. Recently, bundle drawn annealed stainless steel fibers with diameter 5–100 μm are available for applications in structural composites and they possess both high stiffness and high strain-to-failure properties. The stiffness is comparable to carbon fiber (193 GPa) while the strain-to-failure is similar to a silk fiber (up to 20 %). Again, the ductility can be tailored by modifying the heat treatment [8] (Fig. 3).

Titanium possesses a remarkable set of properties as compared to other metals. It has combined properties of high strength and corrosion resistance, high melting point and low density. It is as strong as steel but much lighter. Hence, it is used in number of structural applications, such as aircraft structures, chemical plants, etc. At elevated temperatures, titanium becomes chemically reactive with other materials requiring special manufacturing processes to be developed, which increases the cost of the metal [9]. Only fine wires of titanium and its alloys are available at present.

Aluminum is the major material used in light weight vehicle structures such as aircrafts. Aluminum alloys are used in most of the aerospace components. They are of low density, and possess high electrical and thermal conductivities and good resistance

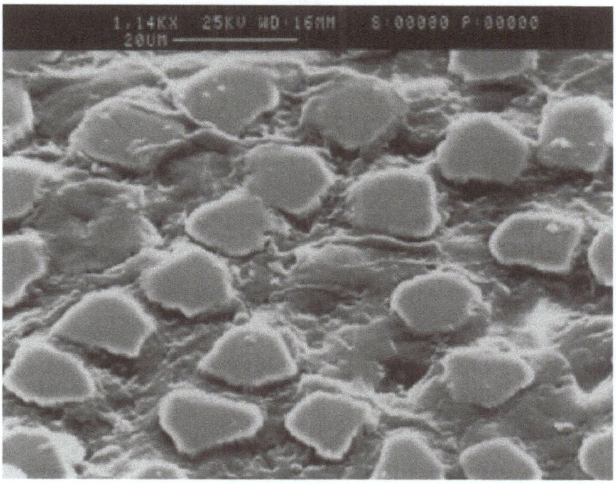

Fig. 3 SEM image of impregnated fiber mat in glass matrix composite. Ref. [7]

to chemical corrosion. Aluminum is also recyclable and properties do not degrade with the recycling process. Aluminum fiber composites showed improved mechanical, thermal and electrical properties than glass fiber reinforced composites [5].

Among the other fibers used, copper has been used in a number of applications as it is tough, ductile, malleable and has exceptionally high electrical and thermal conductivities. Although generally resistant to corrosive environments, copper is nevertheless susceptible to oxidizing agents. Nickel is mostly used in corrosive resistant alloys such as stainless steels, Inconel, Hastelloys, and Nichrome. Pure nickel and its alloys are available as bundle-drawn fibers. Pure tantalum is a heavy, dense, malleable and ductile metal. Additionally, it has exceptional resistance to corrosion and acid attack and these properties allow the metal to be used as a substitute for platinum in laboratory equipments in the chemical industry and for surgical implants. Finally, tantalum is used as an alloying element to produce alloys of good corrosion resistance, strength and ductility. Tungsten has the highest elastic modulus and melting point among all pure metals. Tungsten fiber has been used as reinforcements for superconductor filaments due to its tolerance to large tensile stresses. Much research on these fibers in composites has been focused on their use in metal matrices for improvement in toughness [32].

As reinforcement, metal fibers have many advantages. They are easily produced using several fabrication processes and are more ductile, apart from being not too sensitive to surface damage and possess high strengths and temperature resistance. However, their weight and the tendency to react each other through alloying mechanisms are major disadvantages. Ceramic fibers improve vastly in performance when a fine metal outline is incorporated with refractory ceramics by improving their thermal shock and impact resistance properties. Metal wires can also be used to reinforce polymers or plastics. Such combinations ensure high strength, light weight and good fatigue resistance. Continuous metal fibers can be easily handled. Better flexural properties are observed in some metal fiber reinforced plastic composites which also offer improved strength and weight than glass fibers. However, their poor tolerance to high temperature and the resultant steep variations in thermal expansion coefficient limit their application. Oxide fibers have high temperature tolerance, but they lack in ductility [34]. Hence, metal fibers are often coated with ceramics (oxides and carbides) to overcome their limitations as reinforcements.

Boron-tungsten fibers are obtained by allowing hot tungsten filament to pass through a mixture of gases. Boron is deposited on tungsten and the process continues until the required thickness is achieved. Properties of boron coated tungsten fibers change with the diameter, as the ratio of boron to tungsten and the surface defects change according to size. They are known for their remarkable stiffness and strength (modulus almost four to five times than glass). Boron coated carbons are much cheaper to make than boron tungsten fiber. But it has low modulus of elasticity.

Silicon carbide can be coated over a few metals and their tensile strength and modulus are better than boron–tungsten, especially at higher temperature (only 35 % loss of strength at 1350 °C). The advantages of silicon carbide–tungsten are more than boron tungsten fibers. Silicon carbide-tungsten fibers are dense compared to boron–tungsten fibers of the same diameters. They are prone to surface damage and careful, delicate handling is essential during fabrication of the composite. It is difficult to process in high temperature matrix as tungsten and silicon carbide bonds become weak above 930 °C. Silicon carbide on carbon substrates is lighter than silicon carbide tungsten and is easier to process.

Aluminum oxide (Alumina) fibers are considered a potential reinforcement for metal and polymer matrix composites. They offer good compressive strength though the tensile strength is not very good. They have high melting point of about 2000 °C and the composite can be successfully used at temperature up to about 1000 °C without loss of strength and stiffness properties. The commercial grade alumina fiber developed by Du Pont is known as alumina FP (polycrystalline alumina) fiber. They exhibit high compressive strengths, when they are introduced in a matrix. Typical longitudinal compressive strengths of alumina FP/epoxy composites vary from 2.27 to 2.41 GPa. Alumina whiskers exhibit tensile strength of 20.7 GPa and tensile modulus of 427 GPa. Magnesium and aluminum matrices are frequently used in alumina fiber reinforced composites as they do not damage the fiber even in the liquid state.

3 Properties of Metallic Fibers

Though fibers can be amorphous (glass), polycrystalline (carbon, boron, alumina, etc.) or single crystals (silicon carbide, alumina, beryllium and other whiskers), metallic fibers are mostly polycrystalline. In general, strength and stiffness properties of a fiber are significantly higher compared to the bulk material due to less crystal defects and higher orientation of crystallites along fiber length direction. The orientation of crystallites along the fiber direction also helps considerably in

Table 1 A summary of various fiber forming metal properties [1]

Metal	Density (g/cm^3)	Tensile strength at break (MPa)	Elongation (%)	Resistivity ($\times 10^{-6}$ Ω cm)	Coefficient of thermal expansion (μm/[m °C])
Aluminium (7000)	2.8	76	50	2.7	25.5
Copper	9.0	221–455	55	1.7	17.7
Nickel	8.9	317	30	6.4	13
Stainless steel:					
304	8.0	505–840	>40	72	17.8
316	8.0	460–860	>60	70–78	17.8
Tantalum	16.7	310	–	13	7
Titanium	4.5	241–552	15–25	60	9.2
Tungsten	19.3	530–1920	8	5.4	4.5

improving the strength properties. A whisker, being a single crystal, is not prone to crystal defects unlike polycrystalline fibers and provides very high strength and stiffness. Table 1 listed tensile, electrical and thermal expansion properties of different fiber forming metals. They show good tensile properties with reasonable ductility (except tantalum, titanium and tungsten) and exceptionally high electrical conductivity.

Due to high electrical conductivity, metal fibers impart significant electrical conductivity at very low fiber concentrations. Hence, the material consumption for successful EMI shielding is significantly lower than other fibers. For a 7 micron fiber, only 5–7 wt% is required for EMI shielding of 35–50 dB. Further, as metal fibers are used in low amount, they have minimum adverse effect on the mechanical properties. Other advantages of metal fibers are the similarity of their shrinkage to unfilled resins, excellent abrasion and corrosion resistance, minimal alteration of base resin properties (long life), and cost effectiveness [30].

Metal fiber dispersed composites appear to be light grey in color, hence molded articles cannot be colored with all colors. Mostly orange, beige and some other colors are possible. Because stainless steel fiber is chemically stable and non-oxidizing, fiber filled components do not age (increase resistivity with time), or lose shielding effectiveness due to abrasion, chipping or peeling. This is important in critical applications where shield degradation can create EMI leaks.

Most of the common fibers for structural applications are brittle in nature. Both inorganic and organic fibers are used in making structural composites. Inorganic fibers (including ceramic fibers) such as metal, glass, boron, carbon, silicon carbide, silica, alumina, etc. are most commonly used. The structural grade organic fibers are comparatively very few in number. Aramid and high strength polyethylene fiber (Spectra 900) are the most popular organic fibers. Organic fibers are cheaper, lighter and more flexible. Carbon fibers are often considered as ceramic fibers, though sometimes classified as organic fibers. Inorganic ceramic fibers in general are strong, stiff, thermally stable and insensitive to moisture. They exhibit good fatigue resistant properties, but low energy absorption characteristics. They possess high strength and better impact resistant properties. Metal fibers have minimal adverse effect on mechanical properties of composites as compared to other fillers which are abrasive when used at high loading levels. Metallic fibers are several times finer than a human hair. Most of these fibers are used to produce non-shielding items such as filter media, abradable seals; aircraft sound insulation, air bag filters and anti–static textiles; however, they are extensively used as composites for EMI shielding and structural applications [9]. Typical mechanical and thermal properties of common fibers are listed in Tables 2 and 3, respectively. From the tables it can be

Table 2 Typical mechanical properties of selected fibers [25]

Fiber material	Density (kg/m^3)	Tensile strength (MPa)	Tensile modulus (GPa)	Diameter (μm)
Glass	2550	3450–5000	69–84	7–14
Boron	2200–2700	2750–3600	400	50–200
Carbon	1500–2000	2000–5600	180–500	6–8
Kevlar	1390	2750–3000	80–130	10–12
Polyethelyne	970	2590	117	38
Silica (SiO_2)	2200	5800	72	35
Boron carbide (B_4C)	2350	2690	425	102
Boron nitride	1910	1380	90	6.9
Silicon carbide (SiC)	2800	4500	480	10–12
TiB_2	4480	105	510	–
TiC	4900	1540	450	–
Zirconium oxide	4840	2070	345	–
Borsic(SiC/B/W)	2770	2930	470	107–145
Alumina(Al_2O_3)	3150	2070	210	17
Alumina FP	3710	1380	345	15–25
Steel	7800	4140	210	127
Tungsten	19300	3170	390	361
Beryllium	1830	1300	240	127
Molybdenum	1020	660	320	127
Quartz whisker	2200	4135	76	9
Fe whisker	7800	13800	310	127
SiC whisker	3200	21000	840	0.5–10
Al_2O_3 whisker	4000	20700	427	0.5–10
BeO whisker	2851	13100	345	10–30
B_4C whisker	2519	13790	483	–
Si_3N_4 whisker	3183	13790	379	1–10
Graphite whisker	2100	20800	1000	–

Table 3 Typical thermal properties of selected fibers [25]

Fiber	Melting point (°C)	Heat capacity (kJ/[kg K])	Thermal conductivity (W/mK)	Coefficient of thermal expansion (10^{-6} m/mK)
Glass	840	0.71	13	5
Boron	2000	1.30	38	5
Carbon	3650	0.92	1003	−1.0
Kelvar 49	250	1.05	2.94	−4.0
SiC	2690	1.2	16	4.3
Steel	1575	0.5	29	13.3
Tungsten	3400	0.1	168	4.5
Beryllium	1280	1.9	150	11.5
Molybdenum	2620	0.3	145	4.9
Fe whisker	1540	0.5	29	13.3
Al_2O_3 whisker	2040	0.6	24	7.7
Quartz whisker	1650	0.963	10	0.54

observed that both tensile and thermal properties of metal fibers are very good; however, due to their high density the specific strength and modulus are lower as compared to high performance fibers like Kevlar and carbon.

4 Types of Matrix

The history of metal reinforced composites is long. Virtually all types of matrices have been tried with metal (rod, wire, fiber or cord form). As discussed in introduction, composite matrices are classified into three categories namely polymer, metal and ceramic. Considering the high thermal expansion coefficient of most of the metal fibers, thermoplastic polymer and metal matrix composites are most suitable for various applications. However, polymer matrix has the disadvantage of lower working temperature.

4.1 Polymers Matrix Composites

Polymers are the most popular matrix material for composites. Almost all reinforcements, inorganic and organic, can be used with polymers to produce a wide range of reinforced plastics or polymer composites. The major advantages of polymers are their densities are usually very low and they can be easily processable. The processing and curing temperature are in lower ranges which bring down the manufacturing cost substantially due to a low energy input. Both thermoplastics and thermosets are used to make reinforced plastics.

4.2 Thermoplastics

Thermoplast is a collection of high molecular weight linear or branched molecules. It softens upon heating at temperature above the glass transition temperature, but regains its strength upon cooling. The process of softening at higher temperature and regaining rigidity upon cooling is reversible. Although thermoset polymers are commonly used in structural composites due to their higher strength and stiffness properties, there is a growing interest in recent years to use thermoplastics. The advantages of thermoplastic polymers are numerous; they can be repeatedly molded, casted and reused several times. The repair of a damaged part also becomes simpler and the scrap rate is also less. All these make thermoplasts very much cost effective. Table 4 provides the typical thermo-mechanical properties of a couple of structural grade thermoplastic resins which are thermally stable at higher temperatures. They are strong, tough, stiff, less sensitive to moisture and exhibit high creep resistance.

PEEK and polyphenylene sulfide are crystalline polymers while other polymers mentioned in the table are amorphous. PEEK has a Tg of 143 °C and a melting point of 332 °C. The chemical resistance of PEEK is also good (soluble only in concentrated sulfuric acid). The processing temperature ranges from 300–400 °C. The moisture absorption limit is very low. The fracture toughness is comparatively higher. All these features of PEEK make it a highly attractive thermoplastic resin for application in reinforced composites. Graphite/PEEK composite prepregs are commercially available. Polysulfones reinforced with glass, aramid and carbon fibers have also found several applications.

Table 4 Typical properties of some high performance thermoplastics [28]

Properties	Polyether ether ketones (PEEK)	Polyamide-imide	Polyether-imide	Polysulfone	Polyphenylene sulfide
Density (kg/m^3)	1300	1400	1270	1240	1340
Tensile strength (MPa)	104	138	115	70	76
Tensile modulus (GPa)	4.21	4.48	3.38	2.48	3.31
Poisson's ratio	0.35	0.35	0.35	0.35	0.35
Coefficient of thermal expansion (10^{-6} m/mK)	–	56	50	86.40	88
Maximum service temperature (K)	630	–	490	490	565

4.3 Thermosets

Thermoset polymers are formed from relatively low molecular weight precursor molecules. The polymerization process in a thermoset resin is irreversible. Once cured, they do not soften upon heating. At high temperature, the covalent bonds may break leading to destruction of the network structure and the polymer decomposes. Thermosetting resins are listed in Table 5. Thermosetting resins vary widely with Tg values varying from 45 to 300 °C and elongations ranging from 1 % to more than 100 %.

The most commonly used thermosets are epoxy, polyester and phenolic resins, among which polyester resins are most widely used in various common engineering goods and composite applications. Epoxy resins are stronger and stiffer and constitute the major group of thermoset resins used in composite structures and adhesives. Phenolic resins possess good thermal properties and are normally used in high temperature applications. Silicone, bismaleimide, polyimide, polybenzimidazol, etc. are high temperature resistant polymers for temperature range from 200 to 450 °C.

Epoxy resins in general possess good thermo-mechanical, electrical and chemical resistant properties. They contain two or more epoxide groups in the polymer before cross-linking. Epoxy resins are cured using suitable curing agents or appropriate catalysts. The major curing agents are aliphatic amines, aromatic polyamines and polyanhydrides. Aliphatic amines are relatively strong bases and therefore react with aromatic amines to achieve cure at room temperature. This epoxy resin is useful for contact molding, but not for prepregging and filament winding. Aromatic polyamines are normally solids and require high temperature (100–150 °C) for mixing and curing. Anhydrides need higher thermal exposure (150–200 °C) for a longer duration (8–16 h) for proper curing. Both polyamines and anhydrides are suitable for prepreg manufacturing and filament winding. These epoxy resins are characterized by comparatively high thermal stability and chemical resistance.

Polyester resin is comprised of an unsaturated backbone polymer dissolved in a reactive monomer. The polyester backbone polymer is formed by condensation of a mixture of dibasic acids (saturated and unsaturated) and one or more glycols.

Table 5 Typical properties of some thermosetting resins [28]

Properties	Epoxies	Polyesters	Phenolics	Polyimides
Density (kg/m^3)	1100–1400	1200	1200–1300	1400
Tensile strength (MPa)	35–100	50–60	50–60	100–130
Tensile modulus (GPa)	1.5–3.5	2–3	5–11	3–4
Poisson's ratio	0.35	0.35	0.35	0.35
Coefficient of thermal expansion (10^{-6} m/mK)	50–70	40–60	40–80	30–40
Service temperature (K)	300–370	330–350	440–470	550–750

The components of the most commonly used polyester resin are phthalic anhydride (saturated acid), maleic anhydride (unsaturated acid) and propylene glycol. The process of curing is initiated by adding a source of free radicals (e.g., benzoyl peroxide or hydroperoxide) and catalysts (e.g., organic peroxides such as cobalt naphthenate or alkyl mercaptans). Several types of polyester resins are commercially available. One class of high performance polyester resins is vinyl–ester resins (acrylic esters of epoxy resins dissolved in styrene monomer). Polyester resins are cheaper and more versatile, but inferior to epoxy resins in some respects. Their use in advanced structural composites is therefore limited. However, polyester resins can be reinforced with almost all types of reinforcements to make polyester composites. They have been widely used in boat hulls, civil engineering structures, automobile industries and various engineering products and appliances.

The commonly used phenolic (phenol–formaldehyde) resins are divided into two groups: resoles and novolacs. Resoles are one-stage resins which are synthesized with formaldehyde/phenol ratio greater than one (1.25:1) in presence of an alkaline catalyst. The novolacs are two-stage resins made with an acid catalyst. The ratio of formaldehyde to phenol is about 0.8:1. In the first stage, the reaction is completed to yield an unreactive thermoplastic oligomer which is dehydrated and pulverized. A curing agent (e.g., hexamethylenetetramine) is added in the second stage. Resoles are used for prepregs and structural laminates while novolacs are used as molding compounds and friction products. Phenolic resins provide good dimensional stability as well as excellent chemical, thermal and creep resistance, and exhibit low inflammability. Phenolics char form a layer of carbon which protects the underlying composite. This characteristic has made phenolic resins an ideal candidate for high temperature applications and thermal shielding.

Several other high temperature thermosetting polymers are presently available such as polyimides, bismaleimides, polybenzimidazole, silicone, etc. for composite applications.

4.4 Metal Matrix Composites (MMC)

MMCs are inherently resistant to problems like micro cracking during thermal cycling and radiation exposure and possess electromagnetic interference (EMI) shielding property. Metal matrices can widen the scope of using composites over a wide range of temperatures. Besides, metal matrix composites allow tailoring of several useful properties that are not achievable in conventional metallic alloys. High specific strength and stiffness, low thermal expansion, good thermal stability and improved wear resistance are some of the positive features of metal matrix composites. The metal composites also provide better transverse properties and higher toughness compared to polymer composites. Metal matrices like aluminum, magnesium are often modified with materials like glass, ceramic, boron, silicon–carbide and graphite to prepare cost effective composites for aerospace and other commercial applications [14]. MMCs are made by dispersing a reinforcing material into a metal matrix. The reinforcement surface can be coated to prevent a chemical reaction with the matrix. For example, carbon fiber are commonly used in aluminium matrix to synthesize composites showing low density and high strength. However, carbon reacts with aluminium to generate a brittle and water-soluble compound Al_4C_3 on the surface of the fiber. To prevent this reaction, carbon fibers are coated with nickel or titanium boride.

Table 6 provides the list of some metal matrices and associated reinforcing materials. The reinforcements can be in the form of either particles, or short fibers or continuous fibers. The thermo-mechanical properties of some common matrices are given in Table 7.

The aluminum matrix composites are relatively lightweight, but their applications are limited to the lower temperature range because of its low melting point. Titanium and nickel can be used at a service temperature of up to 1000–1100 °C. Ti-6AL-4 V is commonly used as titanium matrix material. The other alloys of titanium include A-40Ti, A-70Ti, etc. Nickel matrices are comprised of a series of Ni-Cr-W-Al-Ti alloys. Super alloys, NiCrAlY and FeCrAlY are also used as matrices because of their high oxidation resistance properties. Molybdenum is a high temperature matrix. Iron and steel matrices are cheaper and can be used at high temperatures. The high temperature applications of metal matrix composites are mostly for aero engine blades, combustion chamber, thrust chamber, nozzle throat, exit nozzle, engine valves, fins, etc. Most of the metallic structural parts can be

Table 6 Metal matrices and reinforcements [25]

Matrix	Reinforcements
Aluminium and alloys	C, Be, SiO_2, B, SiC, Al_2O_3, Steel, B_4C, Al_3Ni, Mo, W, Z_rO_2
Titanium and alloys	B, SiC, Mo, SiO_2, B_e, ZrO_2
Nickel and alloys	C, B_e, Al_2O_3, SiC, Si_3N_4, steel, W, Mo, B
Magnesium alloys	C, B, glass, Al_2O_3
Molebdenum and alloys	B, ZrO_2
Iron and steel	Fe, Steel, B, Al_2O_3, W, SiO_2, ZrO_2
Copper and alloys	C, B, Al_2O_3, E-glass

Table 7 Typical thermo-mechanical properties of some metal matrices [25]

Matrices	Density (kg/m^3)	Tensile strength (MPa)	Tensile modulus (GPa)	Coefficient thermal expansion (10^{-6} m/mk)	Thermal conductivity (W/mk)	Heat capacity (KJ/ [kg k])	Melting point (° C)
AA6061	2800	310	70	23.4	171	0.96	590
Nickel	8900	760	210	13.3	62	0.46	1440
Ti-6AL-4 V	4400	1170	110	9.5	7	0.59	1650
Magnesium	1700	280	40	26	100	1.00	570
Steel	7300	2070	206	13.3	29	0.46	1460
Copper	8900	340	120	17.6	391	0.38	1080

replaced with metal matrix composite parts; however, cost of the material is the major problem for such uses.

4.5 Ceramic Matrix Composites

Ceramics provide strength at very high temperature (above 1500 °C) with considerable oxidation resistance. They possess several advantageous properties like high elastic modulus, high Peierls yield stress, low thermal expansion, low thermal conductivity, high melting point, good chemical and weather resistance as well as excellent electromagnetic transparency [7]. The cutting rate of an alumina-SiC whisker cutting tool is ten times higher than that of conventional tools. However, the major drawback of ceramics is that they are brittle and exhibit low plasticity which often leads to catastrophic failure. Hence, they are not considered as dependable structural materials. To overcome such limitations, suitable reinforcements are introduced in ceramic matrix composites, which lead to desirable mechanical properties. The ceramic matrices are usually carbon, glass, glass ceramics (lithium aluminosilicates), carbides (SiC), nitrides (SiN_4, BN), oxides (Al_2O_3, Zr_2O_3, Cr_2O_3, Y_2O_3, CaO, TiO_2) and borides (ZrB_2, TiB_2). For ceramic matrix composites, commonly used fibers are carbon, silicon carbide, silica, alumina, steel and other metals. They have excellent resistance to heat and wear, creep, low and high cycle fatigue, corrosion and impact combined with high specific strength at high temperatures [22]. The use of ceramic composites in aero-engine and automotive engine components can reduce their weight and thereby enhance the engine performance due to high specific strength at high temperatures.

5 Fabrication Technique

Composites can be fabricated using different techniques. The fabrication technique used depends on the type of reinforcement and matrix material used to produce the composite material. In addition, it also depends on the application of composite material. Composite materials are so called as one or more materials of different properties or characteristics are integrated together. In this respect, different types of metal fiber composite materials are reviewed in this section ranging from composite yarn, woven fabric, knitted fabric, fiber metal laminates, multilayered fabric laminates, etc. Many fabrication techniques such as hollow spindle spinning, weaving, knitting, slurry casting, lamination, molding, etc. adopted by researchers have been summarized.

Sinmazçelik et al. [27] have reviewed the bonding techniques and test methods of fiber metal laminates. Fiber metal laminates are classified as hybrid composite materials meant for aerospace and automotive applications. They have categorized the Aluminium based fiber metal hybrid composite materials into three different categories namely, Aramid Reinforced Aluminium Laminate (ARALL), Glass Reinforced Aluminium Laminates (GLARE) and Carbon Reinforced Aluminium Laminates (CARALL). The aluminium based fiber metal laminates fabrication has been initiated with the motivation of achieving lightweight and better properties such as toughness, strength, elasticity, etc. than the conventional metal alloys. In the early 80s, National Aerospace Laboratory and Delft University of Technology, Netherland started their research focusing to improve the performance of metal alloys. In mid of 1980s, the first ever fiber metal laminate (FML) has been fabricated with the incorporation of aramid fiber and named, ARALL. Another attempt has been made to produce carbon based FML, which has shown poor fatigue performance due to their low strain characteristics (0.5–2.0 %). In 1990, the fatigue performance has been improved with glass reinforced metal laminate (GLARE). The following flow chart shows the major activities involved in the production of fiber metal laminates (Fig. 4). The fiber—metal composite preparation and characterization are highly emphasized in many research studies and applied in aerospace and automotive applications; thereby, the literature pertaining to fabrication of fiber metal laminates meant for aerospace application is discussed in this section.

The main process involved in the fabrication of fiber metal composite is the bond formation between fiber and metal. It is done by surface treatments followed by adhesive bonding technique. Surface treatments are carried out in five stages

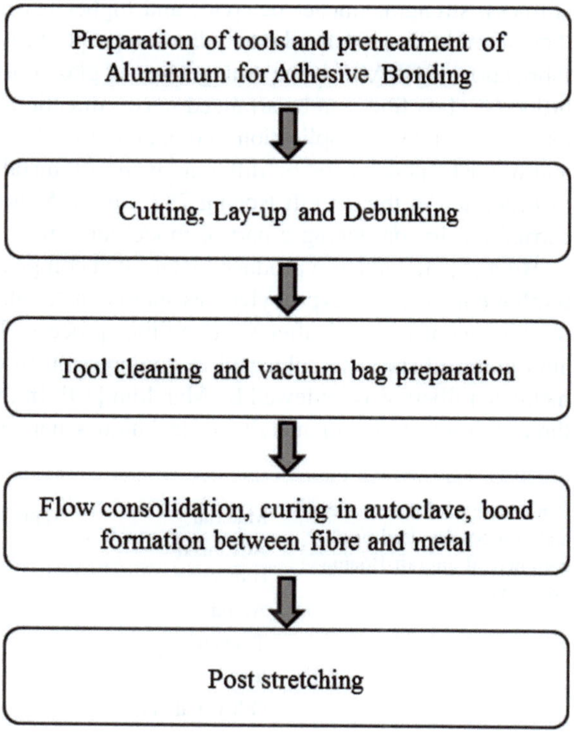

Fig. 4 Steps involved in production of fiber metal laminates

namely, mechanical, chemical, electrochemical, coupling agent and dry surface treatments. The mechanical treatment is provided by grit blasting using alumina or silica grit or glass beads to enable rougher surface or "peak and valley effect" and to remove the initial oxide layer [11, 21]. In chemical treatment, acid etching is given in three steps with sulphuric acid and its mixtures in general. Acid etching is usually the intermediate process between the mechanical and alkaline treatment [6, 10, 21]. Mechanical and chemical treatment alone may not be sufficient for further bonding processes. In order to repair corrosive portion, if any, electrochemical treatment is required, in which a thin oxide film with micro-roughness [12] is coated on the surface of metal. Coupling agent treatment is given to the metal or metal alloy to improve the structural aluminium bonds. Environment friendly chemicals such as silane or sol-gel are used as coupling agents, which form strong covalent bonds between metal oxide and silane, enabling high intermediate modulus between metal and polymer. The completion of mechanical treatment ensures the ready to bond state of the aluminium metal with fibers. ARALL is the unidirectional oriented aramid fibers sandwiched with the aluminium metal layers using epoxy adhesive. The sandwiched panels will be fabricated either in mechanical press or autoclave at preset temperature ranging from 120 to 175 °C for the required time. The fibers are aligned in the direction of load applied. The studies show that the incorporation of aramid fiber hinders the crack propagation developed in the metal layers by reducing the crack tip stress intensity. The next generation of fiber metal laminate is the glass fiber reinforced metal laminate (GLARE). Glass fiber has better specific stiffness, strength, impact behavior and higher resistance to compression loading than aramid fiber and also better adhesion property. Another major difference in the fabrication of GLARE is the arrangement of glass fibers. Because of better adhesion property, glass fiber can be arranged in two directions or multidirections, leading to its wider range of applications compared to ARALL [27]. Table 8 shows the approximate percentage of different types of metallic elements and composites incorporated in the aircraft Boeing 787. Figure 5 shows the resin infusion process carried out for producing a part of model aircraft.

Bigg [5] attempted to produce metal filled composites using aluminium fibers as reinforcement and polypropylene as matrix material. Glass fiber was also used as reinforcement material after short milling process. The chopped glass fibers and aluminium fibers were subjected to compression molding process. Application of metalized fibers was reviewed by Marchini [19]. In the discussed, he has discussed three different types of metalizing techniques namely, physical process, galvanic

Percentage of different metal and composites used in a commercial aircraft Boeing 787 [38]

Material	Approximate percentage used in aircraft
Advanced composite	50 %
Aluminium	20 %
Titanium	15 %
Steel	10 %
Other materials	5 %

Fig. 5 Resin infusion process. Ref. [36]

coating and chemical coating. In physical process, vacuum sprayed metal, preferably aluminium, was suggested to coat the fibers to get the conductivity of around 10^0–10^4 Ω cm^{-1}, but poor processability was observed. Galvanic coating was preferred only for the conductive fibers such as carbon and graphite in which the conductivity of greater than 10^4 Ω cm^{-1} was obtained. Chemical coating was preferred for any sort of fibers which can give 10^2–10^4 Ω cm^{-1} without affecting the properties and processability of the fibers. Boccaccini et al. [7] attempted to produce metal fiber reinforced glass matrix composites. They have demonstrated the composite manufacturing using metal fiber fabric as reinforcement and soda–lime glass as matrix. They have followed two steps in the production of composite. The first step was the infiltration by an electrophoretic deposition (EDP) process. In the deposition process, a small charged particle was attracted towards the conducting fiber perform which was a stainless steel fabric used by the researcher. The movement of the ceramic sol particle was governed by three factors namely, field strength, the pH of the solution and the ionic strength of the solution. The second step was the preparation of composite using EDP infiltrated metallic fiber fabrics by uniaxial cold pressing method at a temperature of 670 °C. Soda lime glass matrix powder was sandwiched between two metallic fiber fabrics and composite was produced.

Conductive metallic yarns can be fabricated using yarn doubling technique. Tezel et al. [29] have fabricated conductive composite yarns. They have used four different types of metallic yarns made of two different metals namely, stainless steel and copper. The former was produced with the diameter of 35 and 50 μm and the later with 50 and 70 μm. They were doubled with cotton yarn using hollow spindle twisting machine to form a conductive composite yarn. Barburski et al. [4] have produced stainless steel fibers by bundle drawing technique. They have enveloped bundle of fibers in a metal pipe and subsequently wire drawing was done followed by a heat treatment to form composite materials of reduced diameter. Due to wire drawing, the metallic wires were converted to thin fibers, around 500–1500 in numbers. At later stage the covering material was removed and the thin fibers were stretch-broken to form small fibers which have the average diameter of around 12 μm and staple length between 30 and 150 mm. The produced fibers were spun into metallic yarn using conventional spinning technology and weft knitted to form a fabric for automotive applications.

Hwang et al. [15, 16] have produced composite fabrics made of stainless steel, silver and polyester yarns. They have produced two different types of yarns in which the type of wrapping materials differs for each yarn. In one of the yarns silver filaments were used as wrapper and in another yarn polyester filament was used as the wrapper. In both the yarns, the stainless steel filament was used in the core structure. The yarns were produced by hollow spindle spinning system with winding speed of 8000 rpm. The commingled yarns were knitted to form a fabric using flat knitting machine.

Pemberton et al. [22] have elaborated the production of composite by slurry casting. Stainless steel fibers were packed in a wooden mould in which the slurry was poured. The slurry consists of ceramic powder mixed with 10 % weight of water. The casting was done with vibration for 12 h at 30 °C and further heat treatment for 12 h at 450 °C during which the water was completely removed from the system and composite was formed. Yu et al. [35] have produced metal composite knitted fabrics using stainless steel filaments. They have produced yarns by wrap spinning technique with stainless steel yarn in core and antibacterial nylon as inner wrapper and polyester filament as outer wrapper yarn. Another set of yarn was also produced with alternate combination of wrapping. The produced yarns were knitted to form a fabric. Such fabrics were laminated together at different angles to form metal composite fabrics.

6 Properties of Composites

Many researchers have analyzed different properties of composite materials with respect to their applications. Such properties may include hardness, tensile strength, elongation, wear resistance (or) corrosion resistance, fracture toughness, electro magnetic shielding efficiency, surface resistivity, FIR emissivity, anion density emissivity, anti bacterial activity, electrical conductivity, thermal conductivity, etc. In this section, the properties of composites analyzed by various researchers are discussed.

Bigg [5] has made an attempt to analyze electrical, thermal and mechanical properties of composite produced using aluminium and glass fibers as reinforcement material and polypropylene as matrix material. Different types of aluminum fibers were selected with variation in length and diameter and composites were produced with varying fiber aspect ratio. The electrical resistivity was studied using the voltage drop between the produced sample and the sample of known resistance. The thermal conductivity was studied using hot plate method with the hot plate temperature of 90 °C. It was reported that the electrical resistivity decreased with the increase in the volume of metal content in the composite. Similarly, thermal conductivity of the composite has shown the increasing trend with the increase in the volume of metal in the total material. The tensile strength of composite was reported higher for polypropylene–glass fiber composite than polypropylene–aluminium fiber composite. Pemberton et al. [22] have produced composites using stainless steel fibers as reinforcement. They have attempted to analyze single fiber tensile testing, single fiber pullout testing and composite fracture energy. The tensile testing was performed using the conventional system of tensile property measurement. Single fiber pullout test was used to study the interface between matrix and reinforcement. The fiber was allowed to partially protrude for the length of 5 mm from the composite material and the frictional and shear behaviors of the fiber in composite was analyzed. It was observed that some of the fibers were broken before actually it came out of the composite system and few fibers were completely pulled out from the composite. The nominal fiber stress was reported to reduce from the range of 400–500 MPa to 0 MPa gradually, as the fiber pulled out of the composite material. The fracture energy was studied using the conventional pendulum Izod apparatus. The energy absorbed during fracture was studied. Fractured composite specimens, shown in Figs. 6 and 7, consists of coarse and fine fibers respectively. It was observed that of 10–15 % of the fibers were ruptured and the formation of neck was observed in the specimen during fracture. It was also reported that there was some loss of matrix adjacent to crack during fracture study.

Tezel et al. [29] analyzed the properties of single jersey knitted fabrics produced using stainless steel or copper metallic yarn doubled with cotton yarn. They have studied the electromagnetic shielding efficiency of doubled metallic yarn fabrics using two different techniques namely, free space measurement technique and coaxial transmission line measurement technique. It was found that the filament made of copper provided higher electromagnetic shielding efficiency than the filament made of stainless steel at low frequency by both coaxial and free space measurement technique. It was also suggested that the fabric made of metallic filaments which are reactive such as copper may cause allergic reaction to skin and discomfort to the human body. Barburski et al. [4] have produced stainless steel

Fig. 6 Optical image of fractured composite material consists of coarse fiber. Ref. [22]

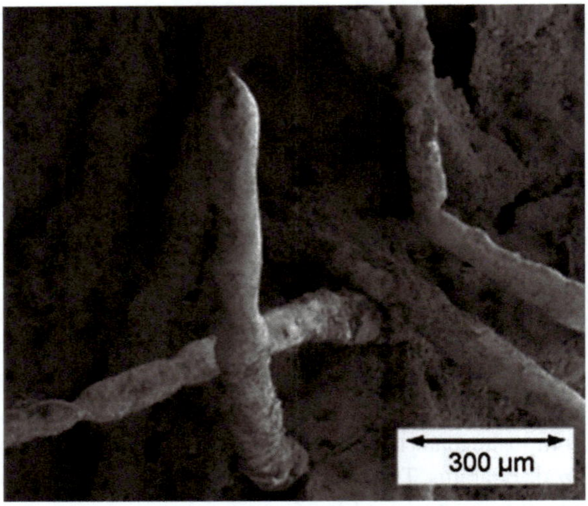

Fig. 7 SEM image of fractured composite material consists of fine fiber. Ref. [22]

fiber knitted fabric for automotive applications. They have observed that the produced yarn out of the stainless steel fiber shown hariness due to short fibers and high rigidity of the fiber to form yarn or fabric. They have attempted to analyze the compressibility of the biaxially sheared strained fabrics. The biaxial shear was introduced in the increment of 5° from 5° to 25° beyond which the fabric started to wrinkle. Strain was applied to the fabric simultaneously in wale and course direction ranging from 0 to 20 %. The deformed fabrics were kept in card board for conducting the compression test. Instron tester was used for compression with 70 mm diameter of compression head, 1 kN load cell and at the speed of 1 mm/min. It was reported that the reduction in thickness was in the range of 8–14 % for different types of knitted fabrics made of stainless steel filament. It was also reported that the maximum change in thickness was observed at 15 % of strain in both wale and course directions. Hwang et al. [15, 16] studied the electromagnetic shielding efficiency (EMSE) and air permeability of the knitted fabrics produced from stainless steel—silver—polyester commingled yarn. EMSE was studied from the frequency range of 300 kHz to 3 GHz. It was reported that the wale wise surface resistance was higher than course wise resistance due to the discontinuous structure in the wale direction compared to the continuous structure in the course direction. It was also reported that the fabrics having higher metal content were observed to give higher shielding efficiency at lower frequency range due to the magnetic and conductive effects. But at higher frequencies the waves were leaked out due to the interspace among knitted fabrics. Yu et al. [35] have analyzed the comfort characteristics namely, wicking, absorbency, water vapour transmission, drying capability and other properties namely, EMSE and antibacterial activity of the composite knitted metal fabrics produced using stainless steel, polyester and nylon yarns. EMSE study was performed by coaxial transmission line and antibacterial activity was performed by qualitative assessment using AATCC 90-2011 standard. It was observed that the fabric consisting of polyester filament yarn showed better vertical wickability than the antibacterial nylon yarn due to the presence of organosilicone which reduced the wickability of the material. Also the fabrics with higher number of wrappings showed less wicking height due to the longest path of water to wick to the required height. It could be also understood that the drying rate and water absorbency properties of the composite fabric was dependant not on the nature of stainless steel filaments but on the other constituent fibers or filaments incorporated in the structure. They have studied the EMSE of the developed fabric in the frequency range of 300 to 3 GHz. It was reported that the developed fabric was able to provide the attenuation of −10 to −20 dB in the frequency range of 45 to 400 MHz and −5 to −10 dB in the frequency range of 0.4 to 2 GHz. Laminated fabrics showed better attenuation of −20 dB in the frequency range of 300 MHz to 1.47 GHz. Similar studies were conducted by many researchers on electromagnetic shielding properties for different applications [15–17, 20, 23, 24].

7 Examples of Application

The metallic fiber reinforced fabrics and composites can be used in wide range of applications such as aircraft engine, parts of motor car, lightning strike protection in aircraft panels, conductive fabrics, radar reflectors and sailing boats, electrically heated blankets or apparel, ablation, military clothing, industrial dust collector bags, filtration, etc. Recent research studies on metal fiber incorporated fabric composites are reviewed and their applications in various products are discussed in this section.

The aramid fiber reinforced aluminium laminate hybrid composite (ARALL) is used in aerospace and automotive applications. The incorporation of aramid fiber with in metal structures reduces the crack propagation behavior by factor of 100 or more when compared to the conventional aluminium metal alloy structures. It finds wide range of applications such as lower wing skin panels in Fokker 27 aircraft, cargo door in Boeing C 17 and fuselage. Even though these materials have been initially developed for aerospace applications, later they found applications also in ballistic materials in military products [2, 31]. Glass fiber reinforced aluminium laminate hybrid composite (GLARE) is also mainly used in aerospace applications. The GLARE panels are not only used in main fuselage skin but also in the leading edges of tails of planes located both in horizontal and vertical direction, particularly in new versions of Air bus such as A380. One such application in Bombardier Fuselage is shown in Fig. 8. The use of GLARE is also explored in various other parts of aircraft such as cockpit crown, forward bulkheads, fire walls, cargo liners, etc. [3, 26].

Marchini [19] reviewed the applications of metalized fibers for technical applications. It was mentioned in the review that the conductivity of the fibers can be increased by coating the fibers with metals. He has suggested that metallized carbon or graphite fibers showing more than 10^4 Ω cm^{-1} conductivity finds application in aerospace as a composite. It was reported that the PAN fibers which are metallized with copper and nickel are used for a range of applications such as

Fig. 8 Picture showing Bombardier C series Fuselage—GLARE. Ref. [37]

antistatic carpets, antistatic blankets, airliners and EMI shielding. A single jersey fabric was produced using copper and stainless steel filaments doubled with cotton yarn by Tezel et al. [29]. The study showed that the produced single jersey fabric can have electro magnetic shielding efficiency of more than 7 dB or 80 % of the radiation produced by GSM 850 or GSM 900 cellular phone. It was categorized as "AA" in "Class II – General Use" by the "Committee for Conformity Assessment of Accreditation and Certification on Functional and Technical Textiles" (CCAACFTT).

Pemberton et al. [22] have developed a stainless steel incorporated composite material which can be used in corrosion resistant and thermally stable environment. Barburski et al. [4] have produced stainless steel fabrics for automotive glass applications. They have studied the application of stainless steel fabric pre-strained at different level of strains for automotive glass with minimum distortions. It has been observed that the produced glass with the pre-strained fabric showed much lower distortion than the glass produced without the pre-strained fabric. Hwang et al. [15, 16] have fabricated a knitted fabric consisting of stainless steel—silver—polyester commingled yarn for electromagnetic shielding applications in the frequency range of 300 kHz–3 GHz. It has been reported that the silver wrapping significantly improved the shielding efficiency compared to polyester warp yarn. Five layer composite fabric structure was proposed in the study conducted by Hwang et al. [15, 16]. Alumina has good compressive strength and high modulus along with electrically insulating property which makes it more efficient with polymer matrix composite applications. Alumina with aramid/epoxy material can be used for many defence applications such as radar transparent structure, circuit boards, and antenna applications [13]. A typical example for EMI shielding composite junction box is shown in Fig. 9. Some other applications of metal fiber reinforced composite applications are shown in Figs. 10 and 11.

Fig. 9 EMI shielding junction box. Ref. [40]

Fig. 10 Metal ceramic composite material. Ref. [41]

Fig. 11 Metal fiber braided sleeve preform for producing composite. Ref. [39]

8 Conclusion

Metallic fibers are gaining importance from the past few decades. The techniques emerged for spinning of yarn, fabric formation and fabrication of preforms for composites allowed metallic fibers or filaments to form any required shape and product. In this chapter, different types of metallic fibers presently used by the researchers and in commercial applications are elaborated and their properties are discussed in detail. The matrix systems that can be used for the production of composites are summarized. Various fabrication techniques adopted by the researchers are reviewed in detail. Properties and applications of the metal or metallic fiber based composites are also discussed. The technological advancements in the textile and composite manufacturing sector ensure that the metallic fibers will certainly find many more applications in defence, aerospace, automotive and other industrial sectors.

References

1. Ahmed JT (2009) Hybrid composite structures: multifunctionality through metal fibers, PhD thesis, TU Delft. http://repository.tudelft.nl/view/ir/uuid%3A85e91d70-dcf2-47c2-892a-cad116fe845f/. Accessed 17 June 2015
2. Alderliesten R (2009) On the development of hybrid material concepts for aircraft structures. Recent Pat Eng 3:25–38
3. Asundi A, Alta CYN (1997) Fiber metal laminates: an advanced material for future aircraft. J Mat Process Technol 63:384–394
4. Barburski M, Lomov S, Lanckmans F, De Ridder F (2014) Steel fiber knitted fabric for automotive glassforming: variations ofthe fabric thickness on the mould and glassoptical quality. J Ind Text. doi:10.1177/1528083714538685
5. Bigg DM (1979) Mechanical, thermal, and electrical properties of metal fiber-filled polymer composites. Polym Eng Sci 19(1):1188–1192
6. Bishopp A (2005) Handbook of adhesives and sealants. Elsevier, Amsterdam
7. Boccaccini AR, Ovenstone J, Trusty PA (1997) Fabrication of woven metal fiber reinforced glass matrix composites. Appl Compos Mater 4:145–155
8. Callens MG, Gorbatikh L, Verpoest I (2014) Ductile steel fiber composites with brittle and ductile matrice. Compos: Part A 61:235–244
9. Callister WD (1994) Materials science and engineering: an introduction, 3rd edn. Wiley, New York
10. Critchlow GW, Brewis DM (1996) Review of surface pretreatments for aluminium alloys. Int J Adhes Adhes 16:255–275
11. Critchlow GW, Yendall KA, Bahrani D, Quinn A, Andrews F (2006) Strategies for the replacement of chromic acid anodising for the structural bonding of aluminium alloys. Int J Adhes Adhes 26:419–453
12. Davis M, Bond D (1999) Principles and practices of adhesive bonded structural joints and repairs. Int J Adhes Adhes 19:91–105
13. Department of Defense Handbook, Composite Materials Handbook, vol 3. Polymer matrix composites, materials usage, design and analysis, USA, 2002
14. Hoskin BC, Baker AA (eds) (1986) Composite materials for aircraft structure, AIAA Education Series. American Institute of Aeronautics and Astronautics Inc., New York
15. Hwang PW, Chen AP, Li TT, Lou CW, Lin JH (2014) Structure design and property evaluation of silver/stainless steel composite fabric. J Ind Text. doi:10.1177/1528083714538683
16. Hwang PW, Chen AP, Lou CW, Lin JH (2014) Electromagnetic shielding effectiveness and functions of stainless steel/bamboo charcoal conductive fabrics. J Ind Text 44:477–494
17. Krishnasamy J, Alagirusamy R, Das A, Basu A (2014) Electromagnetic shielding behaviour of conductive filler composites and conductive fabrics—a review. Ind J fiber Text Res 39:329–342

18. Lee SM (ed) (1990) Encyclopedia of composites, vol 1–4. VCH Publications, New York
19. Marchini F (1991) Advanced applications of metallized fibers for electrostatic discharge and radiation shielding. J Coat Fabr 20:153–166
20. Ozen MS, Sancak E, Beyit A, Usta I, Akalin M (2013) Investigation of electromagnetic shielding properties of needle-punched nonwoven fabrics with stainless steel and polyester fiber. Text Res J 83:849–858
21. Park SY, Choi WJ, Choi HS, Kwon H, Kim SH (2010) Recent trends in surface treatment technologies for airframe adhesive bonding processing: a review (1995–2008). J Adhes 86:192–221
22. Pemberton SR, Oberg EK, Dean J, Tsarouchas D, Markaki AE, Marston L, Clyne TW (2011) The fracture energy of metal fiber reinforced ceramic composites (MFCs). Compos Sci Technol 71:266–275
23. Rajendrakumar K, Thilagavathi G (2012) A study on the effect of construction parameters of metallic wire/core spun yarn based knitted fabrics on electromagnetic shielding. J Ind Text 42:400–416
24. Roh JS, Chi YS, Kang TJ, Nam SW (2008) Electromagnetic shielding effectiveness of multifunctional metal composite fabrics. Text Res J 78:825–835
25. Sinha P K (1995), Composite materials and structures, Composite Centre of Excellence, AR & DB, Department of Aerospace Engineering, I.I.T. Kharagpur, Chapter 2, ebook, http://www.ae.iitkgp.ernet.in/ebooks/chapter2.html. Accessed 17 June 2015
26. Shima DJ, Alderliesten RC, Spearing SM, Burianek DA (2003) Fatigue crack growth prediction in GLARE hybrid laminates. Compos Sci Technol 63:1759–1767
27. Sinmazçelik T, Avcu E, Bora MÖ, Çoban O (2011) A review: fiber metal laminates, background, bonding types and applied test methods. Mater Des 32:3671–3685
28. Smith R (1990) Resin systems, in processing and fabrication technology. In: Bader MG, Smith W, Isham AB, Rolston JA, Metzner AB (eds) Delaware composites design encyclopedia, vol 3. Technomic Publishing Co., Inc. Lancaster, pp 15–84
29. Tezel S, Kavuşturan Y, Vandenbosch Guy AE, Volski V (2014) Comparison of electromagnetic shieldingeffectiveness of conductive single jersey fabrics with coaxial transmission line and free space measurement techniques. Text Res J 84:461–476
30. Toon JJ (1990) Metal fibers and fabrics as shielding materials for composites, missiles and airframes. Electromagnetic compatibility, pp 5–7. In: IEEE International Symposium Record, Washington, DC. http://ieeexplore.ieee.org/xpl/articleDetails.jsp?arnumber=252722. Accessed 17 June 2015
31. Vogelesang LB, Vlot A (2000) Development of fiber metal laminates for advanced. J Mater Process Technol 103:1–5
32. Weeton JW, Signorelli RA (1966) Fiber-metal composite materials. National Aeronautics and Space Administration, Washington, DC. http://www.dtic.mil/cgi-bin/GetTRDoc?AD=ADA400428. Accessed 17 June 2015
33. Weeton JW, Peters DM, Thomas KL (eds) (1987) Engineers' guide to composite materials. American Society for Metals, Metals Park
34. Wilson DM, Visser LR (2000) High performance oxide fibers for metal and ceramic composites. In: Processing of fibers & composites Conference, Barga, Italy. http://www.3m.com/market/industrial/ceramics/pdfs/High_Performance_Oxide_Fibers.pdf. Accessed 17 June 2015
35. Yu ZC, Zhang JF, Lou CW, Lin JH (2015) Investigation and fabrication of multifunctional metal composite knitted fabrics. Text Res J 85:188–199
36. http://img.deusm.com/designnews/2011/11/235863/112810_235522.jpg. Accessed 22 Sept 2015
37. http://s3.e-monsite.com/2011/01/15/05/resize_550_550//bombardier-cseries-fuselage.jpg. Accessed 22 Sept 2015
38. http://www.boeing.com/commercial/aeromagazine/articles/qtr_4_06/article_04_2.html. Accessed 22 Sept 2015

39. http://www.cabletiesandmore.com/images/braidedsleeving/SSNtopimage3.jpg. Accessed 22 Sept 2015
40. http://www.glenair.com/composite/pdf/e/140_105.pdf. Accessed 22 Sept 2015
41. https://www.ceramtec.com/mmc-metal-ceramic-composite-materials/. Accessed 22 Sept 2015

Carbon Nanofibres and Nanotubes for Composite Applications

Maria C. Paiva and José A. Covas

Abstract Carbon nanotubes and nanofibers are now commercially viable making possible a number of effective applications. This chapter provides a brief but didactic revision of the polymer/nanotube or nanofiber mixing methods, with major focus on melt mixing. The nanotube or nanofiber dispersion parameters are discussed, as well as the role of chemical functionalization. As a framework for this discussion the general properties of carbon nanotubes and nanofibers, as well as their polymer composites, are summarized. A market perspective is presented demonstrating the growing interest of these materials. The effective market growth will depend on the efficiency in tackling dominant factors such as price, material quality/purity and consistency, health and safety aspects and, especially in the case of polymer nanocomposites, dispersibility and compatibility with the matrix.

1 Introduction

The seminal work of Ijima [33] brought carbon nanotubes to the attention of both the scientific community and society. The unique mechanical, thermal and electrical properties of these materials rapidly triggered a wide scope of anticipated advanced technical applications, often with a prospective outstanding societal impact.

In particular, as comprehensively presented by Coleman et al. [20], early theoretical and experimental work on carbon nanotubes (CNT) and nanocomposites production and characterization carried out between the late 1990s and the middle of the 2000s revealed the potential of CNT as reinforcement material. The potential of these nanoparticles for electrical and thermal conductivity was demonstrated, whilst the main requirements for effective mechanical reinforcement were identified

M.C. Paiva (✉) · J.A. Covas
Institute for Polymers and Composites/i3N, University of Minho, Guimarães, Portugal
e-mail: mcpaiva@dep.uminho.pt

J.A. Covas
e-mail: jcovas@dep.uminho.pt

© Springer Science+Business Media Singapore 2016
S. Rana and R. Fangueiro (eds.), *Fibrous and Textile Materials for Composite Applications*, Textile Science and Clothing Technology,
DOI 10.1007/978-981-10-0234-2_7

to be the large CNT aspect ratio, their good dispersion in the matrix and the efficient polymer-reinforcement interfacial stress transfer. High interfacial shear stress associated with the large interfacial contact area of the CNTs were expected to improve the capacity for energy dissipation, and thus to induce a large toughness enhancement of the nanocomposites (Wichmann et al. [96]). Suhr and Koratkar [86] reported an increase in loss modulus in the order of 1000 % for polycarbonate (PC) composites with 2 wt% of single wall carbon nanotubes (SWNT), confirming this ability. The improvement in composite properties was observed to vary with the processing technique, but also with the type of polymer matrix, emphasizing the role of interfacial energy and polymer morphology. Carponcin et al. [15] reported large variations of the electrical percolation with CNT content for composites prepared by melt mixing, depending on the polymer matrix. While for amorphous thermoplastics the percolation was reported near 0.3 wt% CNT, for semi-crystalline polymers it ranged from 0.9 to 2.5 wt% CNT.

While CNT were subjected to intense research (often supported by relatively abundant programs put up by funding agencies), a few manufacturers launched their grades and more and more applications were envisaged (some of these were near to visionary, such as an elevator to space, or the global replacement of silica in electronics). However, during this same period it was progressively demonstrated that attaining the full (theoretical) potential of these materials in terms of performance was not easy, namely due to difficulties in producing sufficiently clean (free of impurities such as metal catalyst particles and amorphous carbon) and dimensionally uniform nanotubes and, in the case of polymer nanocomposites, in achieving the required dispersion levels in the matrix. Only a very limited number of products actually attained the commercial stage (mostly automotive and electronic parts, sporting goods, sensors and battery electrodes). More recently, the better understanding of the chemical-physical characteristics and dispersion mechanisms of CNT, together with the advances in manufacturing technologies and in surface functionalization routes, originated a new upsurge in the practical interest in these materials.

These fluctuations in the visibility of CNT and in their real-world applications are typical of a hype cycle of an emerging technology, as illustrated in Fig. 1. Exaggerated expectations were initially created, as the manufacture, characteristics and engineering properties of CNT had not been sufficiently investigated. This created a trough of disillusionment, which affected the interest to conduct (and the capacity to fund) new research. The entrance of graphene derivatives as competitors of CNT for many applications in the early 2000s further complicated the situation. At present, there is renewed optimism concerning the commercial sustainability of CNT. Market studies from 2013 asserting that the global CNT market is expected to grow at an estimated 4 % compound annual growth rate from 2011 to 2016 [53] were revised in 2014 to 15.5 % between 2013 and 2018 [54].

Nonetheless, a significant gap between supply and demand currently still exists. According to De Volder et al. [24], worldwide CNT production increased more than one order of magnitude between 2006 and 2011 (typically, from 200 ton in 2005 to

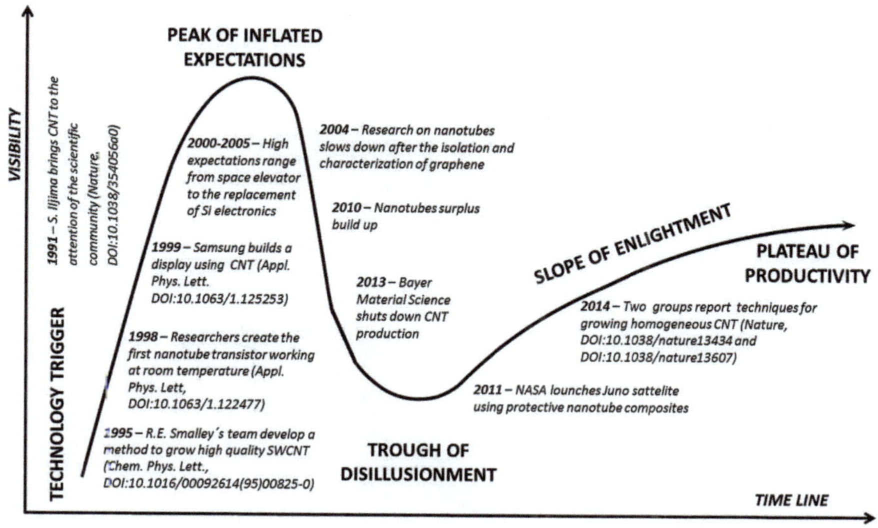

Fig. 1 The hype cycle of carbon nanotubes. Adapted from Davenport [23]

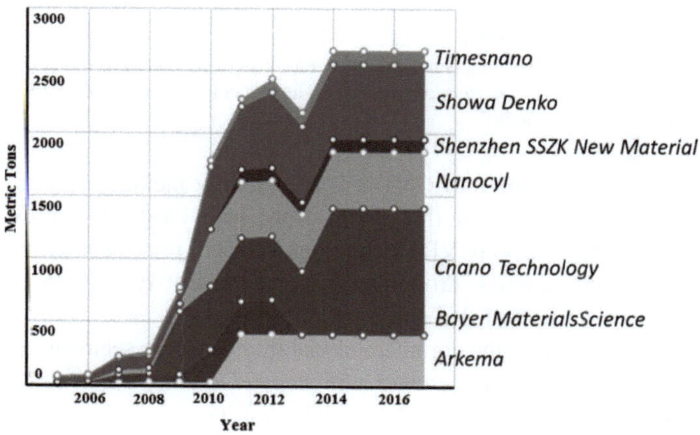

Fig. 2 Main CNT manufacturers (Adapted from: http://cenm.ag/nanotubes2015. Accessed 15 July 2015. *Source* Lux Research)

approximately 4600 ton in 2011). Figure 2 reveals that CNT manufacture is dominated by a few companies (Showa Denko, CNano Technology, Nanocyl S.A and Arkema), whilst Bayer MaterialScience exited production in 2013, apparently due to excessive market fragmentation (at the time, their capacity was over 200 tons per year). In contrast, there is growing optimism in terms of demand (see Fig. 3), even if data can differ significantly between sources. It is also clear that CNT have been and will continue to be primarily utilized in polymers (at least 60 % of the total

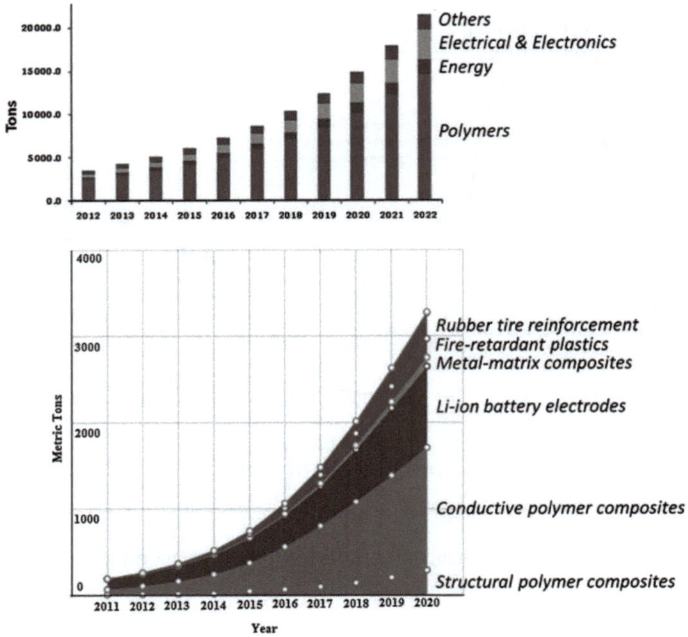

Fig. 3 Market estimates for CNT per area of application. (*Top* adapted from http://www.grandviewresearch.com/industry-analysis/carbon-nanotubes-cnt-market. Accessed 21 July 2015. *Source* Grand View Research; *bottom* adapted from http://cenm.ag/nanotubes2015. Accessed 21 July 2015. *Source* Lux Research

applications), as consumption of CNT will be certainly facilitated by the observed continuous decrease in prices that result from technological improvements in Chemical Vapor Deposition (CVD) technology. Indeed, prices started at more than 1,000 US$/kg in the nineties, dropping to approximately 100 USD$ in 2011 and probably to less than half of this value by 2016.

In conclusion, although the commercial viability of CNTs seems guaranteed, the effective growth of practical demand will depend on the efficiency in tackling dominant factors such as price, material quality/purity and consistency, health and safety aspects and, especially in the case of polymer nanocomposites, dispersibility and compatibility with the matrix.

2 Synthesis and Properties of Carbon Nanotubes and Nanofibers

The element carbon (C) with atomic number 6 may form hybridized sp, sp^2, or sp^3 atomic orbitals, providing such a versatility for chemical bonding that makes it the major element in a variety of materials, from feedstock gases to all organic matter,

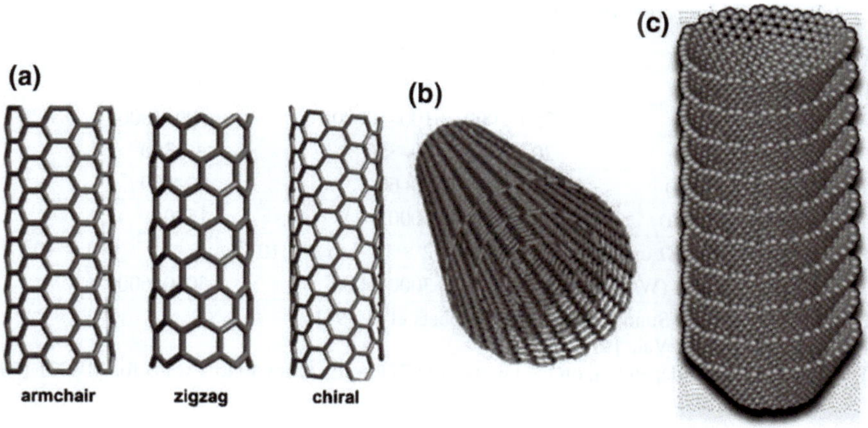

Fig. 4 Representation of the hexagonal graphene layers arrangement to form single wall carbon nanotubes (**a**), multi wall carbon nanotubes (**b**) and carbon nanofibers (**c**)

and to inorganic matter such as diamond and aluminium carbide. Sp^2 C may bond to form a bi-dimensional hexagonal molecular array of C atoms, designated as graphene. Graphene layers may arrange in different shapes, originating diverse carbon allotropes. For example, graphite results from stacking of graphene layers with an interplanar spacing of 0.335 nm; a single rolled up sheet of graphene forms a cylinder designated as single walled carbon nanotube (SWNT) (Fig. 4a); several concentrically stacked graphene cylinders with an interlayer spacing of 0.344 nm (as reported by Saito et al. [77] are designated as multi walled carbon nanotubes (MWNT) (Fig. 4b). A diversity of SWNT may be formed depending on the roll-up angle of the graphene sheet relative to the nanotube axis (Fig. 4a), presenting different electronic properties ranging from metallic to semiconducting behaviour [76]. Carbon nanofibers (CNF) are also graphite-based cylindrical hollow fibres, however not only the graphene layers stack in truncated cones, or "stacked cups", as illustrated in Fig. 4c, but they also present larger diameter compared to typical MWNT.

The synthesis of CNT can be achieved through high temperature methods (>1700 °C) such as arc discharge or laser ablation, although presently the larger production rates are achieved at low temperature (<800 °C) by chemical vapour deposition (CVD) methods [21, 72, 89]. Arc discharge methods require high temperature but produce CNT with few structural defects. Laser ablation also provides good quality CNT free from catalyst contamination, but the large number of process parameters involved make it difficult to control the properties of the particles produced. In both cases the CNT are typically contaminated with less structured carbon. Catalytic chemical vapour deposition (CCVD) methods are based on the reaction of an hydrocarbon-rich gas mixture in the presence of a catalyst, activated by heat or other stimuli such as plasma enhancement (PE), microwave plasma or other methods [72]. The process has been extensively studied

Table 1 Typical properties of CNF, MWNT and SWNT

Property	CNF[a]	MWNT[b]	SWNT[b]
Diameter (nm)	50–200	5–50	0.6–1.8
Aspect ratio	250–2000	100–10000	100–10000
Surface area (m^2/g)	20–30	50–850[c]	~1300[c]
Tensile strength (GPa)	2.9	10–60	50–500
Tensile modulus (GPa)	240	1000	1500
Electrical resistivity (Ω cm)	1×10^{-4}	2×10^{-3} to 1×10^{-4}	1×10^{-3} to 1×10^{-4}
Thermal conductivity (W/m K)	1950	3000–6000	3000–6000

[a]Based on Breuer and Sundararaj [13] and Tibbets et al. [91]
[b]Based on Winey and Vaia [97]
[c]Theoretical estimate depending on the number of CNT walls, from 40 to 2 walls for MWNT [70]

and the CNT growth parameters are well identified and controlled yielding high purity CNT, however containing residual metal catalyst and a less perfect structure, compared to CNT produced by high temperature methods.

Carbon nanofibers have been known and produced for a long time [7], (Oberlin [64]). Their synthesis is based on catalytic CVD in a process that is similar to CNT production, but their structure and dimensions differ considerably [37, 59, 91]. Although they are formed by sp^2 C, their surface chemical activity is higher compared to CNT, while their mechanical and electrical properties are lower. Nevertheless, their lower cost, relatively easier incorporation in matrices and interesting set of mechanical, electrical and thermal properties, make them competitive for a wide range of composite applications. Table 1 presents typical values of physical properties for SWNT, MWNT and CNF, as reported in the literature for nanoparticles produced by different methods.

3 Carbon Nanotube and Nanofiber Polymer Composites: Properties and Applications

The unique set of mechanical, electrical and thermal properties of CNT, as well as their large aspect ratio and high surface area, make CNT remarkable fillers for polymer matrices. However, although their excellent properties are expected to translate into high performance composites, other characteristics such as nanoparticle agglomerate formation (arising from CNT or CNF entanglement during growth) may strongly influence composite properties. The as-produced MWNT and CNF are typically entangled and agglomerated, as represented in Fig. 5a, b, while the SWNT form "ropes", or nanotube strands, strongly attached to each other through Van der Waals interactions. These "ropes" are observed as long fibers with a diameter that may exceed 20 nm (Fig. 5c). The stable aggregates greatly decrease

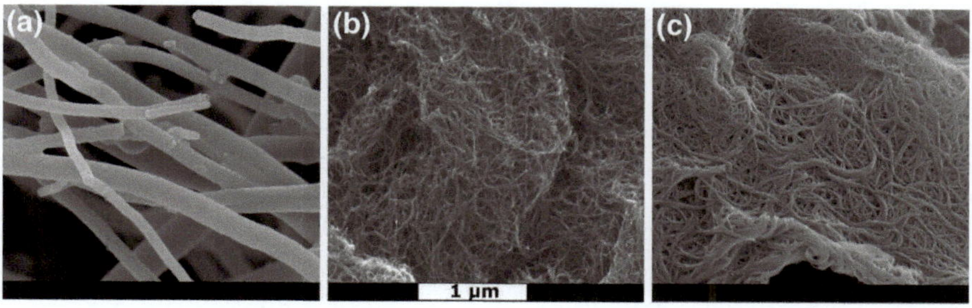

Fig. 5 Scanning electron micrographs of CNF (**a**), MWNT (**b**) and SWNT "ropes" (**c**) observed under the same magnification (same scale bar)

the nanoparticle surface available for interfacial interactions and hinder the dispersion of the individual CNT or CNF in the polymer matrix.

Molecular mechanics and molecular dynamics simulations provide a good insight into the local interactions among individual atoms of CNT and the polymer matrix, helping to understand the load transfer and mechanical behaviour of carbon nanotube/polymer composites. For example, Arash et al. [5] used molecular dynamics simulations to estimate the elastic properties of the interfacial region in CNT/poly (methyl methacrylate) (PMMA) matrix composites under tensile loading. Their simulations estimate that the Young's modulus of a PMMA composite reinforced by an infinite length (5,5) CNT increases 16 times relative to pure PMMA. They also reported that the CNT/PMMA interfacial strength increases with increasing CNT aspect ratio.

Classical continuum mechanics has been used to study the mechanical response of carbon nanotube/polymer composites with SWNT and MWNT (for example, Wagner [95] and Tu and Ou-Yang [93], respectively), adding to the understanding of interfacial stress-transfer in CNT composites. However, modelling of nanocomposites still provides limited information as it cannot account for all the CNT features, such as structural defects on the nanotube surface, ropes or bundles of nanotubes, and waviness of the nanotubes in the nanocomposites. Thus, real composites present a (quite unpredictable) range of mechanical properties that are dependent on the CNT type and structure, level of entanglement and aggregation. Table 2 provides examples of tensile properties experimentally measured for CNT composites prepared by melt mixing methods, for CNT incorporation levels ranging from 0.1 to 5 wt%. Generally, the addition of CNT has a positive effect on the composite Young's modulus, and this property is sensitive to the wt% of CNT and to interfacial effects induced by CNT surface functionalization. Composite strength trends are less predictable, since strength is highly affected by interfacial quality, CNT defects, presence of CNT agglomerates, etc. CNT functionalization frequently results in higher composite strength, associated to stronger interfacial strength, as illustrated also in Table 2.

The electrical conductivity of CNT and CNF make them suitable for the preparation of conductive polymer composites. The large aspect ratio of these

Table 2 Representative results for tensile moduli and strength of CNT/polymer composites produced by melt mixing methods

Matrix/CNT type	CNT wt%	% increase Tensile strength	% increase Young's modulus	CNT % at electrical percolation	Reference
PP/pristine MWNT	2	−0.25	15	1.1	Micusik et al. [55]
PP-MA/pristine MWNT	2	−88	50	1.6	
PP/PP-modified MWNT	4	3	42	4	Novais et al. [62]
PP/pristine MWNT	4	0	6.5	<4	
PA6/purified MWNT	1	−19	13	–	Xia et al. [99]
PA6/polymer encapsulated MWNT	1	−5	34.5	–	
PA6/purified MWNT	1	115	71	–	Zhang et al. [101]
	2	162	214	–	Liu et al. [48]
PA6/pristine MWNT	1	164	220	–	Mahfuz et al. [51]
PA6/SWNT	1.5	−24	15	–	Bhattacharyya et al. [11]
PA6/SMA-modified SWNT	1.5	−17	−16	–	
PA6/pristine MWNT	1.5	38	37	<4.5	Ferreira et al. [27]
PA6/Pyrrolidine functionalized. MWNT	1.5	10	10	>4.5	
HDPE/pristine MWNT	5	12	10	–	Tang et al. [87]
HDPE/PEG-SiO2 modified MWNT	1	20	50	–	Zou et al. [102]
LDPE/MWNT	0.5	−27	6	–	Yang et al. [100]
LDPE/MWNT	2	−27	20	–	
LDPE/PE-grafted MWNT	0.5	28	61	–	
LDPE/PE-grafted MWNT	2	23	92	–	
SBBS/purified MWNT	3	23	102	–	Li et al. [45]
EVA/purified MWNT	3	−20	70	–	Peeterbroeck et al. [68]
PVA/pristine SWNT	2.5	−5	35	–	Paiva et al. [66]
PVA/modified SWNT	2.5	17	40	–	
PVA/modified SWNT	5	54	55	–	

(continued)

Table 2 (continued)

Matrix/CNT type	CNT wt%	% increase Tensile strength	% increase Young's modulus	CNT % at electrical percolation	Reference
PLA/pristine MWNT	0.5	8	6	0.5	Novais et al. [63]
PLA/PLA-modified MWNT	0.5	20	17	1.5–2.0	
PET/pristine SWNT	0.1	8	20	1	Anand et al. [4]
	1	24	58		

Whenever available, the wt% of CNT required to reach electrical percolation is included

nanoparticles allows the modification of the electrical properties of their composites even at low filler content. The electrical percolation threshold varies with the CNT or CNF type, but also with the polymer nature, dispersion level, dispersion method, etc. Bauhofer and Kovacs [9] analyzed a large amount of published data concerning electrical conductivity of polymer/CNT composites produced by solution and melt mixing methods. They concluded that the minimum percolation threshold and maximum conductivity are mainly dependent on the type of polymer matrix and dispersion method and emphasized that non-entangled MWNT originated higher composite conductivity (50 times higher) compared to highly entangled MWNT. Figure 6 [9] presents the values of maximum conductivity as a function of CNT concentration for composites with different types of CNT and polymers.

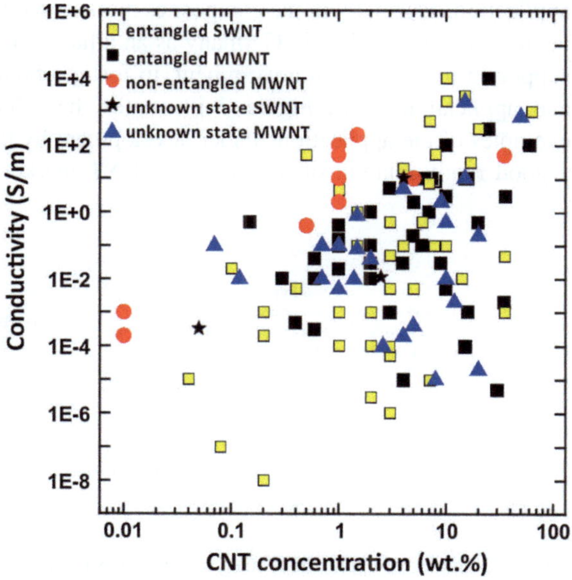

Fig. 6 Maximum conductivity reported for composites produced with different matrices and CNT types, at respective CNT concentration (experimental results selected from data published between 1998–2007. Adapted from Bauhofer and Kovacs [9])

The electrical conductivity of CNF/polymer composites was also observed to be significantly affected by the polymer type and physical properties (surface tension, crystallinity, polarity and molecular weight), as reported by Al-Saleh et al. [2]. The authors underlined the influence of the processing method on the CNF electrical percolation concentration, as well as the importance of selecting dispersion conditions that do not reduce the CNF aspect ratio. Percolation threshold for CNF/polymer composites reported in the literature range from 2 to 18 wt%, varying with CNF, polymer type and mixing method.

CNT and CNF are also expected to increase the thermal conductivity of their polymer composites, although only minor increments were reported so far. Thermal properties are largely dependent on the CNT length and alignment. High thermal conductivity values were measured for aligned CNT fibers produced from forests, such as reported by Myhew and Prakash [58]. The authors measured the thermal conductivity of fibres produced with CNT to be 448 ± 61 W/(m K) and for CNT/polymer composite fibres to be 225 ± 15 W/(m K).

Polymer nanocomposites with good mechanical and electrical properties already find a number of applications. They are used to manufacture fuel lines and filters that dissipate electrostatic charge and parts for electrostatic-assisted painting for the automotive industry, electromagnetic interference (EMI)—shielding packages and wafer carriers [24], touch panels and displays, diodes and transistors for the microelectronics and electronics market, electrical wire and cable and sporting goods (such as racquets, golf clubs, surfboards and ice hockey sticks). In the form of arrays, films, filaments or yarns, CNTs have also been utilized in battery/capacitor electrodes, membranes, sensors, heat exchangers, wind turbine blades and in biomedical applications (drug delivery, biosensors).

The continuous drop in price of CNT will certainly contribute to their rising application, especially in the composites market (see Fig. 3). In 2011, the Inno.CNT venture was established in Germany as an alliance between academia and industry, supported by the federal government, to foster a breakthrough in CNT technology and applications [34]. Figure 7 illustrates the selected fields and offers a few examples of the applications under development by this initiative, thus constituting a good representation of the emergent CNT market. Another stimulating field is

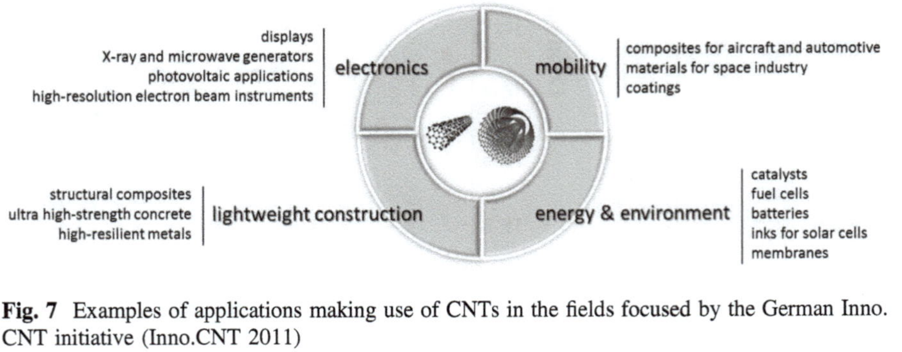

Fig. 7 Examples of applications making use of CNTs in the fields focused by the German Inno. CNT initiative (Inno.CNT 2011)

smart textiles, where CNT can be used to manufacture different types of sensors, such as for movement (torsion, displacement) and health and safety (temperature, liquid and gas exposure), or wearable electronics. Conductive adhesives, as well as biomedical detection, imaging and therapeutics are also potential areas of product development.

4 Preparation of Nanocomposites

4.1 General Preparation Methods

In order to obtain polymer/CNT or polymer/CNF nanocomposites with optimal performance, it is generally necessary to ensure extensive dispersion of the filler in the polymer matrix (but the required dispersion levels to attain the best mechanical properties or electrical conductivity may be different). Dispersion involves the disruption of the initial agglomerates (the process is sometimes denoted as deagglomeration) until the dominant size of the filler becomes nanoscopic (ideally, until disintegration into isolated particles/fibers is achieved). This process must be accompanied by adequate distributive mixing, in order to assure spatial compositional homogeneity. The improvement of the interaction between filler and polymer can be approached by adding surfactants, or by functionalizing the surface of the as-produced fillers [12, 18, 49]. In practice, the dispersion of CNT and CNF is quite difficult [3, 36, 40, 83] due to three main causes:

- They grow as highly entangled agglomerates of several microns or even millimeters in size;
- Van der Waals interactions between individual tubes or fibers promote significant aggregation; together with the physical entanglement, the outcome are highly cohesive agglomerates;
- Chemical inertia of the nanoparticle surface creates weak interfaces with most polymers (the process can be further aggravated by surface contamination resulting from the manufacturing technique), which may prevent reaching the hydrodynamic stresses required for dispersion, as well as the load transfer necessary for enhanced mechanical response under service conditions.

Several methodologies have been developed for the manufacture of polymer/CNT and polymer/CNF composites [13, 79]. In situ polymerization of the monomer in the presence of the filler (e.g. Lin et al. [46], Wu and Chen [98]), solution processing (e.g. Qian et al. [74], Safadi et al. [75], Chen et al. [16], Huang and Terentjev [31], Thomassin et al. [90]) and melt mixing (e.g. Socher et al. [85], Novais et al. [61, 62], Sathyanarayana et al. [80], Jamali et al. [35], Novais et al. [63]) are the most utilized, but other routes have also been explored [79].

In situ polymerization of the monomer in the presence of an initiator and CNT or CNF allows the production of composites with high filler loadings. The technique is particularly suitable for insoluble and thermally unstable polymers. Depending on the target molecular weight (MW) and MW distribution, chain transfer, radical, anionic and ring-opening polymerization routes can be pursued. Covalent grafting of polymers to the nanotubes can be made either as "grafting from" (i.e., initial immobilization of the initiators onto the filler surface followed by the attachment of the monomers and subsequent growth into polymers) or as "grafting to" (when previously end-functionalized polymer macromolecules are attached to functional groups on the filler), leading to distinct grafting densities. Concurrently, the low viscosity levels of the starting monomer enable its efficient infiltration into the filler bundles, thus contributing to an easier deagglomeration of the latter.

Solution processing is widely used due to its simplicity: filler, surfactant and polymer are mixed in a suitable solvent, followed by evaporation of the latter with or without vacuum to (frequently) form a film. Agitation of the low viscosity medium (ultrasonication is often employed) facilitates infiltration and deagglomeration. The former is easier if the CNT/CNF are first dispersed in the solvent, whereas the increased viscosity of a polymer solution will be more efficient for deagglomeration. Although solution processing is not easily scalable and total removal of the solvent may be problematic, it is often adopted for thermoset matrices (e.g. epoxy), which do not require significant shear magnitudes [79]. Conversely, it is unsuitable for insoluble polymers.

4.2 *Carbon Nanotube and Nanofiber Functionalization for Interfacial Enhancement*

The smooth CNT surface (at the atomic level) is prone for physical adsorption of specific molecules and may be chemically modified, or functionalized, by non-covalent chemistry [25]. In this process, the CNT physically adsorb molecules that are constituted by a CNT-compatible part, such as an aromatic moiety and a solvent-compatible part. This type of functionalization is most efficient in the preparation of stable CNT suspensions in solvents, including water [6, 25] and find applications in solution-based composites.

Strong physical interactions help enhancing CNT/matrix interface; however, covalent bonding leads to the formation of much stronger interfaces and thus to higher stress transfer efficiency. It is generally recognized that adequate covalent functionalization of the CNT or CNF surface leads to improved mechanical properties of their polymer composites. A considerable number of organic reactions may be used to functionalize the sp^2-C of CNT surfaces, as described by Hirsch [30] and Tasis et al. [88], for example. The more common approach is based on CNT or CNF oxidation using oxidizing inorganic acids, mixture of acids and other oxidative reagents [22]. These methods are often used as cleaning procedure to remove disordered carbon and purify pristine CNT. Depending on the reaction conditions, oxidative methods bond oxygen-containing groups to the CNT surface,

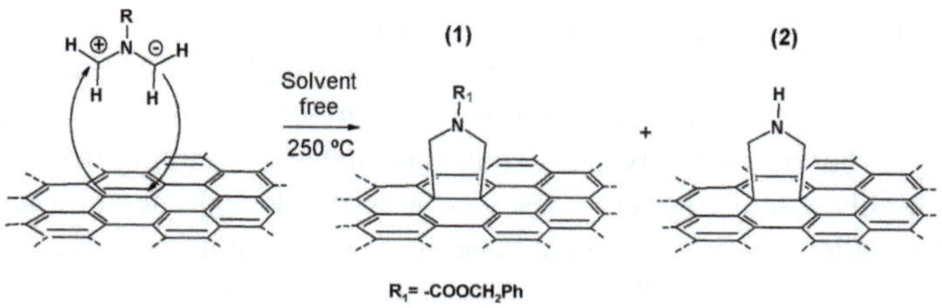

Fig. 8 Functionalization scheme for the 1,3-dipolar cycloaddition of azomethine yilides and products formed under solvent-free reaction conditions. Adapted from Paiva et al. [63]

which may further react through tailored reaction routes. However, extensive oxidation also induce CNT breakage and a considerable decrease of the CNT aspect ratio, which is detrimental for composite properties.

An interesting class of covalent functionalization routes is based on addition reactions to the CNT or CNF surface π-electrons. Cycloaddition reactions were successfully applied to these materials, such as carbene [2 + 1] cycloaddition [17], nitrene functionalization [30], and Diels-Alder cycloaddition [38, 73]. The 1,3-dipolar cycloaddition of azomethine ylides, first used for the functionalization of CNT by Georgakilas et al. [28], originates pyrrolidine functionalized CNT (or CNT with substituted pyrrolidines), as represented in Fig. 8 (2 and 1, respectively). This reaction was carried out in the solid state [65], reducing the reaction time from several days to 3 h and eliminating the use of dangerous solvents.

The CNT or CNF/polymer interactions may be maximized if the nanoparticle surface is modified by covalently bonding polymer molecules. Covalent functionalization of CNT with polymers was achieved by in situ polymerization [19, 44]. A simple route for the covalent bonding of polymer molecules to CNT was achieved under melt processing conditions, thus avoiding in situ polymerization procedures. This method uses solid-state pyrrolidine-functionalized CNT [65], as they were observed to bond to specific polymers under melt processing conditions. The functionalized CNT are thermally stable up to approximately 300 °C, allowing for melt mixing with several polymers without thermal degradation of the functional groups. At these high temperatures and in the polymer melt the pyrrolidine groups were observed to be highly reactive towards ester-based polymers, as well as polycarbonates and maleic anhydride modified polymers. Reaction in solution was only observed with maleic anhydride-modified polymers. This enabled reactive extrusion, forming composites with polymer functionalized CNT during the melt mixing step [62, 63].

4.3 Melt Mixing Methods

Melt mixing (also known as compounding) is probably the most common route to prepare thermoplastic polymer/CNT or polymer/CNF nanocomposites. It consists in the physical/mechanical mixing of the filler with amorphous polymers above their glass transition temperature, or with semicrystalline polymers above their melting temperature. The application of intensive hydrodynamic stresses (usually of the shear type, but occasionally also extensional stresses) resulting from the laminar flow of a highly viscous system causes progressive deagglomeration and disintegration of the filler bundles, with simultaneous distribution in the polymer melt. The method is simple, adaptable to the specific characteristics of different polymer/carbon-based filler systems (tunable process parameters include type and geometry of the equipment and operating conditions) and attractive within an industrial context, as it uses commercial machines (often twin screw extruders, that are capable of a continuous high yield) and can be scaled-up directly. It is also viable for those polymers that cannot be utilized in solution processing. A recent practical trend consists in the preparation (by material manufacturers) of well dispersed masterbatches where the filler is present at high concentration, that are subsequently diluted by processors or end users to the required filler content by mixing with more polymer. This approach brings two major benefits: (i) processors or end users are no longer exposed to direct contact with nanoparticles, which brings about significant health and safety gains, as well as logistical simplifications; (ii) the masterbtach can be directly used in standard shaping techniques, such as extrusion and injection molding. A proposed variant consists in preparing polyethylene/CNT concentrates (with CNT loadings typically in the range 24–44 wt %) followed by dilution with other polymers (for instance, polycarbonate, polyamide). As a consequence of the high interfacial energy between CNT and polyethylene, the filler migrates into matrix polymers with lower interfacial energy [71].

Due to its practical importance, melt mixing of CNT or CNF with thermoplastic matrices will be analyzed below in greater detail. The effect of material and process parameters will be discussed and illustrated with a few examples. A general dispersion mechanism will be presented.

Thermoplastic polymer nanocomposites are usually prepared in batch mixers or extruders. The former are useful for laboratorial R&D and include internal intensive mixers consisting of two counter-rotating rotors (also known as Haake or Brabender mixers, two well-known equipment manufacturer brands), micro-compounders and small-scale prototype devices (for example, Maric and Macosko [52], Lin et al. [47], Novais et al. [61]). At the industrial production scale, intermeshing conjugated co-rotating twin screw extruders are the most popular machines. These are also widely utilized for other plastics compounding operations, such as additivation and polymer modification and blending, especially due to two dominant features: (i) modular construction, i.e., the geometry of both screw and barrel can be changed, enabling to define and adjust the sequence and configuration of process steps such as melting, mixing, devolatilization, secondary feeding, pressure generation, etc. (ii) the operator has independent control over output, screw speed and temperature, i.e., over residence time and shear intensity. Thus, the machine can be adapted both in terms of geometry and operating conditions to the needs of a particular system. The screws usually consist of a series of adjacent individual elements that are hold together at the tips. A typical screw profile includes conveying elements separated by mixing elements that are restrictive in terms of forward transportation and, depending on their geometry, induce more or less intensive distributive or distributive/dispersive mixing. In conveying sections, the material follows a figure-of eight pattern along the helical channels of the partially filled screws, producing some degree of distributive mixing. Once the material attains a restrictive element (typically, a kneading block comprising disks staggered at a negative or neutral angle, or a conveying element with a negative helix angle), it accumulates immediately upstream, in order to generate the pressure required to continue the flow in the axial direction. The higher the pressure required, the longer the length of the conveying section that works fully filled. Restrictive elements create a complex 3D flow and can induce significant shear and extensional stresses, according to their arrangement (see also Kohlgruber [41]). Generally, when preparing polymer nanocomposites, the first restrictive section upstream prompts polymer melting due to the combined contribution of local frictional forces, pressure and heat transfer. The filler is then added and the material flows through various mixing zones downstream, which are generally designed to induce dispersion and distribution without causing excessive filler damage (e.g. fiber breakage) and/or viscous dissipation, which could cause thermal degradation of the matrix and/or surfactant. Prior to generating pressure for extrusion through the die, the material may flow through a devolatilization zone.

It is also important to note that nanocomposites are typically subjected to two thermomechanical sequences, compounding and shaping. During the first, specialty equipment such as the twin screw extruders presented above are capable of generating during sufficient time the stress levels and flow patterns assuring adequate filler dispersion and distribution in the matrix. Subsequent extrusion or injection molding (the two most industrially relevant polymer processing techniques) will generate much lower stress levels, simpler flows and, particularly in the case of extrusion, may subject the composite to significant stretching/orientation. This means that: (i) probably, the necessary dispersion levels should be achieved during compounding, as processing will bring about a small additional contribution; (ii) the viscoelastic properties and the thermal stability of the as-compounded nanocomposite are essential for trouble-free shaping; (iii) specific processing steps, such as axial stretching, may cause an increase in the distance between neighboring individual filler particles, thus affecting, for example, the electrical performance (i.e., creating the need either to achieve higher dispersion levels, or to use higher filler concentrations than initially anticipated).

CNT or CNF dispersion in a molten matrix is a complex process that is influenced by a number of material and processing factors, at various length scales (see Fig. 9). Material parameters encompass agglomerate size, structure and cohesion, surface energy and purity of the filler, as well as viscosity levels, affinity with the filler and thermal stability of the polymer matrix [40]. Processing factors mainly involve velocity fields, intensity and type of the exerted hydrodynamic stresses (i.e., shear versus extensional) and residence time [35, 39, 47]. Indeed, it has been shown both for immiscible fluid systems and solid suspensions (including polymer-clay

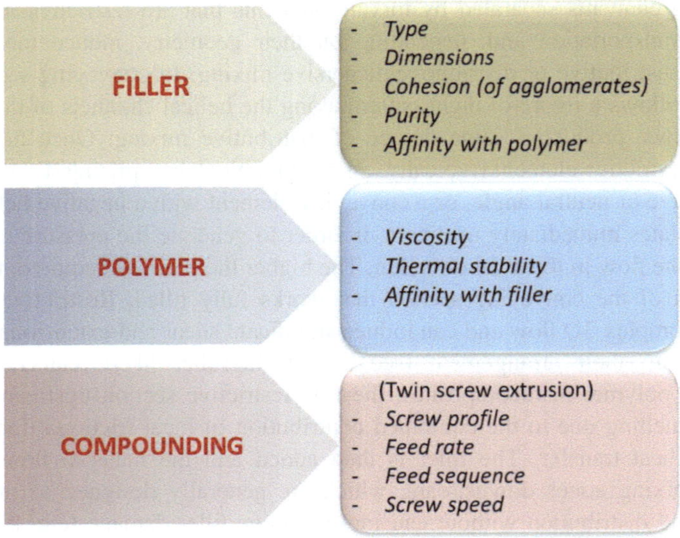

Fig. 9 Main parameters affecting the dispersion of carbon-based fillers in thermoplastic polymer matrices by melt mixing

nanocomposites) that extensional flow promotes dispersion more efficiently than pure shear flow [29, 92].

Since the various manufacturers produce carbon-based fillers with distinct characteristics, it is not surprising that, under identical compounding conditions, nanocomposites with different morphologies will be obtained. This is exemplified in Fig. 10 for the electric conductivity of various Polyamide12/CNT nanocomposites, each containing a different commercial CNT (the figure also includes, for comparison purposes, a carbon black (CB) alternative). The electrical percolation threshold is attained for separate incorporation levels of the various CNT (with carbon black requiring much higher concentrations) [84]. Pötschke and co-workers [42, 43, 69, 84] found a correlation between dispersibility of CNT and bulk density of the initial agglomerates). Likewise, Salzano de Luna et al. [76] showed that synthesized CNT particles in the form of small and loosely packed clusters made by interwoven bundles of combed yarns of nanotubes were easier to disperse then the reference denser commercial counterparts. Matrix viscosity is also a primary parameter. Socher et al. [85] revealed that dispersion of CNT agglomerates increased with increasing matrix viscosity, due to the higher input of mixing energy. In addition, different degrees of nanotube shortening during mixing were found when using matrices with varying viscosities. However, the lowest electrical percolation thresholds were always found in the composites based on the lower viscosity matrix.

It has been extensively demonstrated that polymer/CNT and polymer/CNF nanocomposites prepared by melt mixing typically contain aggregates (this

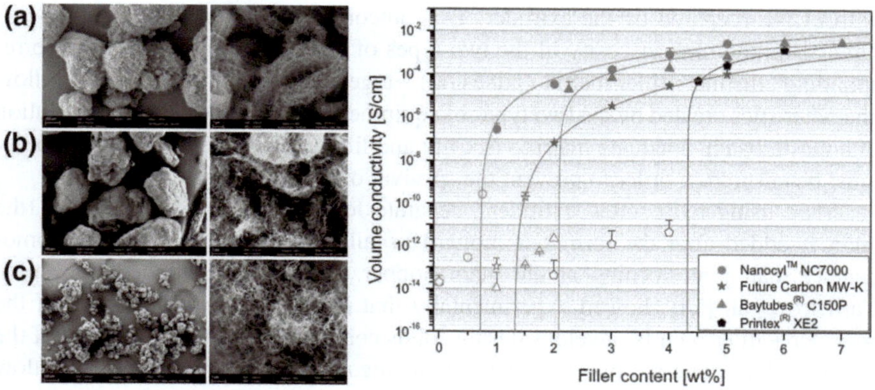

Fig. 10 Commercial CNT powders and electrical volume conductivity of the corresponding composites with Polyamide 12. *Left* CNT as seen by scanning electron microscopy at two magnifications: **a** Nanocyl™ NC7000, **b** Baytubes C150P, **c** FutureCarbon CNT-MW-K. *Right* Effect of filler content on the electrical volume conductivity of PA12/MWNT and PA12/carbon black (Printex XE2 from Evonik Degussa GmbH, Germany). Adapted from Socher et al. [84]

designation is used here for clusters of particles that are smaller than the initial agglomerates), together with well dispersed particles, and that the size and number of the former strongly depend on the melt mixing method and procedure [10, 32, 67]. For instance, Fig. 11 depicts representative optical microscopy and SEM micrographs of polypropylene/CNF nanocomposites (containing either as-received or functionalized CNF (FCNF)) prepared by twin screw extrusion and by a prototype mixer generating a strong extensional flow component [62]. The difference in the levels of dispersion of the various systems is evident, smaller particles being present in the samples containing CNF and produced by the prototype mixer and in the composite

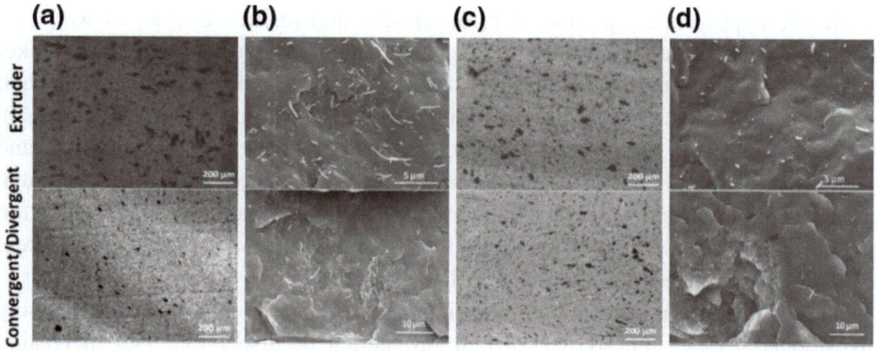

Fig. 11 Optical (**a** and **c**) and scanning electron microscopy (**b** and **d**) micrographs of polypropylene/CNF composites manufactured by twin screw extrusion and by a prototype extensional mixer. The first two columns correspond to composites containing CNF, the remaining refer to composites with functionalized CNF. Adapted from Novais et al. [61]

with FCNF prepared by the extruder. This outcome probably results from the differences in the melt viscosity of the two types of composites (caused by the corresponding distinct filler/matrix adhesion), together with the distinctive flow characteristics created by the two types of equipment. Even in the case of the dilution of a masterbatch, adequate matrix viscosity and filler/matrix chemical compatibility must be ensured in order to prevent the survival of large aggregates [55].

When using twin screw extruders, separate feeding of polymer and filler (the latter is added after the former is molten) usually yields better composite homogeneity than joint feeding at the main hopper. Using a Polyamide/CB system, Carneiro et al. [14] showed experimentally that in joint feeding de-mixing of the polymer + filler premix develops during solids conveying and that at the start of the first kneading zone the two materials are segregated to a point where during flow along this zone there is only limited opportunity for the efficient intermingling of the two components. Conversely, in the case of separate feeding, Carbon black falls directly onto a stream of molten polymer, so that distributive mixing immediately develops and, upon reaching the upcoming kneading zone, dispersion and distribution can be very efficient. Nevertheless, Muller et al. [60] observed that filler type may influence the choice of the best feeding mode. Thus, in order to obtain better dispersion of CNT in polypropylene or in polycarbonate, as well as lower electrical resistivity and good mechanical properties, it was found that separate feeding should be selected for CNT Nanocyl^{TM}, which exhibits a loosely packed structure and a lower agglomerate strength, whereas the denser structure of CNT Baytubes^{TM} advised the adoption of joint feeding (Fig. 10 portrays the morphology of these two commercial products). In turn, setting an optimum extruder screw profile may not only guarantee the appropriate and efficient sequence of the necessary process steps for compounding (typically, as seen above, polymer feeding and melting, feeding of the filler, mixing, devolatilization and pressure generation), but also the adequate dispersion and distribution of the filler in the matrix. For this purpose, the number, length and geometry of the restrictive screw elements (particularly the staggering angle and the length of the individual disks) must be carefully chosen, in order to create the appropriate type and level of stresses and residence time, as well as a suitable flow kinematics. Neutral and negative staggering angles, as well as thicker disks, will create both dispersive and distributive mixing, whilst positive angles and thin disks will induce distributive mixing. However, a too restrictive screw can suffer from limited throughput capacity, considerable energy consumption and insufficient devolatilization, all due to the high average degree of screw filling, and can also cause material degradation owing to viscous dissipation. Conversely, a too mild profile will face difficulties in melting the polymer and in creating the stress levels and residence time required for dispersion.

The effect of the operating conditions on dispersion has been extensively investigated [3, 32, 39, 61, 67, 94]. Improvement of dispersion with increasing screw speed is often reported and attributed to the accompanying higher stress levels. However, the viscous dissipation associated to screw speed may induce degradation and deterioration of the composite performance (it seems worth reminding that viscous dissipation is proportional to viscosity and to the square of the shear rate). Also, high screw speeds diminish the degree of screw fill and the residence time (when the remaining parameters are kept constant). High throughputs entail a reduction of the residence time and an increase of the degree of screw fill, both hindering dispersion. Barrel set temperatures essentially influence the viscosity of the system. In principle, lower melt viscosities facilitate infiltration into the CNT agglomerates, whereas higher viscosities should bring about the higher hydrodynamic stresses required for deagglomeration. The Specific Mechanical Energy consumption (SME) is a global process parameter that is often used to characterize the intensity of the thermo-mechanical environment created inside a twin screw extruder during a particular operation. It is defined as the amount of power input by the motor into each kilogram of material being processed. Multiple attempts to correlate dispersion of nanocomposites with SME have been made, but they obtained limited success. This is probably due to the contribution of factors that are not taken in by this parameter, such as the axial residence time, degree of fill and melt temperature profiles. In conclusion, for each nanocomposite, an optimum combination of screw design and operating conditions will probably exist, albeit being difficult to forecast without performing some preliminary testing.

Even if there is obvious useful value in correlating material and process parameters with final dispersion levels (in turn, correlations between the latter and the final properties are also essential), studying the spatial or temporal evolution of the process should provide important insights into the dispersion mechanisms of carbon-based fillers and thus assist practical process setup and optimization. Surprisingly, this type of studies is relatively scarce [3, 35, 61, 62]. Figure 12 shows the development of the dispersion of polypropylene/CNT composites along a prototype extensional mixer (same equipment as the one mentioned above and used to produce the material depicted in Fig. 11) and along a co-rotating twin screw extruder (Fig. 12a, b, respectively) [35, 62]. In the first case, the graphs illustrate the progress of the average agglomerate area and of the electrical resistivity of the composite through a series of flow channels (labelled as pairs of rings) that create a sequence of repetitive convergent/divergent flows, when processing at 100 and 3000 s^{-1}. The far from linear evolution of dispersion is obvious. Since the area of the smallest agglomerates of the as-received in powder form CNT is approximately 3.58×10^5 μm^2, and the initial average value of approximately 1000 μm^2 is measured at the first pair of rings when processing at 100 s^{-1}, a significant dispersion took place upon flow through the first convergence. Thus, even at this relatively low shear rate, the hydrodynamic stresses generated overcome the cohesive strength of the agglomerates. After this initial step, dispersion seems to proceed gradually, but after a certain residence time (corresponding to attaining pair 4) faster dispersion is again triggered. This behavior is matched by an equivalent decline of the electrical resistivity between pairs numbers 7 and 8. Once smaller aggregates were formed, further progress in dispersion seems to require higher hydrodynamic stresses and/or longer residence times, as inferred from the little evolution in dispersion beyond pair 6 (and in electrical resistivity after pair 8). Processing at a high wall shear rate (3000 s^{-1}) reduces the size of the particle clusters present after the first pair of rings down to roughly half the size of those

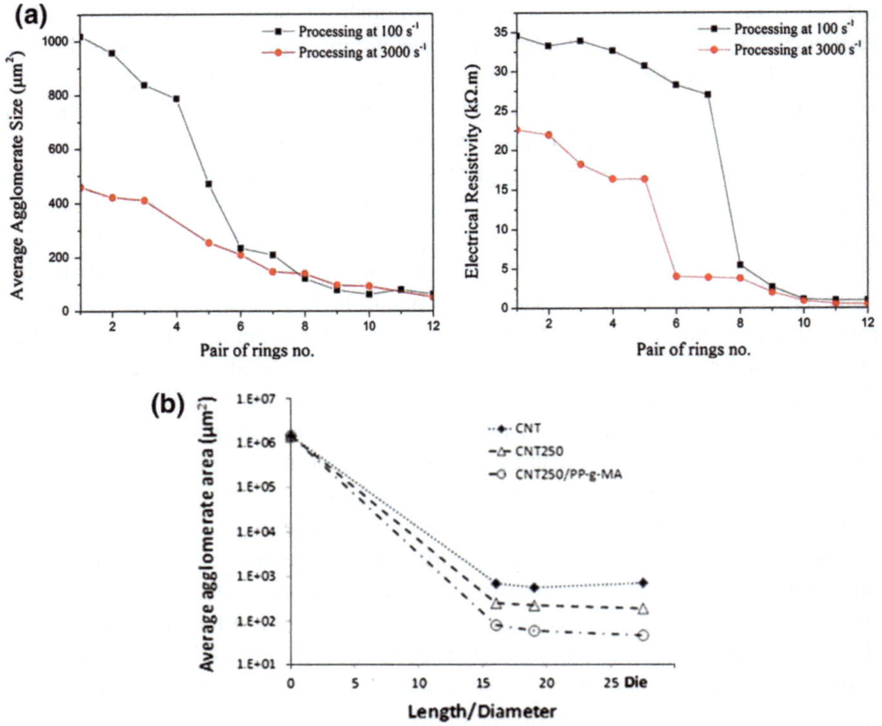

Fig. 12 Evolution of the dispersion of polypropylene/CNT composites along the length of the mixing device; **a** average agglomerate area (*left*) and electrical resistivity (*right*) along a prototype extensional mixer, at two shear rates (adapted from Jamali et al. [35]); **b** average agglomerate area along a twin screw extruder, using three types of CNT: as-received (CNT), functionalized by the 1,3-dipolar cycloaddition reaction of azomethine ylides at 250 °C (CNT250) and CNT250 further functionalized with PP-g-MA and with PP molecules bonded to the CNT surface (CNT250/PP-g-MA) Adapted from Novais et al. [62]

seen at 100 s^{-1}. This is followed by a gradual dispersion until reaching dispersion levels similar to those obtained at 100 s^{-1}. Still, the electrical resistivity also shows a sharp decline, now between pair of rings 5 and 6, i.e., earlier than in the preceding condition. Since data on level of dispersion and electrical conductivity were made available, the existence of a correlation between both could be investigated. Indeed, an electrically conductive network seemed to be formed when at least 50 % of the surviving aggregates had an area smaller than 2000 μm^2 [35]. When using an extruder (Fig. 12b), Novais et al. [62] observed that most of the decrease in agglomerate size took place along the first half of the extruder, up to the first kneading zone. Here, the screw channels work fully filled due to the flow restriction created and temperatures are still low, as polymer melting is ongoing. Hence, substantial hydrodynamic stresses may develop, together with a relatively large local residence time and a complex 3D flow. A comparable behavior has been reported for other types of nanocomposites and polymer blends [8, 50]. Although higher stresses and/or longer mixing times are probably needed to break-up the aggregates formed upstream, temperature is much higher in the second part of the extruder and thus the process is difficult. Anyway, dispersion should depend on the cohesive strength and interfacial bonding of each system, differences being perceived in Fig. 12b for the three types of CNT that were used in the experiments: (i) as-received (CNT), (ii) chemically functionalized CNT by the 1,3-dipolar cycloaddition reaction of azomethine ylides (discussed earlier in this chapter) performed at 250 °C (CNT250) and (iii) CNT250 further functionalized with PP-g-MA and with PP molecules bonded to the CNT surface by reaction of the NH-pyrrolidine groups with the anhydride grafted on the PP-g-MA (CNT250/PP-g-MA) (the aim was to create covalent bonding between the pyrrolidine groups at the CNT surface with the maleic anhydride on the PP-g-MA molecules).

The evolution of the dispersion of a polypropylene/CNF composite along the length of the extensional mixer, as depicted in Fig. 13 [61], can be distinct from that pictured in Fig. 12a for a PP/CNT system. Data for composites incorporating as-received CNF and functionalized CNF (using the 1,3-dipolar cycloaddition reaction cited above) are presented in terms of Area Ratio (A_R), that is, the ratio of the sum of the areas of all surviving clusters to the total area of composite analyzed.

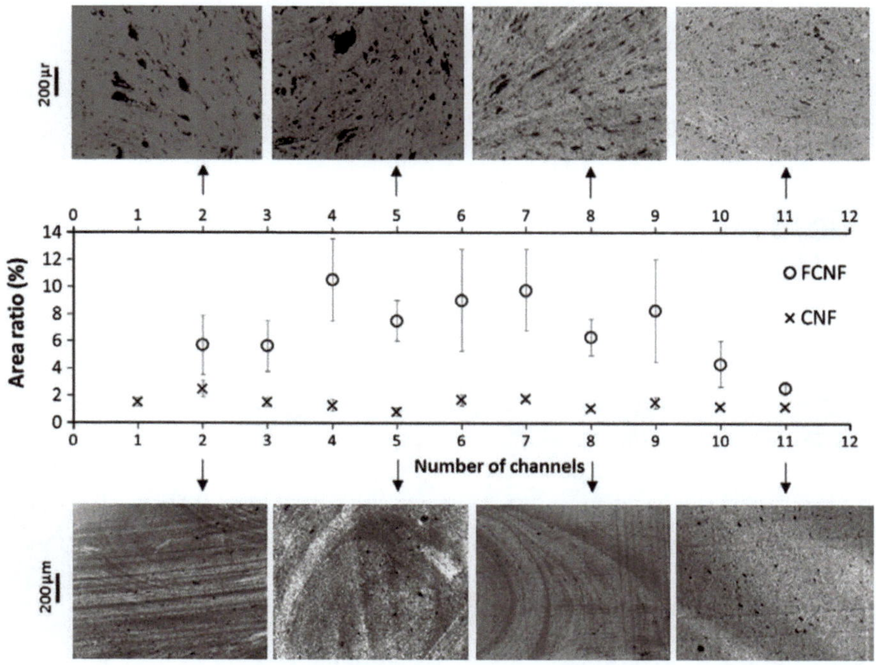

Fig. 13 Evolution of the dispersion of polypropylene/CNF composites along the length of the prototype extensional mixer, containing as-received (CNF) or functionalized (FCNF) fillers. Adapted from Novais et al. [61]

The low molecular weight PP grafted with maleic anhydride has greater mobility relative to the high molecular weight unmodified PP, as well as good chemical compatibility with the FCNF surface. The PP/CNF composite evidenced a relatively similar low agglomerate area ratio along the mixer, which indicates that significant filler dispersion was achieved during flow through the first converging/diverging sequence. Equivalent scanning electron microscopy micrographs (not shown) revealed that impregnation of the FCNF agglomerates was easier, resulting in the formation of PP-g-MA/FCNF clusters. However, these are not compatible with the non-polar, high molecular weight PP and, therefore, remain difficult to break into smaller fragments. The combined effect of these two factors could explain the slight increase in Area Ratio (A_R) measured for the FCNF filler along the first part of the mixer. The subsequent gradual decrease in A_R probably took place by erosion. Towards the exit of the mixer, sufficiently low values of A_R will correspond to increased viscosities and so to higher hydrodynamic stresses, generating higher dispersion rates.

The body of experimental knowledge accumulated on the dispersion by melt mixing of CNT and CNF in various thermoplastic polymer matrices, using different types of equipment and processing conditions, enabled the progressive identification of various common features that were used to gradually build-up a general phenomenological dispersion model. As illustrated in Fig. 14, dispersion starts with the polymer melt wetting the agglomerates of filler and then infiltrating them, thus

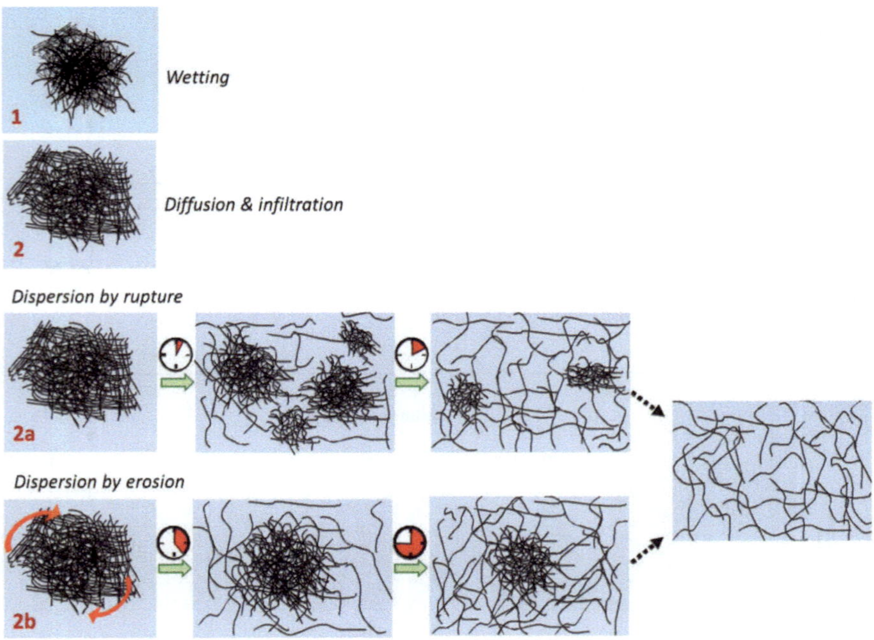

Fig. 14 Proposed general dispersion mechanism for fiber-like carbon-based fillers. Adapted from Kasaliwal et al. [40] and Sathyanarayana and Hübner [79]

reducing their strength. The application of sufficiently intensive hydrodynamic stresses (shear/extensional) during flow will cause a decrease in the size of the initial agglomerates by rupture and/or erosion. Rupture involves the successive break-up of the agglomerates into smaller aggregates and, eventually, into the individual particles. As the name implies, erosion involves the continuous detachment of particles or small aggregates from the surface of bigger clusters.

Effective wetting of the agglomerates by the polymer melt requires a small interfacial tension. This is not problematic when the hydrophilic CNTs are to be mixed with polar polymers (such as polyamide, polycarbonate, or polyimide). In the case of non-polar polymers (for example, polyolefins), it is necessary to incorporate surface functionalities to the carbon-based fillers, as discussed in detail in a separate section of this chapter. Infiltration of the polymer melt obviously depends on the agglomerate density and size (in the two cases, the lower the better [3], as well as on local surface tension gradients [79], but it unquestionably requires good mobility of the polymer chains, i.e., low viscosity values. Viscosity and elasticity increase steadily with filler content. Even at low filler concentration, the formation of a filler–polymer chain network produces a significant change in the rheological behavior, with a transition from liquid like to solid like behavior [1]. Dispersion will only develop when the magnitude of the hydrodynamic stresses acting on the agglomerates becomes higher than their cohesive strength. Estimates of inter-tube binding forces and shear forces suggest that only at sufficiently high shear energy density complete separation of CNTs can be achieved, but the considerable energy density input may also induce unwanted fiber breakage [31]. Conversely, if the hydrodynamic stresses are small, interfacial forces become dominant, similarly to what is well-known for colloids and immiscible polymer blends (where droplet coalescence is a commonly observed phenomenon). In these systems, the balance between the two types of stresses is usually quantified by the capillary number [29, 57]. Likewise, in the case of suspensions containing clusters of solid particles, Scurati et al. [82] defined a fragmentation number (Fa), given by the ratio between the hydrodynamic stresses and the cohesive strength of the agglomerate. In the case of a specific polymer/silica particles system, it was experimentally shown that for $Fa < 2$ dispersion does not occur, for $2 \leq Fa < 5$ the agglomerates erode and that for $Fa \geq 5$ rupture becomes the predominant mode of dispersion. Moreover, while rupture is a quick process, once erosion starts it continues for long times. Even for sufficiently high Fa, there is a finite probability associated to the break-up process, which is proportional to the fractional change in the agglomerate surface area with respect to the initial surface area [26]. Again, a parallelism can be made with the time for droplet break-up for colloids and immiscible polymer blends, which decreases with increasing stresses and is greater for smaller droplets [57]. Although critical Fa have not yet been estimated for CNT or CNF (this is not an easy challenge, as agglomerate strength depends on the extent of infiltration which, in turn, is governed by the various factors identified above), the concept appears to be applicable to these fillers [39, 79].

A demonstration that when the applied hydrodynamic stresses are small interfacial forces become dominant is clearly given in Fig. 15 (see detailed description

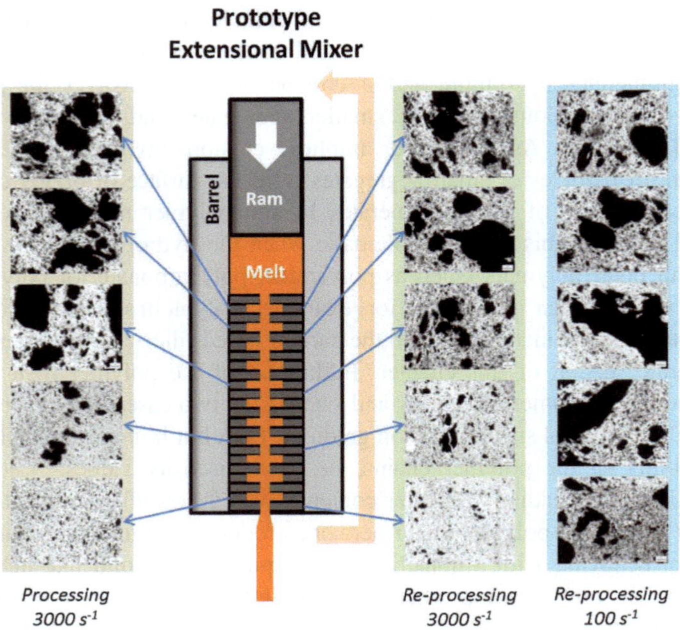

Fig. 15 Evolution of the morphology of a polypropylene/CNT nanocomposite prepared with the prototype extensional mixer at a shear rate of 3000 s^{-1} (*left* column) and then re-processed at identical (*central* column) and lower (*right* column) shear rates. Adapted from Jamali et al. [35]

and discussion in [35]. Using the prototype extensional mixer recurrently cited in this chapter, processing of a polypropylene/CNT composite was performed at 3000 s^{-1}. As shown by the micrographs in the left column of the figure, the system achieved a fine dispersion after passing through the 12 pairs of rings that created repetitive converging/diverging flows. Once extruded and cooled down to room temperature, the material was pelletized and re-heated in the reservoir of the capillary rheometer on the bottom of which the mixer is positioned, i.e., under quiescent conditions. Then, it was reprocessed at the same shear rate. As seen in the central column of Fig. 15, the evolution of morphology upon re-processing closely mimics that of processing under identical conditions. This means that significant re-agglomeration of the CNTs took place during the re-heating stage. Remarkably, if re-processing is performed at a much lower shear rate than that used for processing (in this experiment, 100 s^{-1}), dispersion becomes less efficient and the extrudate presents a coarser morphology—see right column of Fig. 15. Indeed, processing at 100 s^{-1} generated higher dispersion levels (not shown) than those obtained after processing at 3000 $s^{-1}$and then reprocessing at 100 s^{-1}. Thus, it appears that re-agglomeration is a fast process during which CNTs recreate entities with high cohesive strength. In these experiments, equivalence can be made between processing and re-processing and practical industrial compounding and processing. Compounding involves high shear rates and should be able to yield fine morphologies, but apparently these are not stable. During processing (i.e., shaping of a product), the morphology can become coarser due to reagglomeration, the magnitude of the phenomenon depending on the operating conditions. In the same way, Alig et al. [3] reported that by thermal annealing a quiescent melt, a secondary agglomeration (or cluster formation) of nanotubes took place, with the formation of a conductive network of interconnected agglomerates. Although this last result has significant practical importance, it could not be reproduced in the experiments discussed above [35].

5 Conclusions

Carbon-based nanotubes and nanofibers with excellent physical properties are now synthesized at large scale and descending price. Since their discovery the scientific knowledge on these nanomaterials has been growing steadily. In contrast, the hope for immediate applications has dropped drastically along the first decade of 2000. This "trough of disillusionment" did not affect significantly the potential markets that have been consolidating slowly but steadily.

Polymer nanocomposites incorporating CNT and CNF may be presently produced under controlled conditions and the expectations are rising again, envisaging new applications targeting at mechanical reinforcement, toughening, electrical conductivity, sensor integration, EMI shielding, and other prospects that will be presenting in the near future.

References

1. Abdel-Goad M, Potschke P (2005) Rheological characterization of melt processed polycarbonate multiwalled carbon nanotube Composites. J Non-Newton Fluid Mech 128:2–6
2. Al-Saleh MH, Sundararaj U (2009) A review of vapor grown carbon nanofiber/polymer conductive composites. Carbon 47:2–22
3. Alig I, Pötschke P, Lellinger D, Skipa T, Pegel S, Kasaliwal G, Villmow T (2012) Establishment, morphology and properties of carbon nanotube networks in polymer melts. Polymer 53:4–28
4. Anand A, Agarwal U, Rani J (2007) CNT–reinforced PET nanocomposite by melt compounding. J Appl Polym Sci 104:3090–3095
5. Arash B, Wang Q, Varadan VK (2014) Mechanical properties of carbon nanotube/polymer composites. Sci Rep 4:6479
6. Araújo RF, Silva CJ, Paiva MC, Melle-Franco M, Proença MF (2013) Efficient dispersion of multi-walled carbon nanotubes in aqueous solution by non-covalent interaction with perylene bisimides. RSC Adv 3:24535–24542
7. Baker RTK, Barber MA, Harris PS, Feates FS, Waite RJ (1972) Nucleation and growth of carbon deposits from the nickel catalyzed decomposition of acetylene. J Catal 26(1):51–62

8. Barbas JM, Machado AV, Covas JA (2014) Processing conditions effect on dispersion evolution in a twin screw extruder: polypropylene-clay nanocomposites. Chem Eng Technol 37:1–11
9. Bauhofer W, Kovacs JZ (2009) A review and analysis of electrical percolation in carbon nanotube polymer composites. Comp Sci Technol 69:1486–1498
10. Bhattacharyya AR, Sreekumar TV, Liu T, Kumar S, Ericson LM, Hauge RH (2003) Crystallization and orientation studies in polypropylene/single wall carbon nanotube composite. Polymer 44:2373–2377
11. Bhattacharyya AR, Potschke P, Haubler L, Fischer D (2005) Reactive compatibilization of melt mixed PA6/SWNT composites: mechanical properties and morphology. Macromol Chem Phys 206:2084–2095
12. Bose S, Khare RA, Moldenaers P (2010) Assessing the strengths and weaknesses of various types of pre-treatments of carbon nanotubes on the properties of polymer/carbon nanotubes composites: a critical review. Polymer 51:975–993
13. Breuer O, Sundararaj U (2004) Big returns from small fibers: a review of polymer/carbon nanotube composites. Polym Compos 25:630–645
14. Carneiro OS, Covas JA, Reis R, Brulé B, Flat JJ (2012) The effect of processing conditions on the characteristics of electrically conductive thermoplastic composites. J Thermoplast Compos Mat 25:607–629
15. Carponcin D, Dantras E, Aridon G, Levallois F, Cadiergues L, Lacabanne C (2012) Evolution of dispersion of carbon nanotubes in Polyamide 11 matrix composites as determined by DC conductivity. Compos Sci Technol 72:515–520
16. Chen L, Pang XJ, Qu MZ, Zhang QT, Wang B, Zhang BL, Yu ZL (2006) Fabrication and characterization of polycarbonate/carbon nanotubes composites. Compos A 37:1485–1489
17. Chen J, Hamon MA, Hu H, Chen Y, Rao AM, Eklund PC, Haddon RC (1998) Solution properties of single-walled carbon nanotubes. Science 282:95–98
18. Chen J, Y-Y Shi, J-H Yang, Zhang N, Huang T, Wang Y (2013) Improving interfacial adhesion between immiscible polymers by carbon nanotubes. Polymer 54:464–471
19. Clavé G, Delport G, Roquelet C, Lauret J-S, Deleporte E, Vialla F, Langlois B, Parret R, Voisin C, Roussignol P, Jousselme B, Gloter A, Stephan O, Filoramo A, Derycke V, Campidelli S (2013) Functionalization of carbon nanotubes through polymerization in micelles: a bridge between the covalent and noncovalent methods. Chem Mater 25 (13):2700–2707
20. Coleman JN, Khan U, Blau WJ, Gun'ko YK (2006) Small but strong: a review of the mechanical properties of carbon nanotube–polymer composites. Carbon 44:1624–1652
21. Dai H (2002) Carbon nanotubes: synthesis, integration and properties. Acc Chem Res 35:1035–1044
22. Datsyuk V, Kalyva M, Papagelis K, Parthenios J, Tasis D, Siokou A, Kallitsis I, Galiotis C (2008) Chemical oxidation of multiwalled carbon nanotubes. Carbon 46:833–840
23. Davenport M (2015) Much ado about small things. Chem Eng News 93:11–15
24. De Volder MFL, Tawfick SH, Baughman RH, Hurt AJ (2013) Carbon nanotubes: present and future commercial applications. Science 339:535–539
25. Di Crescenzo A, Ettorre V, Fontana A (2014) Non-covalent and reversible functionalization of carbon nanotubes. Beilstein J Nanotechnol 5:1675–1690
26. Domingues N, Gaspar-Cunha A, Covas JA, Camesasca M, Kaufman M, Manas-Zloczower I (2010) Dynamics of filler size and spatial distribution in a plasticating single screw extruder —modeling and experimental observations. Int Polym Proc XXV:188–198
27. Ferreira T, Paiva MC, Pontes AJ (2013) Dispersion of carbon nanotubes in polyamide 6 for microinjection moulding. J Polym Res 20:301–310
28. Georgakilas V, Kordatos K, Prato M, Guldi DM, Holzinger M, Hirsch A (2002) Organic functionalization of carbon nanotubes. J Am Chem Soc 124:760–761
29. Grace HP (1982) Dispersion phenomena in high viscosity immiscible fluid systems and application of static mixers as dispersion devices in such systems. Chem Eng Commun 14:225–277

30. Hirsch A (2002) Functionalization of single-walled carbon nanotubes. Angew Chem Int Ed 41:1853–1859
31. Huang Y-Y, Terentjev EM (2012) Dispersion of carbon nanotubes: mixing, sonication, stabilization, and composite properties. Polymers 4:275–295
32. Hwang TY, Kim HJ, Ahn Y, Lee JW (2010) Influence of twin screw extrusion processing condition on the properties of polypropylene/multi-walled carbon nanotube nanocomposites. Korea-Aus Rheol J 22:141–148
33. Iijima S (1991) Helical microtubules of graphitic carbon. Nature 354:56–58
34. Inno.CNT (2011) Innovation Alliance CNT—Carbon Nanomaterials Conquer Markets, Innovationsallianz CNT. http://www.inno-cnt.de. Accessed 22 July 2015
35. Jamali S, Paiva MC, Covas JA (2013) Dispersion and re-agglomeration phenomena during melt mixing of polypropylene with multi-wall carbon nanotubes. Polym Test 32:01–707
36. Jogi BF, Sawant M, Kulkarni M, Brahmankar PK (2012) Dispersion and performance properties of carbon nanotubes (CNTs) based polymer composites: a review. J Encapsulation Adsorpt Sci 2(69):78
37. De Jong KP, Geus JW (2000) Carbon nanofibers: catalytic synthesis and applications. Catal Rev-Sci Eng 42(4):481–5107
38. Delgado JL, de la Cruz P, Langa F, Urbina A, Casado J, Lopez-Navarrete JT (2004) Microwave-assisted sidewall functionalization of single-wall carbon nanotubes by Diels-Alder cycloaddition. Chem Commun 15:1734–1735
39. Kasaliwal GR, Pegel S, Goldel A, Potschke P, Heinrich G (2010) Analysis of agglomerate dispersion mechanism of multiwalled carbon nanotubes during melt mixing in polycarbonate. Polymer 51:2708–2720
40. Kasaliwal G, Villmow T, Pegel S, Pötschke P (2011) Influence of material and processing parameters on carbon nanotube dispersion in polymer melts. In: McNally T, Pötschke P (eds) Polymer carbon nanotube composites: preparation, properties and applications. Woodhead Publishing Limited, Cambridge, pp 92–132
41. Kohlgruber K (ed) (2008) Co-rotating twin screw extruders. Carl Hanser Verlag, Munich
42. Krause B, Petzold G, Pegel S, Pötschke P (2009) Correlation of carbon nanotube dispersability in aqueous surfactant solutions and polymers. Carbon 47:602–612
43. Krause B, Mende M, Pötschke P, Petzold G (2010) Dispersability and particle size distribution of CNTs in an aqueous surfactant dispersion as a function of ultrasonic treatment time. Carbon 48:2746–2754
44. Li Y, Yang D, Adronov A, Gao Y, Luo X, Li H (2012) Covalent functionalization of single-walled carbon nanotubes with thermoresponsive core cross-linked polymeric micelle Macromolecules 45(11):4698–4706
45. Li Y, Shimizu H (2007) High-shear processing induced homogenous dispersion of pristine multiwalled carbon nanotubes in a thermoplastic elastomer. Polymer 48:2203–2207
46. Lin TS, Cheng LY, Hsiao CC, Yang ACM (2005) Percolated network of entangled multiwalled carbon nanotubes dispersed in polystyrene thin films through surface grafting polymerization. Mat Chem Phys 94:438–443
47. Lin B, Sundararaj U, Potschke P (2006) Melt mixing of polycarbonate with multiwalled carbon nanotubes in miniature mixers. Macromol Mater Eng 291:227–238
48. Liu T, Phang IY, Shen L, Chow SY, Zhang YD (2004) Morphology and mechanical properties of MWCNT reinforced nylon-6 composites. Macromolecules 37:7214–7222
49. Ma PC, Siddiqui NA, Marom G, Kim JK (2010) Dispersion and functionalization of carbon nanotubes for polymer-based nanocomposites: a review. Compos A 41:1345–1367
50. Machado AV, Covas JA, Walet M, Van Duin M (2001) Effect of composition and processing conditions on the chemical and morphological evolution of PA-6/EPM/ EPM-g-MA blends in a corotating twin-screw extruder. J Appl Polym Sci 80:1535–1546
51. Mahfuz H, Adnan A, Rangari VK, Hasan MM, Jeelani S, Wright WJ et al (2006) Enhancement of strength and stiffness of Nylon 6 filaments through carbon nanotubes reinforcement. Appl Phys Lett 88:083119

52. Maric M, Macosko CW (2001) Improving polymer dispersions in mini-mixers. Polym Eng Sci 41:118–130
53. MarketsandMarkets (2013) Global CNT Market—SWCNT, MWCNT, Technology, Applications, Trends & Outlook (2011–2016)". http://www.marketsandmarkets.com/Market-Reports/carbon-nanotubes-139.html. Accessed 15 July 2015
54. MarketsandMarkets (2014) Carbon Nanotubes (CNTs) Market by Type (SWCNTS & MWCNTS), Application (Electronics & Semiconductors, Chemical & Polymers, Batteries & Capacitors, Energy, Medical, Composites, & Aerospace & Defense) & Geography—Global Trends & Forecasts to 2018. http://www.marketsandmarkets.com/Market-Reports/carbon-nanotubes-139.html. Accessed 15 July 2015
55. Micusik M, Omastova M, Krupa I, Prokes J, Pissis P, Logakis E, Pandis P, Potschke P, Pionteck (2009) A comparative study on the electrical and mechanical behaviour of multi-walled carbon nanotube composites prepared by diluting a masterbatch with various types of polypropylenes. J Appl Polym Sci 113:2536–2551
56. Menzer K, Krause B, Boldt R, Kretzschmar B, Weidisch R, Pötschke P (2011) Percolation behaviour of multiwalled carbon nanotubes of altered length and primary agglomerate morphology in melt mixed isotactic polypropylene-based composites. Compos Sci Technol 71:1936–1943
57. Manas-Zloczower I (2009) Mixing and compounding of polymers: theory and practice, 2nd edn. Hanser, Munich
58. Mayhew E, Prakash V (2014) Thermal conductivity of high performance carbon nanotube yarn-like fibers. J Appl Phys 115:174306
59. Monthioux M, Noé L, Dussault L, Dupin J-C, Latorre N, Ubieto T, Romeo E, Rovo C, Monzón A, Guimon C (2007) Texturising and structurising mechanisms of carbon nanofilaments during growth. J Mater Chem 17:4611–4618
60. Muller MT, Krause B, Kretzschmar B, Potschke P (2011) Influence of feeding conditions in twin-screw extrusion of PP/MWCNT composites on electrical and mechanical properties. Compos Sci Technol 71:1535–1542
61. Novais RM, Covas JA, Paiva MC (2012) The effect of flow type and chemical functionalization on the dispersion of carbon nanofiber agglomerates in polypropylene. Compos A 43:833–841
62. Novais RM, Simon F, Paiva MC, Covas JA (2012) The influence of carbon nanotube functionalization route on the efficiency of dispersion in polypropylene by twin-screw extrusion. Compos A 43:2189–2198
63. Novais RM, Simon F, Pötschke P, Villmow T, Covas JA, Paiva MC (2013) Poly(lactic acid) composites with poly(lactic acid)-modified carbon nanotubes. J Polym Sci Part A: Polym Chem 51:3740–3750
64. Oberlin A, Endo M, Koyama T (1976) Filamentous growth of carbon through benzene. J Crystal Growth 32(3):335–349
65. Paiva MC, Simon F, Novais RM, Ferreira T, Proença MF, Xu W, Besenbacher F (2010) Controlled functionalization of carbon nanotubes by a solvent-free multicomponent approach. ACS Nano 4(12):7379–7386
66. Paiva MC, Zhou B, Fernando KAS, Lin Y, Kennedy JM, Sun Y-P (2004) Mechanical and morphological characterization of polymer–carbon nanocomposites from functionalized carbon nanotubes. Carbon 42:2849–2854
67. Pan Y, Chan SH, Zhao J (2010) Correlation between dispersion state and electrical conductivity of MWCNTs/PP composites prepared by melt blending. Compos A 41:419–426
68. Peeterbroeck S, Alexandre M, Nagy JB, Pirlot C, Fonseca A, Moreau N et al (2004) Polymer-layered silicate–carbon nanotube nanocomposites: unique nanofiller synergistic effect. Compos Sci Technol 64:2317–2323
69. Pegel S, Pötschke P, Petzold G, Alig I, Dudkin SM, Lellinger D (2008) Dispersion, agglomeration, and network formation of multiwalled carbon nanotubes in polycarbonate melts. Polymer 49:974–984

70. Peigney A, Laurent Ch, Flahaut E, Bacsa RR, Rousset A (2001) Specific surface area of carbon nanotubes and bundles of carbon nanotubes. Carbon 39:507–514
71. Potschke P, Pegel S, Claes M, Bonduel D (2008) A novel strategy to incorporate carbon nanotubes into thermoplastic matrices. Macromol Rapid Commun 29:244–251
72. Prasek J, Drbohlavova J, Chomoucka J, Hubalek J, Jasek O, Adam V, Kizek R (2011) Methods for carbon nanotubes synthesis—review. J Mater Chem 21:15872–15884
73. Proença MF, Araújo RF, Paiva MC, Silva CJ (2009) The Diels-Alder cycloaddition reaction in the functionalization of carbon nanofibers. J Nanosci Nanotechnol 9:6234–6238
74. Qian D, Dickey EC, Andrews R, Rantell T (2000) Load transfer and deformation mechanisms in carbon nanotube-polystyrene composites. Appl Phys Lett 76:2868–2871
75. Safadi B, Andrews R, Grulke EA (2002) Multiwalled carbon nanotube polymer composites: synthesis and characterization of thin films. J Appl Polym Sci 84:2660–2669
76. Saito R, Fujita M, Dresselhaus G, Dresselhaus M (1992) Electronic structure of graphene tubules based on C_{60}. Phys Rev B 46(3):1804–1811
77. Saito Y, Yoshikawa T, Bandow S, Tomita M, Hayashi T (1993) Interlayer spacings in carbon nanotubes. Phys Rev B 48(3):1907–1909
78. Salzano de Luna M, Pellegrino L, Daghetta M, Mazzocchia CV, Acierno D, Filippone G (2013) Importance of the morphology and structure of the primary aggregates for the dispersibility of carbon nanotubes in polymer melts. Compos Sci Technol 85:17–22
79. Sathyanarayana S, Hübner C (2013) Thermoplastic nanocomposites with carbon nanotubes. In: Njuguna J (ed) Structural nanocomposites, perspectives for future applications. Springer, Heidelberg, pp 19–60
80. Sathyanarayana S, Olowojoba G, Weiss P, Caglar B, Pataki B, Mikonsaari I, Hübner C, Henning F (2013) Compounding of MWCNT with PS in a twin-screw extruder with varying process parameters: morphology, interfacial behaviour, thermal stability, rheology and volume resistivity. Macromol Mat Eng 298:89–105
81. Scott CE, Macosko CW (1991) Model experiments concerning morphology development during the initial stages of polymer blending. Polym Bull 26:341–348
82. Scurati A, Feke DL, Manas-Zloczower I (2005) Analysis of the kinetics of agglomerate erosion in simple shear flows. Chem Eng Sci 60:6564–6573
83. Skipa T, Lellinger D, Böhm W, Saphiannikova M, Alig I (2010) Influence of shear deformation on carbon nanotube networks in polycarbonate melts: interplay between build-up and destruction of agglomerates. Polymer 51:201–210
84. Socher R, Krause B, Boldt R, Hermasch S, Wursche R, Potschke P (2011) Melt mixed nano composites of PA12 with MWNTs: influence of MWNT and matrix properties on macrodispersion and electrical properties. Compos Sci Technol 71:306–314
85. Socher R, Krause B, Müller MT, Boldt R, Pötschke P (2012) The influence of matrix viscosity on MWCNT dispersion and electrical properties in different thermoplastic nanocomposites. Polymer 53:495–504
86. Suhr J, Koratkar NA (2008) Energy dissipation in carbon nanotube composites: a review. J Mater Sci 43:4370–4382
87. Tang W, Santare MH, Advani SG (2003) Melt processing and mechanical property characterization of multi-walled carbon nanotube high density polyethylene composite films. Carbon 41:2779–2785
88. Tasis D, Tagmatarchis N, Bianco A, Prato M (2006) Chemistry of carbon nanotubes. Chem Rev 106:1105–1136
89. Tessonnier J-P, Su DS (2011) Recent progress on the growth mechanism of carbon nanotubes: a review. ChemSusChem 4:824–847
90. Thomassin J-M, Vuluga D, Alexandre M, Jérôme C, Molenberg I, Huynen I, Detrembleur C (2012) Convenient route for the dispersion of carbon nanotubes in polymers: application to the preparation of electromagnetic interference (EMI) absorbers. Polymer 53:169–174
88. Tibbetts GG, Lake ML, Strong KL, Rice BP (2007) A review of the fabrication and properties of vapor-grown carbon nanofiber/polymer composites. Comp Sci Technol 67:1709–1718

92. Tokihisa M, Yakemoto K, Sakai T, Utracki LA, Sepehr M, Li J, Simard Y (2006) Extensional flow mixer for polymer nanocomposites. Polym Eng Sci 41:1040–1050
93. Tu ZC, Ou-Yang ZC (2002) Single-walled and multiwalled carbon nanotubes viewed as elastic tubes with the effective Young's moduli dependent on layer number. Phys Rev B 65:233407
94. Villmow T, Potschke P, Pegel S, Haussler L, Kretzschmar B (2008) Influence of twin-screw extrusion conditions on the dispersion of multi-walled carbon nanotubes in a poly(lactic acid) matrix. Polymer 49:3500–3509
95. Wagner HD (2002) Nanotube–polymer adhesion: a mechanics approach. Chem Phys Lett 361:57–61
96. Wichmann MHG, Schulte K, Wagner HD (1998) On nanocomposite toughness. Compos Sci Technol 68 (1):329–331
97. Winey KI, Vaia RA (2007) Polymer nanocomposites. MRS Bull 32(4):314–322
98. Wu TM, Chen EC (2008) Preparation and characterization of conductive carbon nanotube-polystyrene nanocomposites using latex technology. Comp Sci Technol 68:2254–2259
99. Xia H, Wang Q, Qiu G (2003) Polymer encapsulated CNTs prepared through ultrasonically initiated in situ emulsion polymerization. Chem Mater 15:3879–3886
100. Yang BX, Pramoda KP, Xu GQ, Goh SH (2007) Mechanical reinforcement of polyethylene using polyethylene-grafted multiwalled carbon nanotubes. Adv Funct Mater 17:2062–2069
101. Zhang WD, Shen L, Phang IY, Liu T (2004) CNT reinforced nylon-6 composite prepared by simple melt compounding. Macromolecules 37:256–259
102. Zou Y, Feng Y, Wang L, Liu X (2004) Processing and properties of MWNT/HDPE composites. Carbon 42:271–277

Natural Nanofibres for Composite Applications

Carlos F.C. João, Ana C. Baptista, Isabel M.M. Ferreira, Jorge C. Silva and João P. Borges

Abstract Cellulose and chitin are the two most abundant natural polysaccharides. Both have a semicrystalline microfibrillar structure from which nanofibres can be extracted. These nanofibres are rod-like microcrystals that can be used as nanoscale reinforcements in composites due to their outstanding mechanical properties. This chapter starts by reviewing the sources, extraction methods and properties of cellulose and chitin nanofibres. Then, their use in the fabrication of structural and functional nanocomposites and the applications that have been investigated are reviewed. Nanocomposites are materials with internal nano-sized structures. They benefit from the properties of the nanofillers: low density, nonabrasive, nontoxic, low cost, susceptibility to chemical modifications and biodegradability. Diverse manufacturing technologies have been used to produce films, fibres, foams, sponges, aerogels, etc. Given their natural origin and high stiffness, these polymers have attracted a lot of attention not only in the biomedical and tissue engineering fields but also in areas such as pharmaceutics, cosmetics, agriculture, biosensors and water treatment.

1 Introduction

Nanocomposites are polyphasic[1] materials where the reinforcement has at least one dimension in the nanometre range (1–100 nm). Nanoscale reinforcements present high aspect ratio and high surface-to-volume ratio, which ensure optimum strength

[1]Usually biphasic: comprising a soft phase (the matrix) and a strong and stiff phase (the reinforcement).

C.F.C. João · A.C. Baptista · I.M.M. Ferreira · J.P. Borges (✉)
CENIMAT/I3N, Departamento de Ciência dos Materiais,
Faculdade de Ciências e Tecnologia, FCT, Universidade Nova de Lisboa,
2829-516 Caparica, Portugal
e-mail: jpb@fct.unl.pt

J.C. Silva
CENIMAT/I3N, Departamento de Física, Faculdade de Ciências e Tecnologia, FCT,
Universidade Nova de Lisboa, 2829-516 Caparica, Portugal

© Springer Science+Business Media Singapore 2016
S. Rana and R. Fangueiro (eds.), *Fibrous and Textile Materials for Composite Applications*, Textile Science and Clothing Technology,
DOI 10.1007/978-981-10-0234-2_8

Table 1 Properties of cellulose and chitin and several reinforcement materials. Adapted from [4]

Material	ρ (g cm^{-3})	σ_f (GPa)	E_A (GPa)	E_T (GPa)	Reference
Kevlar-49 fibre	1.4	3.5	124–130	2.5	[4]
Carbon fibre	1.8	1.5–5.5	150–500	–	
Steel wire	7.8	4.1	210	–	
Clay nanoplatelets	–	–	170	–	
Carbon nanotubes	–	11–63	270–950	0.8–30	
Boron nanowhiskers	–	2-8	250–360	–	
Cellulose nanofibres	1.6	7.5–7.7	110–220	10–50	
Chitin nanofibres	1.425	–	150	15	[6, 7]

ρ density; σ_f ultimate tensile strength (UTS); E_A elastic modulus along the crystal axial direction; E_T elastic modulus along the crystal transverse direction

and maximum tolerance to flaws (robustness) of nanocomposites [1]. Nanoscale reinforcements can be particles (e.g., minerals), sheets (e.g., nanoclays) or fibres (e.g., carbon nanotubes, electrospun fibres or polysaccharide nanofibres) [2]. Polysaccharide nanofibres have received increased attention because of their low density, nonabrasive, nontoxic, low cost and biodegradable properties [3]. Cellulose and chitin are the two most abundant polysaccharides in nature and are increasingly being used for the preparation of nanofibres. The reason for this lies not only in the biocompatible and biodegradable character of these biopolymers but also in the fact that they compete well in terms of mechanical strength/weight performance with other reinforcements (Table 1) [4, 5].

Cellulose and chitin nanofibres are highly crystalline particles that are extracted from the corresponding polysaccharides. This is only possible due to the hierarchical structure and semicrystalline nature of these polysaccharides. Native cellulose[2] and chitin fibres are composed of smaller long, thin and stiff filaments, called microfibrils, from which nanofibres are extracted. In fact, hierarchical structures are common in natural materials. Trees and the exoskeleton of arthropods, which are the most abundant sources of cellulose and chitin, are typical examples of structures with a hierarchical design (Fig. 1). Despite the complicated hierarchical structures, one feature is common: the smallest "building blocks" in such structures are on the nanometre length scale (the nanofibres). This chapter will focus on the extraction of these "building blocks" and their use in the production of composites. The main applications of cellulose and chitin-based nanocomposites will also be highlighted.

[2]Native cellulose is the designation of the cellulose produced by trees, plants, tunicates, algae, fungi and bacteria.

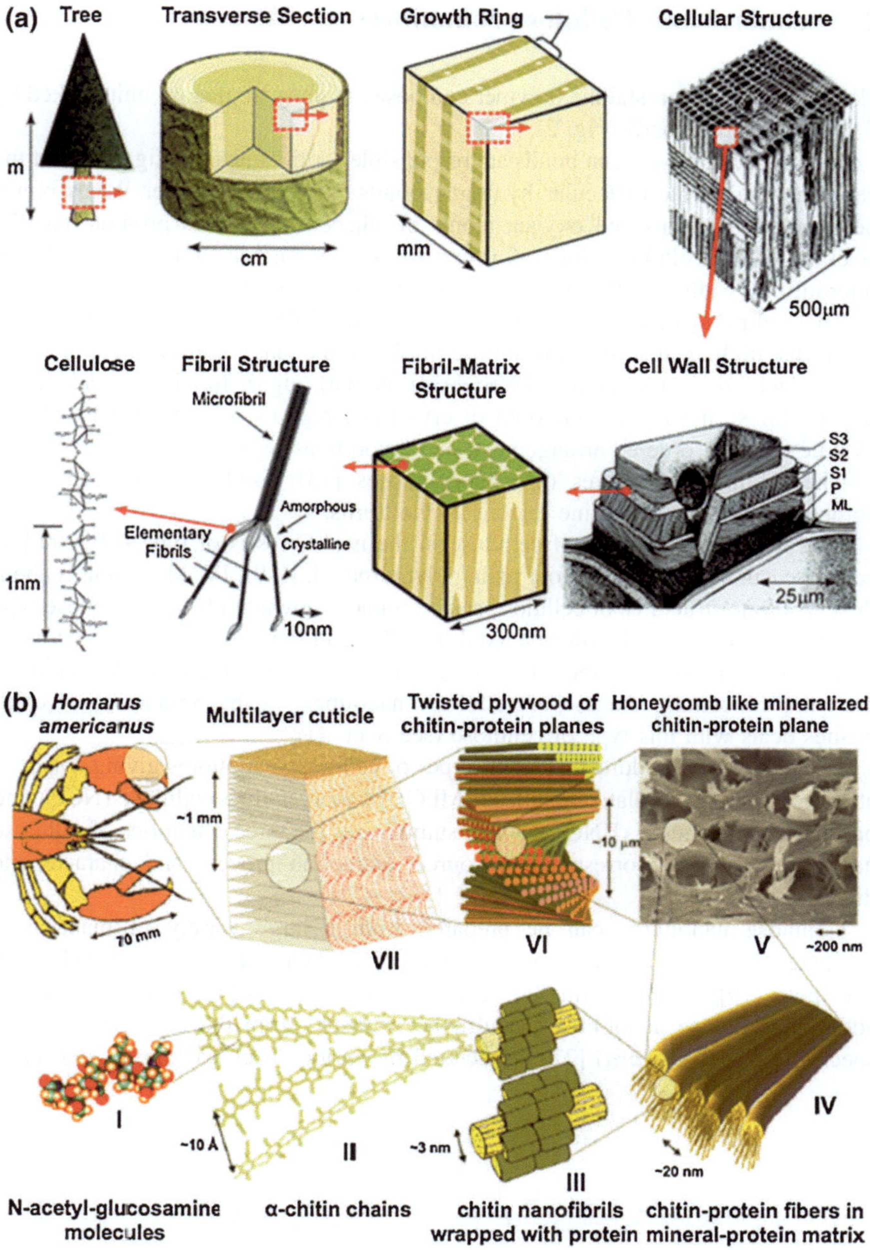

Fig. 1 Hierarchical design in: **a** trees (Reprinted with permission from Ref. [8]); **b** in the exoskeleton of *Homarus americanus*, the american lobster (Reprinted with permission from Ref. [9] copyright © 2009 John Wiley & Sons.)

2 Cellulose and Cellulose Nanofibres

Cellulose is a semicrystalline polymer composed of anhydroglucose units linked by β 1–4 glucosidic bonds (Fig. 2).

Intramolecular hydrogen bonds are responsible for the linear configuration of the cellulose chain. Intermolecular hydrogen bonds as well as van der Waals bonds between –OH groups and oxygen atoms of adjacent molecules promote parallel stacking of the cellulose chains forming fibrils that further aggregate into larger microfibrils (width: 5–50 nm; length of several microns) (Fig. 1a) [4].

These intra- and intermolecular bonds are responsible for cellulose' stability and for fibrils' high axial stiffness. Cellulose fibrils are the main reinforcement phase of trees, plants, tunicates (marine invertebrate animal), algae, fungi and bacteria [4, 10]. Cellulose fibrils are composed of crystalline regions, where cellulose chains assume a highly ordered arrangement, and amorphous regions [11].

The cellulose nanofibres (or nanocelluloses [13]) used as reinforcement in composites are the crystalline regions of the fibrils.

The crystalline structure of cellulose in the nanofibres is of type I. Crystalline cellulose can be divided into four main polymorphs: I, II, III, IV. More details about the crystalline structures of cellulose can be found elsewhere [4, 10]. Cellulose type I is the structure with the highest axial modulus and the one that can be extracted from native cellulose. For these reasons the great majority of the works found in the literature concerning the use of cellulosic nanofibres in the production of composites deals with this type of cellulose (see Sect. 4)

This chapter will address different types of cellulose nanofibres giving a special attention to microfibrillated cellulose (MFC), nanocrystalline cellulose (NCC) and bacterial nanocellulose (BNC). Table 2 summarizes the three categories of cellulose nanofibres with the corresponding source, extraction process and characteristic dimensions.

Cellulose nanofibres can be prepared from a large variety of plant-based resources (Fig. 3) such as cotton [14, 15], flax [16], hemp [16], sisal [17, 18], sugar beet pulp [19], soybean hull [20, 21] and banana rachis [18] among others. In addition, cellulose nanofibres can also be extracted from the cell wall of algae species (e.g. *Cladophora*) [22] or secreted by various bacteria (e.g. *Acetobacter*)

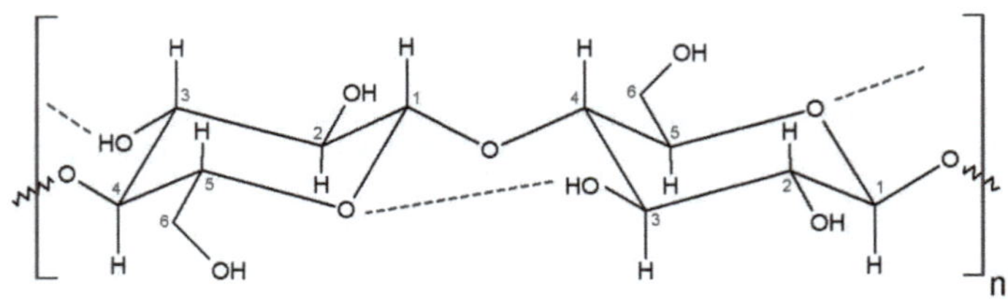

Fig. 2 Structure of a single cellulose chain repeat unit with the glycosidic bond (between C1 and C4) and intramolecular hydrogen bonds represented. The degree of polymerization (n) depends on the cellulose source (n = 10 000 in wood cellulose and 15 000 in cellulose cotton)

Table 2 Cellulose nanofibres—extraction methods and fibrils' characteristics

Cellulose nanofibres						
Particle type	Other denomination	Source	Extraction method	Particle dimensions		Ref.
				Length (nm)	Width (nm)	
MFC	Microfibrillated cellulose, nanofibrils and microfibrils, nanofibrillated cellulose	Banana rachis	Alkaline and bleaching treatments followed by acid hydrolysis	>1000	18 ± 7	[18]
		Sisal			13 ± 4	
		Kapok			16 ± 3	
		Pineapple leaf			22 ± 3	
		Coir			14 ± 5	
		Spruce	Mechanical homogenization	~1x10^{6}	–	[41]
		Wood pulp	Grinding	>1000	1–100	[27]
		Palm leaf	Chemical pre-treatment (with HCl and $NaHCO_3$) and mechanical microfluidization	~0.5x10^{6}	10–30	[42]
		Poplar wood	Chemical pre-treatment and high-intensity ultrasonication	>1000	5–20	[28]
		Spruce	Combination of severe shearing in a refiner, followed by high-impact crushing under liquid nitrogen (cryocrushing).	–	100–1000	[26]
NCC	Cellulose nanocrystals, crystallites, whiskers, rodlike cellulose microcrystals	Tunicate	Acid hydrolysis (H_2SO_4)	1187 ± 1066	9.4 ± 5.0	[43]
		Rice straw		117 ± 39	11 ± 4	[44]
		Pineapple leaf	Pressure streaming followed by rapid decompression (steam explosion) combined with acid treatment	200–300	5–60	[45]
		Bacteria	Acid hydrolysis (H_2SO_4)	>1000	~ 50	[46]
				1103 ± 698	14 ± 7	[43]
		Softwood pulp	Acid hydrolysis (HCl)	180 ± 75	3.5	[47]
		Sugarcane bagasse		250–480	20–60	[48]
		Cotton linter pulp	TEMPO-mediated oxidation with ultrasonic treatment	100–400	5–10	[49]
		Bamboo pulp		400–800	5–15	[15]
		Softwood pulp		400–800	5–15	
		Cotton linter pulp		200–400	15–25	
		Tunicate	Enzymatic hydrolysis	>1000	17 ± 3	[50]
			TEMPO-mediated oxidation	1590 ± 759	16 ± 2	[50]
			Acid hydrolysis (H_2SO_4)	694 ± 312	20 ± 3	[50]

(continued)

Table 2 (continued)

Cellulose nanofibres						
Particle type	Other denomination	Source	Extraction method	Particle dimensions		Ref.
				Length (nm)	Width (nm)	
BNC	Bacterial cellulose, microbial cellulose, biocellulose	*A. xylinum*	Bacterial synthesis (carbon source: sucrose)	<100	2–4	[23]
		G. medellensis	Bacterial synthesis (carbon source: glucose)	–	40–70	[51]
		G. xylinus	Bacterial synthesis (carbon source: mannitol)	–	–	[52]
		G. xylinus	Bacterial synthesis (carbon source: fructose)	–	~30	[53]
		G. xylinus	Bacterial synthesis (carbon source: glucose)	~1000	64 ± 19	[54]

[23]. Several approaches based on mechanical and chemical treatments for isolating cellulose nanofibres can be found in the literature [24]. Microfibrilated cellulose is characterized for their long and flexible fibres composed of alternating crystalline and amorphous domains (width of 5–60 nm and length of several micrometres). MFC are generally produced by delamination of wood pulp using mechanical processes, such as high-pressure homogenizers [25], cryocrushing [26], grinding [27] and high intensity ultrasonic treatments [28]. The main action of the reported methodologies to isolate MFC from the fibres is the high shear force that makes the separation of fibrils, which form the fibres, possible. Since the previous processes can be highly energy consuming, pre-treatments of the fibres prior to isolation of the nanofibres have been studied including acid hydrolysis, enzymatic pretreatment, and the introduction of charged groups through carboxymethylation or 2,2,6,6-tetramethylpiperidine-1-oxyl (TEMPO)-mediated oxidation [24, 29].

Nanocrystalline cellulose exhibits elongated crystalline rodlike shapes and limited flexibility, when compared to MFC, since it does not contain amorphous regions (width of 5–70 nm and lengths from 100 to several micrometres) [30]. Using strong acid hydrolysis, native NCC suspensions can be prepared from a variety of sources (e.g. wood, tunicates, algae and bacteria) under specific synthesis conditions (e.g. temperature, time and acid concentrations). During the acid hydrolysis process, the H_3O^+ ions penetrate the cellulose chains in the amorphous regions promoting the hydrolytic cleavage of the glycosidic bonds and releasing individual crystallites after mechanical treatment (e.g. sonication). Additional processes to prepare NCC have been reported allowing the release of crystalline domains from cellulosic fibres by using enzymatic hydrolysis treatment, TEMPO oxidation, hydrolysis with gaseous acid, and treatment with ionic liquids [29]. The variety of dimensions, morphologies and degree of crystallinity of NCCs strongly depends on the source of cellulosic material and conditions under which they were extracted [30].

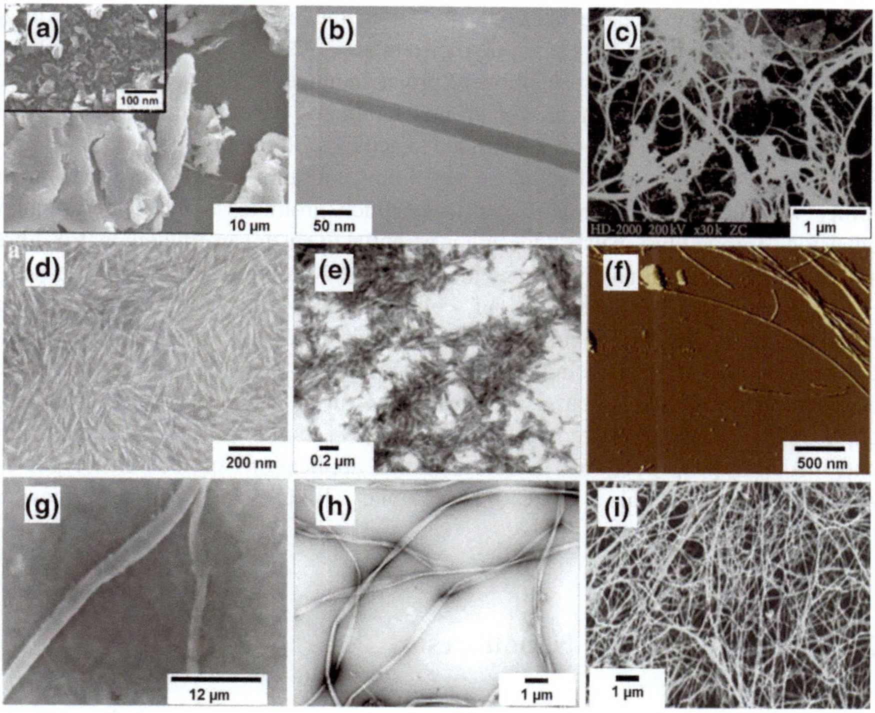

Fig. 3 Images of microfibrillar, nanowhiskers and bacterial cellulose obtained from different sources and methods: **a** banana rachis—alkaline and bleaching treatment followed by acid hydrolysis, SEM image adapted with permission from [18] copyright© 2015 Springer; **b** spruce—mechanical homogenization, TEM image adapted with permission from [41] copyright© 2007 Springer; **c** wood pulp—grinding, TEM image adapted with permission from [27] copyright© 2012 S. Pahnthapulakkal and M. Sain; **d** rice Straw—acid hydrolysis, TEM image adapted with permission from [44] copyright© 2012 Elsevier; **e** cotton linter pulp—TEMPO-mediated oxidation and ultrasonication, TEM image adapted from [49]; **f** tunicate—enzimatic hydrolysis, AFM image adapted with permission from [50] copyright© 2015 Elsevier; **g** *A. Xylinum*—bacterial cellulose, TEM image adapted from [23]; **h** *G. Medelensis*—bacterial cellulose, TEM image adapted with permission from [51] copyright© 2012 Elsevier; **i** *G. Xylinus*—Bacterial cellulose, SEM image adapted with permission from [53] copyright© 2013 Elsevier

Bacterial nanocellulose is a network structure produced by bacteria in a reverse way, synthesizing cellulose and building up bundles of nanofibres, with 2–4 nm in width and less than 100 nm in length. The synthesis of cellulose by bacteria generally consists in the formation of β-1,4 glucan chain with polymerization of glucose units, followed by assembly and crystallization of cellulose chain. Concerning this process, *Acetobacter* species cultivated in a culture medium containing carbon and nitrogen sources can convert glucose or other organic substrates into pure cellulose. One of the major advantages of such process is the ability to adjust the culturing conditions to modify the formation of nanofibres and crystallization [23, 31].

Different strategies have been considered in the literature to homogeneously disperse cellulose nanofillers in a polymeric matrix [32, 24]. Cellulosic nanofibres, like parental cellulose, are susceptible to broad chemical or physical modifications. The reactivity of surface hydroxyl groups facilitates the attachment of different functional groups onto nanofibres' surface to achieve different properties [5, 10, 33–36]. Surface modification of cellulose nanofibres is often used to improve compatibility with non-polar polymeric matrices, which promotes a better fibre-matrix adhesion. A good fibre-matrix adhesion is essential to ensure an efficient matrix-fibre stress transfer and ultimately defines the mechanical properties of the composite.

Given the small dimensions of cellulose-based nanostructures and their remarkable physical and chemical properties, it is clear that these building blocks can be used both as reinforcing elements in composite materials and in high-value added products. Cellulose-based nanomaterials can be found in different fields of application, including automotive applications [37], building [38], packaging [39] and medicine [40]. Section 4 addresses some applications of the three types of nanocellulose materials (MFC, NCC and BNC) in high-performances nanocomposites.

3 Chitin and Chitin Nanofibres

Chitin is the second most abundant biopolymer in nature, after cellulose. Chitin appears within cell wall of fungi, as essential component in all arthropods (cuticle of insects, arachnids and crustaceous), in the eggshells of nematods, in the jaws of rotifers, in the cell walls of green algae and in shells of mollusks [53, 55].

Chitin is a polysaccharide similar to cellulose, in which the glucose units are substituted by *N*-acetylglucosamine and glucosamine units linked by β 1–4 glucosidic bonds (Fig. 4). The fraction of N-glucosamine units (x) in the copolymer is defined as the degree of deacetylation (DD). In chitin, DD is lower than 0.5 (typically 0.1–0.3).

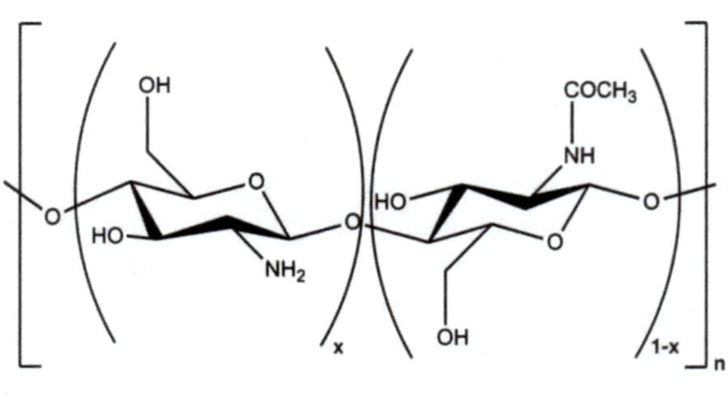

Fig. 4 Chitin and chitosan single chain repeating unit showing the copolymer character of the biopolymers. The DD defines whether the biopolymer is called chitin ($x < 0.5$) or chitosan ($x > 0.5$)

When the glucosamine units are predominant ($x > 0.5$) the biopolymer is termed chitosan. Chitosan is the principal derivative of chitin presenting different solubility and nitrogen content. Chitin is soluble in 5 % lithium chloride/N, N-dimethylacetamide solvent and insoluble in aqueous acetic acid while the reverse is true for chitosan. The nitrogen content in purified samples is less than 7 % for chitin and more than 7 % for chitosan [56].

Like cellulose, chitin has a structural function. Cellulose is the organic matrix of skeletal structures in plants and chitin is cellulose's analogue that appears within cell wall of fungi, as essential component in all arthropods (cuticle of insects, arachnids and crustaceous), in the eggshells of nematods, in the jaws of rotifers, in the cell walls of green algae and in shells of mollusks [55]. Also, like cellulose, chitin generates fibrillar structures from which chitin nanofibres[3] (also termed whiskers, nanowhiskers and nanocrystals) can be extracted.

Different chitin chains associate through N–H¯O = C hydrogen bonds giving rise to three polymorphic crystalline forms designated as α-, β- and γ-chitin [56, 57]. α-chitin is the most abundant form and has a highly ordered crystalline structure resultant from the extensive intra- and inter-molecular hydrogen bonds. Inter-molecular bonds in α-chitin are possible due to the anti-parallel packing of the adjacent polymeric chains and are responsible for the rigid and intractable physical properties of the biopolymer. In β-chitin the chains are stacked in a parallel disposition. This arrangement of the polymeric chains does not favour the inter-chain hydrogen bonding making β-chitin more soluble in water than α-chitin. β-chitin extracted from squid pen is completely soluble in water. However, the availability of β-chitin is limited and not much interesting for the commercial isolation and production of chitin in a hydrophilic form [56]. γ-chitin is the less common polymorhic structure and can be found in cocoon fibres of the *Ptinus* beetle and the stomach of *Loligo*. Results as the combination of both α and β-chitin with a three-chain unit structure in which two chains are parallel and the third one is arranged in an anti-parallel fashion [58].

The assembly of these fibre-forming polymeric chains results in the formation of the chitin's microfibrillar structure, which is responsible for physico-chemical properties of tissues.

The microfibrillar arrangement—chitin nanofibres orientational packing- has raised growing interest in scientific community in order to apply chitin intrinsic properties as structural and/or functional reinforcements in nanocomposites.

In almost all biological species where it is present, chitin does not occur in a pure form, but is linked (covalently or not) to other molecules. In the shells of crabs and shrimps (main source of chitin in industry, since they can be easily recovered from food wastes (Fig. 5) there are proteins (30–40 %), calcium carbonate (30–50 %), lipids and astaxanthin (<1 %), and only 20–30 % chitin. Therefore, the first step in the production of nanochitin is the isolation of the biopolymer from the source material. This involves demineralization, normally with acids such as hydrochloric,

[3] Also termed whiskers, nanowhiskers and nanocrystals.

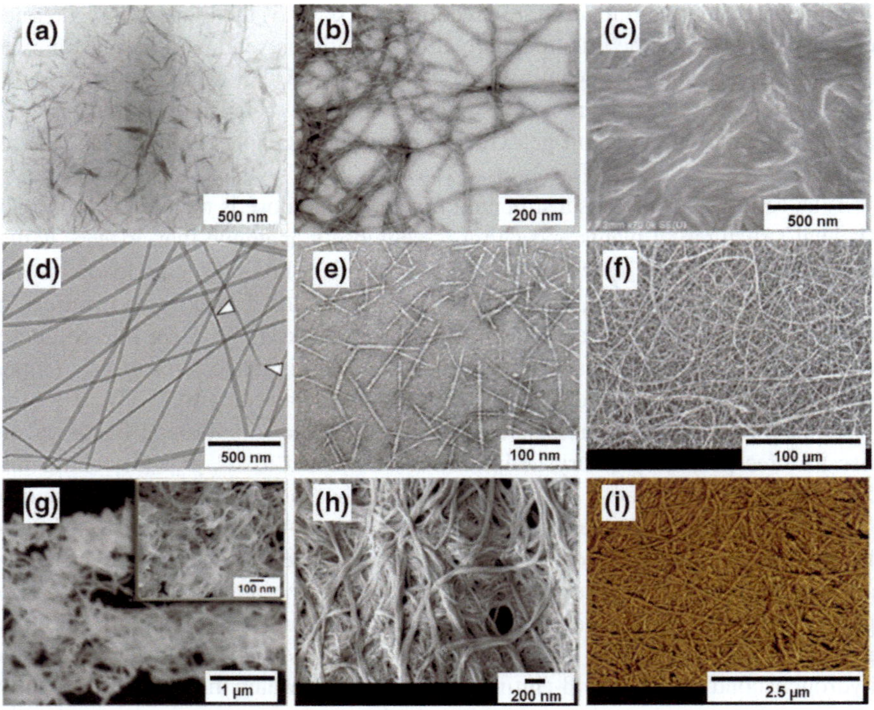

Fig. 5 Images of chitin nanofibres and nanowhiskers obtained from different sources and methods: **a** shrimp—acid hydrolysis (HCl), TEM image adapted with permission from [115] copyright© 2005 Elsevier; **b** lobster—acidic medium (acetic acid) followed by mechanical grinding, STEM image adapted with permission from [104] copyright© 2014 Elsevier; **c** crab—gelation and regeneration, SEM image adapted with permission from [66] copyright© 2011 Elsevier; **d** Tubeworm—TEMPO-mediated oxidation, TEM image adapted with permission from [64] copyright© 2009 Elsevier; **e** Crab—deacetylation and surface cationization, TEM image adapted with permission from [94] copyright© 2010 Elsevier; **f** crab—electrospinning, SEM image adapted with permission from [97] copyright© 2006 Elsevier; **g** Prawn—Ultrasonication, SEM image adapted with permission from [100] copyright© 2013 Elsevier; **h** crab—grinding, FE-SEM image adapted with permission from [92] copyright© 2009 American Chemical Society; **i** lobster—high pressure homogenizer, AFM image adapted with permission from [105] copyright© 2015 Elsevier

deproteinization (with a strong base such as sodium hydroxide), decolorization (with organic solvents) and bleaching (treatment with sodium hypochlorite) [6].

At this point, chitin powders or solution/dispersions can be obtained, which with further treatment can give rise to a multiplicity of fibrillar and/or crystalline nanochitin with wide variety of sizes, shapes, crystallinity, aspect ratio and morphology [55]. Table 3 lists methods of nanofibres and nanowhiskers production and the characteristics of the products obtained. The differences observed in nanochitin morphology are dependent on extraction method, conditions (time/temperature/pressure/pH/agitation, etc.) and chitin source.

Table 3 Chitin nanowhiskers and nanofibres, production methods and fibrils' characteristics

Chitin Nanowhiskers and Nanofibres						
Particle dimensions						
Source	Extraction Method	Key Conditions	Length (nm)	Width (nm)	Aspect Ratio	Ref.
Shrimp	Electrospinning	–	–	670	–	[67]
	Acid hydrolysis (HCl)	80–90 °C	150–800	10–50	–	[68]
		90 °C/90 min	200–500	10–15	20–33	[69]
		104 °C/90 min	417	33	17	[70]
			150–400	15–20	–	[71]
		105 °C/180 min	200–560	18–40	18	[72]
			–	–	16	[73]
		104 °C/360 min	427	43	10	[74]
			307.7	27.1	11	[75]
			343	46	8	[76]
		120 °C/360 min	549	31	18	[77]
		-/360 min	300	20	15	[78]
Crab		80 °C/90 min	400	10–30	–	[79]
		90 °C/90 min	240	15	16	[80]
		95 °C/90 min	240	18	13	[81, 82]
		104 °C/90 min	500 ± 50	50 ± 10	10	[83]
			200	15	13	[84]
			200–500	5–20	15–20	[85]
			100–200	6–8	–	[86]
			300	20	15	[87, 88]
			100–350	5–30	16	[89]
		120 °C/360 min	255 ± 56	31 ± 6	8	[90]

(continued)

Table 3 (continued)

Chitin Nanowhiskers and Nanofibres

Particle dimensions

Source	Extraction Method	Key Conditions	Length (nm)	Width (nm)	Aspect Ratio	Ref.
	Acidic medium (acetic acid) + Mechanical treatment	–	–	25–40	–	[62]
		–	10–20	–	–	[91, 92]
	Mechanical treatment (4 methods)	–	200 ± 20	8 ± 1	–	[93]
	TEMPO-mediated oxidation	Room °C	340	8	–	[65]
	Deacetylation + cationization	90 °C/120–240 min	250 ± 140	6, 2 ± 1, 1	–	[94]
	Gelation + regeneration	100 °C/2880 min	>100	20–60	–	[66, 95]
	Electrospinning	–	–	152 ± 70	–	[96]
	Irradiation + electrospinning	Co60	–	163	–	[97]
		Co60	–	40–60	–	[98]
Prawn	Ultrasonication	1000 W/40 min	–	30–50	–	[99]
		300 W/30 min	300	2–5	–	[100]
Mushroom	Mechanical treatment	Grinder/pH = 7	–	10–100		[101]
Lobster	Acidic medium (acetic acid) + mechanical treatment	1500 rpm	–	10	–	[102]
		pH 3–4	–	10–20	–	[103]
			1000–1500	3–4	–	[104]
	High pressure homogenizer	1000 bar	>1000	80–100	–	[105]
		1000 bar	5000	90	–	[106]
	Acid hydrolysis (HCl)	100 °C/90 min	300	60	–	[106]

(continued)

Table 3 (continued)

Chitin Nanowhiskers and Nanofibres						
Particle dimensions						
Source	Extraction Method	Key Conditions	Length (nm)	Width (nm)	Aspect Ratio	Ref.
Riftia		104 °C/90 min	2200	18	120	[107]
Squid			150	10	18	[108]
	Acidic medium (acetic acid) + mechanical treatment	–	–	3–4	–	[63]
	Ultrasonication	pH 3–4	>1000	3–10	–	[109, 110]
	Self-assembly	Solvent evaporation (HFIP)	>10000	2.8 ± 0.7	–	[111, 112]
		Precipitation (DMAC)	>10000	10.2 ± 2.9	–	[113, 114]
Tubeworm	TEMPO-mediated oxidation	Room ℃	–	–	–	[64]
			>1000	20–50	–	[64]

Different nanocrystalline/ nanofibrous chitin extraction methods have been proposed either by acid hydrolysis, 2.2.6.6-tetramethylpiperidine-1-oxyl radical (TEMPO) mediated oxidation, high-pressure homogenizer, mechanical grinding or surface cationization of partially deacetylated chitin. These methods enable access to the nano sub-level within chitin's hierarchic structure (Fig. 1b) and isolation of the rod/needle-like shaped and highly crystalline regions (Fig. 1b stage II-III)—chitin nanowhiskers (also called chitin nanocrystals, chitin whiskers)—or the basic fibrillar system (Fig. 1b stage IV–VI)—chitin nanofibres (also called chitin nanofibrils, chitin fibrils).

Of the procedures mentioned above, acid hydrolysis (as in cellulose whisker extraction) remains the key method. This process comprises the treatment of chitin powders with an acid bath, usually 3 M HCl, at the boiling point (104 °C) for a period of 90 min. This procedure causes the continuous depolymerization of chitin, through the removal of the amorphous regions, until the highly crystalline chitin regions are extracted. After hydrolysis, the samples are washed (dialyzed) and/or centrifuged several times until a pH of 2 is reached. At this point, a colloidal suspension arises naturally, characterized by a whitish appearance [59]. Depending on concentration, these suspensions reveal birefringence under cross polarizers, similar to textures found in living tissues [60, 61]. Other methods, based on acidic medium (acetic acid) combined with mechanical treatment, have been reported for nanofibres extraction from crab and squid pen [62, 63]. In this approach, chitin powders are suspended in an acidic bath (pH = 3–4) and the resultant slurry is fibrillated with mechanical grinders or ultrasonication, until highly viscous suspensions are obtained.

TEMPO mediated oxidation method has been successfully adapted from cellulose nanowhisker extraction being capable of nanowhisker/nanofibre production from crab and squid [64]. This method comprises a suspension of chitin (1 g) in water containing TEMPO (0.016 g, 0.2 mmol) and sodium bromide (0.1 g, 1 mmol) with further addition of NaClO (0–10 mmol per gram of chitin) in order to start oxidation. The process is performed at room temperature and in a basic medium (addition of NaOH to maintain pH = 10). After stopping the reaction with ethanol addition and neutralizing the solution until pH = 7 is reached, the mixture is centrifuged and the insoluble fraction containing the chitin nanofibres is reaped [65].

Chitin nanocrystals can also be produced with a different approach involving a process of gelation with 1-allyl-3-methylimidazolium bromide (AMIMBr) followed by regeneration with methanol. In this method, chitin powder from commercial sources (1.23 mmol, 0.250 g) is drenched, for 24 h at room temperature, with AMIMBr (12.3 mmol, 2.50 g). Subsequently, chitin is subjected to a heating process at 100 °C for 48 h, resulting in chitin/AMIMBr gel. After cooling down the system until room temperature, the regeneration stage begins with dropwise addition of methanol (40 mL) followed by sonication, in order to obtain suspensions of chitin nanocrystals [66].

Individual chitin nanofibres were obtained using a new process of α-chitin deacetylation and fibrillation by surface cationization. The process consisted in treating chitin with NaOH and $NaBH_4$, in order to promote deacetylation without depolymerization and weight loss. The treatment with NaOH promotes the increase of C2-primary amino groups at chitin's surface. After a period of 1–4 h and 90 °C heating, the deacetylated chitin is washed and centrifuged several times before a final adjustment of the medium with acetic acid to pH = 3–4 followed by sonication. When acetic acid is added to the medium, chitin nanofibres' surface becomes protonated with positive charges which by electrostatic repulsion results in nanofibres separating from each other [94].

A simple method based on ultrasonication has been demonstrated as capable of producing nanofibres with diameters ranging from 30 to 120 nm. In this approach, chitin or other natural fibre-forming polymers are suspended in a water solution (0.05 g in 100 ml) and subjected to ultrasonication at 20 kHz with different powers and durations for different fibres. The impact of sonication waves breaks the weak interactions at the interfaces between nanofibres causing gradual disintegration of the microfibers and individualization into nanofibres [116].

Chitin nanofibres have been produced based on the dynamic high-pressure homogenization method. This process makes use of high pressure (1000 bar) on fluids to induce subdivision into dispersed particles. To obtain nanoparticles, chitin is first dispersed in water and pre-treated with ultrasonication to facilitate the process. Then the suspension is passed several times in a dynamic high pressure homogenizer in order to obtain well-dispersed and individualized nanofibres [105].

4 High-Performance Nanocomposites Based on Cellulose and Chitin

Both cellulose and chitin present a variety of advantages for being used in structural and functional nanocomposites applications. Properties such as nontoxicity, biocompatibility, biodegradability, antimicrobial effect, low density, easy chemical modification and vast availability are features desired by researchers and developers in all kinds of industries. Due to their biological nature and high stiffness, these natural polymers have attracted a lot of attention not only in the biomedical and tissue engineering fields but also in areas such as pharmaceutics, cosmetics, agriculture, biosensors and water treatment [6, 101, 117, 118]. Manufacturing technologies have been used to produce films [119–121], electrospun [67, 122, 123] and wet-spun fibres [124, 125], freeze-dried sponges [100, 126], freeze-cast foams [71], aerogels [127, 128], calcinated replicas [129], magnetic nanoparticles [130, 131] and biomimetic hybrids [132], nanocomposites that make use of cellulose and chitin nanofibres and nanowhiskers to improve their performance.

In the following subsections, functional and structural nanocomposites based on cellulose and chitin will be presented highlighting the potential for application in different fields (Fig. 6).

Fig. 6 Areas of application of cellulose and chitin based nanocomposites. Images adapted from *freeforcommercialuse.net*

4.1 Structural Applications

Both cellulose and chitin have been introduced for structural reinforcement of chitosan, starch, soy, rubber, polycaprolactone (PCL), poly(vinyl alcohol) (PVA) and polyurethane (PU) based nanocomposites [55, 133, 134].

As aforementioned, the concept of using nanocellulose as reinforcement comes from the possibility of exploring the stiffness and strength of cellulose crystals. The strong reinforcing effects of small amounts of nanocellulose have been studied and reported by several authors [133, 135–137]. The preparation of high-strength elastomeric nanocomposite has been reported by Wu et al. [136]. The nanocomposite based on microcrystalline cellulose and polyurethane showed that the cellulosic material is acting as a reinforcement additive since the average true strength for this composition was 257 MPa, compared with 39 MPa for the neat polyurethane. Bendahou and co-workers [138] have compared the role of NCC and MFC as reinforcing agents in natural rubber. The tensile modulus showed to be systematically higher for a given composition for MFC reinforced nanocomposites since entangled MFC provide a higher stiffness compared to cellulose whiskers. Iwatake et al. [137] investigated the reinforcement of poly(lactic acid) (PLA) using MFC. The tensile strength of PLA increased from about 50–70 MPa and the average Young's modulus of MFC/PLA composites attained 4 GPa. Later, Suryanegara and collaborators [135] described the effect of PLA crystallization on thermal and mechanical properties of MFC/PLA nanocomposites. They concluded that MFC in PLA improved the storage modulus of crystallized PLA at 120 °C from 293 to 1034 MPa. This study has demonstrated that MFC could extend the application of PLA, particularly for products exposed to high temperatures.

Studies about the use of BNC as reinforcement in high performance structural materials can also be found in literature [139–142]. Martins et al. [140] concluded that the Young modulus of a starch thermoplastic (TPS) matrix reinforced with BNC was 30 times higher than that of the unfilled TPS. Besides, it was observed that the addition of low amounts of BNC nanofibres to the TPS matrix resulted in a slight increase in the thermal stability of the composites, which can be explained by the higher stability of the cellulose substrate. Further studies on BNC/starch composites (Fig. 7) have found that the presence of BC nanofibres improved not only the tensile properties (tensile modulus of 361 MPa compared with 155 MPa of the pure matrix) but also the resistance to moisture and microorganism attack [141].

The investigation carried out by Trovatti and co-workers [142] demonstrated that a composite composed of random thermoplastic copolymers of butyl acrylate and methyl methacrylate and BNC displayed enhanced mechanical properties (tensile strength nearly double that of the pure matrix) and thermal stability compared with those of the acrylic copolymers.

Although there are only a few studies on nanochitin's mechanical properties, it is expected that (as in cellulose) due to the nanosized and crystalline structures, chitin nanowhiskers and nanofibres will show high Young's modulus, high fracture strength and low thermal expansion [101]. Zeng et al. [6] reported chitin nanowhiskers with longitudinal modulus of 150 GPa and transverse modulus of 15 GPa. Hassanzadeh et al. [143] tested a single chitin nanofibre obtained through the self-assembly method and determined an elastic modulus in the range of 5–7 GPa. Mushi et al. [104] produced an entirely nanostructured chitin nanowhisker (length: 1000–1500 nm; width: 3–4 nm) membrane (86 % acetylation degree) and recorded a tensile strength of 153 MPa, a Young modulus of 7.3 GPa and a strain to failure of 8 %.

Expectations of higher performances are indirectly confirmed by studies on nanocomposites containing chitin, often indicating mechanical properties improvement. As in every nanocomposite, mechanical properties can also decrease when critical chitin nanofiller concentration is reached leading to agglomeration of the "reinforcement" within the matrix, causing mechanical disruptions in final structures. To support these claims, various examples can be found in thermoplastic composites reinforced with chitin nanofillers based on starch [106], styrene/acrylamide [108], soy protein [83], polymethyl methacrylate [93],

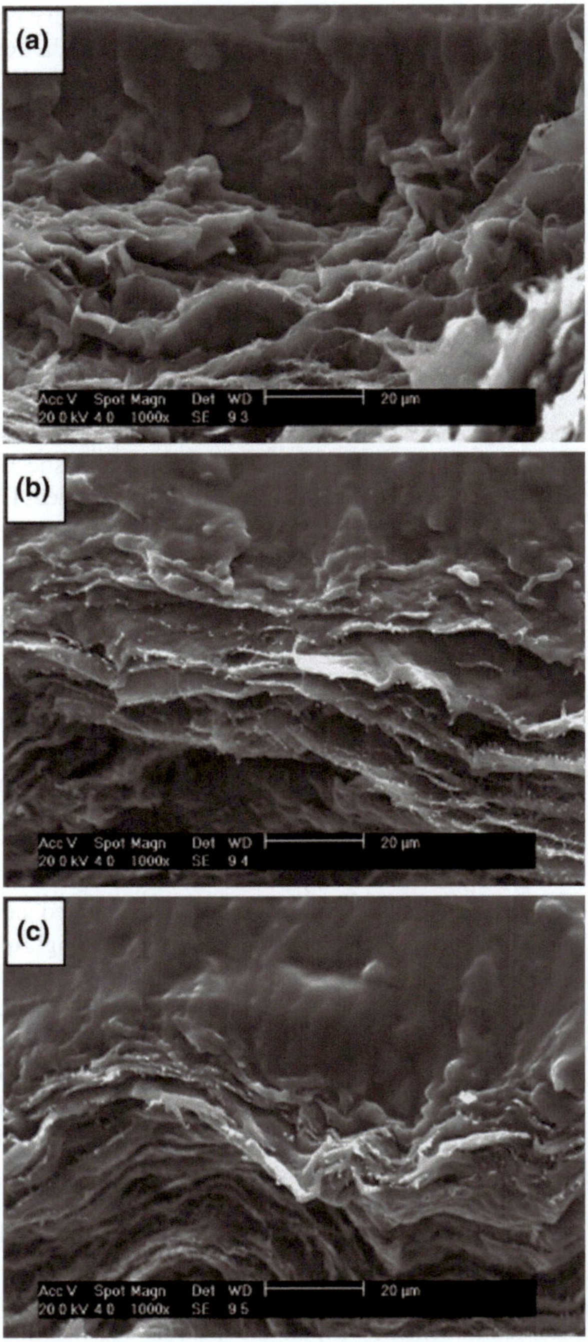

Fig. 7 Scanning electron micrographs of tensile fracture surfaces of BNC/starch biocomposites with different BNC contents **a** 7.8 wt%, **b** 15.1 wt%, and **c** 22.0 wt%. Image reprinted from [141] copyright © 2009 Elsevier

polylactide [144], poly (3-hydroxybutyrate-co-3-hydroxyvalerate) [145] and PVA. Rujiravanit and co-workers highlighted mechanical properties improvement in PVA films and nanofibers reinforced with chitin whiskers [77, 115], even verifying Young modulus improvement by 4 to 8 times in electrospun nanofibres. Working with the same composite composition, Deng et al. [99] report a small decrease of light transmittance from 92 to 86 % but a strong interaction between chitin fibers and PVA, with an increase of 100–140 % in Young modulus and substantial decrease in thermal expansion coefficient, from 124 to 25 ppm K^{-1}.

Nair and collaborators studied the effects of chitin whiskers (length: 240nm; aspect ratio: 16) and chemically modified chitin whiskers on the reinforcement of natural rubber, regarding swelling behavior, mechanical properties [80, 146, 147]. Products obtained through film casting and hot pressing showed that final properties were dependent upon manufacturing technique and also that whiskers formed a 3D percolating network only in the film cast samples that was responsible for mechanical properties enhancement and swelling resistance. Morin and Dufresne [107] also found the same percolating mechanism in PCL composite films reinforced with chitin whiskers from riftia tube (length: 120 nm; width: 18 nm). Above 5wt% whisker content and at high temperatures the percolating behavior was even more evident with the stabilization of the mechanical properties.

The combination embodying chitin nanofillers most used remains the chitosan/chitin nanocomposite either in the form of hydrogels [86], membranes [88], spheres [89] or films [148]. As example, Ma et al. [87] emphasized the excellent biocompatibility and affinity between chitin and chitosan which allows chitosan based films to achieve 2.8 times higher tensile strength values and increase the inhibitory effect against typical bacteria. Also in chitosan based films, Rubentheren et al. [149] observed that addition of chitin improves the nanocomposite films mechanical properties up to 137 % compared to neat chitosan, with a reduction of moisture content by 294 % and water solubility by 13 %.

4.2 Functional Composites

4.2.1 Tissue Engineering Applications

An important feature of tissue engineering is the development of suitable biodegradable scaffolds with a 3-dimensional (3-D) porous network structure and inter-connected pores. Bacterial nanocellulose displays an interesting network nanostructure that closely resembles the structure of native extracellular matrices (ECM). Compared with other natural biodegradable polymers such as collagen, chitosan, chitin, and gelatin, BNC presents much better mechanical properties, which are required in many cases where it is used as scaffold in tissue engineering. For that reason, it has been recently used for the development of nanocomposites for tissue-engineered constructs [23].

A composite made of BNC and PVA has been proposed by Millon and Wan [150] to mimic the role of collagen and elastin, respectively, in heart valve replacement. PVA physically crosslinked by low temperature thermal cycling is one of the few materials that exhibits a stress-strain relationship similar to soft tissues, such as cardiovascular tissue. The combination of this property with the high elastic modulus and degree of crystallinity characteristic of BNC fibres has originated a composite structure with mechanical properties similar to those of a porcine aorta.

The preparation of BNC/hydroxyapatite (HA) nanocomposites for bone healing applications was described by Zimmermann [151]. Since BNC is an attractive alternative to mimic collagen, the authors reported a new approach for the development of a biomineralized BNC scaffold. This biomimetic approach involves the use of carboxymethyl cellulose (CMC) to activate the surface of BC scaffolds and initiate the nucleation of calcium-deficient hydroxyapatite. Due to its ability to mimic the basic composition of bone, HA is one of the most frequently used biocompatible ceramics for bone and dental tissues reconstitution exhibiting excellent biocompatibility with hard tissues and high osteoconductivity and bioactivity. As a result, the surface modification and biomineralization processes used in this study were shown to be effective methods for activating and forming calcium-deficient HA crystals on the surface of BC scaffolds, respectively. Biomineralized scaffolds yielded a better cellular response when compared with pure BNC scaffolds, which alone did not promote osteoblast adhesion on their surface. Also, Tazi and co-workers [152] have investigated the incorporation of HA in BNC scaffolds to support osteoblast growth and bone formation. For this purpose, these researchers have successfully enriched BNC with HA by performing alternating incubation cycles with calcium and phosphate solutions. By incorporating HA into BNC it was possible to design a new nanocomposite with improved osteoblast adhesion, proliferation and mineralization when compared to the BNC polymer alone (Fig. 8).

In the same year, Zhijiang et al. [153] have prepared a biocomposite based on poly(3-hydroxubutyrate-co-4-hydroxubutyrate) (P(3HB-co-4HB)) and BNC by freeze-drying using trifluoroacetic acid (TFA) as co-solvent. The PHB biopolymer provides a means to target a wide range a tissues with potential applications for the cardiovascular system, cornea, pancreas, kidney, musculoskeletal, nervous system, skin, teeth and oral cavity. However, its utilization has been limited due to its inherent physical and chemical properties such as brittleness and hydrophobicity. The combination of PHB copolymer with BNC resulted in a composite scaffold with enhanced biocompatibility, hydrophilicity and mechanical properties when compared with pure P(3HB-co-4HB) scaffolds (UTS = 46 ± 5 MPa and E = 0.88 ± 0.24 GPa).

All-cellulose composites have also been introduced to tissue engineering. The dispersed phase and matrix used in these composites are made from the same material, which brings benefits such as recyclability and better interface adhesion. Mathew and co-workers [154] aimed at preparing all-cellulose artificial ligaments and tendons with mechanical properties similar or better than the natural tissues. In the human body the main function of the tendon is to transfer the force of muscle

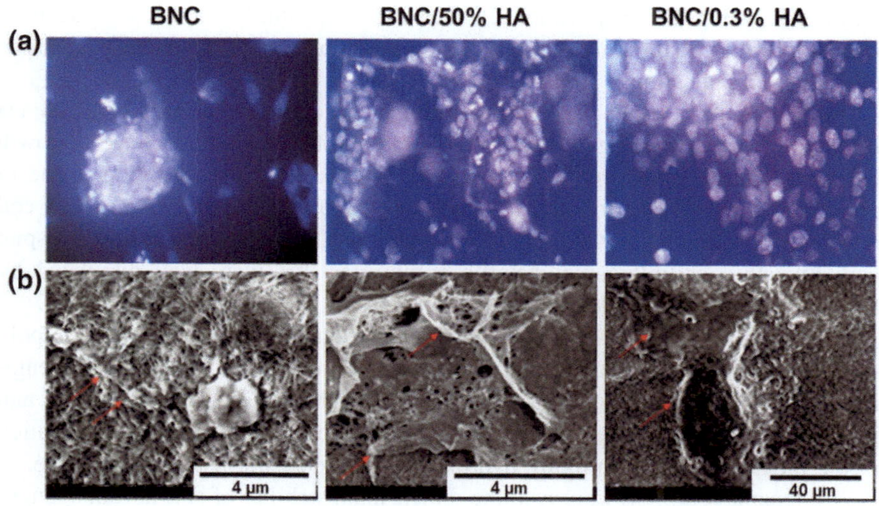

Fig. 8 Interaction between the osteoblasts and the different BNC polymers. Osteoblasts were cultured on the BNC polymers for two days. **a** Adherent cells were subjected to Hoechst staining. **b** SEM analysis of the BNC polymers containing osteoblasts cultured for 2 days, showing cell adhesion (arrows). Image reprinted with permission from [152] copyright © 2012 Springer

contraction to the bones whereas ligaments stabilise the joints preventing abnormal movements. The authors described the preparation of a fibrous nanocomposite by a partial dissolution method. This method consists in the preparation of a cellulose nanofibre network (from Norway spruce) by mechanical fibrillation followed by a partial dissolution using an ionic liquid (1-butyl-3-methylimidazolium chloride—[C4mim]Cl). This method allows the production of homogeneous and uniform structures, leading to an enhancement of fibre-matrix interaction and, consequently, better mechanical properties. The prepared all-cellulose composites showed the potential to be used as ligament/tendon substitutes since their mechanical properties (TS = 25–30 MPa) are in the range required for ligaments or tendons (TS = 28–38 MPa).

Recently, an electrospun fibrous bio-nanocomposite scaffold reinforced with NCC was proposed by Zhou et al. [123]. The use of the electrospinning technique to generate continuous one dimensional fibres is one of the most recent methods of processing nanofibre composites [155–158]. Nanocomposite suspensions of maleic anhydride (MAH) grafted PLA, designated by MPLA, and NCCs were electrospun and characterized. The effective improvement of interfacial adhesion between CNCs and polymer matrix through MAH grafting on PLA inhibited their phase separation and facilitated the efficient stress transfer from MPLA to CNCs. The main results demonstrated that the addition of NCCs successfully improved the thermal stability and the mechanical properties of the composite scaffolds by more than 10 MPa.

Chitin nanofibres also have the ability of mimicking the extracellular matrix of human tissues and, because of that, researchers have been intensively studying new ways of boosting their application. Besides biocompatibility and biodegradability, similarity with collagen fibrils makes chitin nanofibres good candidates for tissue engineering applications [118].

Noh et al. [97] tested the cytocompatibility of chitin nanofibres obtained via electrospinning by seeding of keratinocytes and fibroblasts. Cell attachment, growth and proliferation were superior when comparing with chitin microfibers, due to nanofibres high surface area availability that promotes a 3D environment for cell attachment/proliferation. Xin et al. [159] found out the same relation in elecrospun carboxymethyl chitin/organic rectorite composites that were well populated by mouse lung fibroblasts after 24 h of incubation.

Shalumon et al. [160] worked in the combination of carboxymethyl chitin/poly (vinyl alcohol) and produced electrospun nanofibrous scaffolds for tissue engineering. Besides proving cell viability with mesenchymal stem cells adhesion and proliferation assays, the researchers were also able to induce in vitro mineralization of the membranes by identifying hydroxyapatite deposits at nanofibres surface.

Wongpanit et al. [74] improved the dimensional stability of silk fibroin sponges by adding chitin whiskers (length: 427 nm; width 43 nm) as nanofillers. The nanocomposite sponges, exhibiting pore size of 150 μm, were biologically evaluated by seeding with L929 cells, with good results in cell spreading caused by the presence of chitin whiskers.

Hassanzadeh et al. [112] explored the engineering of cell sheets by the fabrication of substrates with controlled micro and nanopatterns as support for cell growth, cellular elongation and orientation via contact guidance. For that, the replica-molding method was applied in the production of chitin nanofibre flexible micropatterns. NIH-3T3 cells seeded on the substrates aligned and proliferated along the major axis, forming ultra-thin (<10 μm) and ordered cell sheets. The investigators additionally suggest that this advance in chitin nanofibre application could be used in myocardial repair or even in retinal regeneration where the damaged tissue could be mechanically supported by the chitin substrate while new tissue develops.

Hariraksapitak and Supaphol [90] constructed nanocomposite scaffolds for bone tissue engineering based on hyaluronan-gelatin reinforced with chitin whiskers (length: 255 ± 56 nm; width: 31 ± 6 nm; aspect ratio: 8). Applying the freeze-drying technique they constructed cylindrical and disc shaped products with transverse section pores of 139 μm and longitudinal section pores of 166 μm. After testing diverse matrix/reinforcement compositions, the researchers found that 2 % whisker content composites had twice the tensile strength; however 10 % whisker content scaffolds had better thermal stability, resistance to biodegradation and enhanced human osteosarcoma cells adhesion and proliferation.

Yamamoto et al. [85] described a biomimetic approach to produce biomineralized hydrogels using chitin nanowhiskers. After acid hydrolysis of chitin powders, they prepared suspensions starting from 5 % whiskers content and posteriorly added calcium chloride ($CaCl_2$) for mineralization. Under cross-polarized optical microscopy the suspensions revealed isotropic, biphasic and anisotropic textures with the appearance of liquid crystalline mesophases. When the suspensions were put in contact with ammonium vapors, they converted in gels maintaining liquid crystalline character, with the emergence of fingerprint textures (space fringes of 20 to 30 μm) characteristic of cholesteric liquid crystals. Chitin gel immersion in $CaCl_2$ and further contact with the ammonium vapors lead to the formation of the hybrid ceramic/polymeric hydrogel with the appearance of calcium carbonate crystals at the gels surface.

Cooper et al. [114] developed chitin nanofibre templates by the self-assembling method. After deacetylation of nanofibre's surface, substrates with different morphology (4 nm and 12 nm fibre width) were used to examine mouse cortical neuron cultures. The nanofibrillar templates promoted neuron viability up to 39 %, a huge improvement when comparing with PDL coated surfaces. With this result the chitin nanofibre template proves to be an efficient substrate for in vitro primary neuron cultures with the possibility of being applied as artificial neural networks for diagnostics and therapeutics.

4.2.2 Wound Dressing and Drug-Delivery Systems

The replacement or regeneration of damaged tissues by new ones may be referred to as wound healing process. In order to developed wound dressing films with controllable drug release properties, Bajpai and co-workers [161] investigated the preparation NCC/poly(sodium acrylate) hydrogel films. They concluded that the release of an antibiotic drug (minocycline, Mic) from the composite hydrogel can be controlled mainly by the variation of NCCs content. Moreover, the Mic-loaded films showed fair antifungal and antibacterial properties, which are fundamental properties for wound healing applications.

A further approach was described by Ul-Islam [162] with the incorporation of nanoclays into a BNC porous structure. The impregnation of montmorillonite (MMT) nanoparticles into the empty spaces of BNC network resulted in a composite structure with antibacterial activity and improved mechanical properties (UTS = 209 MPa and E = 6.1 GPa). This cellulose-based nanocomposite exhibit excellent properties for wound healing applications.

Nanowhisker-reinforced all-cellulose composite gels were prepared by Wang and Chen [163] envisaging a high-performance porous material for drug delivery systems. The authors describe a thermally induced phase separation as the most appropriate and rapid method to prepare physically crosslinked NCC/cellulose gels. In this structure, cellulose nanowhiskers act as a 'bridge' to facilitate the crosslinking of cellulose chains providing a good support to the gel network formation improving the mechanical properties of the final composite. With the increase of NCC content, the shear modulus of the neat cellulose gel increased from 81 to 160 kPa, while a higher tensile stress was also observed, indicating an obvious reinforcing effect. Moreover, in vitro assays have demonstrated that these composites can be used for the controlled release of macromolecules in simulated body fluids.

Recently, Mauricio [164] described the preparation of a microhydrogel composite of NCCs and starch via radical polymerization in an ultrasound assisted-emulsion. Starch and NCCs were chemically modified enabling the creation of a microhydrogel composite in which NCC played a role as a covalent crosslinker and also acting as an emulsifying agent for emulsion, improving both sphericity and homogeneity of the microparticles. The drug release profile of the composite was studied by using Vitamin B_{12} as a model drug. It was verified that the introduction of NCC to starch microparticles leads to a sustainable profile of drug release. It was also found that the release rate become approximately 2.9 times slower when NCC is added so that when combined with starch, it acts as a retardation factor for drug release.

Chitin presents a lot of properties that are extremely important for wound healing such as biodegradability, anti-bacterial and anti-fungal activity, hemostatic action and water adsorption. These properties, combined with high tensile strength and elasticity, have been harnessed by researchers in the fabrication of various nanocomposites.

Madhumathi et al. [165] and Kumar et al. [166] developed nanocomposites of chitin and silver for wound dressing applications. The investigators developed freeze-dried scaffolds based on hydrogels of β-chitin from squid and incorporated silver nanoparticles with minimum inhibitory and bactericidal concentration (Fig. 9).

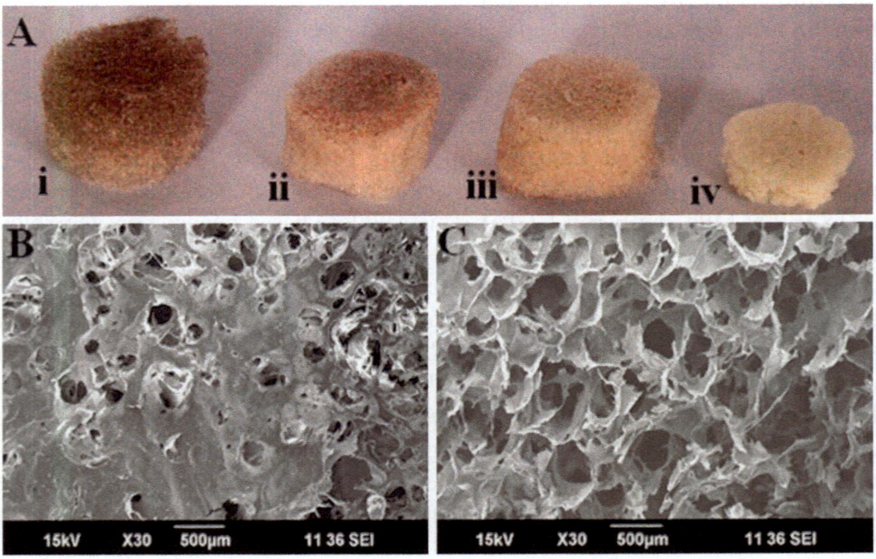

Fig. 9 (**a**) Images of composite scaffolds. (i) β-Chitin + 0.006 % nanosilver; (ii) β-chitin + 0.003 % nanosilver; (iii) β-chitin + 0.001 % nanosilver; (iv) β-chitin control. (**b**) SEM image of β-chitin control scaffold (**c**) SEM image of β-chitin with nanosilver (0.006 %). Images reprinted with permission from [166] copyright© 2010 Elsevier

Scaffolds' biological characterization showed that alongside with no toxicity and with good cell adhesion (Vero cells), this nanocomposite was bactericidal against Escherichia coli and Staphylococcus aureus and allowed quick blood-clotting.

Watthanaphanit and co-workers reported a wetspun method to produce nanocomposite fibres of alginate and chitin nanowhiskers [76, 124]. Using acid hydrolysis, they extracted chitin nanowhiskers (length: 110–975 nm; width: 8–73 nm; aspect ratio: 7.5) and included them in alginate solutions for further extrusion and coagulation in $CaCl_2$ and MeOH baths. The nanoparticles had a significant effect in the wet-spun composite with mechanical and thermal properties improving alongside with accelerated biodegradation in Tris-HCl buffer solution containing lysozyme.

Combining chitin nanofibrils extracted from crustacean, chitosan glycolate and chlorhexidine, Muzzarelli et al. [167] manufactured composite spray, gel and gauze for wound healing. These three products were tested using mice models and the gauze was clinically applied in 75 patients. Results suggested different usages for the products: spray was suitable as first aid tool in abrasions with little bleed; the gel had the ability to promote physiological repair in areas with thin epidermal layers; the gauze lead to better epithelial differentiation, keratinization and reorganization of the basal lamina, resulting in scarless skin repair.

Ang-atikarnkul et al. [75] developed a freeze-dried bionanocomposite of cellulose nanofibre (length: >400 nm; width: 7.3 nm; aspect ratio: >55)/chitin whisker (length: 308 nm; width: 27 nm; aspect ratio: 11)/silk sericin with potential to become a wound care product. The macroporous structure obtained by the freeze-drying process is advantageous to the extent that it facilitates oxygen diffusion resulting in better wound repair. The presence of chitin nanowhiskers attributes wound healing ability to the scaffolds. When immersed in Tris-HCl buffer at physiological pH with and without lysozyme, the nanocomposite is able to release silk sericin to the medium and it is expected that this release may prevent cell damage by oxidation and provide wound-moisturizing effect.

Yoo et al. [168] developed two types of electrospun membranes to mimic the human extracellular matrix for epidermal keratinocytes. Through this process the researchers obtained bicomponent scaffolds of chitin nanofibres (140 nm) combined with silk fibroin fibres (1260 nm) and also a blend membrane of chitin/silk fibroin (340–920 nm). In both fibre types, cell morphology, attachment and proliferation were evaluated and after 1-hour incubation the chitin/silk fibroin membrane demonstrated a better performance.

Shelma et al. [169] suggested composite films of chitosan and chitin nanowhiskers (length: 300 nm; width 20 nm) for wound healing applications, where the nanofiller allows increase in mechanical properties without alteration in water vapor permeability.

Naseri et al. [79] also produced wound dressing materials by electrospinning chitosan based mats with reinforcing chitin nanowhiskers. Nanocomposite fibres with diameters of 223–966 nm were obtained and assembled in membranes that showed whisker influence in tensile strength (64.9 MPa) and modulus (10.2 GPa). Along with high surface area, the membranes allowed high water vapor transmission rate (1290–1548 $gm^{-2}day^{-1}$) similar to skin wounds. The chitosan/nanochitin membranes demonstrated no toxicity and no inhibition in adipose derived stem cells growth.

Ifuku et al. [170] examined the preventive effects of chitin nanofibres in a mouse model of dextran sulfate sodium (DSS)-induced acute ulcerative colitis. Oral administration of chitin nanofibres improved clinical symptoms, colon inflammation and histological tissue injury. Severe erosions, crypt destruction and edema were almost completely suppressed by chitin's action. Chitin nanofibres were responsible for anti-inflammatory effect by suppressing NF-kB activation and antifibrosis effects in DSS-induced UC mice models.

Lin et al. [171] developed composite alginate/polysaccharide based microspheres and studied their structure, properties and drug release kinetics. Applying the acid hydrolysis method, nanocrystals of cellulose (length: 200–300 nm; width: 10–20 nm; aspect ratio: 20), chitin (length: 300–400 nm; width: 10–20 nm; aspect ratio: 30) and starch (length: 40–60 nm; width 15-30 nm) were obtained and used to produce alginate nanocomposite microspheres of (3500 ± 50) μm in the wet state and (900 ± 20) μm after drying. The presence of the polysaccharide nanocrystals not only improved microspheres' mechanical properties but also enhanced the swelling behavior consistency and increased the encapsulation efficiency. The analysis of the drug release profile revealed that nanocrystals' presence probably restricted the motion of the alginate polymer chains, inhibiting drug diffusion and slowing microspheres structural disintegration, pointing to an improved drug load and release effect.

Zhang et al. [84] produced injectable cyclodextrin hydrogels reinforced with polysaccharide nanocrystals (cellulose and chitin whiskers, platelet-like starch) and obtained structures with superior elastic modulus, where cellulose whiskers were able to offer 50 times higher strength than native hydrogel. The reinforcements allowed accelerated hydrogel gelation enabling higher mechanical stability but at the same time maintaining essential shear-thinning and temperature sensitive properties for an injectable material. The nanocomposite hydrogels showed a more sustained release profile of bovine serum albumin and no cytotoxicity.

4.2.3 Electronics

Along the years, nanofibre based sensors have been proposed as very useful for enhanced detection in food industry, clinical and environmental areas. They present unique advantages such as high surface area/volume ratio that increases binding site availability, faster nanofibre mass transfer rates that allow lower limit and faster detection, easy fabrication of nanofibre membranes with different morphologies (size, shape, pore size) and mechanical properties [118].

Cellulose-based electroconductive nanocomposites can be prepared by combining conducting electroactive materials with biocompatible nanocellulose. The deposition of conductive polymers, such as polyaniline (PANI) and polypyrrole (PPy), on the surface of fibres used in fabrics and yarns have been widely investigated in the last few years due to its importance to the fabrication of new functional devices, including bio-batteries [172], sensors and biosensors [173].

Hu et al. [174] studied the oxidative polymerization of aniline using the three-dimensional structure of BNC as a template. The resulting PANI-coated bacterial nanocellulose composite formed a uniform and flexible membrane with a high conductivity (5.0×10^{-2} S/cm) and good mechanical properties (UTS = 95.7 MPa and E = 5.6 GPa) which could be applied in sensors and flexible electrodes. The development of multilayer PANI/BNC composites (Fig. 10) as flexible electrode materials has been also reported by Lin and co-workers [175]. They proposed a new synthetic route of interfacial polymerization by which aniline was polymerized in situ on the BNC surface. The surface resistivity of the BNC fibres covered with PANI significantly decreased from 108 to 40 Ωcm.

Another approach proposed by Valentini and collaborators [176] regards the preparation of electrically conductive cellulose-based nanocomposites. A composite film of NCC and graphene oxide (GO) was produced by drop casting of a water dispersion of GO in the presence of NCCs. The composite conductivity was investigated after thermal reduction of GO and the composite demonstrated a significant transition from an electrically insulating state to a conductive one with the increase of GO content.

Nakorn et al. [177] proposed a chitin nanowhisker (300 nm) and chitosan nanoparticle (39 nm) based system targeting protein immobilization, in order to develop a system suitable for glucose detection.

Stephan et al. [178] reported a chitin-incorporated poly(ethylene oxide) based nanocomposite electrolytes for lithium batteries. Chitin nanowhiskers produced by acid hydrolysis were incorporated in PEO nanocomposite prepared by hot pressing. After incorporation, an increase in ionic conductivity and lithium transference number were observed.

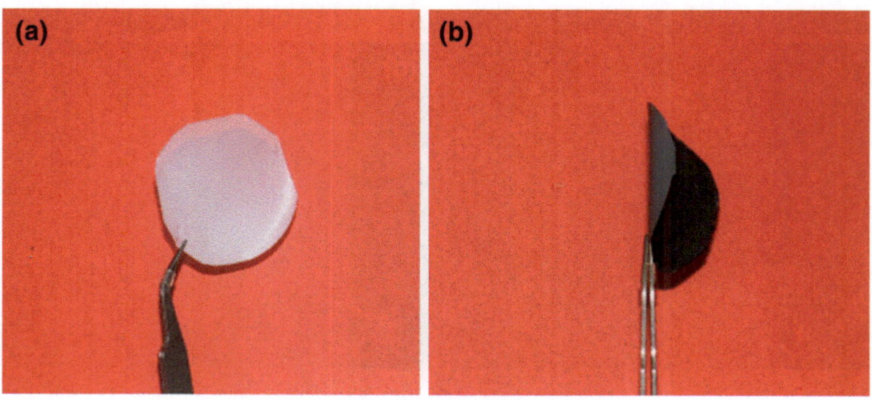

Fig. 10 Optical images of **a** pure BNC film and **b** BNC/PANI nanocomposite film containing 60.6 wt% PANI. Reprinted with permission from [175] copyright © 2013 American Chemical Society

4.2.4 Cosmetics

Recently, efforts have been made in order to develop facial masks for revitalizing, healing, and refreshing facial skin. Aramwit and co-workers [179] proposed the fabrication of a cellulose-based nanocomposite with improved biological properties. For such purpose, a BNC gel was absorbed with silk sericin, a biocompatible protein with interesting antioxidant properties and often used as bioadhesive and bioactive agent.

The authors concluded that the nanocomposite gel was less adhesive and non-cytotoxic (Fig. 11) than the commercially available paper masks, which is an advantage since it may be removed more easily and with reduced pain.

Chitin nanofibres and nanowhiskers have a huge potential for cosmetic applications. To raise awareness about this area, Morganti and collaborators contributed with a wide variety of works and reviews where chitin properties are explored for the benefit of skin and hair treatment [180–185].

Nanochitin can bind with compounds used in cosmetics (anti-oxidants such as lutein, melatonin and lipoic acid and immune modulators such as ecotoin and beta-glucan) and enhance their biologic activity and effectiveness [185]. Although

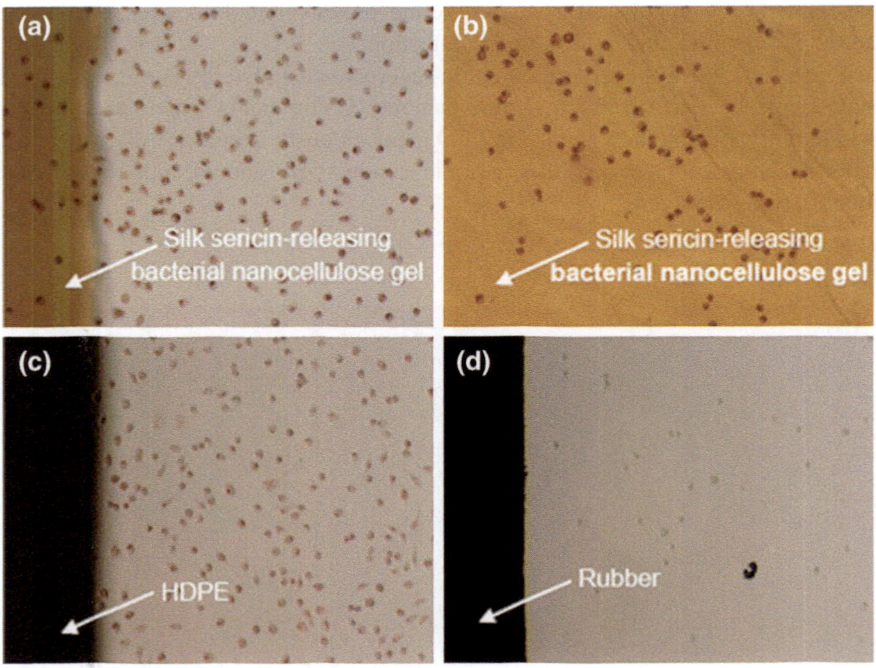

Fig. 11 Morphology of L929 mouse fibroblast cells after cultured on materials for 48 h. **a**, **b** silk sericin-releasing bacterial nanocellulose gel, **c** high-density polyethylene (HDPE, negative control), and **d** rubber (positive control). Image reprinted [179] copyright © 2014 Aramwit and Bang

chitin is easily metabolized by human body's endogenous enzymes (lysozyme, lipases and N-acetyl-D-glucosaminidase), chitin nanofibrils do not penetrate cells nor the circulatory system due to their high molecular weight, but they can reach different levels of skin permeability, being in that way good candidates to deliver active compounds to the treatment site. The nanofibres can be posteriorly catabolized to glucose and glutamic acid and introduced in the cell energy production cycle [186].

Chitin nanofibrils have the ability to integrate growth factors such as heparin binding factor, TGF-β1 and FGF-2 through their hydrophilic property and consequently intervene in cell behavior as microenvironmetal stimulus. Chitin nanofibres and nanowhiskers can control skin epithelium-mesenchyma molecular relationships and the hair follicle cycle by making strong bonds with keratin due to their electropositive surface charges [187]. Enhancing the ability of tissue to uptake and fix water, nanochitin can be used in association with moisturizers such as PCA and hyaluronic acid, allowing skin hydration for long periods, skin barrier repairing and free radicals neutralization [186].

4.2.5 Absorbents

Making use of cellulose and chitin nanofibrils morphology, surface charge and hydrophilicity, some researchers have proposed their use as absorbents/physical barriers in different situations. Hatakeyama and colleagues [188] have described the development of compostable and environmentally compatible polymeric foams for water absorption. A polyurethane (PU) composite foam was prepared using polyols derived from molasses and lignin filled with NCCs. During water immersion, it was observed that sorption time markedly decreases by adding NCCs as fillers and the PU matrix restrains a large amount of water for a long period. Later, a study conducted by Gabr et al. [189] focused on the water absorption behavior of a novel material comprised of MFC and laponite clay. The composite showed a considerable improvement of its mechanical properties with the addition of nanoclay. This reinforcement decreased the water absorption property of the nanocomposite, acting as a water barrier, which could be a great advantage for packaging applications.

Ma et al. [190] developed thin nanofibrous membranes of chitin and cellulose nanofibres and nanowhiskers. Fibres obtained through the TEMPO-mediated oxidation and presenting diameters between 5 and 10 nm were assembled in membranes for water ultrafiltration. These membranes had high surface/volume ratio and negatively charged surfaces, which allowed high virus adsorption capacity verified by MS2 bacteriophage testing (Fig. 12).

When comparing with commercially available filters, these membranes stand as a better choice because they are able to maintain equal rejection ratio while increasing water flux and reduce particles clogging due to smooth nanofibres surface.

Dolphen et al. [68] produced chitin nanowhiskers with 150–800 nm length and 10–50 nm and suggested their application in melanoidin (or other pigments)

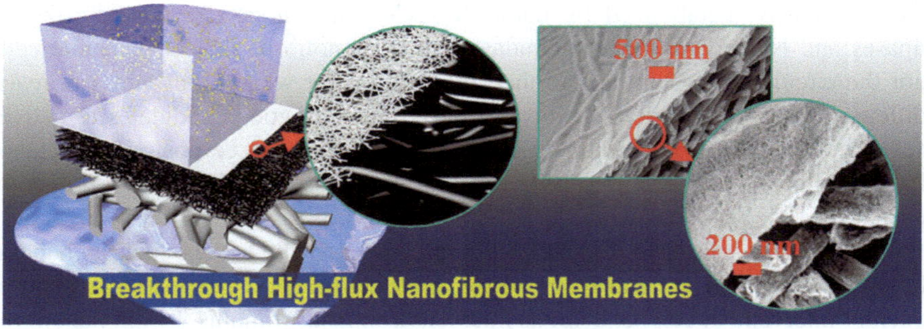

Fig. 12 Graphical abstract of chitin and cellulose nanofibres membrane for water purification developed by Ma et al. Image reprinted with permission from [190] copyright© 2011 American Chemical Society

adsorption in sugar syrup. The researchers realized that chitin nanofibres were capable of establishing electrostatic and chemical interaction with melanoidin and showed increase adsorption with increasing temperature.

Huang et al. [191] also proposed the use of chitin whiskers in non-aqueous systems. To achieve that, chitin whiskers where modified with long chain alkyl group by reacting with bromohexadecane, and their hydrophobicity was evaluated in sunflower oil. The introduction of alkyl group chains decreased surface energy and also provoked surface roughness, leading to high hydrophobicity. Besides being well dispersed in sunflower oil, chitin whiskers were also able to remain stable for at least 6 months. Nanoparticles dimension, high aspect ratio and alkyl side chain were pointed as responsible for that occurrence.

5 Conclusions

The hierarchical and semicrystalline structure of cellulose and chitin is particularly suited for the extraction of mechanically strong crystalline nanofibres. The use of these nanofibres as reinforcements in the production of nanocomposites with unique properties has been the subject of intense research in recent years. The availability, biocompatibility and biodegradability of these polysaccharides, associated with green fabrication methods and environmentally friendly polymeric matrices imparts these nanocomposites with properties that are much sought after by researchers in many fields. Both cellulose and chitin have been used as structural reinforcements in PU, natural rubber, PLA, starch, polymethyl methacrylate, soy protein, PVA and many others with significant improvement in the elastic modulus and the ultimate tensile strain. Applications of these structural composites to the automotive, building and packaging fields have been studied. Functional composites based on cellulose and chitosan have also been developed for usage in tissue engineering and other biomedical applications such as wound healing and drug delivery, as well as in the electronics, cosmetics and absorbents sectors. The high performance of cellulose and chitin based nanocomposites makes them promising materials for novel applications and solutions for current societal needs and certainly warrants the pursuit of research and development on these amazing nanocomposites.

Acknowledgments This work work was partially funded by FEDER funds through the COMPETE 2020 Programme and National Funds through the Portuguese Foundation for Science and Technology (FCT - MEC) under the project UID/CTM/50025/2013. Carlos F. C. João and Ana C. Baptista also acknowledge FCT- MEC for SFRH/BD/80860/2011 and SFRH/BPD/104407/2014 grants, respectively.

References

1. Gao H, Ji B, Jäger IL et al (2003) Materials become insensitive to flaws at nanoscale: lessons from nature. Proc Natl Acad Sci USA 100:5597–5600
2. Ray D, Sain S (2014) Nanocellulose-reinforced polymer matrix composites fabricated by In-Situ polymerization technique. In: Nanocellulose polymer nanocomposites, Wiley, pp 131–161
3. Kalia S, Dufresne A, Cherian BM et al (2011) Cellulose-based bio- and nanocomposites: a review. Int J Polym Sci 2011
4. Moon RJ, Martini A, Nairn J et al (2011) Cellulose nanomaterials review: structure, properties and nanocomposites. Chem Soc Rev 40:3941–3994
5. Borges JP, Godinho MH, Martins AF et al (2004) Tensile properties of cellulose fiber reinforced hydroxypropylcellulose films. Polym Compos 25:102–110
6. Zeng J-B, He Y-S, Li S-L et al (2012) Chitin whiskers: an overview. Biomacromolecules 13:1–11
7. Carlstrom D (1957) The crystal structure of alpha-chitin (poly-N-acetyl-D-glucosamine). J Biophys Biochem Cytol 3:669–683
8. Michael TP, András V, John D et al (2011) Development of the metrology and imaging of cellulose nanocrystals. Meas Sci Technol 22:024005
9. Nikolov S, Petrov M, Lymperakis L et al (2010) Revealing the design principles of high-performance biological composites using Ab initio and multiscale simulations: the example of lobster cuticle. Adv Mater 22:519–526
10. Klemm D, Heublein B, Fink H-P et al (2005) Cellulose: fascinating biopolymer and sustainable raw material. Angew Chem Int Ed 44:3358–3393
11. Borges JP, Canejo JP, Fernandes SN et al (2014) Cellulose-based liquid crystalline composite systems. In: Nanocellulose polymer nanocomposites, Wiley, pp 215–235
12. Sjostrom E (1993) Chapter 2—introduction to carbohydrate chemistry. In: Sjostrom E (ed) Wood chemistry, 2nd edn. Academic Press, San Diego, pp 21–50
13. Klemm D, Kramer F, Moritz S et al (2011) Nanocelluloses: a new family of nature-based materials. Angew Chem Int Ed 50:5438–5466
14. Ibrahim M, El-Zawawy W (2015) Extraction of cellulose nanofibers from cotton linter and their composites. In: Pandey JK, Takagi H, Nakagaito AN et al (eds) Handbook of polymer nanocomposites. Processing, performance and application, Springer Berlin, pp 145–164
15. Qian Y, Qin Z, Vu N-M et al (2012) Comparison of nanocrystals from TEMPO oxidation of bamboo, softwood and cotton linter fibers with ultrasonic-assisted process. BioResources 7 (4):4952–4964
16. Kopania E, Wietecha J, Ciechanska D (2012) Studies on isolation of celluose fibres from waste plant biomass. Fibres Text East Eur 20:167–171

17. Morán J, Alvarez V, Cyras V et al (2008) Extraction of cellulose and preparation of nanocellulose from sisal fibers. Cellulose 15:149–159
18. Deepa B, Abraham E, Cordeiro N et al (2015) Utilization of various lignocellulosic biomass for the production of nanocellulose: a comparative study. Cellulose 22:1075–1090
19. Li M, L-j W, Li D et al (2014) Preparation and characterization of cellulose nanofibers from de-pectinated sugar beet pulp. Carbohydr Polym 102:136–143
20. Alemdar A, Sain M (2008) Isolation and characterization of nanofibers from agricultural residues—wheat straw and soy hulls. Bioresour Technol 99:1664–1671
21. Flauzino Neto WP, Silvério HA, Dantas NO et al (2013) Extraction and characterization of cellulose nanocrystals from agro-industrial residue—Soy hulls. Ind Crops Prod 42:480–488
22. Mihranyan A (2011) Cellulose from cladophorales green algae: From environmental problem to high-tech composite materials. J Appl Polym Sci 119:2449–2460
23. Keshk S (2014) Bacterial cellulose production and its industrial applications. Bioprocess Biotechniques 4:1–10
24. Dufresne A (2013) Nanocellulose: a new ageless bionanomaterial. Mater Today 16:220–227
25. Nakagaito AN, Yano H (2008) The effect of fiber content on the mechanical and thermal expansion properties of biocomposites based on microfibrillated cellulose. Cellulose 15:555–559
26. Chakraborty A, Sain M, Kortschot M (2005) Cellulose microfibrils: a novel method of preparation using high shear refining and cryocrushing. Holzforschung 59:102
27. Panthapulakkal S, Sain M (2012) Preparation and characterization of cellulose nanofibril films from wood fibre and their thermoplastic polycarbonate composites. Int J Polym Sci 2012:6
28. Chen W, Yu H, Liu Y et al (2011) Individualization of cellulose nanofibers from wood using high-intensity ultrasonication combined with chemical pretreatments. Carbohydr Polym 83:1804–1811
29. Habibi Y (2014) Key advances in the chemical modification of nanocelluloses. Chem Soc Rev 43:1519–1542
30. Brinchi L, Cotana F, Fortunati E et al (2013) Production of nanocrystalline cellulose from lignocellulosic biomass: Technology and applications. Carbohydr Polym 94:154–169
31. Klemm D, Schumann D, Kramer F et al (2009) Nanocellulose materials—different cellulose, different functionality. Macromol Symp 280:60–71
32. Dufresne A (2010) Processing of polymer nanocomposites reinforced with polysaccharide nanocrystals. Molecules 15:4111
33. Borges JP, Godinho MH, Figueirinhas JL et al (2011) All-cellulosic based composites. In: Kalia S, Kaith BS, Kaur I (eds) Cellulose fibers: bio- and nano-polymer composites. Springer, Berlin, pp 399–421
34. Habibi Y (2013) Chemical modification of nanocelluloses. In: Biopolymer nanocomposites, Wiley, pp 367–390
35. Missoum K, Belgacem M, Bras J (2013) Nanofibrillated cellulose surface modification: a review. Materials 6:1745
36. Ifuku S, Nogi M, Abe K et al (2007) Surface modification of bacterial cellulose nanofibers for property enhancement of optically transparent composites: dependence on acetyl-group DS. Biomacromolecules 8:1973–1978
37. Faruk O, Sain M, Farnood R et al (2014) Development of lignin and nanocellulose enhanced bio PU foams for automotive parts. J Polym Environ 22:279–288
38. MacVicar R, Matuana LM, Balatinecz JJ (1999) Aging mechanisms in cellulose fiber reinforced cement composites. Cement Concr Compos 21:189–196
39. Khan A, Huq T, Khan RA et al (2012) Nanocellulose-based composites and bioactive agents for food packaging. Crit Rev Food Sci Nutr 54:163–174
40. Lin N, Dufresne A (2014) Nanocellulose in biomedicine: current status and future prospect. Eur Polym J 59:302–325
41. Stenstad P, Andresen M, Tanem B et al (2008) Chemical surface modifications of microfibrillated cellulose. Cellulose 15:35–45

42. Ferrer A, Filpponen I, Rodríguez A et al (2012) Valorization of residual Empty Palm Fruit Bunch Fibers (EPFBF) by microfluidization: production of nanofibrillated cellulose and EPFBF nanopaper. Bioresour Technol 125:249–255
43. Sacui IA, Nieuwendaal RC, Burnett DJ et al (2014) Comparison of the properties of cellulose nanocrystals and cellulose nanofibrils isolated from bacteria, tunicate, and wood processed using acid, enzymatic, mechanical, and oxidative methods. ACS Appl Mater Interfaces 6:6127–6138
44. Lu P, Hsieh Y-L (2012) Preparation and characterization of cellulose nanocrystals from rice straw. Carbohydr Polym 87:564–573
45. Cherian BM, Leão AL, de Souza SF et al (2010) Isolation of nanocellulose from pineapple leaf fibres by steam explosion. Carbohydr Polym 81:720–725
46. Grunert M, Winter W (2002) Nanocomposites of cellulose acetate butyrate reinforced with cellulose nanocrystals. J Polym Environ 10:27–30
47. Araki J, Wada M, Kuga S et al (1998) Flow properties of microcrystalline cellulose suspension prepared by acid treatment of native cellulose. Colloids Surf A 142:75–82
48. Kumar A, Negi YS, Choudhary V et al (2014) Characterization of cellulose nanocrystals produced by acid-hydrolysis from sugarcane bagasse as agro-waste. J Mater Phys Chem 2:1–8
49. Qin Z-Y, Tong G, Chin YCF et al (2011) Preparation of ultrasonic-assisted high carboxylate content cellulose nanocrystals by TEMPO oxidation. BioResources 6(2):1136–1146
50. Zhao Y, Zhang Y, Lindström ME et al (2015) Tunicate cellulose nanocrystals: preparation, neat films and nanocomposite films with glucomannans. Carbohydr Polym 117:286–296
51. Castro C, Zuluaga R, Álvarez C et al (2012) Bacterial cellulose produced by a new acid-resistant strain of gluconacetobacter genus. Carbohydr Polym 89:1033–1037
52. Nguyen V, Flanagan B, Gidley M et al (2008) Characterization of cellulose production by a gluconacetobacter xylinus strain from kombucha. Curr Microbiol 57:449–453
53. Nimeskern L, Martínez Ávila H, Sundberg J et al (2013) Mechanical evaluation of bacterial nanocellulose as an implant material for ear cartilage replacement. J Mech Behav Biomed Mater 22:12–21
54. Wesarg F, Schlott F, Grabow J et al (2012) In Situ synthesis of photocatalytically active hybrids consisting of bacterial nanocellulose and anatase nanoparticles. Langmuir 28:13518–13525
55. Salaberria AM, Labidi J, Fernandes SCM (2015) Different routes to turn chitin into stunning nano-objects. Eur Polym J 68:503–515
56. Khor E (2001) Chitin: fulfilling a biomaterials promise. Elsevier Science Limited, Oxford
57. Martínez JP, Falomir MP, Gozalbo D (2014) Chitin: a structural biopolysaccharide with multiple applications. eLS. Wiley, Chichester
58. Jang M-K, Kong B-G, Jeong Y-I et al (2004) Physicochemical characterization of α-chitin, β-chitin, and γ-chitin separated from natural resources. J Polym Sci Part A Polym Chem 42:3423–3432
59. Gupta NS (2010) Chitin. Springer
60. Belamie E, Giraud-Guille MM (2004) Liquid-crystalline behavior in aqueous suspensions of elongated chitin microcrystals. Springer, Berlin
61. Belamie E, Mosser G, Gobeaux F et al (2006) Possible transient liquid crystal phase during the laying out of connective tissues: α-chitin and collagen as models. J Phys Condens Matter 18:S115–S129
62. Shervani Z, Taisuke Y, Ifuku S et al (2012) Preparation of gold nanoparticles loaded chitin nanofiber composite. Adv Nanopart 01:71–78
63. Fan Y, Saito T, Isogai A (2008) Preparation of chitin nanofibers from squid pen β-chitin by simple mechanical treatment under acid conditions. Biomacromolecules 9:1919–1923
64. Fan Y, Saito T, Isogai A (2009) TEMPO-mediated oxidation of β-chitin to prepare individual nanofibrils. Carbohydr Polym 77:832–838
65. Fan Y, Saito T, Isogai A (2008) Chitin nanocrystals prepared by TEMPO-mediated oxidation of α-chitin. Biomacromolecules 9:192–198

66. J-i K, Takegawa A, Mine S et al (2011) Preparation of chitin nanowhiskers using an ionic liquid and their composite materials with poly(vinyl alcohol). Carbohydr Polym 84:1408–1412
67. Barber PS, Griggs CS, Bonner JR et al (2013) Electrospinning of chitin nanofibers directly from an ionic liquid extract of shrimp shells. Green Chem 15:601–607
68. Dolphen and Thiravetyan (2011) Adsorption of melanoidins by chitin nanofibers. Chem Eng J 166:6
69. Goodrich JD, Winter WT (2007) α-Chitin nanocrystals prepared from shrimp shells and their specific surface area measurement. Biomacromolecules 8:252–257
70. Sriupayo S, Supaphol P, Blackwell J et al (2005) Preparation and characterization of @a-chitin whisker-reinforced poly(vinyl alcohol) nanocomposite films with or without heat treatment. Polymer 46:8
71. Zhou Y, Fu S, Pu Y et al (2014) Preparation of aligned porous chitin nanowhisker foams by directional freeze–casting technique. Carbohydr Polym 112:277–283
72. Phongying S, S-i A, Chirachanchai S (2007) Direct chitosan nanoscaffold formation via chitin whiskers. Polymer 48:393–400
73. Lertwattanaseri T, Ichikawa N, Mizoguchi T et al (2009) Microwave technique for efficient deacetylation of chitin nanowhiskers to a chitosan nanoscaffold. Carbohydr Res 344:331–335
74. Wongpanit P, Sanchavanakit N, Pavasant P et al (2007) Preparation and characterization of chitin whisker-reinforced silk fibroin nanocomposite sponges. Eur Polym J 43:4123–4135
75. Ang-atikarnkul P, Watthanaphanit A, Rujiravanit R (2014) Fabrication of cellulose nanofiber/chitin whisker/silk sericin bionanocomposite sponges and characterizations of their physical and biological properties. Compos Sci Technol 96:88–96
76. Watthanaphanit A, Supaphol P, Tamura H et al (2008) Fabrication, structure, and properties of chitin whisker-reinforced alginate nanocomposite fibers. J Appl Polym Sci 110:890–899
77. Junkasem J, Rujiravanit R, Supaphol P (2006) Fabrication of α-chitin whisker-reinforced poly(vinyl alcohol) nanocomposite nanofibres by electrospinning. Nanotechnology 17:4519–4528
78. Ji, Wolfe, Rodriguez, et al. (2012) Preparation of chitin nanofibril/polycaprolactone nanocomposite from a nonaqueous medium suspension. Carbohydr Polym 87:7
79. Naseri N, Algan C, Jacobs V et al (2014) Electrospun chitosan-based nanocomposite mats reinforced with chitin nanocrystals for wound dressing. Carbohydr Polym 109:7–15
80. Gopalan Nair K, Dufresne A (2003) Crab shell chitin whisker reinforced natural rubber nanocomposites. 1. Processing and swelling behavior. Biomacromolecules 4:657–665
81. Tzoumaki MV, Moschakis T, Biliaderis CG (2010) Metastability of nematic gels made of aqueous chitin nanocrystal dispersions. Biomacromolecules 11:175–181
82. Tzoumaki MV, Moschakis T, Kiosseoglou V et al (2011) Oil-in-water emulsions stabilized by chitin nanocrystal particles. Food Hydrocolloids 25:1521–1529
83. Lu Y, Weng L, Zhang L (2004) Morphology and properties of soy protein isolate thermoplastics reinforced with chitin whiskers. Biomacromolecules 5:1046–1051
84. Zhang X, Huang J, Chang PR et al (2010) Structure and properties of polysaccharide nanocrystal-doped supramolecular hydrogels based on cyclodextrin inclusion. Polymer 51:4398–4407
85. Yamamoto Y, Nishimura T, Saito T et al (2010) CaCO3/chitin-whisker hybrids: formation of CaCO3 crystals in chitin-based liquid-crystalline suspension. Polym J 42:583–586
86. Araki J, Yamanaka Y, Ohkawa K (2012) Chitin-chitosan nanocomposite gels: reinforcement of chitosan hydrogels with rod-like chitin nanowhiskers. Polym J 44:713–717
87. Ma B, Qin A, Li X et al (2014) Structure and properties of chitin whisker reinforced chitosan membranes. Int J Biol Macromol 64:341–346
88. Ma B, Qin A, Li X et al (2014) Bioinspired design and chitin whisker reinforced chitosan membrane. Mater Lett 120:82–85
89. Pereira AGB, Muniz EC, Hsieh Y-L (2014) Chitosan-sheath and chitin-core nanowhiskers. Carbohydr Polym 107:158–166

90. Hariraksapitak P, Supaphol P (2010) Preparation and properties of α-chitin-whisker-reinforced hyaluronan-gelatin nanocomposite scaffolds. J Appl Polym Sci 117:3406–3418
91. Ifuku S, Morooka S, Morimoto M et al (2010) Acetylation of chitin nanofibers and their transparent nanocomposite films. Biomacromolecules 11:1326–1330
92. Ifuku S, Nogi M, Abe K et al (2009) Preparation of chitin nanofibers with a uniform width as alpha-chitin from crab shells. Biomacromolecules 10:1584–1588
93. Chen C, Li D, Hu Q et al (2014) Properties of polymethyl methacrylate-based nanocomposites: Reinforced with ultra-long chitin nanofiber extracted from crab shells. Mater Des 56:1049–1056
94. Fan Y, Saito T, Isogai A (2010) Individual chitin nano-whiskers prepared from partially deacetylated α-chitin by fibril surface cationization. Carbohydr Polym 79:1046–1051
95. Hatanaka D, Yamamoto K, J-i K (2014) Preparation of chitin nanofiber-reinforced carboxymethyl cellulose films. Int J Biol Macromol 69:35–38
96. Schiffman JD, Stulga LA, Schauer CL (2009) Chitin and chitosan: Transformations due to the electrospinning process. Polym Eng Sci 49:1918–1928
97. Noh HK, Lee SW, Kim J-M et al (2006) Electrospinning of chitin nanofibers: degradation behavior and cellular response to normal human keratinocytes and fibroblasts. Biomaterials 27:3934–3944
98. Min B-M, Lee SW, Lim JN et al (2004) Chitin and chitosan nanofibers: electrospinning of chitin and deacetylation of chitin nanofibers. Polymer 45:7137–7142
99. Deng Q, Li J, Yang J et al (2014) Optical and flexible Î ± -chitin nanofibers reinforced poly (vinyl alcohol) (PVA) composite film: fabrication and property. Compos A 67:55–60
100. Lu Y, Sun Q, She X et al (2013) Fabrication and characterisation of α-chitin nanofibers and highly transparent chitin films by pulsed ultrasonication. Carbohydr Polym 98:1497–1504
101. Ifuku S, Shervani Z, Saimoto H (2013) Preparation of chitin nanofibers and their composites. In: Biopolymer nanocomposites, Wiley, pp 11–31
102. Nogi M, Kurosaki F, Yano H et al (2010) Preparation of nanofibrillar carbon from chitin nanofibers. Carbohydr Polym 81:919–924
103. Ifuku S, Nomura R, Morimoto M et al (2011) Preparation of chitin nanofibers from mushrooms. Materials 4:1417–1425
104. Mushi NE, Butchosa N, Salajkova M et al (2014) Nanostructured membranes based on native chitin nanofibers prepared by mild process. Carbohydr Polym 112:255–263
105. Salaberria AM, Fernandes SCM, Diaz RH et al (2015) Processing of α-chitin nanofibers by dynamic high pressure homogenization: characterization and antifungal activity against A. niger. Carbohydr Polym 116:286–291
106. Salaberria AM, Labidi J, Fernandes SCM (2014) Chitin nanocrystals and nanofibers as nano-sized fillers into thermoplastic starch-based biocomposites processed by melt-mixing. Chem Eng J 256:356–364
107. Morin A, Dufresne A (2002) Nanocomposites of chitin whiskers from riftia tubes and Poly (caprolactone). Macromolecules 35:2190–2199
108. Paillet M, Dufresne A (2001) Chitin whisker reinforced thermoplastic nanocomposites. Macromolecules 34:6527–6530
109. Nata IF, Wang SS-S, Wu T-M et al (2012) β-chitin nanofibrils for self-sustaining hydrogels preparation via hydrothermal treatment. Carbohydr Polym 90:1509–1514
110. Nata IF, Wu T-M, Chen J-K et al (2014) A chitin nanofibril reinforced multifunctional monolith poly(vinyl alcohol) cryogel. J Mater Chem B 2:4108
111. Zhong C, Kapetanovic A, Deng Y et al (2011) A chitin nanofiber ink for airbrushing, replica molding, and microcontact printing of self-assembled macro-, micro-, and nanostructures. Adv Mater (FRG) 23:4776–4781
112. Hassanzadeh P, Kharaziha M, Nikkhah M et al (2013) Chitin nanofiber micropatterned flexible substrates for tissue engineering. J Mater Chem B 1:4217–4224
113. Zhong C, Cooper A, Kapetanovic A et al (2010) A facile bottom-up route to self-assembled biogenic chitin nanofibers. Soft Matter 6:5298

114. Cooper A, Zhong C, Kinoshita Y et al (2012) Self-assembled chitin nanofiber templates for artificial neural networks. J Mater Chem 22:3105
115. Sriupayo J, Supaphol P, Blackwell J et al (2005) Preparation and characterization of α-chitin whisker-reinforced chitosan nanocomposite films with or without heat treatment. Carbohydr Polym 62:130–136
116. Zhao H-P, Feng X-Q, Gao H (2007) Ultrasonic technique for extracting nanofibers from nature materials. Appl Phys Lett 90:073112
117. João CFC, Silva JC, Borges JP (2015) Chitin based Nanocomposites: Biomedical Applications. In: Thakur VK, Thakur MK (eds) Eco-friendly polymer nanocomposites. Springer, India, p 576
118. Ding F, Deng H, Du Y et al (2014) Emerging chitin and chitosan nanofibrous materials for biomedical applications. Nanoscale 6:9477–9493
119. Fan Y, Fukuzumi H, Saito T et al (2012) Comparative characterization of aqueous dispersions and cast films of different chitin nanowhiskers/nanofibers. Int J Biol Macromol 50:69–76
120. Ifuku S, Ikuta A, Izawa H et al (2014) Control of mechanical properties of chitin nanofiber film using glycerol without losing its characteristics. Carbohydr Polym 101:714–717
121. Shams MI, Ifuku S, Nogi M et al (2011) Fabrication of optically transparent chitin nanocomposites. Appl Phys A 102:325–331
122. Ji Y, Liang K, Shen X et al (2014) Electrospinning and characterization of chitin nanofibril/polycaprolactone nanocomposite fiber mats. Carbohydr Polym 101:68–74
123. Zhou C, Shi Q, Guo W et al (2013) Electrospun Bio-Nanocomposite Scaffolds for Bone Tissue Engineering by Cellulose Nanocrystals Reinforcing Maleic Anhydride Grafted PLA. ACS Appl Mater Interfaces 5:3847–3854
124. Watthanaphanit A, Supaphol P, Tamura H et al (2010) Wet-spun alginate/chitosan whiskers nanocomposite fibers: Preparation, characterization and release characteristic of the whiskers. Carbohydr Polym 79:9-9
125. Yudin VE, Dobrovolskaya IP, Neelov IM et al (2014) Wet spinning of fibers made of chitosan and chitin nanofibrils. Carbohydr Polym 108:176–182
126. Valo H, Arola S, Laaksonen P et al (2013) Drug release from nanoparticles embedded in four different nanofibrillar cellulose aerogels. Eur J Pharm Sci 50:69–77
127. Heath L, Zhu L, Thielemans W (2013) Chitin Nanowhisker Aerogels. ChemSusChem 6:537–544
128. Tsutsumi Y, Koga H, Qi Z-D et al (2014) Nanofibrillar chitin aerogels as renewable base catalysts. Biomacromolecules 15:4314–4319
129. Belamie E, Boltoeva MY, Yang K et al (2011) Tunable hierarchical porosity from self-assembled chitin–silica nano-composites. J Mater Chem 21:16997
130. Chatrabhuti S, Chirachanchai S (2013) Single step coupling for multi-responsive water-based chitin/chitosan magnetic nanoparticles. Carbohydr Polym 97:441–450
131. Galateanu B, Bunea M-C, Stanescu P et al (2015) In Vitro studies of bacterial cellulose and magnetic nanoparticles smart nanocomposites for efficient chronic wounds healing. Stem Cells Int 2015:10
132. Malho J-M, Heinonen H, Kontro I et al (2014) Formation of ceramophilic chitin and biohybrid materials enabled by a genetically engineered bifunctional protein. Chem Commun 50:7348–7351
133. Aspler J, Bouchard J, Hamad W et al (2013) Review of nanocellulosic products and their applications. In: Biopolymer Nanocomposites, Wiley, pp 461–508
134. Lee K-Y, Aitomäki Y, Berglund LA et al (2014) On the use of nanocellulose as reinforcement in polymer matrix composites. Compos Sci Technol 105:15–27
135. Suryanegara L, Nakagaito AN, Yano H (2009) The effect of crystallization of PLA on the thermal and mechanical properties of microfibrillated cellulose-reinforced PLA composites. Compos Sci Technol 69:1187–1192
136. Wu Q, Henriksson M, Liu X et al (2007) A high strength nanocomposite based on microcrystalline cellulose and polyurethane. Biomacromolecules 8:3687–3692

137. Iwatake A, Nogi M, Yano H (2008) Cellulose nanofiber-reinforced polylactic acid. Compos Sci Technol 68:2103–2106
138. Bendahou A, Kaddami H, Dufresne A (2010) Investigation on the effect of cellulosic nanoparticles' morphology on the properties of natural rubber based nanocomposites. Eur Polym J 46:609–620
139. Lee K-Y, Tang M, Williams CK et al (2012) Carbohydrate derived copoly(lactide) as the compatibilizer for bacterial cellulose reinforced polylactide nanocomposites. Compos Sci Technol 72:1646–1650
140. Martins IMG, Magina SP, Oliveira L et al (2009) New biocomposites based on thermoplastic starch and bacterial cellulose. Compos Sci Technol 69:2163–2168
141. Wan YZ, Luo H, He F et al (2009) Mechanical, moisture absorption, and biodegradation behaviours of bacterial cellulose fibre-reinforced starch biocomposites. Compos Sci Technol 69:1212–1217
142. Trovatti E, Oliveira L, Freire CSR et al (2010) Novel bacterial cellulose–acrylic resin nanocomposites. Compos Sci Technol 70:1148–1153
143. Hassanzadeh P, Sun W, de Silva JP et al (2014) Mechanical properties of self-assembled chitin nanofiber networks. J Mater Chem B 2:2461–2466
144. Rizvi R, Cochrane B, Naguib H et al (2011) Fabrication and characterization of melt-blended polylactide-chitin composites and their foams. J Cell Plast 47:283–300
145. Wang B, Li J, Zhang J et al (2013) Thermo-mechanical properties of the composite made of poly (3-hydroxybutyrate-co-3-hydroxyvalerate) and acetylated chitin nanocrystals. Carbohydr Polym 95:100–106
146. Nair KG, Dufresne A (2003) Crab shell chitin whisker reinforced natural rubber nanocomposites. 2. Mechanical behavior. Biomacromolecules 4:666–674
147. Nair KG, Dufresne A, Gandini A et al (2003) Crab shell chitin whiskers reinforced natural rubber nanocomposites. 3. Effect of chemical modification of chitin whiskers. Biomacromolecules 4:1835–1842
148. Li X, Li X, Ke B et al (2011) Cooperative performance of chitin whisker and rectorite fillers on chitosan films. Carbohydr Polym 85:747–752
149. Rubentheren V, Ward TA, Chee CY et al (2015) Processing and analysis of chitosan nanocomposites reinforced with chitin whiskers and tannic acid as a crosslinker. Carbohydr Polym 115:379–387
150. Millon LE, Wan WK (2006) The polyvinyl alcohol–bacterial cellulose system as a new nanocomposite for biomedical applications. J Biomed Mater Res B Appl Biomater 79B:245–253
151. Zimmermann KA, LeBlanc JM, Sheets KT et al (2011) Biomimetic design of a bacterial cellulose/hydroxyapatite nanocomposite for bone healing applications. Mater Sci Eng C 31:43–49
152. Tazi N, Zhang Z, Messaddeq Y et al (2012) Hydroxyapatite bioactivated bacterial cellulose promotes osteoblast growth and the formation of bone nodules. AMB Express 2:61
153. Zhijiang C, Chengwei H, Guang Y (2012) Poly(3-hydroxubutyrate-co-4-hydroxubutyrate)/bacterial cellulose composite porous scaffold: preparation, characterization and biocompatibility evaluation. Carbohydr Polym 87:1073–1080
154. Mathew AP, Oksman K, Pierron D et al (2012) Fibrous cellulose nanocomposite scaffolds prepared by partial dissolution for potential use as ligament or tendon substitutes. Carbohydr Polym 87:2291–2298
155. Baptista AC, Martins JI, Fortunato E et al (2011) Thin and flexible bio-batteries made of electrospun cellulose-based membranes. Biosens Bioelectron 26:2742–2745
156. Franco PQ, João CFC, Silva JC et al (2012) Electrospun hydroxyapatite fibers from a simple sol–gel system. Mater Lett 67:233–236
157. Baptista AC, Ferreira I, Borges JP (2013) Electrospun fibers in composite materials for medical applications. J Compos Biodegrad Polym 1:56–65
158. Baptista AC, Botas AM, Almeida APC et al (2015) Down conversion photoluminescence on PVP/Ag-nanoparticles electrospun composite fibers. Opt Mater 39:278–281

159. Xin S, Li Y, Li W et al (2012) Carboxymethyl chitin/organic rectorite composites based nanofibrous mats and their cell compatibility. Carbohydr Polym 90:1069–1074
160. Shalumon KT, Binulal NS, Selvamurugan N et al (2009) Electrospinning of carboxymethyl chitin/poly(vinyl alcohol) nanofibrous scaffolds for tissue engineering applications. Carbohydr Polym 77:7-7
161. Bajpai SK, Pathak V, Soni B (2015) Minocycline-loaded cellulose nano whiskers/poly (sodium acrylate) composite hydrogel films as wound dressing. Int J Biol Macromol 79:76–85
162. Ul-Islam M, Khan T, Park JK (2012) Nanoreinforced bacterial cellulose–montmorillonite composites for biomedical applications. Carbohydr Polym 89:1189–1197
163. Wang Y, Chen L (2011) Impacts of nanowhisker on formation kinetics and properties of all-cellulose composite gels. Carbohydr Polym 83:1937–1946
164. Mauricio MR, da Costa PG, Haraguchi SK et al (2015) Synthesis of a microhydrogel composite from cellulose nanowhiskers and starch for drug delivery. Carbohydr Polym 115:715–722
165. Madhumathi K, Sudheesh Kumar PT, Abhilash S et al (2009) Development of novel chitin/nanosilver composite scaffolds for wound dressing applications. Journal of materials science. Mater Med 21:807–813
166. Kumar PTS, Abhilash S, Manzoor K et al (2010) Preparation and characterization of novel β-chitin/nanosilver composite scaffolds for wound dressing applications. Carbohydr Polym 80:761–767
167. Muzzarelli RAA, Morganti P, Morganti G et al (2007) Chitin nanofibrils/chitosan glycolate composites as wound medicaments. Carbohydr Polym 70:274–284
168. Yoo CR, Yeo I-S, Park KE et al (2008) Effect of chitin/silk fibroin nanofibrous bicomponent structures on interaction with human epidermal keratinocytes. Int J Biol Macromol 42:324–334
169. Shelma R, Paul W, Sharma CP (2008) Chitin nanofibre reinforced thin chitosan films for Wound healing application. Trends Biomater Artif Organs 22:111–115
170. Ifuku S, Saimoto H, Azuma K, et al. (2015) Preparation of Chitin Nanofibers for Biomedical Application. *(null).* CRC Press, 169–179
171. Lin N, Huang J, Chang PR et al (2011) Effect of polysaccharide nanocrystals on structure, properties, and drug release kinetics of alginate-based microspheres. Colloids Surf B Biointerfaces 85:270–279
172. Li S, Guo ZP, Wang CY et al (2013) Flexible cellulose based polypyrrole-multiwalled carbon nanotube films for bio-compatible zinc batteries activated by simulated body fluids. J Mater Chem A 1:14300–14305
173. Yoon H (2013) Current trends in sensors based on conducting polymer nanomaterials. Nanomaterials 3:524
174. Hu W, Chen S, Yang Z et al (2011) Flexible electrically conductive nanocomposite membrane based on bacterial cellulose and polyaniline. J Phys Chem B 115:8453–8457
175. Lin Z, Guan Z, Huang Z (2013) New bacterial cellulose/polyaniline nanocomposite film with one conductive side through constrained interfacial polymerization. Ind Eng Chem Res 52:2869–2874
176. Valentini L, Cardinali M, Fortunati E et al (2013) A novel method to prepare conductive nanocrystalline cellulose/graphene oxide composite films. Mater Lett 105:4–7
177. Na NP (2008) Chitin nanowhisker and chitosan nanoparticles in protein immobilization for biosensor applications. J Met Mater Miner 18:73–77
178. Stephan AM, Kumar TP, Kulandainathan MA et al (2009) Chitin-incorporated poly(ethylene oxide)-based nanocomposite electrolytes for lithium batteries. J Phys Chem B 113:1963–1971
179. Aramwit P, Bang N (2014) The characteristics of bacterial nanocellulose gel releasing silk sericin for facial treatment. BMC Biotechnol 14:104
180. Morganti P, Muzzarelli RAA, Muzzarelli C (2006) Multifunctional use of innovative chitin nanofibrils for skin care. J Appl Cosmetol 24:105–114

181. Morganti P, Morganti G, muzzarelli RAA et al (2007) Chitin nanofibrils: a natural compound for innovative cosmeceuticals. Cosmet Toilet 122:81–88
182. Morganti P, Fabrizi G, Palombo P et al (2008) Chitin-nanofibrils: a new active cosmetic carrier. J Appl Cosmetol 26:113–128
183. Morganti P, Morganti G (2008) Chitin nanofibrils for advanced cosmeceuticals. Clin Dermatol 26:334–340
184. Morganti P (2010) Chitin nanofibrils for cosmetic delivery. Cosmet Toilet 125:36–39
185. Morganti P (2011) Chitin nanofibrils and their derivatives as cosmeceuticals. In: Kim S-K (ed) Chitin, chitosan, oligosaccharides and their derivatives: biological activities and applications, CRC Press, New York
186. Morganti P (2015) Chitin-nanofibrils in skin treatment. J Appl Cosmetol 27:251–270
187. Biagini G, Zizzi A, Giantomassi F (2008) Cutaneous absorption of nanostructured chitin associated with natural synergistic molecules (lutein). J Appl Cosmetol 26:69–80
188. Hatakeyama H, Kato N, Nanbo T et al (2012) Water absorbent polyurethane composites derived from molasses and lignin filled with microcrystalline cellulose. J Mater Sci 47:7254–7261
189. Gabr MH, Phong NT, Abdelkareem MA et al (2013) Mechanical, thermal, and moisture absorption properties of nano-clay reinforced nano-cellulose biocomposites. Cellulose 20:819–826
190. Ma H, Burger C, Hsiao BS et al (2011) Ultrafine polysaccharide nanofibrous membranes for water purification. Biomacromolecules 12:970–976
191. Huang Y, He M, Lu A et al (2015) Hydrophobic modification of chitin whisker and its potential application in structuring oil. Langmuir 31:1641–1648

Surface Preparation of Fibres for Composite Applications

Mohammad S. Islam and Anup K. Roy

Abstract This chapter discusses about the surface preparation methods of fibres prior to use in composite materials to improve their performance or to introduce other essential characteristics. Chemical coatings, plasma and chemical surface modification techniques as well as mechanical treatments are presented. Also, advanced surface tailoring using nano-coating technology is included in this chapter.

1 Introduction

As composite material systems become increasingly sophisticated to meet ever-increasing performance requirements, it has become necessary to develop more advanced methods for controlling the manner in which the reinforcing fibres interact with the matrix material. Compatibility between the reinforcing fibre and the matrix polymer is paramount factor in achieving required composite properties. To increase the compatibility between the fibre and matrix, depending on the fibre type (whether the fibre is natural or man-made synthetic) various surface coating or surface modification processes have been adopted by researchers. The article also illustrates nanotechnology applications to improve and/or to induce some of these properties. Improvement of these properties can give the fibres an important position between the textile fibres which make them more convenient in different uses.

M.S. Islam (✉)
School of Aerospace Mechanical and Mechatronic Engineering,
The University of Sydney, Sydney, NSW 2006, Australia
e-mail: saifulctp@yahoo.com

A.K. Roy
Laboratory for Sustainable Technology, School of Chemical and Biomolecular Engineering, The University of Sydney, Sydney, NSW 2006, Australia

© Springer Science+Business Media Singapore 2016

S. Rana and R. Fangueiro (eds.), *Fibrous and Textile Materials for Composite Applications*, Textile Science and Clothing Technology,
DOI 10.1007/978-981-10-0234-2_9

Processing of polymer composites by using green source of fibres as reinforcement has increased dramatically in recent years. Advantages of using natural fibre over man-made fibres include low density, low cost, recyclability and biodegradability. These benchmark properties make natural fibre a potential replacement for synthetic fibres in composite materials opening up further industrial possibilities. However, high level of moisture absorption by the fibre leads to poor wettability and insufficient adhesion within the matrix (interfacial adhesion) resulting degradation of composite properties. These properties hinder the potential of these fibres in providing successful reinforcement for polymer composites. In order to expand the use of natural fibres as successful reinforcement in polymer composites the fibre surface needs to be modified to enhance fibre-matrix adhesion.

In this study, surface coating methods for synthetic fibres for advanced composites applications have been discussed along with fibre surface preparation methods for natural fibres in the composite production.

2 Surface Coatings

Since there are a limited number of available types of fibres that have the requisite properties for high-temperature applications, it has become necessary to develop coatings and surface treatments that facilitate the use of these fibres in an ever-widening selection of matrix systems. The application of a coating to the fibre surface is one of the most versatile methods [9, 26] for controlling the fibre matrix interaction. Other methods include providing additional elements to the matrix that will either form precipitates that segregate at or react with the fibre surface during fabrication and in situ modification of the fibre surface during fibre manufacturing. In this section fibre coatings will be emphasized. Coatings offer the possibility of tailoring the fibre-matrix interfacial properties so that optimum composite properties can be achieved. Fibre coatings can have a profound effect on the composite material at all stages of its existence from fabrication to in-service use. Fibre coatings perform the following general functions including controlling fibre-matrix bond strength, improving strength by reduction of surface stress concentrations, altering the wettability of fibres by the matrix, improving the chemical compatibility with the matrix, providing diffusion barriers, protecting fibres from damage during consolidation and fabrication and protecting the fibre and fibre-matrix interface from environmental degradation during service. There are many different methods for applying coatings to fibres. The most frequently used techniques include electrodeposition, CVD, metallorganic coating, polymer precursor coating, and line-of-sight vacuum deposition techniques [40].

2.1 Electrodeposition

Electrodeposition has proven to be one of the most convenient methods for applying metal coatings to fibres. Electroplating is used for depositing metals onto conducting substrates such as carbon fibres. In the electroplating process, continuous fibres are pulled through a plating solution while an electrical bias is applied between the fibre, which usually serves as the cathode and an anode structure in the plating bath. During this process the fibre passes through several plating baths to build up the desired thickness of metal. The process provides good penetration of the fibre tow and usually results in uniform coating thickness [40]. A schematic representation of electroplating process is shown in Fig. 1.

2.2 Chemical Vapor Deposition

Chemical vapor deposition (CVD), previously discussed as a method of fibre synthesis, is also used for coating. It can be broadly defined as a materials synthesis method in which the constituents of a gaseous phase react with one another and with the substrate surface to produce a solid film having the desired composition on the substrate. It is most favorable that the reactions be heterogeneous, that is, occur at the substrate surface. This usually results in high-quality films that adhere well to the substrate. If the reactions are homogeneous and take place in the gas phase, the resulting films exhibit poor properties and adhesion as a result of the formation of solid particles in the gas that eventually contact the substrate surface producing a film [40]. A schematic representation of a continuous CVD fibre-coating process is shown in Fig. 2.

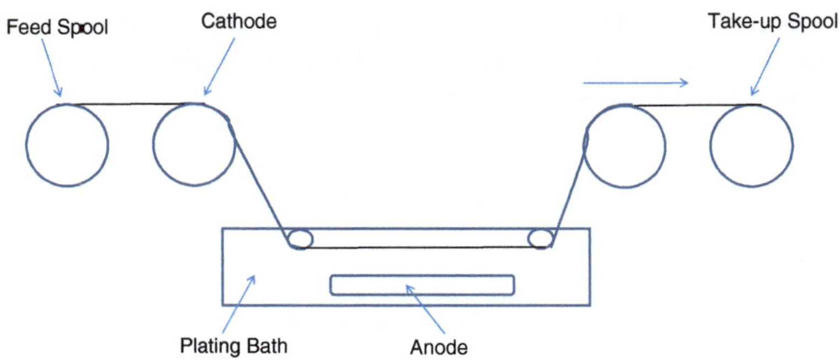

Fig. 1 Schematic of an electroplating process of continuous fibre

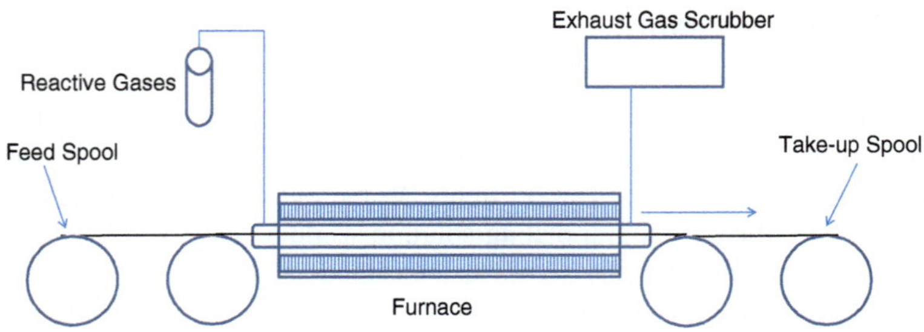

Fig. 2 A schematic of a continuous CVD fibre-coating process

2.3 Metallorganic Deposition

Metallorganic deposition is an entirely nonvacuum technique for the preparation of thin films. In this process a liquid precursor that contains a metallorganic species dissolved in an organic solvent is deposited onto the substrate surface. The substrate is then subjected to drying and heating steps that result in removal of the solvent and decomposition of the organometallic compound to produce the desired coating [40]. A schematic representation of metalorganic deposition process is shown in Fig. 3.

2.4 Vacuum Deposition

This category of fibre coatings includes sputtering, physical vapor deposition, e-beam evaporation, plasma-assisted CVD, and ion-plating techniques. With the exception of the plasma and ion plating, these processes are line-of-sight deposition techniques, making deposition of uniform film on a multifilament tow very difficult to achieve due to shadowing effects by other filaments in the tow. The film deposition rates are usually slower than with CVD methods, and hence the throughput of these processes is limited [40].

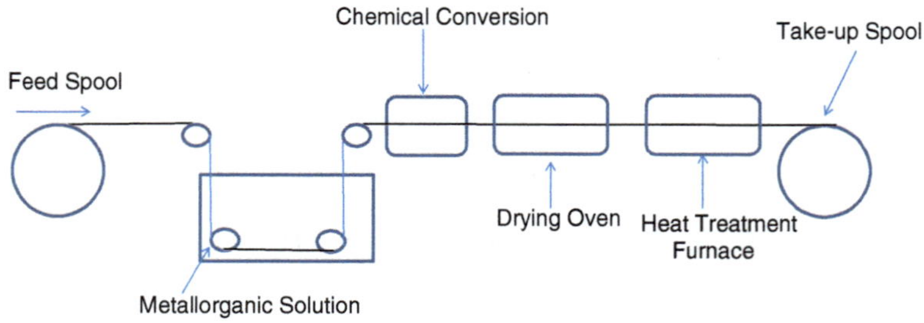

Fig. 3 A schematic of a metalorganic deposition process

2.5 *Future Fibre-Coating Technology*

As the demand for structural materials that can operate at higher temperatures increases, it is imperative that a thorough understanding of the role of fibre coatings be developed. Given that there is a limited set of existing fibres that can be utilized at elevated temperatures, and that these fibres are usually unstable either relative to the matrix material or the environment during fabrication of the composite or in-service use, it becomes necessary to consider the use of coatings as a means of moderating the interaction of the matrix with the fibre.

3 Plasma Surface Modification

The main purpose of plasma surface treatment of fibres or whiskers used as reinforcements in composite materials is to modify the chemical and physical structures of their surface layer, tailoring fibre-matrix bonding strength. A simple but most effective method for improving mechanical properties of carbon fibre reinforced polymer is to impart/promote strong adhesion between the fibre surfaces and the polymer matrix [11]. However, the nonpolar nature of carbon fibres makes them difficult to wet and chemically bond to general polymer matrices. Surface treatments of the fibre by oxidation can improve adhesion property, as oxidation of the surface introduces reactive groups onto the fibre surfaces so that they can react with matrices as well as increase the surface energy for improved wetting. Plasma surface modification technique is attractive for this application due to its environmental compatibility and high treatment effect without affecting the textural characteristics of the bulk materials, and thus has been extensively studied [22]. Some of the techniques include batch treatment of carbon fibre using atmospheric pressure dielectric barrier discharge for improving adhesion property [32]. Furthermore, atmospheric pressure plasmas have been used to treat glassy carbon plates, which are thought to be ideal model specimens for fundamental studies of adhesive properties of carbon fibres due to the structural similarity and easier handling than carbon fibres [27, 33]. Apart from that, low pressure plasma treatments are also being used for continuous treatment of carbon [34], imparting higher interfacial shear strength with epoxy resin than batch-treated fibres, possibly because moving the fibres in the plasma gives a more uniform level of treatment. However, for continuous treatment atmospheric pressure plasmas are better choice, as vacuum equipment can be avoided, and it permits large-scale development more easily than low pressure plasmas.

The plasma treatment system generally consists of four parts: power supply, discharge, exhaust and vacuum detection. To establish the conditions for plasma treatment of glass fibres, glass with a composition similar to that of glass fibre usually subjected preliminarily to plasma treatment. Plasma was generated from a plasma generator while supplying the reaction gas (e.g. oxygen). The next step to the plasma treatment is the measurement of contact angle to determine the wettability and surface energy in order to set the optimal conditions for plasma treatment. Under the optimal conditions set through the wettability analysis, a composite material was prepared by combining the plasma-treated glass fibres and epoxy resins [29].

Atmospheric air pressure plasma (AAPP) is a promising surface modification method to reduce, as far as possible, the negative effects on the environment of other commonly used methods, such as caustic treatment (NaOH), acetylation, silanisation and polymer grafting [3]. Plasma treatment techniques remove the need of exposing lignocellulosic fibres to solvents and solutions that otherwise need to be disposed of after treatment. It is usually accepted that during any plasma treatment at least two combined effects take place: an etching process that sputters the surface and removes weakly attached layers, such as oils and waxes adhering to the material being treated, and the formation of new functional groups. The nature of these functional groups created depends on the feed gas used to produce the plasma. Of particular interest is the possibility of improving the wetting behaviour and the adhesive properties of lignocellulosic fibres using AAPP treatments to remove noncellulose substances attached to the surfaces of the plant cell walls. We characterised the effectiveness of AAPP treatment on different lignocellulosic fibres by means of the capillary rise and the streaming potential measurements [3].

The enhancement of the adhesion between a polymer matrix and a plasma-treated fibre is caused by both physical and chemical modifications [50]. The physical modification is the surface roughening of the fibre by the sputtering effect, producing an enlargement of contact area which increases the friction between the fibre and the polymer matrix. The chemical modification is the implantation of active polar groups on the fibre surface, reducing the surface energy and promoting chemical bonding between the fibre and the polymer matrix. According to the gas temperature difference of various plasmas, they are generally divided into: high-temperature plasma and low-temperature plasma.

The plasma surface treatments at low temperature can be further divided into plasma polymerization and plasma ablation, which are two totally opposite effects, depending largely on the plasma gas used [10]. If the plasma gas has high proportions of carbon and hydrogen atoms in its composition, such as methane, ethylene and ethanol, the plasma will result in plasma polymerization. If the plasma gas has a strong electron affinity, such as oxidizing gas, air, CF_4 or SF_6, the plasma will have a strong ablation tendency. Among various cold plasma treatment techniques, corona discharge has a remarkable effect on roughening a fibre surface, and it also has an advantage of enabling the treatment to be carried out in open atmosphere [49]. However, corona is not real cold plasma but a very fine arc moving randomly on the surface of a substrate. It can melt surface material and harm its mechanical properties. To minimize local melting of the material, corona discharge is normally used to conduct fast treatment at a speed of about 1 m/s for two-dimensional materials such as fabrics or plastic films, not being suitable for dimensional fibres or zero-dimensional particles. Figure 4 shows a schematic of a cold spray plasma fibre coating process.

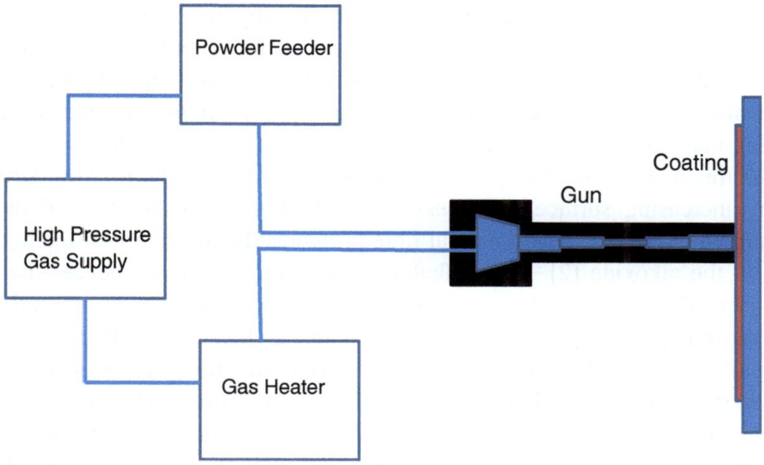

Fig. 4 A schematic of a cold spray plasma coating process

4 Chemical Surface Modification

The fibre-matrix interface is the diffusion or reaction zone, in which fibre and matrix phases are chemically and/or mechanically combined. Interfacial adhesion between fibre and matrix plays a predominant part in characterizing the mechanical properties of the composites. If there is a poor adhesion across the phase boundary, then relatively weak dispersion of force occurs and resulting poor mechanical properties of the composite [54, 55].

Most of the problems in natural fibre composite originate from the hydrophilic nature of the fibre and hydrophobic nature of the matrix. As a result, there is an inherent incompatibility between fibre and matrix. Chemical treatment on reinforcing fibre can reduce hydrophilic tendency and thus improve compatibility with the matrix. Several research activities have been conducted to improve fibre properties through different chemical treatments and their effects on composites properties. The following reviews the different treatment methods and their effects on the properties of fibres and composites.

4.1 *Alkaline Treatment*

Alkaline treatment is one of the most widely used chemical treatments of natural fibres for reinforcing with polymeric materials. Treatment of fibres by alkali changes the orientation of highly packed crystalline cellulose order and forms amorphous region by swelling the fibre cell wall providing more access to penetration of chemicals. Thus, the alkali sensitive hydrogen bonds that exist among the fibres break down and at the same time new reactive hydrogen bonds form between the cellulose molecular chains. Alkali treatment partially removes the hydrophilic hydroxyl groups of natural fibres improving their moisture removal property. This treatment also removes certain portion of hemicelluloses, lignin, pectin, wax and oil covering materials [12, 28, 55]. The important modification which can be done by alkaline treatment is the disruption of hydrogen bonding in the network structure, thereby increasing surface roughness. On top of this, the addition of aqueous sodium hydroxide (NaOH) to natural fibre promotes the ionization of the hydroxyl group to the alkoxide [2]. Thus, alkaline processing directly influences the cellulosic fibril, the degree of polymerization and the extraction of lignin and hemicellulosic compounds [20]. Alkali treatment makes the fibre surface cleaner and more uniform due to the elimination of micro voids and thus stress transfer capacity between the ultimate cells improves. In addition to this, this treatment reduces fibre diameter and thereby increase aspect ratio, which results in better fibre matrix interfacial adhesion [24]. Composite properties such as mechanical and thermal behaviour can be improved significantly by this treatment. If the alkali concentration is higher than the optimum condition, excess delignification of fibre take place, which results in weakness or damage to the fibre [54, 55]. In alkaline treatment, fibres are immersed in NaOH solution for a given period of time. Ray et al. [39] and Mishra et al. [30] treated jute and sisal fibres with 5 % aqueous NaOH solution for a period of 2–72 h at room temperature. For degumming and defibrillation to individual fibres a suitable method with a 2 wt% alkali solution at a temperature of 200 °C and a pressure of 1.5 MPa pressure for 90 s was reported by Garcia et al. [13]. These researchers observed that alkali led to an increase in amorphous cellulose content at the expense of crystalline cellulose. It is reported that alkaline treatment has two effects on the fibre. Firstly, it increases surface roughness resulting in better mechanical interlocking; and secondly, it increases the amount of cellulose exposed on the fibre surface, thus increasing the number of possible reaction sites [52]. Consequently, alkaline treatment has a lasting effect on the mechanical behavior of flax fibres, especially on fibre strength and stiffness [20]. Van de Weyenberg et al. [53] reported that alkaline treatment gave up to a 30 % increase in tensile properties (both strength and modulus) for flax fibre–epoxy composites and coincided with the removal of pectin. Alkaline treatment also significantly improved the mechanical, impact fatigue and dynamic mechanical behaviours of fibre-reinforced composites [19, 23, 44]. Jacob et al. [19] examined the effect of NaOH concentration (0.5, 1, 2, 4 and 10 %) for treating sisal fibre-reinforced composites and concluded that maximum tensile strength resulted from the 4 % NaOH treatment at room temperature. Mishra et al. [31] reported that 5 % NaOH treated sisal fibre-reinforced polyester composite had better tensile strength than 10 % NaOH treated composites. This is because at higher alkali concentration, excess delignification of natural fibre occurs resulting in a weaker or damaged fibre. The tensile strength of the composite decreased drastically after certain optimum NaOH concentration.

4.2 Silane Treatment

Silane is used as coupling agents to modify fibre surface. It undergoes several stages of hydrolysis, condensation and bond formation during the treatment process with the fibre. Silanols forms in the presence of moisture and hydrolysable alkoxy groups. It reacts with cellulose hydroxyl group of the fibre and improves fibre matrix adhesion to stabilize composite properties [55]. The chemical composition of silane coupling agents (bifunctional siloxane molecules) allows forming a chemical link between the surface of the cellulose fibre and the resin through a siloxane bridge. This co-reactivity provides molecular continuity across the interface region of the composite. It also provides the hydrocarbon chains that restrains fibre swelling into the matrix [15, 54]. Natural fibres exhibit micropores on their surfaces and silane coupling agent act as a surface coating which penetrates into the pores and develops mechanically interlocked coating on their surface. Silane treated fibre reinforced composite provides better tensile strength properties than the alkaline treated fibre composites [52]. Seki [45] investigated the effect of alkali (5 % NaOH for 2 h) and silane (1 % oligomeric siloxane with 96 % alcohol solution for 1 h) treatment on the flexural properties of jute epoxy and jute polyester composites. For jute epoxy composites alkali over silane treatment resulted in about 12 and 7 % higher strength and modulus properties compared to the alkali treatment alone. Similar treatment led to around 20 and 8 % improvement for jute polyester composites. Sever et al. [46] applied different concentration of (0.1, 0.3 and 0.5 %) silane (γ-Methacryloxypropyltrimethoxysilane) treatment on jute fabrics polyester composites. Tensile, flexural and interlaminar shear strength properties were investigated and compared with untreated composites. The results for 0.3 % silane treated composites showed around 40, 30 and 55 % improvement in tensile, flexural and interlaminar shear strength respectively. Three aminopropyl trimethoxy silane with concentration of 1 % in a solution of acetone and water (50/50 by volume) for 2 h was also reportedly used to modify the flax surface [53]. Rong et al. [41] soaked sisal fibre in a solution of 2 % aminosilane in 95 % alcohol for 5 min at a pH value of 4.5–5.5 followed by 30 min air drying for hydrolyzing the coupling agent. Silane solutions in a water and ethanol mixture with concentration of 0.033 and 1 % were also carried by Valadez et al. [52] and Agrawal et al. [1] to treat heneque'n fibres and oil palm fibres. Silanes can also be used as coupling agents to let glass fibres adhere to a polymer matrix, stabilizing the composite material.

4.3 Acetylation Treatment

Acetylation treatment on natural fibre is generally known as esterification method for plasticizing of cellulose fibres. Fibres are acetylated with and without an acid catalyst to graft acetyl groups onto the cellulose structure. It reacts with the hydrophilic hydroxyl groups and swells the fibre cell wall. As a result, the hydrophilic nature of the fibre decreases leading to improvement in dimensional stability of the composites [48]. In general, acetic acid and acetic anhydride individually do not react sufficiently with cellulosic fibres. For this, to accelerate the reaction, fibres are initially soaked in acetic acid and consequently treated with acetic anhydride between the time periods of 1–3 h with higher temperature. Moreover, this treatment provides rough surface topography with less number of void contents that give better mechanical interlocking with the matrix [51, 55]. Rowell et al. [42] investigated acetic anhydride treatment on several types of cellulosic fibres to analyse the effects of equilibrium moisture content and reported that, this treatment improved moisture resistance properties. This was due to the removal of hemicellulose and lignin constituents from the treated fibre. Mishra et al. [31] used acetic anhydride treatment on alkali pre-treated (5 and 10 % NaOH solution for 1 h at 300 °C) dewaxed sisal fibre with glacial acetic acid and sulphuric acid, and reported, improved fibre matrix adhesion characteristics of the composites. Bledzki et al. [6] studied different concentration of acetylation treatment on flax fibres and reported 50 % higher thermal properties. Moreover, 18 % acetylated flax fibre polypropylene composites showed around 25 % increase in tensile and flexural properties compared to the untreated fibre composites.

4.4 Benzoylation Treatment

Benzoylation treatment uses benzoyl chloride to decrease the hydrophilicity of the fibre and improve fibre matrix adhesion, thereby increasing the strength of the composite. It also enhances thermal stability of the fibre [35, 55]. During benzoylation treatment alkali pretreatment is used to activate the hydroxyl groups of the fibre. Then the fibre is soaked in benzoyl chloride solution for 15 min. Afterwards ethanol solution is used for 1 h to remove benzoyl chloride that adhered to the fibre surface followed by washing with water and oven dried. Joseph et al. [23] implied benzoyl chloride treatment on alkali pre-treated sisal fibre and reported higher thermal stability compared to the untreated fibre composites. Similar treatment was carried out on flax fibre reinforced low density polyethylene composites by Wang et al. [54] and reported 6 and 33 % improvement on tensile strength and moisture resistance properties.

4.5 Peroxide Treatment

Interface properties of fibre and matrix can be improved by peroxide treatment. The peroxide-induced grafting of polyethylene adheres onto the fibre surface and the peroxide initiated free radicals react with the hydroxyl group of the fibre and matrix. As a result, good fibre matrix adhesion occurs. This treatment also reduced moisture absorption tendency by the fibre and improves thermal stability [25, 54].

Sapieha et al. [43] reported that, treatment with benzoyl peroxide or dicumyl peroxide of the cellulosic fibre led to higher mechanical properties of composites. The mechanism of peroxide treatment involves alkali pre-treatment on the fibres then coated with benzoyl peroxide or dicumyl peroxide (around 6 % concentration) in the acetone solution for 30 min. Complete decomposition of peroxide can be achieved by heating the solution at higher temperature [25, 55]. Joseph et al. [23] investigated optimum concentration of benzoyl peroxide (6 %) and dicumyl peroxide (4 %) treatment on short sisal fibre-reinforced polyethylene composites and reported improved tensile strength.

5 Mechanical Surface Treatment

In comparison with chemical fibre treatments, mechanical treatment is more environmental-friendly. Mechanical fibre treatment such as stretching and twisting modify the structural and surface properties of fibre while keeping the chemical composition of it unchanged, thereby influencing the mechanical interlocking process of the polymer [5, 47]. On twisting of the fibre, an increase in the helix angle of it increases the radial pressure, causing the overall strengthening of fibre yarns [38]. At the same time, the direction of principal stress changes—the higher the helix angle of the fibre, the lower the yarn strength. Because of these two competing effects, a twist level, called optimal twist, exists at which the maximum fibre strength is achieved. Also it is considered that for impregnated fibres with a higher twist level, a higher tension generates greater internal forces preventing the separation of the fibre and resin. Frictional forces between fibres and resin improve the load transfer among yarns [38]. To evaluate the interaction of these factors, optimisation of the fibre twist level for natural fibre reinforced composites is a key parameter [17]. Reinforcement pre-tension has several effects on the polymer-matrix composite's properties. The influence of pre-tension on both the interaction at the interface and matrix toughness was thoroughly investigated by Pang et al. [37] and Foster et al. [14]. Pre-tension improves the flexural strength of matrix polymer by inducing compressive forces, which increases the toughness of the matrix. Compression is conditioned by stretching fibres during composite moulding. When the matrix solidifies due to the release of tension, a reinforcement compressive force is applied to the matrix as the rein-forcement seeks to contract to its original unstretched length. The matrix polymer becomes tougher and will have higher flexural strength [37]. At the same time, compressive forces generate frictional forces, which improve interaction at the interface [14]. Although the pre-tension technique is mainly used to improve the flexural strength of matrix polymer, in the case of natural fibres, pre-tension improves the tensile strength of the composite as well [4]. Vegetable fibres have a lack of orientation compared to that of high-orientated synthetic and man-made fibres. However, it is well known that during the tension process, natural fibre alignment and orientation are improved and its tensile strength increases [21].

Mechanical means such as fibre surface fibrillation by hydrostatic pressure or by sending the fibre surface in a controlled manner may be used to roughen the fibre surface to improve mechanical interlocking with polymer matrices. However, surface roughening in this way can reduce the fibre tensile strength [16].

Mechanical surface treatment by fibre beating—a mechanical process well known in paper industry—and its effect on the performance of the wood fibre cement composite has been studied [7, 8]. Generally, it has been found that beating enhances composites manufactured by the Hatschek process. Fibre beating improves the fibres' ability to retain particles and maintain sufficient drainage rate for processing in the Hatschek machine. However, beating also leads to fibre shortening, as well as fibre fibrillation. From the standpoint of mechanical behaviour, it was found that beating affects flexural strength and toughness. The flexural strength was found to increase with beating to an optimum level (for CSF of about 550) but reduced after that level. On the other hand the flexural toughness was found to be reduced with fibre beating [7]. This might be resulting from the fibres shortening effect and increased fibre-cement bonding due to the beating process.

6 Nano-Coatings

Coatings can make fibre surface compatible to polymer matrices, increase the thermal mechanical or chemical stability of it, increase wear protection, durability or lifetime, decrease friction or inhibit corrosion, or change the overall physico-chemical properties of the fibre. Coatings by means of conventional methods have several issues such as loss of strength, lack of adhesion and poor abrasion resistance and durability [36]. To achieve better surface property and compatibility with polymer matrices, nano-coatings are applied on fibre surfaces by minimizing the coat to weight ratio, developing new functionalities as well as improving the existing functions significantly without losing the texture of the fibre. Nano-coatings are being explored using mainly processes such as plasma-assisted polymerization, sol-gel process, layer-by-layer process along with other processes such as cathode arc vapour technique, electrode-less coating, in situ polymerization and chemical vapour deposition [18].

7 Conclusions

Various fibre surface treatment and coating techniques described in this chapter has been used for years and paved the way for technological advancement and viable solution in the field of composite materials. It enables the tailoring of advanced composite structures to fit into various types of applications. It also helps to produce low cost and economically and environmentally viable and sustainable materials for the future. Even though a lot has been done in the area of surface treatments there are a lot of technological advancements yet to be seen in coming years.

References

1. Agarwal A, Foster SJ, Hamed E, Ng TS (2014) Influence of freeze-thaw cycling on the bond strength of steel-frp lap joints. Compos Part B Eng 60:178–185
2. Agrawal R, Saxena N, Sharma K, Thomas S, Sreekala M (2000) Mater Sci Eng A 277(1): 77–82
3. Baltazar-Y-Jimenez A, Bismarck A (2007) Surface modification of lignocellulosic fibres in atmospheric air pressure plasma. Green Chem 9(10):1057–1066
4. Bekampiene P, Domskiene J, Širvaitiene A (2011) The effect of pre-tension on deformation behaviour of natural fabric reinforced composite. Mater Sci 17(1):56–61
5. Bledzki A, Gassan J (1999) Composites reinforced with cellulose based fibres. Prog Polym Sci 24(2):221–274
6. Bledzki A, Mamun A, Lucka-Gabor M, Gutowski V (2008) The effects of acetylation on properties of flax fibre and its polypropylene composites. Express Polym Lett 2(6):413–422
7. Coutts RSP (1984) Autoclaved beaten wood fibre-reinforced cement composites. Composites 15:139–143
8. Coutts RSP, Kightly P (1982) Microstructure of autoclaved refined wood-fibre cement mortars. J Mater Sci 17:1801–1806
9. Cranmer D (1989) Fibre coatings and characterization. Am Ceram Soc Bull 68:415
10. d'Agostino R, Favia P, Oehr C, Wertheimer MR (2005) Low-temperature plasma processing of materials: past, present, and future. Plasma Processes Polym 2(1):7–15
11. Dilsiz N (2000) Plasma surface modification of carbon fibers: a review. J Adhes Sci Technol 14(7):975–987
12. Dipa R, Sarkar BK, Rana AK, Bose NR (2001) Effect of alkali treated jute fibres on composite properties. Bull Mater Sci 24(2):129–135
13. Dupeyre D, Vignon M (1998) Fibres from semi-retted hemp bundles by steam explosion treatment. Biomass Bioenergy 14(3):251–260
14. Foster JS, Greeno R, Harington R (2007) Structure and fabric. Pearson Education, Harlow
15. George J, Sreekala MS, Thomas S (2001) A review on interface modification and characterization of natural fiber reinforced plastic composites. Polym Eng Sci 41(9): 1471–1485
16. Goda K, Takagi H, Netravali AN (2008) Fully biodegradable green composites reinforced with natural fibres. Old City Publishing, Philadelphia
17. Goutianos S, Peijs T, Nystrom B, Skrifvars M (2006) Development of flax fibre based textile reinforcements for composite applications. Appl Compos Mater 13(4):199–215
18. Gulrajani M (2013) Advances in the dyeing and finishing of technical textiles, edited. Woodhead Publishing, Cambridge
19. Jacob M, Thomas S, Varughese KT (2004) Mechanical properties of sisal/oil palm hybrid fiber reinforced natural rubber composites. Compos Sci Technol 64(7):955–965
20. Jahn A, Schroder M, Futing M, Schenzel K, Diepenbrock W (2002) Spectrochim Acta Part A: Mol Biomol Spectrose 58:2271–2279
21. John MJ, Anandjiwala RD (2008) Recent developments in chemical modification and characterization of natural fiber-reinforced composites. Polym Compos 29(2):187
22. Jones C (1991) Special issue interfaces in composites the chemistry of carbon fibre surfaces and its effect on interfacial phenomena in fibre/epoxy composites. Compos Sci Technol 42(1):275–298

23. Joseph K, Thomas S, Pavithran C (1996) Effect of chemical treatment on the tensile properties of short sisal fibre-reinforced polyethylene composites. Polymer 37(23):5139–5149
24. Joseph P V (2001) Studies on short sisal fibre reinforced isotactic polypropylene composites. Ph.D. thesis, Mahatma Gandhi University
25. Kalaprasad G, Francis B, Thomas S, Kumar CR, Pavithran C, Groeninckx G, Thomas S (2004) Effect of fibre length and chemical modifications on the tensile properties of intimately mixed short sisal/glass hybrid fibre reinforced low density polyethylene composites. Polym Int 53(11):1624–1638
26. Kerans R, Hay R, Pagano N, Parthasarathy T (1989) The role of the fibre-matrix interface in ceramic composites. Ceram Bull 68:529
27. Kusano Y, Mortensen H, Stenum B, Goutianos S, Mitra S, Ghanbari-Siahkali A, Kingshott P, Sørensen BF, Bindslev H (2007) Atmospheric pressure plasma treatment of glassy carbon for adhesion improvement. Int J Adhes Adhes 27(5):402–408
28. Leonard YM, Nick T, Andrew JC (2007) Mechanical properties of hemp fibre reinforced euphorbia composites. Macromol Mater Eng 292(9):993–1000
29. Lim KB, Lee DC (2004) Surface modification of glass and glass fibres by plasma surface treatment. Surf Interface Anal 36(3):254–258
30. Mishra S, Misra M, Tripathy S, Nayak S, Mohanty A (2001) Graft copolymerization of acrylonitrile on chemically modified sisal fibers. Macromol Mater Eng 286(2):107–113
31. Mishra S, Mohanty A, Drzal L, Misra M, Parija S, Nayak S, Tripathy S (2003) Studies on mechanical performance of biofibre/glass reinforced polyester hybrid composites. Compos Sci Technol 63(10):1377–1385
32. Mittal K L (2004) Polymer surface modification: relevance to adhesion. CRC Press, Boca Raton
33. Mortensen H, Kusano Y, Leipold F, Rozlosnik N, Kingshott P, Sørensen B, Stenum B, Bindslev H (2006) Jpn J Appl Phys 45(10B):8506
34. Mujin S, Baorong H, Yisheng W, Ying T, Weiqiu H, Youxian D (1989) The surface of carbon fibres continuously treated by cold plasma. Compos Sci Technol 34(4):353–364
35. Nair KM, Thomas S, Groeninckx G (2001) Thermal and dynamic mechanical analysis of polystyrene composites reinforced with short sisal fibres. Compos Sci Technol 61(16): 2519–2529
36. Othmers K (1930) Encyclopedia of chemical technology. Interscience, New York
37. Pang JW, Fancey KS (2008) Analysis of the tensile behaviour of viscoelastically prestressed polymeric matrix composites. Compos Sci Technol 68(7):1903–1910
38. Porwal P, Beyerlein I, Phoenix S (2007) Statistical strength of twisted fiber bundles with load sharing controlled by frictional length scales. J Mech Mater Struct 2(4):773–791
39. Ray D, Sarkar BK, Rana A, Bose NR (2001) Effect of alkali treated jute fibres on composite properties. Bull Mater Sci 24(2):129–135
40. Report NC (1992) High-performance synthetic fibres for composites. National Reserach Council, Washington
41. Rong MZ, Zhang MQ, Liu Y, Yang GC, Zeng HM (2001) The effect of fiber treatment on the mechanical properties of unidirectional sisal-reinforced epoxy composites. Compos Sci Technol 61(10):1437–1447
42. Rowell RM, Young RA, Rowell JK (2000) Paper and composites from agro-based resources. Carbohydr Polym 41(1):69–78
43. Sapieha S, Allard P, Zang Y (1990) Dicumyl Peroxide-Modified Cellulose/Lldpe Composites. J Appl Polym Sci 41(9–10):2039–2048
44. Sarkar B, Ray D (2004) Effect of the defect concentration on the impact fatigue endurance of untreated and alkali treated jute–vinylester composites under normal and liquid nitrogen atmosphere. Compos Sci Technol 64(13):2213–2219
45. Seki Y (2009) Innovative multifunctional siloxane treatment of jute fiber surface and its effect on the mechanical properties of jute/thermoset composites. Mater Sci Eng A 508(1):247–252

46. Sever K, Sarikanat M, Seki Y, Erkan G, Erdoğan ÜH (2010) The mechanical properties of γ-methacryloxypropyltrimethoxy silane-treated jute/polyester composites. J Compos Mater 44(15):1913–1924
47. Širvaitienė A, Jankauskaitė V, Bekampienė P, Kondratas A (2013) Influence of natural fibre treatment on interfacial adhesion in biocomposites. Fibres and textiles in Eastern Europe 21(4 (100)):123–129
48. Sreekala M, Kumaran M, Joseph S, Jacob M, Thomas S (2000) Oil palm fibre reinforced phenol formaldehyde composites: influence of fibre surface modifications on the mechanical performance. Appl Compos Mater 7(5–6):295–329
49. Sun D, Stylios G (2004) Effect of low temperature plasma treatment on the scouring and dyeing of natural fabrics. Text Res J 74(9):751–756
50. Tang LG, Kardos JL (1997) A review of methods for improving the interfacial adhesion between carbon fiber and polymer matrix. Polym Compos 18(1):100–113
51. Tserki V, Zafeiropoulos N, Simon F, Panayiotou C (2005) A study of the effect of acetylation and propionylation surface treatments on natural fibres. Compos A Appl Sci Manuf 36(8):1110–1118
52. Valadez-Gonzalez A, Cervantes-Uc J, Olayo R, Herrera-Franco P (1999) Chemical modification of henequen fibers with an organosilane coupling agent. Compos B Eng 30(3):321–331
53. van de Weyenberg I, Ivens J, De Coster A, Kino B, Baetens E, Vepoes I (2003) Compos Sci Technol 63:(1241)
54. Wang B, Panigrahi S, Tabil L, Crerar W (2007) Pre-treatment of flax fibers for use in rotationally molded biocomposites. J Reinf Plast Compos 26(5):447–463
55. Xue L, Lope GT, Satyanarayan P (2007) Chemical treatment of natural fibre for use in natural fibre-reinforced composites: a review. Polym Environ 15(1):25–33

Reinforcements and Composites with Special Properties

Arobindo Chatterjee, Subhankar Maity, Sohel Rana and Raul Fangueiro

Abstract This chapter presents an overview of different types of composite reinforcements (fibres, textiles, nanofibres, nanotubes, particles, etc.) with special properties like piezoresistivity, self-sensing property, self-healing capability, conductivity, electromagnetic shielding, heat generation, and so on. The properties of reinforced polymers and composites have also been discussed in detail. Additionally, fundamental aspects behind these special properties are also presented in this chapter. The last section of this chapter is dedicated to novel multi-scale reinforcements and composite materials, and their properties.

1 Introduction

Composites are an important class of engineering material having wide variety of applications. The main advantage of composite is the ability of tailor making of properties to suite different application conditions. The initial development of composite was aimed at having a strong material with high strength: weight ratio and fillers of high l/d ratios are preferred so that the potential for controlled anisotropy offers considerable scope for integration between the processes of material specification and component design.

With the advancement of technology, the need for materials with functional properties is significantly increasing. Materials based on fillers having novel properties and nano-sized materials will represent an adequate solution to many present and future technological demands.

A. Chatterjee (✉) · S. Maity
Department of Textile Technology, Dr. B R Ambedkar National Institute of Technology, Jalandhar 144011, India
e-mail: chatterjeea@nitj.ac.in

S. Rana · R. Fangueiro
Fibrous Materials Research Group, University of Minho, 4800-058 Guimarães, Portugal

© Springer Science+Business Media Singapore 2016

S. Rana and R. Fangueiro (eds.), *Fibrous and Textile Materials for Composite Applications*, Textile Science and Clothing Technology,
DOI 10.1007/978-981-10-0234-2_10

Fillers play an important role in deciding the properties of the composite materials. When electrically conducting filler is added in sufficient quantities to a polymeric resin, a conductive composite is formed. Conductive composites are composite having significant electrical or thermal conductivity. The unique properties of such composites make them technologically superior to or more cost effective than alternative materials in a variety of applications. The inevitable link between conductive composites and the electronics industry has provided much of the impetus for their rapid development. Conductive polymeric composite finds its application in different diverse areas such as sensors, structural health monitoring system, self healing material, EMI shielding, energy storage materials, anticorrosive material, wearable electronics and materials for heat generation etc.

Structural Health Monitoring (SHM) is the process of detection, evaluation and interpretation of the damage in the engineering structure to improve its reliability. Primary goal of Structural Health Monitoring (SHM) is to replace the current inspection cycles with a continuously monitoring system. Piezoelectric and piezoresistive composite materials offer a great scope as sensors for real time monitoring of the health of structures and systems.

With the development of microelectronics, the assessment of failure and repair has become imperative for the reliability of these mass produced miniaturized devices. The high cost and difficulties involved in the repairing of such devices has led to the development of a new material capable of self healing.

The electrical and electronic devices used by us emit electromagnetic waves with frequencies that are potential hazards to health and hence the concept of EMI shielding has become practically important. Similarly sensors for soft electronics based on piezoresistive material have become potentially important.

In the present chapter, an effort has been made to discuss the structure, properties and applications of composite structures having these special properties.

2 Piezoresistive and Self-Sensing Behavior

Piezoresistive sensors translate an induced mechanical displacement into an electrical signal which is useful for monitoring minute structural deformations in composites over time. They comprise of electro-conductive fibers and fillers inside their structure to detect deformations occur. The principle of working of these sensors is the change in resistance due to applied strain between the conducting fibre/filler units. Tensile strain increases the distance between the electro-conductive fibre/filler units, thus increasing the resistivity. And inversely, compression decreases this distance, thus decreasing the resistivity. So, as an embedded circuitry, piezoresistive sensor system enables the sensing of local strain in a composite structure via connection to an external resistance meter.

2.1 Theory of Conduction Mechanism

The piezoresistivity of such composites based on conductive filler has been most successfully modeled by percolation theory and a quantum tunneling mechanism [81]. With a sufficiently large volume fraction of conductive filler material, nano-junctions will form between filler particles in which electrons can tunnel through the insulating barrier. Then an electrical network is created through the composite and it becomes conductive. The resistivity of a nanojunction can be calculated using Eq. (1).

$$\rho = \frac{h^2}{e^e\sqrt{2m\lambda}} ecp\left(\frac{4\pi\sqrt{2m\lambda}\lambda}{h} s\right) \tag{1}$$

where *h* is the Planck constant (Js), *e* is the charge of electron (C), *m* is the mass of electron (kg), λ is the height of tunneling barrier (J), and *s* is the distance between particles (m). Thus the resistivity of a nanojunction is a function of the inter-particle distance. As a composite is strained *s* will change and a piezoresistive signal is obtained.

Figure 1 shows the evolution of logarithm of resistivity ρ with volume fraction of conductive fillers ϕ in a polymer matrix charged with conductive fillers. This $\rho - \phi$ plot can be divided into two distinctive zones viz. the insulator regim and conductor regim. In the insulator regim, due to insufficient conductice filler (ϕ_1), few contacts will create a few charges to flow in the composite, and in the conductor regim (ϕ_2), due to abundant fillers, many contacts will create a healthy

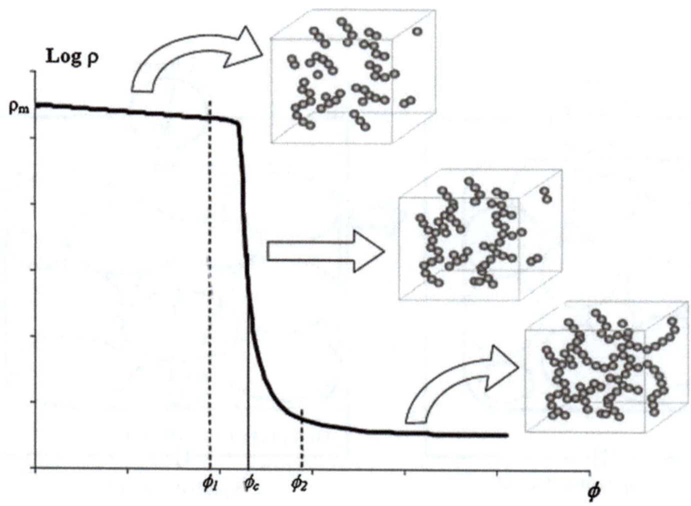

Fig. 1 Evolution of electrical resistivity in a composite with volume concentration of electric fillers [134]

conductive network for the flow of electrons and thus the electric current. The transition between the insulating and conductive regim occurs at a particular volume concentration of conductive filler. This critical concentration of conductive filler (ϕ_c) is called as percolation threshold [24]. Because of the formation of conductive networks of fillers around the percolation threshold, a sudden and rapid drop in resistivity is observed. When filler particles are in direct contact with each other then the electrical conduction is explained by metallic conduction. In metallic conduction the energy bands are overlapped which allows the electrons to flow from one site to another without any input energy. However, when there is gap or barrier between the filler particles then for conduction, the electrons need to jump from one site to another in presence of external input energy. This hopping can be either short range hopping when sites are energetically distant and geographically close or variable range hopping when sites are energetically close and geographically distant [37, 134].

The mechanism of change in electrical resistance due to reorientation of filler particles in composites structure by mechanical deformation is illustated by Fig. 2. Beyond percolation threshold, conductivity of the filler composites obey a power law relationship which is dependent on the concentration of filler as expressed in Eq. (2).

$$\frac{\sigma}{\sigma_0} = (\varphi - \varphi_c)^t \tag{2}$$

where, σ is the electrical conductivity of the composite (S m^{-1}), ϕ is the volume concentration of the conductive filler, ϕ_c is the concntration of filler at the percolation threshold, σ_0 is the effective conductivity of the filler itself, and t is critical exponent [119].

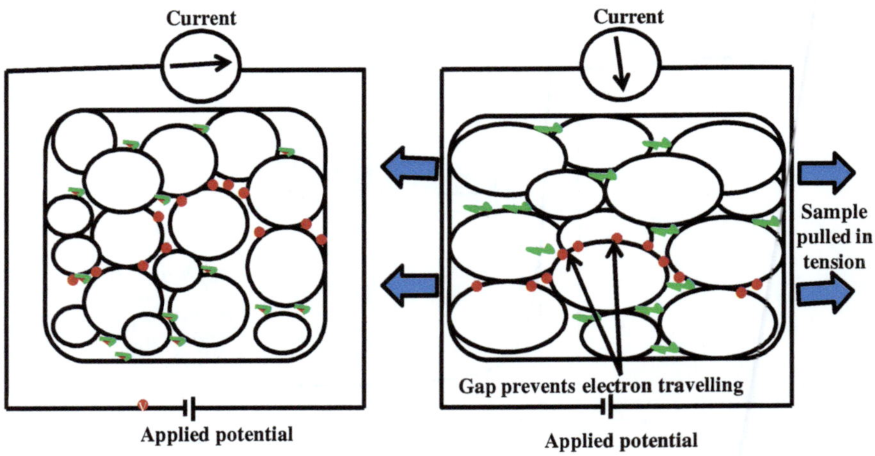

Fig. 2 Schematic representation of change of resistance of filler composites due to application of uniaxial tensile stress [119]

2.2 *Metal-Polymer Composites as Piezoresistive Sensor*

Metal-polymer composites are prepared by reinforcing polydimethylsiloxane (PDMS) polymer with copper (Cu) and nickel (Ni) particles by a method of wet mixing. These composites have high stress sensitivity which show the resistivity changes by more than 10 orders of magnitude when pressure around 4 MPa is applied. These composites are proposed as good piezoresistive sensor for practical applications in various fields like robotics, biomechanics, etc. [3].

2.3 *Carbon Fibres and Carbon Black Composites as Piezoresistive Sensor*

Carbon fiber is itself piezoresistive by nature, and can thus act as a strain gauge. Also, carbon black and metal nanostrand composites possess remarkable piezoresistive response [107]. Self-sensing multifunctional carbon fibre reinforced composite is prepared which has sensing function executed by its change of electrical resistance [186, 187]. But, when they are plied to prepare multifunctional laminate nanocomposites, the transverse plies of them largely short circuit their piezoresistive response. Use of insulators between the layers of the laminate helps to mitigate this problem [82]. It is discovered that direct embedding of nickel nanostrand patch into unidirectional carbon fiber laminates yields significant piezoresistive response [82, 83]. Rana et al. [154, 160] have prepared core reinforced braided composite rods (BCRs) by braiding polyester fibres around an unsaturated polyester resin impregnated core, composed of glass and carbon fibre mixture for self monitoring of deformation and damage. Their sensing behavior is characterized by monitoring the change in electrical resistance with deformation using two terminal DC methods. Three types of BRC samples are prepared altering the core materials viz. BRC1 (77/23 E-glass/carbon), BRC2 (53/47 E-glass/carbon) and BRC 3 (100 % carbon). Their piezoresistive response behavior is shown in Fig. 3. It can be seen that the

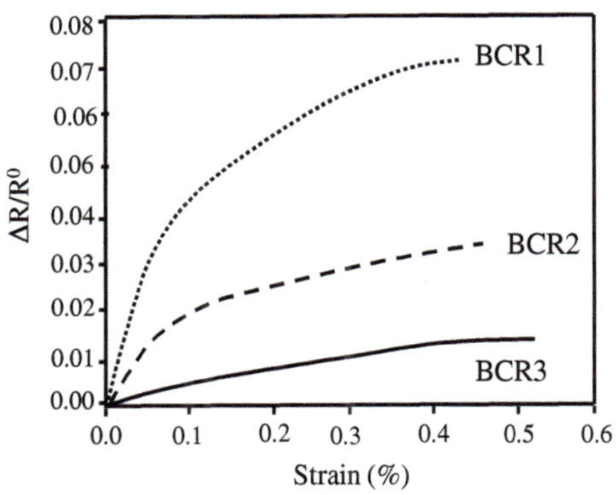

Fig. 3 Strain responses of different BCRs by change of electrical resistance [154, 160]

highest piezoresistive behavior is obtained with BCR1 and the strain sensibility decreases with increase in the carbon fibre content. It is interesting to note that the curve for BCR1 presents more non-linearity than the other BCRs. Its fractional resistance increases sharply with a linear trend up to 0.1 % strain and then more gradually at higher strains due to saturation in the electrical contacts.

An elastomer based filler composite sensor made of evoprene/carbon black nanoparticle is prepared and mounted on a thin and light nylon fabric for self-sensing. Its shows a nonlinear behaviour trend of rise of resistance at low strain level below 15 %. However, above 15 % of strain level, it shows a linear ohmic behavior [37].

Relationship between stress and strain and that between fractional resistance change (ΔR/Ro) and strain during static tensile testing up to failure for a carbon fibre reinforced composite with 2.57 vol % carbon fibers (continuous, 11 mm diameter) is shown in Fig. 4. The stress–strain curve is linear up to a strain of 0.2 %, at which the resistance is increasing steadily. However, after strain level of 0.2 % resistance starts to increase abruptly.

The variation of ΔR/Ro during cyclic loading and unloading for various stress amplitudes within the linear portion of the stress–strain curve of this composite is shown in Fig. 5 [199]. The resistance increases upon loading and decreases upon unloading in every cycle and the resistance increase are not totally reversible. The gage factor, which has been commonly used as a parameter to quantify the pieozoresistive sensitivity (the fractional change in reversible portion of resistance per unit strain), are found to be 28, 21 and 17 for the first, second and third cycles respectively. The decrease in gage factor with increasing cycle number is attributed to the decrease in reversibility with increasing stress amplitude.

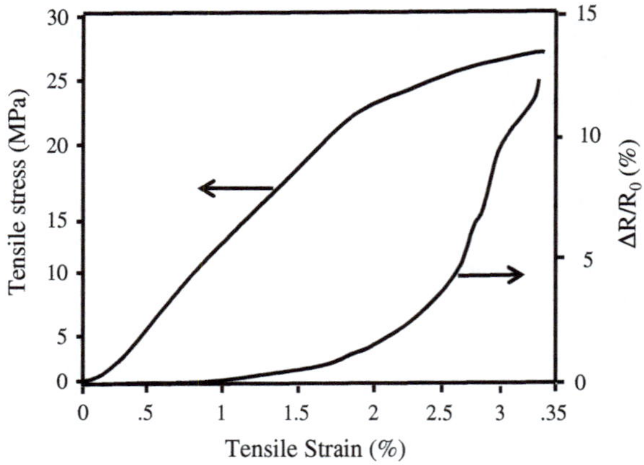

Fig. 4 Relationship between stress and strain and that between fractional resistance change (ΔR/Ro) and strain during static tensile testing for a carbon fibre reinforced composite [199]

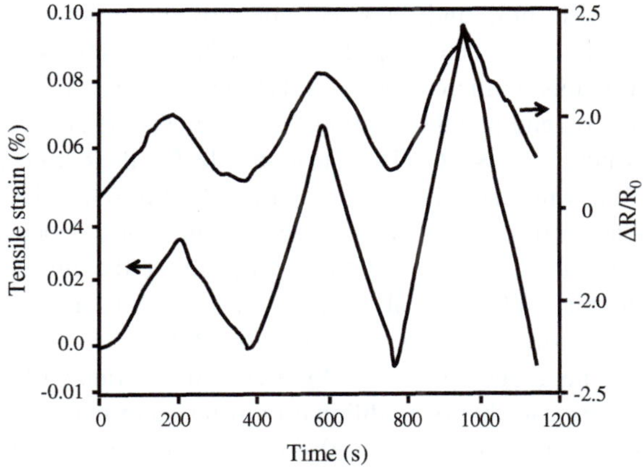

Fig. 5 Variation of Fractional change of resistance during cyclic loading for a carbon fibre reinforced composite [199]

2.4 Carbon Nanotube (CNT) as Piezoresistive Sensor

The incorporation of carbon nanotubes (CNT) into matrix of composites effectively increases the electrical conductivity of the resulting composites by several orders of magnitude and affects various other properties. This is attributed to the formation of a conductive network of the nanofillers inside the structure and it is called a percolated network. Carbon nanotubes can be fabricated in the form of single wall (SWCNT) and multi-wall (MWCNT) carbon nanotubes. SWCNTs consist of one cylindrical lattice of carbon atoms (Fig. 6a) while MWCNTs are structured by multiple helical concentrically positioned lattices (Fig. 6b) [153]. Though both the CNTs appear as hollow tubes of rolled graphene sheets, they are not produced by rolling up the sheets. Their peculiar structure is achieved under precisely controlled conditions in the presence of a catalyst. SWCNTs have diameters in the range of

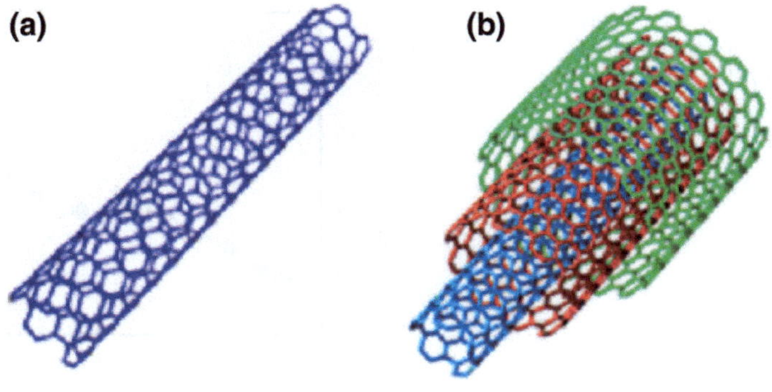

Fig. 6 Structural geometry of **a** SWCNT and **b** MWCNT [153]

0.7–1.0 nm, and they can be grown to long lengths to achieve very high length-to-diameter aspect ratios. CNTs possess very high stiffness with Young's modulus of approximately 1 TPa and a density of about 1.33 g/cm^3. They can bear torsion and bending without breaking. The hexagonally-bonded carbon honeycomb structure of SWCNTs is responsible for their high mechanical strength up to 60 GPa, with maximum strain up to 10 %. From the electrical standpoint, CNTs can be classified as conductors or semi-conductors, depending on the orientation of the carbon atoms in the lattice structure of the tubes. The high strength, stiffness, thermal and electrical conductivities of CNTs make them promising for developing composites for a lot of applications [196].

The gauge factor of SWCNTs is in the range of ~400 to ~2900, which is far superior to that of traditional piezoresistive materials, such as doped single-crystal silicon (gauge factor: 100–170) and metals (gauge factor: 2–5). Stampfer et al. [177] have fabricated pressure sensors based on individual SWNTs as active electromechanical transducer elements. The electromechanical measurements show a piezoresistive gauge factor of approximately 210. The change in resistance of the sensor with respect to applied strain is shown in Fig. 7. A linear response is observed in terms of increase of resistance with increase of strain.

In another study, electrically conductive rigid Poly-Urethane (PUR) composite foams are prepared by dispersing MWCNT at various concentrations. This PUR/MWCNT foams show varying electrical conductivity on a wide range of MWCNT contents and compression. Due to the application of compressive load, the electrical resistance of these foams gradually reduces as shown in Fig. 8 [10].

Composite nanomaterials with micron and submicron layers are prepared by carboxymethyl cellulose (CMC) and MWCNT. The specific bulk conductivity of the composite was found to be ~40 kS/m and the surface conductivity is found to be ~0.2 S. These composites are suggested for different applications viz. flexible electronics, conductive glues, protection of electronic circuits, devices, and many more [78].

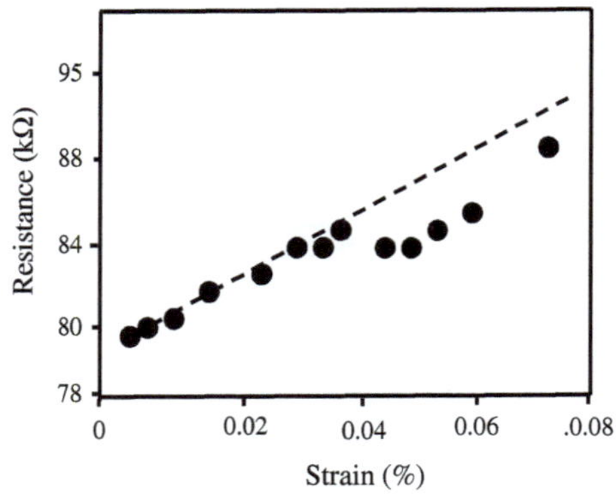

Fig. 7 The resistance change as a function of axial strain of MSWNT [177]

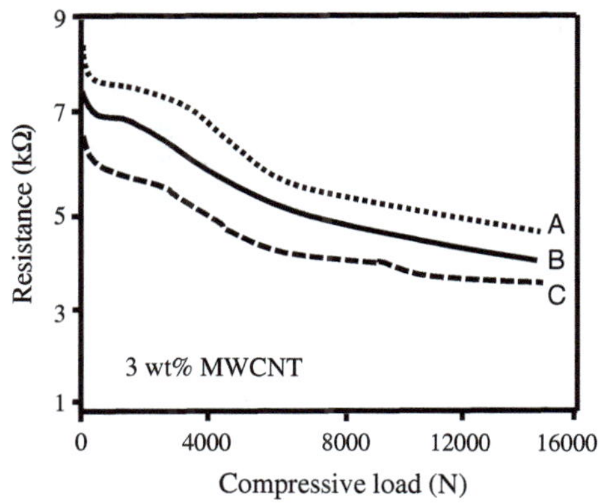

Fig. 8 Change in electrical resistance due to applications of compressive loads [10]

2.5 *Conductive Polymer Composites as Strain Sensor*

Conductive polymer composites are a well known class of composites which are being studied for its application as piezo-resistive strain sensors. Conductive polymers are used as filler material to prepare electro-conductive composites which can respond to structural deformation due to external stress or strain in terms of change in its resistivity [65].

Flexible, polymeric, and electro-conductive nanocomposite films are prepared by dispersing polyaniline (PANI) into a poly(vinyl acetate) (PVA) latex matrix. These films posses a sensitivity of 3 times larger than classical metallic gauges with certain repeatibility [119]. A typical pressure sensor comprising of conductive polymer composites is prepared by sandwiching the composite in between two electro-conductive layers as electrodes. The electrodes are then covered with two layers of insulating polyester films. When a compressive force is applied to the surface of the sensor, the gap between filler particles reduces inside the matrix, resulting an increase in conductivity of composites. This change of conductivity is detected as output of pressure sensitivity [87].

Poly(3,4-ethylenedioxythiophene)-poly(styrene sulfonate) (PEDOT:PSS) based fibrous yarn sensors are developed by Trifigny et al. [189] to directly measure the mechanical stress on fabrics under static or dynamic conditions. These sensors have ability to locally detect mechanical stresses all along the warp or weft yarn. This local detection is undertaken in real time during the weaving process [189]. In another study, polypyrrole coated lycra fibre is explored as strain sensor. The electrical resistance of these fibres increases exponentially with the strain as shown in Fig. 9 [197, 198].

Polypyrrole/polyester textile and PEDOT/polyester textile composites exhibit a monotonic increase of the electrical resistance with the elongation up to 50 %. These elastic textile composites are proposed as a strain sensor for large deformation [99]. Polypyrrole coated cotton yarn show increase of their resistivity due to application of tensile strain and twist [125]. These yarns are proposed for sensory applications of strain and torsion.

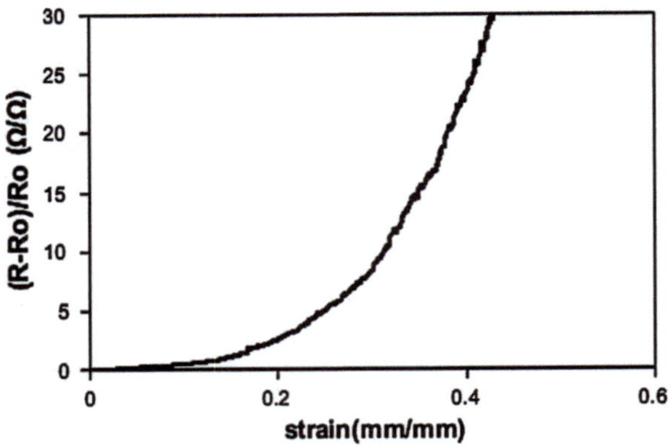

Fig. 9 Typical resistance vs. strain curve of PPy-coated lycra fibers [197, 198]

2.6 Carbon Nanotubes for Structural Health Monitoring (SHM)

Over the past two decades significant efforts have been made by researchers in order to miniaturize electronic devices for SHM applications and optimize their performance. The excellent mechanical properties and the high thermal and electrical conductivities of CNTs play a fundamental role for the development of devices simultaneously showing structural and functional capabilities, including actuation, sensing and generating power. These characteristics allows the development of high performance sensors and actuators operating at low voltage. A number of such sensors have been designed by researchers. A homogeneous SWCNT film obtained through vacuum filtration of SWCNT solution has been used as a strain sensor [41]. A proper design of nanotube concentrations and polyelectrolyte matrices has yielded thin films whose electrical properties are sensitive to strain and pH changes. Layer by layer assembled carbon nanotube composites have been used as strain sensors as well as a wireless transmission systems [122]. Hybrid composites are prepared using carbon nanotubes in combination with fibers for self-sensing properties [16]. Later alternative strategies based on radial in-situ growth of CNTs on the fiber surface are attempted [152]. Also, CNTs can be fabricated as film to develop neural systems in the form of a grid attached to the surface of a structure to make a sensor network. Such an artificial neural system has potential applications in the field of monitoring of large civil structures [153]. Again, CNT/PMMA nanocomposites are prepared for electro-magnetic wave absorption and dynamic strain sensing for structural health monitoring [174]. Change in resistance (normalized resistance) of the strain sensor with respect to change of strain is shown in the Fig. 10. Slope of the curve is the gauge factor of the sensor. This sensor shows consistent piezoresistive behavior under repetitive loading and unloading and has good resistance stability.

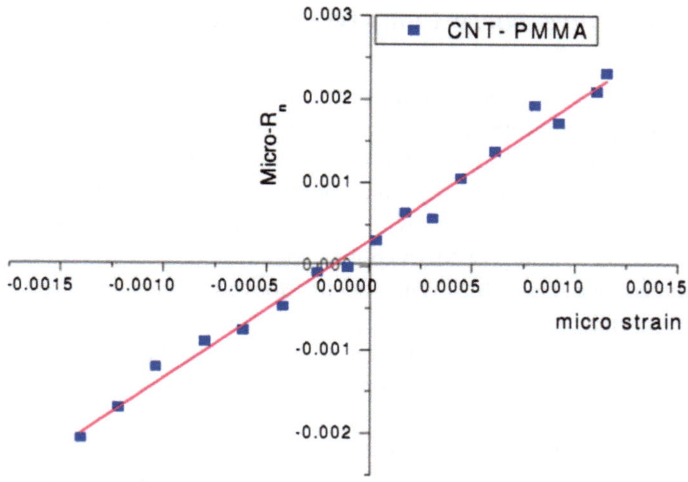

Fig. 10 Strain-Resistance behavior of CNT/PMMA nanocomposite strain sensor [174]

2.7 Piezo-Resistive and Self-Sensing Behavior of Engineered Cementitious Composites

Engineered Cementitious Composites (ECC), like concrete and other cement-based materials, act as a piezoresistive material with bulk resistivity of 101–105 Ω m which is similar to that of a semi-conductor [35]. The microstructure of ECC materials have microscopic pores, partially filled with unbound water and dissolved ions. These ions are mobile under an externally applied electric field to generate electric current. However, there is limited communication between these pores inside ECC structure which increases the resistance of current flow. Also, there exists high contact impedance between various conductive phases of these ECCs. Under applied mechanical strain, changes occur in spatial separation between conductive phases which causes a change in the bulk resistivity of ECCs. This makes the ECCs piezoresistive and self-sensing [162]. The self-sensing ability is achieved in certain ECCs by incorporating a small dosage of carbon black (CB) into the ECC system to enhance its piezoresistive behavior [121]. The significant increase in resistivity is observed when the specimens are subjected to uniaxial tension. Baeza et al. [8] develop multiple sensors based on multifunctional ECC with carbon fibers (CF) and CNF fillers and attach them to a conventional reinforced concrete beam for strain sensing. Both the admixtures, CF and CNF, are found suitable as conductive filler to fabricate cement composites capable of measuring strains on the surface of the structural beam, regardless of local stresses either tensile or compression. In another study, electrically conductive cementitious composites are prepared by adding MWCNTs to the cement matrix. The electrical resistivity of these composites changes with stress conditions with static and dynamic loads [38].

2.8 Piezoresistive Micro-Cantilever Biochemical Sensor

A compact, piezoresistive and low cost polymer nano-composite micro-cantilever sensor is prepared by Patil et al. [142], which is capable of detecting 2,4,6-trinitrotoluene (TNT), pentaerythritoltetranitrate (PETN) and 2,4,6-trinitrotoluene (RDX) vapors at low ppt level within seconds of exposure under ambient conditions. The gold coated micro-cantilevers are modified with self-assembled monolayers (SAM) of 4-mercaptobenzoic acid (4-MBA),6-mercaptonicotonic acid (6-MNA) and 2-mercaptonicotonic acid (2-MNA) in order to achieve sufficient piezo-resistivity. This nano-composite micro-cantilever is ideally suited as "electronic nose" applications. The change in resistance (ΔR/R) as a function of deflection is shown in Fig. 11 The calculated deflection sensitivity is 1.1 ppm/nm and surface stress sensitivity of 7.6×10^{-3} $(Nm^{-1})^{-1}$ [142].

Seena et al. [170] have developed a highly sensitive piezo-resistive SU-8 nano-composite micro-cantilever sensor for detection of explosives in vapor phase. Carbon black (CB) is dispersed by 8–9 vol% in the SU-8/CB nano-composite to prepare it. The resonance frequency and spring constant are found to be 22 kHz and 0.4 N m^{-1} respectively. The force-displacement of the micro-cantilever is found to be linear with surface stress sensitivity of 7.6 ppm $(mNm^{-1})^{-1}$ as shown in Fig. 12. It can detect TNT vapor concentrations down to approximately 6 ppb with an approximate sensitivity value of 1 mV/ppb of TNT as shown in Fig. 13. The output voltage is found to increase with increase in TNT concentration.

2.9 Polymeric Filaments for Piezo-Electric Sensor

Poly(vinylidene fluoride), PVDF, is a widely studied material due to its piezo-electric properties, including in the context of integration into textiles to prepare composites for self-sensing. Piezoelectric PVDF filaments are prepared in co-axial geometry. Multilayered filaments incorporating electrically conductive layers used

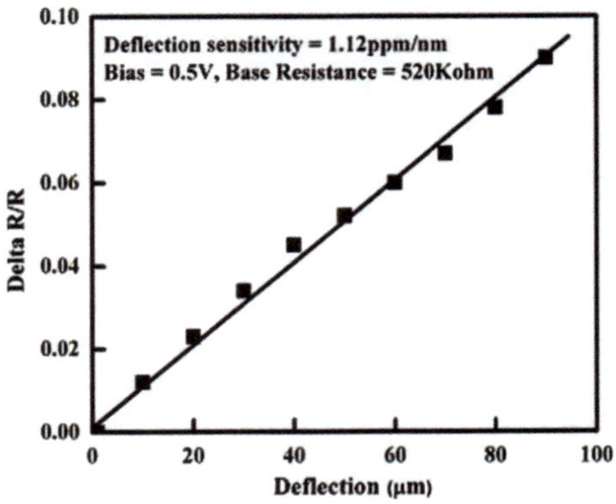

Fig. 11 Change in resistance (ΔR/R) as a function of deflection of polymer nano-composite micro-cantilevers [142]

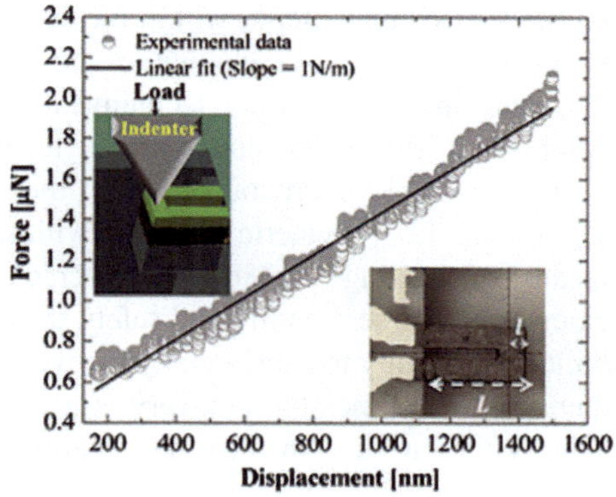

Fig. 12 Load displacement characteristics of the SU-8/CB nano-composite micro-cantilever [170]

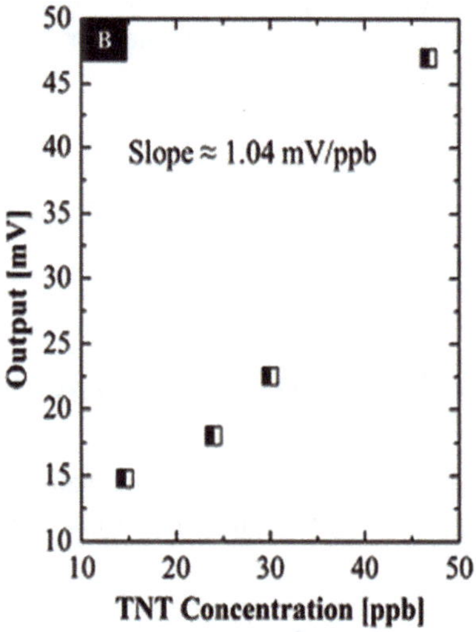

Fig. 13 TNT vapor detection sensitivity plot of SU-8/CB nano-composite micro-cantilever [170]

as electrodes, in co-axial arrangemnt, are produced by using single step co-extrusion technique [165]. These filament sensors are tested under bending and extension stress and found suitable for application as sensor. The sensors have a sensitivity of 2.2 nC/N.

3 Self-Healing Characteristics

Fibre reinforced composites have various applications in diversified fields such as transportation, civil, constructions, geotextiles, sports, etc. These composites are susceptible to damage due to mechanical, chemical, thermal, radiation, or any combination of these factors during their service life. This damage leads to the formation of micro-cracks in the composite structure. The micro-cracks may form on the surface or in depth within the structure. Detection and external intervention to repair these micro-cracks within the composite structure are often difficult.

There are conventional techniques to repair visible cracks/damages of fibre composites. But, these conventional techniques are not suitable or effective for healing invisible microcracks within the structure of the composites during their service life. So, polymeric materials with self-healing capability are introduced as a means of healing invisible micro-cracks within the composite structures for extending their service life and safety [86]. Self-healing polymeric materials have a built-in capability to substantially recover their functional ability automatically after damage. Sometimes the recovery is activated by specific external stimuli such as heat, radiation etc. An ideal self-healing material can be defined as the material which is capable of continuously sensing and recovering induced damages throughout its service life without affecting its performance and base properties. These materials are contributing greatly to the safety and durability of fibre reinforced composites at cheaper costs [57]. Wool and O'Connor [201] proposed a basic method for describing the extent of healing in polymeric material by following Eq. (3).

$$Healing\ Efficiency = 100 \times \frac{Property\ Value_{Healed}}{Property\ Value_{Initial}} \quad (3)$$

3.1 Self-Healing of Thermoplastic Materials

A number of different mechanisms for self-healing of thermoplastic polymers as proposed by various researchers are as follows.

3.1.1 Self-Healing by Molecular Interdiffusion

Molecular interdiffusion is the process of crack healing by molecular diffusion of two similar polymers at the polymer-polymer interface when broken pieces are brought into contact above their glass transition temperature (Tg). Researchers [148, 201] proposed various models to explain the mechanism of crack healing of composites at the thermoplastic interface. The reptation model was proposed by De Gennes [40] and Doi and Edwards [44] to explain chain dynamics for self healing. In this model, motion of a polymer molecule is described as wormlike inside a strongly cross-linked polymeric gel and it is found correlated with molecular mass. Later, Wool and O'Connor [201] suggested a five stages model to explain the crack healing process such as (i) surface rearrangement of molecules, (ii) surface approaching of molecules, (iii) wetting of surfaces, (iv) diffusion between surfaces, and (v) equilibrium and randomization of molecules (Fig. 14). This theory is suitable for crack healing in amorphous polymers and polymer resins during melt

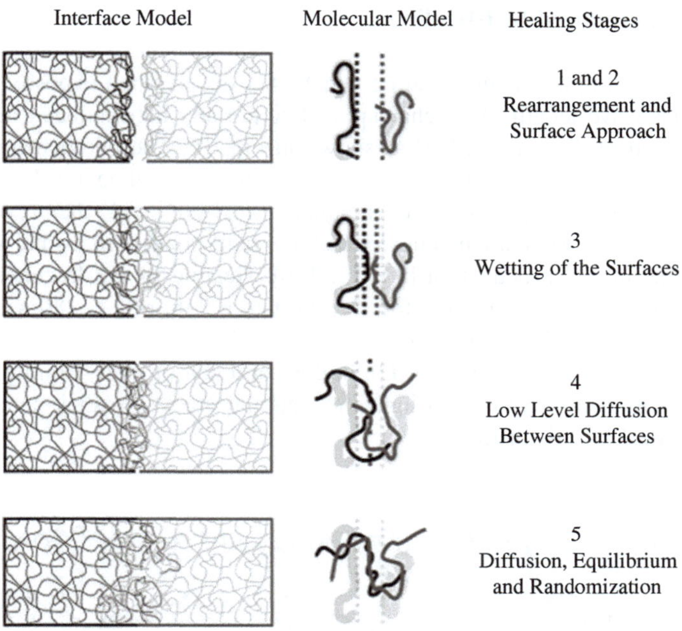

Fig. 14 Five stages model to explain the crack healing by Molecular Interdiffusion of thermoplastic composites [201]

processing. Also, Kim and Wool [102] proposed a microscopic theory of chain diffusion and randomization based on the reptation model. This theory describes the motion of chains at the damaged interface and calculates average interpenetration distance of the polymer segments as a function of time and molecular weight. Later, Kausch and Jud [91] observed that the development of the mechanical strength during the crack healing process of glassy polymers is related to inter diffusion of the molecular chains and subsequent formation of molecular entanglements.

There are various factors which affect the healing process and after healing performance of the composites such as healing time, healing temperature, clamping pressure, idle time after fracturing etc. Lin et al. [120] achieved full recovery of tensile strength of cracked poly (methyl methacrylate) (PMMA) by methanol treatment between 40 and 60 °C. But Boiko et al. [17] observed low levels of adhesion between virgin poly(ethylene terephthalate)/poly(ethylene terephthalate) and poly(ethylene terephthalate)/poly(styrene) joints even after 15 h of treatment at a temperature 18 °C higher than their T_g. Yang and Pitchumani [207] observe 100 % healing efficiency of carbon reinforced polyether-etherketone (PEEK) and polyether-ketone-ketone (PEKK) under non-isothermal conditions. Jud and Kausch [85] have also found some influence of molecular weight and degree of copolymerization on the crack healing behavior of poly (methyl methacrylate) (PMMA) and PMMA- poly (methoxyethylacrylate) (PMEA) copolymers.

3.1.2 Photo-Induced Self-Healing

It is a process of light initiated healing which only occurs upon exposure to a light of particular wave length. A mechanism of fracture and light-induced healing has been proposed by Wu et al. [202] as shown in Fig. 15. The photochemical [2+2] cycloaddition of cinnamoyl groups was chosen as the healing mechanism since photo-cyclo addition produced cyclobutane structure and the reversion of cyclobutane to the original cinnamoyl structure readily occurs in a solid state upon crack formation and propagation [34, 73, 136]. Healing of the fractures in these films is achieved by re-irradiation for 10 min with a light of $\lambda > 280$ nm. Healing efficiencies in flexural strength up to 14 and 26 % are reported using light or a combination of light and heat (100 °C). In this particular process, healing is limited to the surfaces being exposed to light. So, internal cracks or thick substrates are difficult to heal by this process.

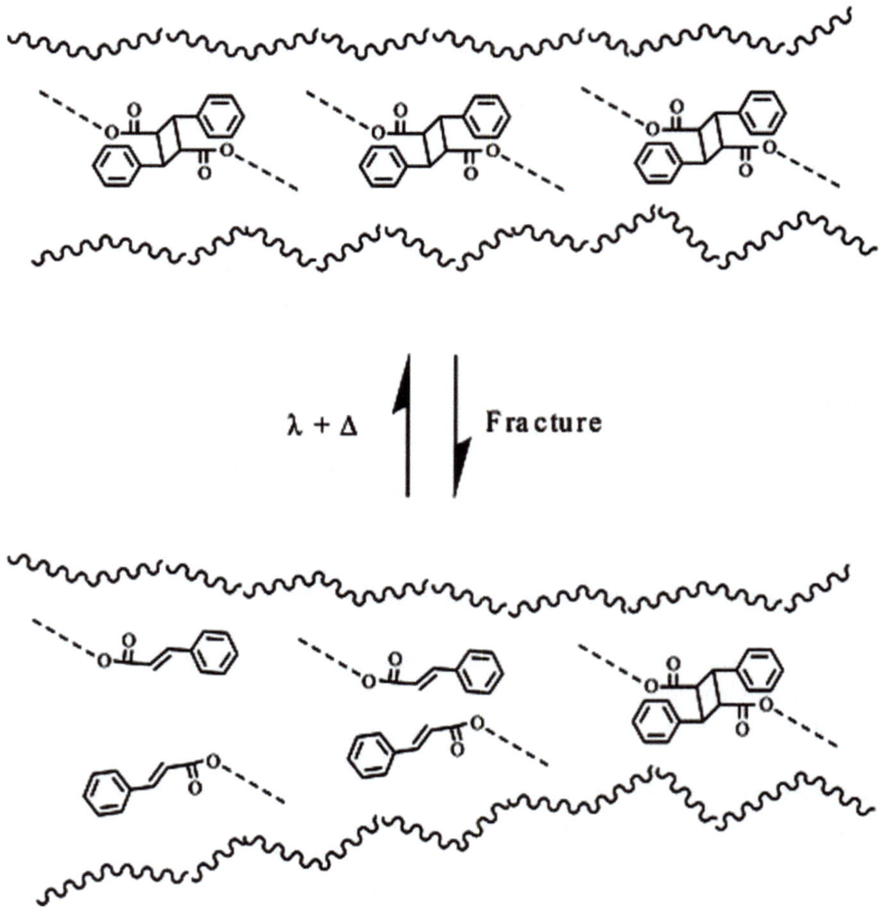

Fig. 15 Mechanism of photo-induced healing [202]

3.1.3 Self-Healing by Recombination of Chain Ends

Recombination of chain ends is a relatively new technique to heal structural and molecular damages in certain thermoplastic composites. This technique does not require external source of energy (e.g. heat, radiation etc.) for healing. Thermoplastic polymeric materials prepared by condensation reactions such as polycarbonate (PC), polybutylene terephthalate (PBT), Polyphenyl ether (PPE), polyether-ketone (PEK), polyethyl-ether-ketone (PEEK), etc. can be healed by a simple reaction that reverses the chain scission [61, 79, 129, 149, 167, 182, 183]. About 98 % of healing efficiency in recovery of tensile strength and molecular weight are achieved for PC material after a healing period of above 600 h [183]. Self healing of PC composites occurs by recombination of the phenolic end groups with phenyl end groups in presence of a small amount (0.1 ppm) of Na_2CO_3 (Fig. 17). A mechanism of the self healing reaction of these polymeric materials is proposed by Takeda et al. [182]. There are various steps for self healing via recombination of chain ends. These steps include (i) chain scissoring due to degradation; (ii) diffusion of oxygen into polymer structure; (iii) re-combination of scissored chain ends by catalytic redox reaction in presence of oxygen and catalyst; and (iv) discharging of water as a result of the self healing reaction. The self-healing process of PC and PPE are shown in Figs. 16 and 17 respectively. The kinetics of the self healing reaction depends on factors such as oxygen concentration, catalyst, chain mobility, reactant end groups etc. As the reaction progresses, the speed of the healing reaction decreases due to a reduction of the polymer chain mobility with increasing molecular weight and a gradual decrease of available hydroxyl (OH) end groups.

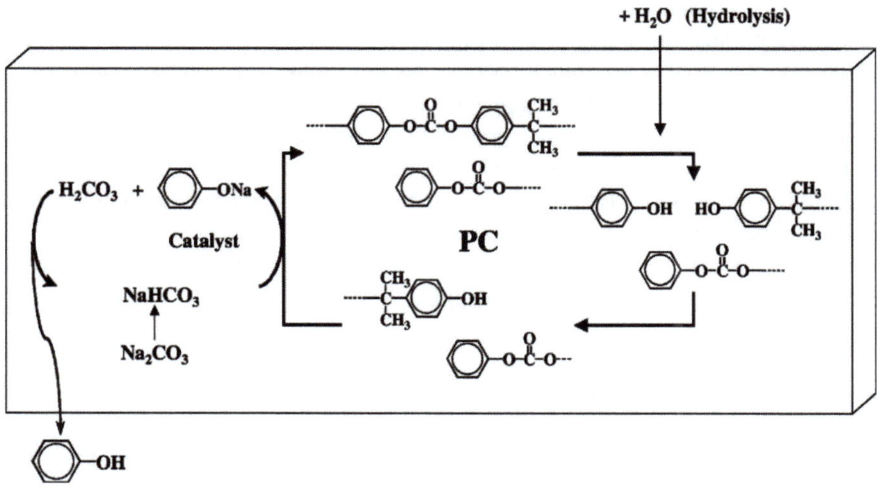

Fig. 16 Self-healing scheme in PC [182]

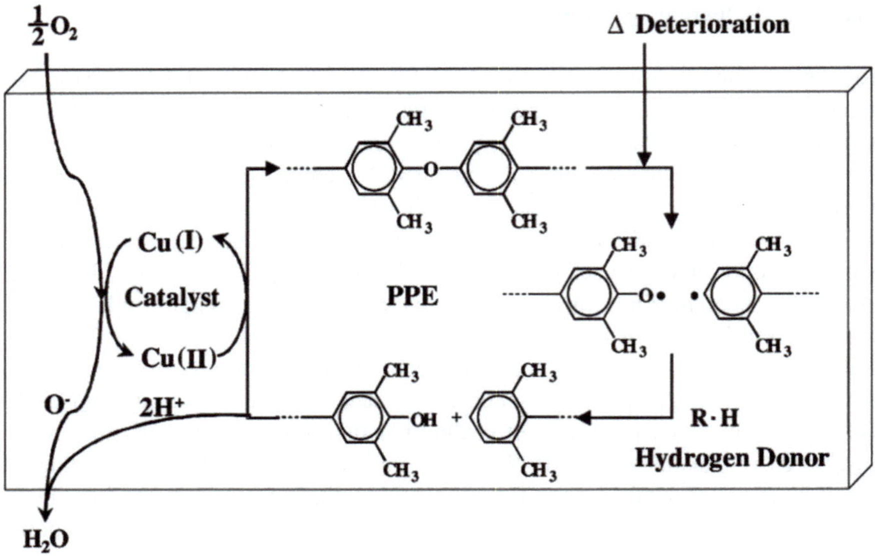

Fig. 17 Self-healing scheme in PPE [182]

3.1.4 Self-Healing by Reversible Bond Formation

Self-healing of certain thermoplastic polymers can be achieved by inclusion of reversible bonds in the polymer matrix at ambient temperatures. This technique utilizes hydrogen or ionic bonds to heal damaged polymer networks. A mechanism of molecular self healing via reversible bond formation is patented by Harreld et al. [71]. The production of polypeptide-polydimethylsiloxane copolymers as self-healing materials is depicted in Fig. 18. The silicon-based primary polymeric networks are grafted or block copolymerized with a secondary network of crosslinking peptides via hydrogen and/or ionic bonding. This hydrogen and/or ionic bondings act as intermediate crosslinks to strengthen the network and provide a good overall toughness to the material while allowing for self healing due to reversible crosslinking. Healing is initiated when the fractured surfaces come in contact either through physical closure or via a solvent induced chain mobility. The healing times can be adjusted by varying the structure of the polymer, degree of crosslinking, and strength of the crosslinks.

3.1.5 Self-Healing of Ionomers

Ionomers are defined as polymers having less than 15 % ionic groups along with polymer backbone [51]. Common ionomers are not suitable for self-healing applications. Suitable ionomers can be synthesized or modified by fillers or fibers for self-healing. Ionic content and its order-disorder transition is the driving force

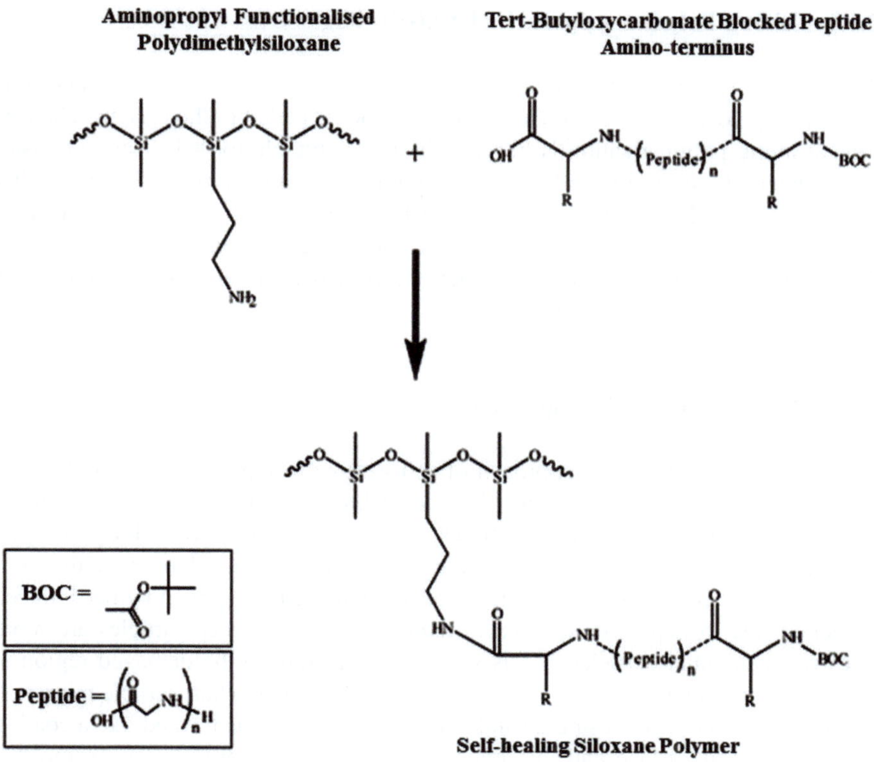

Fig. 18 Self healing of Siloxane polymer [202]

behind the healing process. self-healing behaviors of poly ethylene-co-methacrylic acid (EMAA), carbon nanotube filled EMAA composites, and low density polyethylene (LDPE) are investigated by Kalista et al. [89]. All those samples, except LDPE, exhibit the self healing behavior at room temperature. The lack of self healing in LDPE reveals the existence of the ionic functionality and/or the polar acid groups in the EMAA polymers which are essential to achieve self-healing. A new self-healing composite material is prepared by embedding carbon fibres in Surlyn ionomer [179]. Surlyn is an ethylene/methacrylic acid (E/MAA) copolymer in which the MAA groups are partially neutralized with sodium ions. This sodium ionomer is chosen as the matrix of the self healing composite. The polymer matrix made out of Surlyn melts at 95 °C from the heat generated in the carbon fiber layer and heals any surface and through thickness damage in the polymer matrix. Carbon fibers are selected as a heating element due to its low electrical resistance, resistive heating property, and reinforcement properties [179]. A two-stage theory on healing of ionic polymers by the application of heat, has been proposed by Kalista and Ward [88], and Pingkarawat et al. [145]. As temperature rises, polymer chain mobility increases which helps to complete the healing process.

3.1.6 Living Polymer Approach of Self-Healing

The living polymers are equipped with active groups at their chain ends and capable of resuming polymerization if additional monomer is added to the system. The free radical living polymerization is likely to be more suitable for self-healing considering the high reactivity and stringent conditions required for the ionic living polymerization [29]. Living polymers can be prepared with a number of macro-radicals i.e. polymer chains capped with radicals, by either ionic polymerization or free radical polymerization during which the polymer chains grow without chain transfer and termination [36, 63, 181].

3.1.7 Self-Healing by Nanoparticles

It is an interesting and new approach of self-healing to repair cracks in polymeric composites by using nanoparticles. Instead of breaking and rejoining of polymer chains like other self-healing techniques this technique involves filling and dispersing nanoparticles into cracks and flaws. The nanoparticles have a natural tendency to be driven towards the fractured or damaged area by polymer-induced attraction due to depletion at elevated temperature. The larger particles are more effective than small particles in this sense for migrating to the damaged region at relatively short time span. The system will be cooled down when particle migration will be completed. As a result, a solid nanocomposite layer is formed like a coating on the damaged surface and that effectively repairs the flaws. As an example, a 50 nm thick silicon oxide (SiO_2) layer deposited on top of a 300 nm thick PMMA film embedded with 3.8 nm CdSe/ZnS nanoparticles. The migration of the nanoparticles towards the cracks in the brittle SiO_2 layer is dependent on the enthalpy and entropic interactions between the PMMA matrix and the nanoparticles [191, 202].

3.2 *Self-Healing of Thermoset Materials*

Thermoset materials are rigid and thermally stable due to their cross linked molecular structure. They do not melt and even have no chain mobility due to application of heat. So, there are distinctly different approaches of self-healing of these thermoset materials.

The most common approaches for self-healing of thermoset materials involve incorporation of brittle vessels into the polymeric matrix which contain self-healing agents inside. These vessels fracture upon certain stimuli (load, strain, heat etc.) of the polymer and release low viscosity self-healing agents to the damaged sites for subsequent curing and repairing of the microcracks. This approach may differ slightly depending upon the healing functionality integrated into the composites as shown in Fig. 19. Many different self-healing approaches for thermoset materials

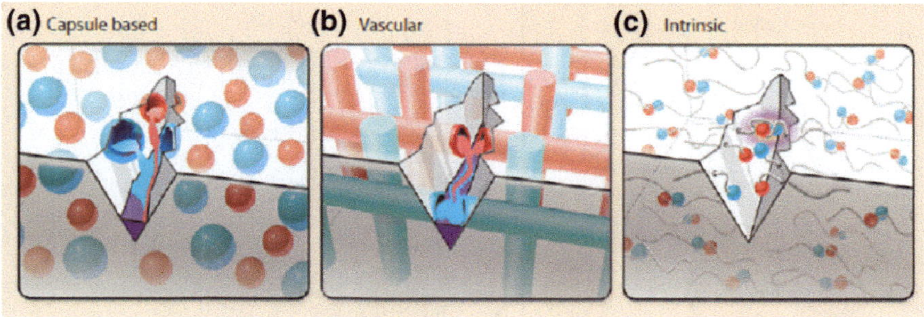

Fig. 19 Self-healing approaches of thermoset materials **a** capsule-based, **b** vascular, and **c** intrinsic methods [50]

have been reported, such as, (a) Capsule-based self-healing approach; the healing agents are stored in capsules and the capsules are incorporated into polymer matrix, which fracture upon stimuli to release self-healing agents in damaged sites (b) Vascular approach; here the healing agent is stored in hollow channels or fibers until damage ruptures the them to release the healing agent. (c) Intrinsic materials contain a latent functionality that triggers self healing of damage via thermally reversible reactions, hydrogen bonding, ionomeric arrangements, or molecular diffusion and entanglement. The exact nature of the self-healing approach depends on (i) the nature and location of the damage; (ii) the type of self-healing agents; and (iii) the operational environment [19, 50].

3.2.1 Self-Healing by Microencapsulation

In recent years, it is the most studied self-healing concept of thermoset materials. This particular approach involves incorporation of micro-capsules filled with healing agent and a dispersed catalyst within a polymer matrix. The microcapsules are fractured or ruptured by the propagating cracks in composites, resulting in release of the healing agent into the cracks by capillary action. Subsequent chemical reaction between the healing agent and the embedded catalyst heals the material and prevents further crack to propagate. These steps of self-healing are shown in Fig. 20 [110, 200]. Self-healing composite is prepared by blending microcapsules (<15 micron) containing various combinations of a 5-ethylidene-2-norbornene (5E2N) and dicyclopentadiene (DCPD) monomers, with (polyMelamine-urea-formaldehyde) (PMUF) shell which react with ruthenium Grubbs' catalyst (RGC) upon emergency. These self-healing materials are mixed with an epoxy resin and single-walled carbon nanotubes (SWNTs) to prepare nanocomposites which possess tremendous potential for space applications under hypervelocity impacts [5, 140]. Poly urea-formaldehyde microcapsules filled with epoxy resin 711 and E-51 healing agents are used in-situ polymerization of urea-formaldehyde in an oil-in-water emulsion has been prepared successfully for corrotion prtection [118].

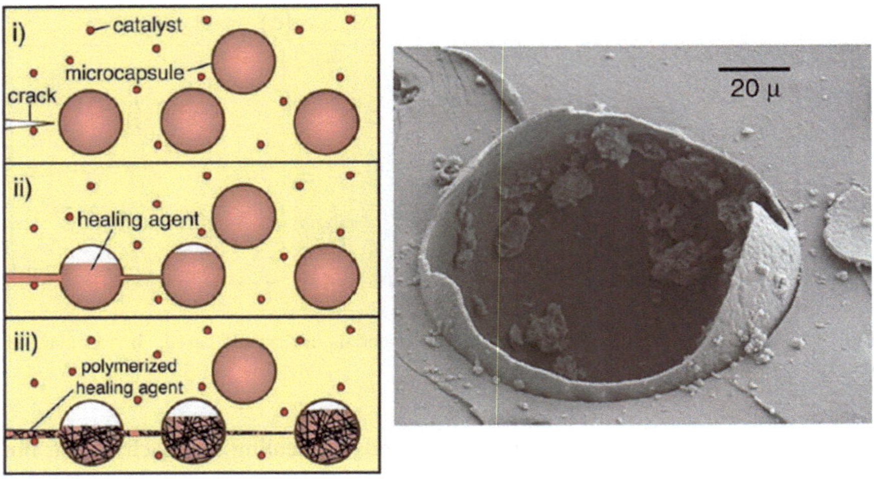

Fig. 20 Self-healing by micro-encapsulation approach, ruptured microcapsules [203]

3.2.2 Hollow Fiber Approach of Self-Healing

Dry [45–48] pioneered the concept of releasing healing chemicals stored in hollow fibers to repair damage in thermoset materials. This concept has been initially applied to cementitious materials to alter the cement matrix permeability, repair cracks, prevent corrosion, and as sensors for remedial actions. The feasibility of this approach has been subsequently extended to polymeric materials [138]. Hollow fibre approaches are used in preference to embedded microcapsules because they offer the advantage of being able to store functional agents for composite self-repair systems as well as acting as reinforcement [188]. A typical hollow fibre for self-healing used within composite structure can take the form of fibres containing a one-part resin system, a two-part resin and hardener system or one part resin system with an encapsulated catalyst or hardener contained within the matrix material as shown in Fig. 21. The exact nature of the self-healing method depends upon the nature and location of the damage, the potential of self-healing resins, and the influence of the operational environment [188].

The tough homogeneous nanofibers and core-shell healing agent-loaded nanofibers can be produced by coaxial electrospinning (co-electrospinning), emulsion electrospinning, and emulsion solution co-blowing, and several other advanced nanofabrication techniques. By these techniques low cost and continuous nanofibers for interfacial toughening and damage self-repairing of high-value advanced structural composites can be prepared. Solution co-blowing method can preferebly be followed for mass production in comparison to other methods due to its higher productivity. The core of these fibers which is filled up by healing agents is surrounded by a polymer shell, which provides them with structural stability [203].

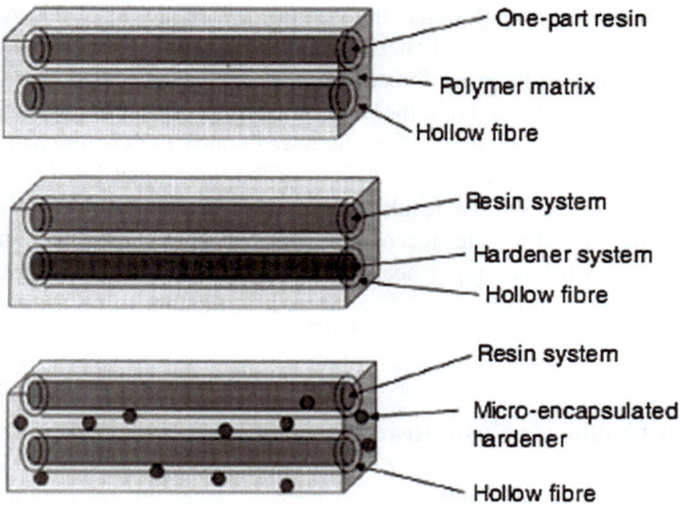

Fig. 21 Different hollow fibre approaches of self-healing [188]

3.2.3 Self-Healing of Thermally Reversible Cross Linked Polymers

Certain cross linked polymers are capable of healing internal cracks through thermo-reversible covalent bonds. This approach of thermally reversible crosslinks eliminates the need of incorporation of micro-capsules or hollow fibres filled with healing agents or catalysts in the polymeric matrix. However, heat is required to initiate the healing [25, 26, 72, 204]. This approach depends upon the nature of the fracture and can repair specific covalent bonds. Two new remendable highly cross-linked polymers, 2ME4F and 2MEP4F, are prepared by Chen et al. [26] without solvent. The study of thermal reversibility of Diels–Alder (DA) cross-linking shows that DA connections and disconnections of both polymers are thermally reversible. Cracks in these polymers can be healed effectively with a simple thermal healing procedure. These thermally reversible cross linked polymers may be suitable to fabricate fiber reinforced polymer composites for structural applications.

3.2.4 Self-Healing via Inclusion of Thermoplastic Additives

Inclusion of thermoplastic additives as self-healing agent inside the thermoset matrices enables them of re-bonding fracture surfaces on application of heat [212]. The feasibility of this technology is demonstrated using up to 40 vol % of thermoplastic epoxy particles (average diameter of 105 μm) in a glass fiber reinforced epoxy composite. Upon heating epoxy particles melted, flowed into internal cracks and healed them.

Jones and Hayes [84] describe the healing system as a “solid solution” of thermoplastic and thermoset polymers instead of the two phase system described above for self-healing fiber reinforced composites. It is specified that the matrix should contain 10–30 % (wt) of a thermoplastic polymer. Thermoplastic preferably forms a homogeneous solution with the thermoset matrix both before and after cure. There should be optimum molecular weight of the thermoplastics. Low molecular weight polymer diffuses faster resulting in quicker healing whilst high molecular weight polymer provides better mechanical properties but slower in action. Hence, there is a need to balance rapid healing and good healed mechanical properties. Healing temperature also has its influence as it is thought to be a diffusion controlled process.

3.2.5 Self-Healing via Chain Rearrangement

Healing of thermoset polymers has also been achieved by rearranging polymer chains at ambient temperature. This technique has similarity to molecular interdiffusion technology of thermoplastics. Self-healing of cracks is achieved by interdiffusion of dangling chains or chain slippage in the polymer network at ambient temperature [206]. This process does not require heating which is essential for thermoplastic additives or the thermally reversible crosslinks approach.

3.2.6 Self-Healing of Hydrogels

Concept of self-healing hydrogels is developed for repair of lightly crosslinked hydrophilic polymer gels via metal-ion-mediated reactions [192, 193]. In this technique, metal-ions are absorbed from an aqueous solution for rearrangement of crosslinked networks and then they are incorporated into the hydrogel. The metal-ion-mediated healed material has an entirely different structure and physical properties from the un-healed material.

One self-healing hydrogels system is demonstrated by Varghese et al. [192] where acryloyl-6-amino caproic acid (A6ACA) gels are synthesized. They contains flexible hydrophobic side chains with a terminal carboxyl group and undergo healing at ambient temperature through the formation of coordination complexes mediated by transition-metal ions. In another system healing of the gels are conducted by placing dried pieces of the gels together in a dilute aqueous solution of 0.1 M $CuCl_2$ at ambient temperature [193]. The tensile strength of the healed gels increased with time and achieved up to 75 % strength recovery after 12 h healing. The factors affecting the healing ability include the metal-binding capacity of the gel, the nature of the complexation, and the ability to deform under stress.

3.2.7 Self-Healing by Incorporating Water Absorbable Materials

Easter [49] developed a low cost cable capable of self-healing damage through expansion action of the water absorbable materials surrounding the conductor. The water absorbable material can be located in any one of the covering layers of cable. When the cable is damaged due to punctures, cracks or voids then water ingress reach to the water absorbable materials. The water absorbable materials expand and fill that cracks, punctures and voids. Thus damages are sealed in the cable. The water absorbable material is comprised of either water absorbable filler such as sodium bentonite or polyethylene oxide dispersed in a non-water absorbable polymer such as polyisobutene or polyisoprene, or a water absorbable polymer, i.e. polyethylene vinyl chloride or polyacrylic resins. This self-healing mechanism is only effective for repairing damage when water is present in the environment.

3.2.8 Self-Healing via Passivation

Passivation is a technique of light-coating on a protective material, such as metal oxide, to create a shell against corrosion. This coating material is developed to automatically heal the damage caused by exposure to UV radiation, oxygen, and in particular atomic oxygen in low earth orbit environment [166]. The self-healing polymer layer may be an organo-silicon material which operates by providing silicon to react with oxygen from the environment to form a SiO_x compound that condenses on micro-cracks, encapsulating impurities and filling the voids, fractures and other flaws. This self-healing structure can be used by itself or applied on top of a UV-sensitive substrate.

3.3 *Other Approaches*

When very fine polyethylene fibres are incorporated in fibre reinforced cementitious composites then self-healing capability of the composites is improved because fine polyethylene fibres form bridges over the crack which helps to attach the crystallization products [74]. Artificial skin is prepared with both pressure sensitivity and mechanical self healing properties with graphene and polymers by integrating them into a thin film which mimics both the mechanical self healing and pressure sensitivity behavior of natural skin without any external power supply [76]. Some of the self-healing materials, their system, mechanism, and assessment techniques are mentioned in Table 1.

Table 1 Self-healing materials, system, mechanism, and assessment techniques [211]

Category of materials to be repaired	Materials to be repaired	Healing system	Trigger mechanism	Healing mechanism	Assessment of healing effect
Thermoplastic	Poly(methylmethacrylate), etc.	Bulk	Heating or solvent induced	Chain interdiffusion and entanglements	Compact tension (CT) test or photography
Thermoplastic	Poly(ethylene-comethacrylic acid)	Bulk	Thermo-mechanically induced melting	Chain interdiffusion and entanglements	Visual inspection after sawing, cutting and puncture
Thermoplastic	Polycarbonate	Bulk (weak alkali/hydrolyzed chains)	Steam	Weak alkali catalyzed polymerization	Molecular weight and mechanical strength
Thermoplastic	Poly(phenylene ether)	Bulk (copper ion/oxygen/scission chains)	Heating	Copper ion catalyzed polymerization	Molecular weight
Thermoset	Epoxy	Bulk	Heating	Post-curing of residual functional groups	Impact strength
Thermoplastic/thermoset semi-interpenetrating network	Poly (bisphenol-Aco-epichlorohydrin)/ epoxy	Bulk	Heating	Chain interdiffusion and entanglements	CT and impact tests; photography
Thermally reversible crosslinking network	Crosslinked multifuran/multimaleimide	Bulk	Heating	Diels-Alder reaction	CT and double cleavage drilled compression
Thermoset	Epoxy, fiber/unsaturated polyester, and fiber/epoxy	Cyanoacrylate, epoxy, unsaturated polyester, etc.	Crack induced breakage of hollow tubes containing healant	Curing of healant	Tensile, flexural and impact tests; photography; ultrasonic C-scan
Thermoset	Epoxy	Dicyclopentadiene/Grubbs' catalyst	Crack induced damage of 3D microvascular networks, releasing healant	Ring-opening metathesis polymerization of healant	Four-point bending

(continued)

Table 1 (continued)

Category of materials to be repaired	Materials to be repaired	Healing system	Trigger mechanism	Healing mechanism	Assessment of healing effect
Thermoset	Unsaturated polyester	Styrene or epoxy	Crack induced rupture of microencapsulated healant	Polymerization or curing of healant	Mechanical strength and visual inspection
Thermoset	Glass fiber/epoxy	Epoxy granules	Heating	Curing of healant	Three-point bending and tensile fatigue
Thermoset	Epoxy and woven glass fiber/epoxy	Epoxy/latent hardener	Crack induced rupture of microencapsulated epoxy	Curing of healant	Single edge notched bend (SENB) and double cantilever beam (DCB) tests
Thermoset	Epoxy, fiber/unsaturated polyester, and fiber/epoxy	Dicyclopentadiene/Grubbs' catalyst	Crack induced rupture of microencapsulated healant	Ring-opening metathesis polymerization of healant	Tapered double cantilever beam (TDCB), DCB and fatigue tests
Thermoset	Unsaturated polyester	Phase-separated polysiloxane droplets/tin catalyst	Crack induced rupture of microencapsulated catalyst	Polycondensation of polysiloxanes	TDCB
Elastomer	Silicone rubber	Polysiloxane/platinum catalyst/initiator	Crack induced rupture of microencapsulated healant and initiator	Polycondensation of polysiloxanes	Tear strength
Thermoset	Epoxy	Solvent (chlorobenzene)	Crack induced rupture of microencapsulated solvent	Solvent induced crosslinking of incompletely cured resin	TDCB

4 Heating Properties

A resistor can generate heat when it is connected with an external voltage supply. This phenomenon of heat generation is well explained by Joul's law. A composite material can act as a resistor with certain level of electrical conductivity to generate heat through Joul's effect. Electrical conductivity in composites can be achieved by incorporation of various conductive materials in their structure by various means. However, self-heating phenomena is noticed in certain composite materials which are particularly used in automobiles and aircraft structures. When dynamic loads are introduced in these composites, energy is starting to dissipate through its matrix, which is transformed into heat. The evolved heat accumulates in the structure if matrix material has small values of heat transfer coefficients. However, the evolved heat can dissipate through the structure when matrix has high heat transfer coefficient. When the evolved heat and the thermal dissipation conditions are balanced, a self-heating effect of the composite in the steady-state can be observed. The self-heating effect could be used in problems of the damage detection and evaluation in polymer-based composite structures [90].

4.1 Metal Fibres and Wires Used in Composites for Heat Generation

Structures prepared with highly conductive metals alone such as copper, steel, aluminum etc. are not suitable for heat generation. Because, for Joul's effect of heat generation, a material requires a moderate electrical resistivity. That's why, metal are blended with textile fibres by various means to develop hybrid electro-conductive composite yarns or fabrics for heat generation [27, 213]. A stainless steel multifilament and a carbon yarn can be produced having moderate resistance that may be suitable for heat generation [106, 168, 169]. It is found that each of the selected electro-conductive yarns/wire exhibits particular brittle characteristics and poor bending properties that are not typical of yarns for textiles applications [147]. A flexible knitted fabric made of silver and elastomeric yarn can generate sufficient heat to warm up the body. This fabric can be used to manufacture personal heating garments that can generate heat on application of external voltage [69].

4.2 Conductive Polymer Coated Composites for Heat Generation

Conductive polymers such as polyaniline, polypyrrole, polythiophene etc. are coated/applied on the surface of textile yarns/fabrics by various means to prepare electro-conductive composites. These non-metallic polymeric composites are found

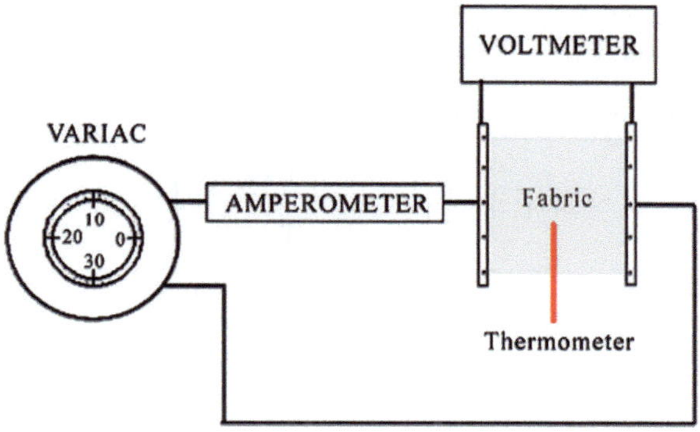

Fig. 22 Experimental set-up diagram for measuring the heating effect of a textile composite [176]

suitable for potential application of heat generation. Compared to other heating materials, the merits of these heating fabrics are their temperature homogeneity, low power density on large surface area, light-weight, their suppleness and fineness. They can be sewed up, cut off or pasted on subtracts for a large range of applications [21].

Dall'Acqua et al. [39] have prepared textile composites by embedding polypyrrole in natural and manmade cellulosic fibres, such as cotton, viscose, cupro and lyocell, by in situ vapour phase polymerization and they are found suitable for application of heat generation. In another study, polypyrrole is incorporated in cotton woven fabrics and various properties such as anti-static, anti-microbial, heat generation is investigated [171]. A possible application of the polypyrrole coated composite fabric is demonstrated as heating devices as shown in Fig. 22 [176]. A square shape fabric (6 cm × 6 cm) is positioned between two pressed electric contacts. The temperature rise is measured using an Omega infrared thermometer, placed at the center of the sample.

The temperature-current (*T-I*) and voltage-current (*V-I*) characteristics of this textile composite are shown in Fig. 23. The *T-I* characteristic follow an exponential trend and *V-I* characteristics follow a power law. According to the power law, maximum theoretical power achieved from the fabrics is: $P = VI$, where P is the power developed and *V*, *I* the voltage and current. In Fig. 24, the power and the impedance as a function of current are shown.

In another study, when a constant voltage of 9 V is applied to cotton/polypyrrole fabric for 10 min, its surface temperature raised up to 90 °C and for many numbers of repeating cycle performance of the fabric does not deteriorated. The voltage-temperature (*V-T*) characteristic of the fabric is shown in Fig. 25. Initially temperature is rising sharply with time at a fixed applied voltage and then level off at a particular temperature. These conductive fabrics can be used as heating pads which are more comfortable to use than metal incorporated fabrics [171].

The behavior of temperature versus voltage of viscose/polypyrrole fabrics prepared by in situ vapor phase polymerization (VPV) with different $FeCl_3$ concentration, are fitted by exponential curves with the following general Eq. (4).

Fig. 23 Behavior of the voltage and temperature of the PPy coated sample as a function of the current [176]

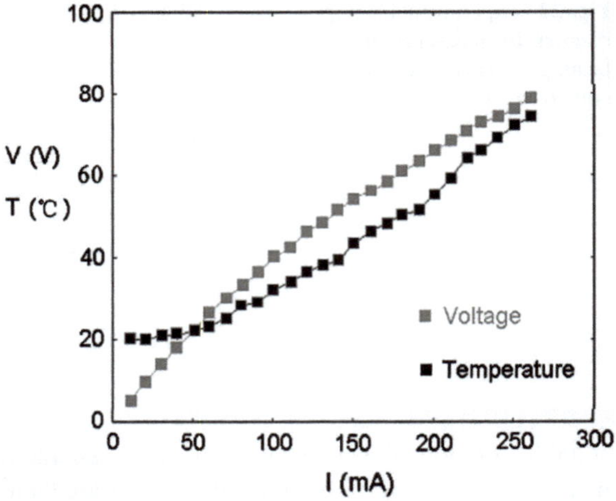

Fig. 24 Behavior of the impedance and power developed by PPy coated textile composite as a function of the current [176]

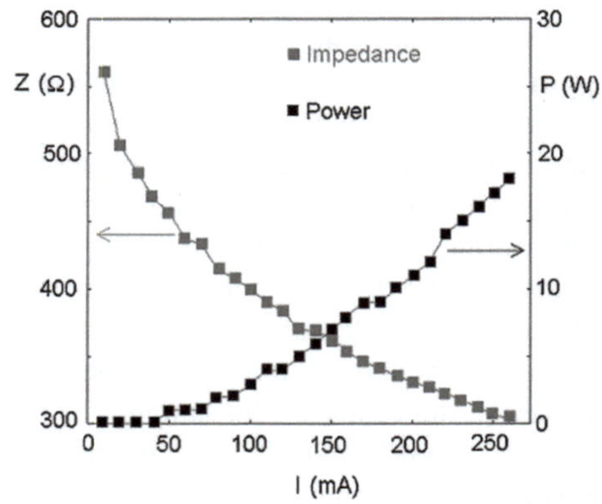

Fig. 25 Voltage-temperature characteristics of Cotton/polypyrrole fabric [171]

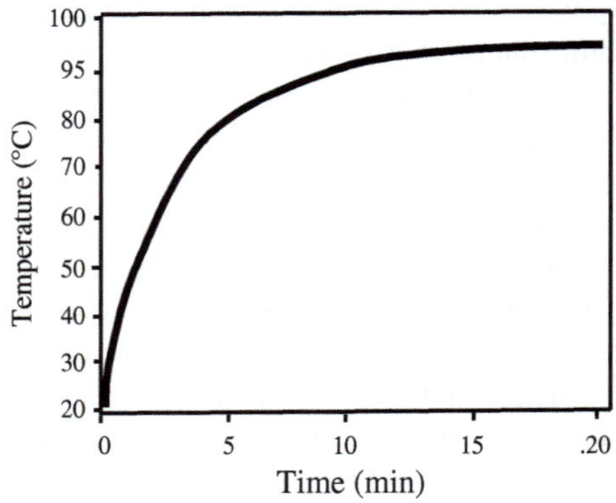

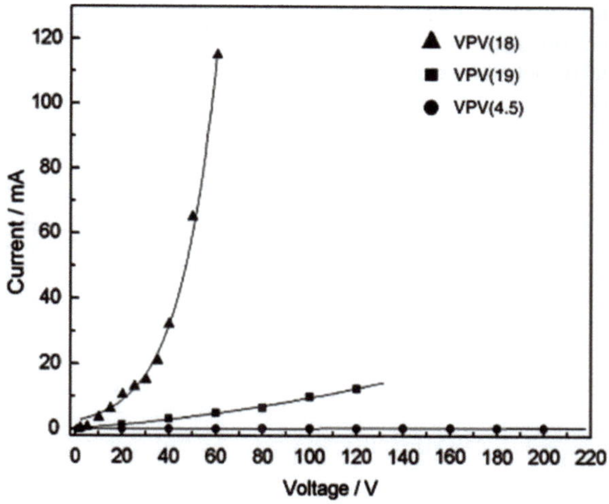

Fig. 26 Plot of temperature vs. applied voltage for different VPV samples [39]

$$T = T^0 + ae^{-kV} \tag{4}$$

where, V is the applied potential, T^0 the initial temperature and T is the final temperature. The resulting current–voltage characteristics of viscose fabric prepared by in situ vapor phase polymerization depends on $FeCl_3$ concentration as shown in Fig. 26. It shows a liner fit until $FeCl_3$ concentration of 9 gl^{-1},and an exponential fit for higher $FeCl_3$ concentrations [39].

It is suggested by Sparavigna et al. [176] and Macasaquit and Binag [123] that 100 % polyester fabrics can easily be made electrically conductive by polypyrrole coating and they are practically useful for many applications, including flexible, portable, surface-heating elements for medical or other applications. Polypyrrole coated polyester/lycra woven composite fabrics exhibit reasonable electrical conductivity and effective heat generation capacity where temperature reaching up to 40.55 °C at 24 V [94]. For all these composites the rate of change of temperature has two distinct phases, an initial sharp rise and followed by a leveling-off to plateau, similar to cotton/PPy composites [68]. Rodriguez et al. [163] have observed that electrical resistivity and heating effect of PPy composites depends on the doping anion present, whether it is chloride (Cl-) or other. Nylon/polypyrrole composite fabric with electrical resistance of 5 Ω/square is prepared by sequencial high temperature high pressure (HTHP) chemical and electrochemical polymerizations [117]. Surface temperature of this fabric is increased very quickly from room temperature to about 55 °C within 2 min due to application of 3.6 V from a commercial battery. The heat generating property of the fabric is so stable that it exhibits similar behaviors for at least 10 repeated cycles [117]. PPy coated E-glass fabric exhibits reasonable electrical stability, and is found effective in heat generation. By application of a constant voltage across the fabric, surface temperature increases whereas power consumption is found to be decreasing [1]. Polypyrrole coated electro-conductive woven, needlepunched nonwoven and spunlace

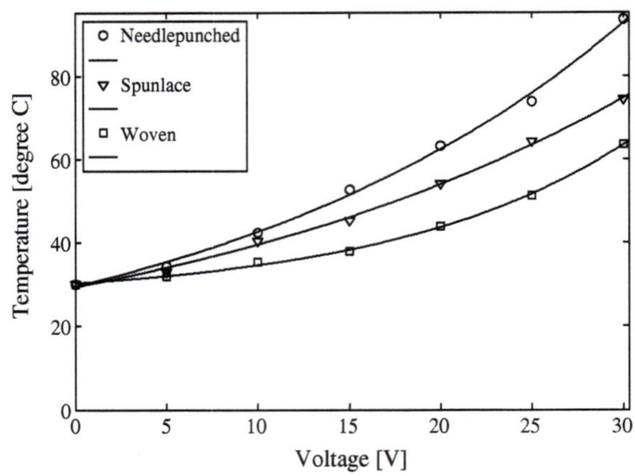

Fig. 27 V-T characteristics of Polyester fabric/polypyrrole composites [126]

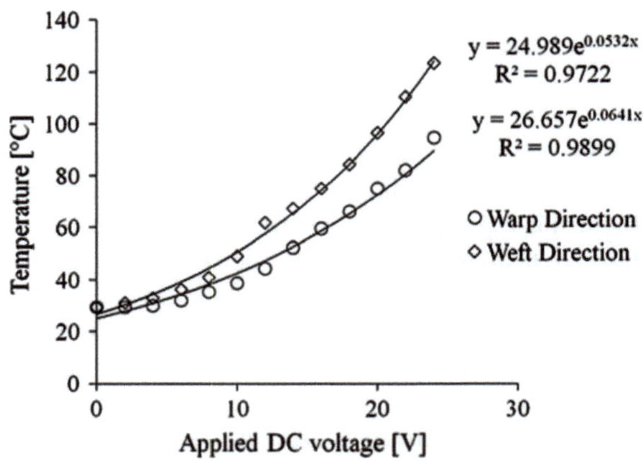

Fig. 28 V-T characteristics of Silk fabric/polypyrrole composite [128]

nonwoven fabric composites are prepared and characterized for heat generation [126]. Their *V-T* characteristics are shown in Fig. 27. All these fabrics follow an exponential trend of heat generation due to application of voltage in the form: $T = ae^{bV} + c$, where T is the measured temperature, V is the applied voltage, a, b and c are coefficients, exponent and constant respectively. Among these fabrics needlepunch nonwoven fabric shows highest rate of heat generation.

Silk/Polypyrrole electro-conductive composites are also prepared for the application of heat generation [20, 128]. Figure 28 shows the heat generation behavior of silk/polypyrrole composite woven fabrics. It can be seen that as voltage is increasing, surface temperature of the composite is increasing exponentially when tested in both warp and weft direction [128].

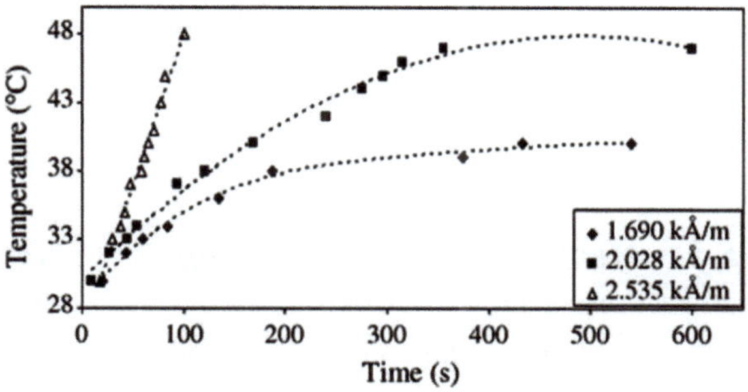

Fig. 29 Induction heat generation behavior of Fe_3O_4/PVA composites [114]

4.3 *Composites for Induction Heating*

Idea of using induction heating technology for the processing of fibre reinforced polymer composites is a recent approach. Induction technology is suitable for the processing of thermoplastic and thermoset polymer materials but requires special additives (conductive materials) either in the form of structured fibres and fabric or particulate that can transform the electromagnetic energy into heat. This approach is found suitable for thermoplastic composite welding, thermoset curing, selective material heating and fast mould heating technologies [11, 56]. The magnetic particles can be used as filler of the composite and are stimulated by the applied electromagnetic field and act as heat sources during the induction heating procedure. Micro-sized Fe_3O_4 particles are dispersed in polyvinyl alcohol hydrogel or other substrate to prepare composites for hyperthermia applications [114, 214]. Induction heating behavior of such composites is shown in Fig. 29. Maximum temperature achieved by applying alternating magnetic field 1.7 kV/m is 40 °C which is sufficient for application in hyperthermia [114].

5 Electromagnetic Shielding

Electromagnetic Interference (EMI) shielding is a process of restricting the penetration of electromagnetic waves into a medium or object by blocking them by a barrier made of electro-conductive material. It is a very popular method of protecting electronic and electrical equipments and even human beings against electromagnetic radiation. The barrier or protector which protects a body, environment or a circuit from harmful electromagnetic radiation is called a shield. It is well known that exposure to long term or acute electromagnetic radiation can have harmful effects on human tissue [4, 59, 105, 112, 124] and furthermore, electromagnetic radiation can interfere with certain bio-electronic devices such as pacemakers [184], and affects the lives of people. The recent proliferation of electronic devices, such as cell phones, and computer equipment, that emit low levels of electromagnetic radiation, or interference, has significantly increased the problem and created a need for everyday shielding garments. In this regard, a variety of electromagnetic shielding composites have been produced to prevent electromagnetic radiations [108, 164, 215].

5.1 Mechanism of Electromagnetic Shielding

When electromagnetic waves pass through a medium or object they interact with molecules of the medium or object and waves are attenuated in certain amount depending uopon the medium or object. This phenomena of interaction can be divided in two major steps such as (a) Attenuation due to absorption, and (b) Attenuation due to reflection [164]. As the waves strike a conductive object, the charges in the object are forced to oscillate at the same frequency of the incident wave. These forced oscillating charges behave like antenna and cause reflection from the surface. The signal wave may reflect in many directions depending upon the pattern associated with a signal oscillating charge. Hence, the signal is scattered and there is some signal loss. This mode of signal loss is called attenuation due to reflection. There are also successive losses of signal into the depth of the layer of the medium. It is called attenuation due to successive internal reflections. Again, the forced oscillating charge losses some energy in the medium in terms of heat. This mode of signal loss is known as attenuation due to absorption. Thus, electromagnetic shields work on the basis of the two above mentioned major electromagnetic mechanisms such as reflection from a conducting surface, and absorption in a conductive volume [130, 144, 164]. The combined effect of these losses (reflection and absorption) determines the effectiveness of the shield as shown in Fig. 30. Reflection from an electromagnetic shield occurs when the impedance of the wave in free space is different from the impedance of the electromagnetic wave in the shield. This phenomenon is independent of the thickness of shield and is a function of conductivity, magnetic permeability, and frequency of the shield [18, 130]. Whereas, electromagnetic shields made of electromagnetic absorbers attenuate undesirable electromagnetic waves and substantially dismiss electromagnetic radiation.

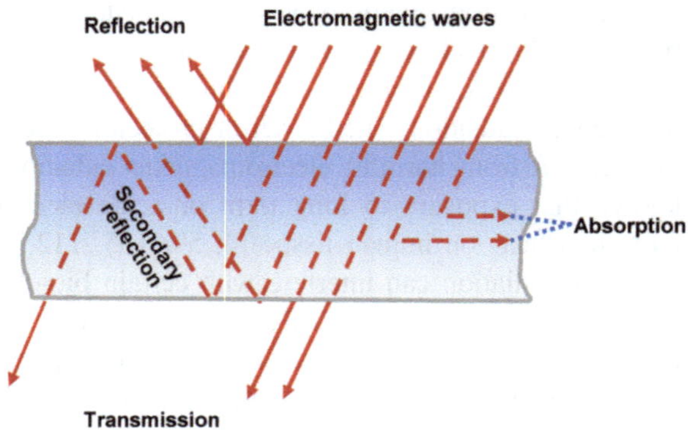

Fig. 30 Schematic representation of shielding phenomena

Shielding efficiency of this type of shield depends upon factors such as type of material, its thickness, size, shape, orientation of apertures etc. [18, 113, 132].

5.2 Electromagnetic Shielding Efficiency

Electromagnetic Interference Shielding Efficiency (EMISE) value expressed in dB is defined as the ratio of the incident to transmitted power of the electromagnetic wave [42, 144]. Mathematically,

$$SE = 10log\left|\frac{P_1}{P_2}\right| = 20log\left|\frac{E_1}{E_2}\right|(decibles, dB) \tag{5}$$

where, $P_1(E_1)$ and $P_2(E_2)$ are the incident power (incident electric field) and the transmitted power (transmitted electric field), respectively.

By measuring the reflectance (R_e) and the transmittance (T_r) of the material, the absorbance (A_b) can be calculated using Eq. (6).

$$A_b = 1 - T_r - R_e \tag{6}$$

where, R_e and T_r are the square of the ratio of reflected (E_r) and transmitted (E_t) electric fields to the incident electric field (E_i), respectively, as representes in Eqs. (7) and (8).

$$R_e = \left|\frac{E_r}{E_i}\right|^2 \tag{7}$$

$$T_r = \left|\frac{E_t}{E_i}\right|^2 \tag{8}$$

5.3 Preparation Methods of Composites for EMI Shielding

We can protect ourselves from electromagnetic radiation by covering us with an electro-conductive media which can generate and transport free charges. Metals are conventional conductive media but metal sheet or foils are not flexible enough to cover easily a complex three dimensional contour of a body or device to be protected. Even if, when metal powder or other conductive particles (such as carbon or conductive polymers) are incorporated as filler material in a polymeric matrix, a rigid or semi flexible composite structure is produced and which will not effectively solve the problem. If textile fabric as a flexible substrate can be made electro-conductive then flexible, moldable, light-weight, composites can be developed for a more versatile application. It is known that synthetic textile fibers such as polyester, polyamide, acrylic, cellulose-acetate etc. exhibit a very poor electrical conductivity and are hydrophobic in nature. When these fiber mass are rubbed, static electricity is generated and accumulates on the fibers [9, 178]. Natural fibers, such as cotton, wool and silk, exhibit a relatively high hydrophilic property and have relatively higher electric conductivity in favor of static charge dissipation. They are suitable for static charge dissipation but do not serve as effective electromagnetic shield [185]. For this purpose, metallic fibers, or fibers coated with a metal, as an electrically conductive material are commonly used [23, 27, 130, 143, 164, 194, 213]. These metallic fibers are not proved to be satisfactory for practical use due to their poor recovery from bending and breakage [108]. The breakage results in a decrease in the conductive effect of the metallic fiber. Also, it is difficult to mix spin, mix weave or mix knit the metallic fibers with organic polymer fibers. In case of textile fiber, yarn or fabric coated with a metal layer by electro-plating technique, it is required to form a uniform and continuous layer on the textile surface. In order to satisfy the above-mentioned requirement the surface of the fiber is required to be smooth and highly polished. Also, intensive care is required during coating operation to coat a continuous layer of metal with uniform thickness leading to substantial enhancement of the cost of the coating. Also, the metal-coated fiber has demerits in terms of easy peeling off from the fiber during processing or use [30]. The fiber coated with a polymer dope containing an electrically conductive material, such as carbon black, CNT, silver particles etc. is also found unsatisfactory because the coating operation is expensive and the coated layer is easily peeled off from the fiber during processing [13, 135]. When carbon black is used in dope mixing with polymer to spin a conductive filament then at least 15 % of carbon black is required based on the weight of the fiber matrix polymer. This large amount of carbon black causes the fiber-forming process to be difficult, complex and expensive. Also, it is impossible to contain the carbon black in the inside of the natural fibers. Many such limitations associated with processability, flexibility, durability etc. could be successfully overcome by coating/applying conducting polymers such as polyaniline, polypyrrole, polythiophene etc. on textile substrates [109]. These conductive polymer coated textile composites, owing to their flexibility, durability, ease of preparation and application, are considered promising for shielding of electromagnetic radiation [58, 108, 111, 176].

5.4 Textile/Metal Composites for EMI Shielding

Textile fabrics have been coated with metals such as aluminium, copper, nickel, silver and combination of them in industrial scale to prepare electro-conductive composites [7, 108]. The coating methods used are electroless plating, laminating with aluminium foils, dyeing with copper sulfide, vapor deposition etc. Electroless metal plating is a non electrolytic method of deposition of metal from solution by process of oxidation and reduction [70]. This method has some advantages such as coherent metal deposition, excellent conductivity and shielding effectiveness. It is also applicable to complex shaped insulating textile materials [70]. Vapor deposition coating of woven or nonwoven fabrics are important step for producing a variety of new thin film composites for new applications. Depending on the evaporation conditions, these coated composite fabrics can present quite different surface properties [43]. The shielding efficiency of these metalized textile fabrics mainly works in principle of energy reflection and not its absorption. In many cases, such a phenomenon is not satisfactory. Hence, there have been searches for materials with greater capability of absorbing electromagnetic radiation. In recent years, some patents have appeared which deal with the problem of producing textile composites with ferromagnetic properties. Such products can be applied as flexible screens for attenuating electromagnetic radiation, to produce cores in transformers, motors, generators, etc., andto produce filters to remove substances showing magnetic properties from air and water [108]. The process of coating on the surface of textile fabrics with layers of ferromagnetic powder can give a way to products of such quality composites. Nonwoven materials with ferromagnetic coating showed EMSE of about 30–35 dB for frequency range of 1000–2000 MHz [108]. Those materials are proposed for applications of the camouflage of military objects, restricting the range of fields emitted by devices such as shortwave and microwave diathermy. Textile materials embedded with metallic wires and fibres are also explored as EMI shield. Knitted fabric forms a good conductive network which can better stop the electromagnetic radiation and achieve the shielding effectiveness [216]. They have good impact energy absorption characteristics. Electro-conductive composites are prepared by using polypropylene (PP) matrix and knitted glass fibers as the reinforcement [28]. Copper wires are incorporated as conductive fillers into knitted structure to provide desired EMISE in composite structure. The EMISE of these knitted composites is greatly influenced by the amount of copper, which can be varied by changing the knit structure, stitch density and linear density of yarns used for knitting [28, 137, 175]. The copper and stainless steel wires are wrapped with polyamide filaments to produce a hybrid conductive yarn which is co-woven-knitted to produce fabrics for EMISE [22]. Here a multilayer structure of various thickness and stainless steel (SS) fiber content are responsible for blocking the electromagnetic interference in different frequency. Also, different compositions of Cu/Zn/Sn coated polyester nonwoven (CNW) and Ni/Cu coated polypropylene nonwoven (SNW) show various EMISE depending on incident frequency as shows in Fig. 31. Coated CNW fabric shows higher EMISE due to its higher thickness than that of the coated SNW fabric [22]. In another study, the effect of single and multilayer fabrics made of stainless steel/polyester (40/60) spun yarn on shielding effectiveness has been investigated and the results are shown in Fig. 32. Also, polypropylene and polyamide filaments are wrapped on copper wire to produce conductive yarn. This yarn is utilized as warp and weft identically for manufacturing woven fabric to obtain isotropic shielding behavior [23]. It is observed that the EMSE of these woven fabrics increases with increase of copper wire content and as well as with the density of warp or weft at all frequencies of incident radiation. The shielding effectiveness of these fabrics of a single layer is barely sufficient for general applications and the multi-layers provide adequate shielding

Fig. 31 Shielding effectiveness of various metalized nonwovens [22]

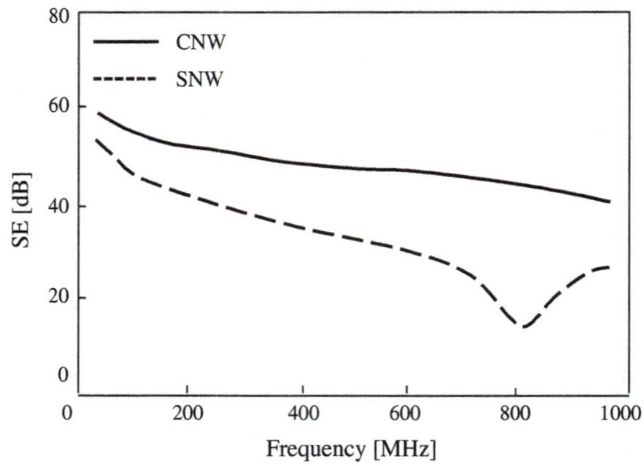

Fig. 32 Shielding effectiveness of laminated SS/PET (40/60) woven fabrics [22]

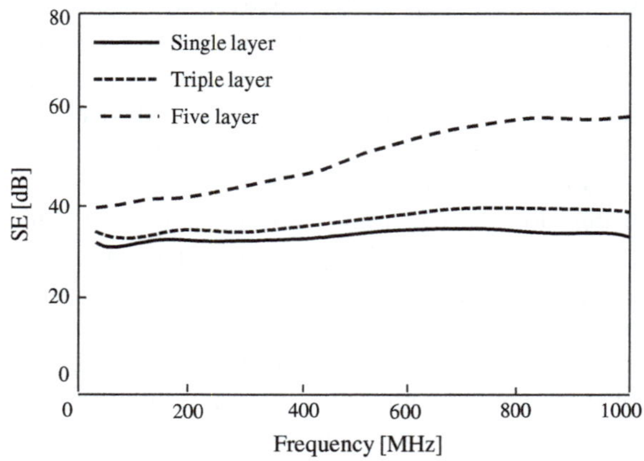

effectiveness (20–55 dB) when the wave is normally incident on fabric of thickness higher than 1.6 mm. The EMISE increases with increase of fabric thickness as the number of conductive layers increases as shown in Fig. 33 [23].

5.5 *Conductive Polymer Coated Textile Composites for EMI Shielding*

Coating of conducting polymers on insulating fabrics produces novel composites for potential applications of EMI shielding [53, 66, 96, 100, 127, 208]. Various composite textiles are prepared by coating different textile substrates, such as cotton, wool, viscose, cupro, lyocell, polyester, nylon etc. with conductive polymers by means of in situ polymerization [1, 15, 31, 39, 54, 92, 93, 97, 133, 141, 197, 198, 205, 209]. These composite textiles show better performance than metal fibre and filler composites which are susceptible to galvanic corrosion or loss of

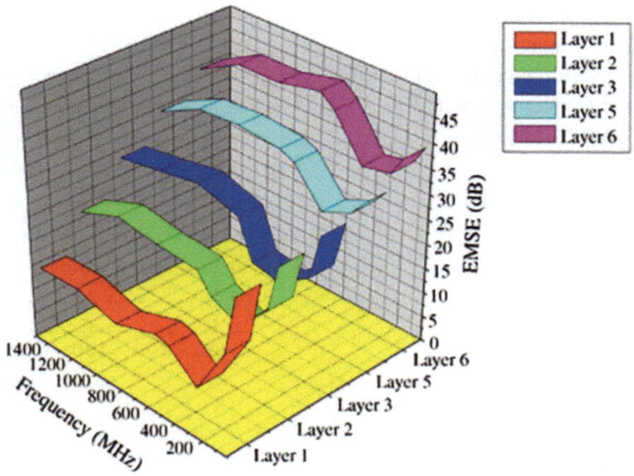

Fig. 33 EMSE of woven fabric at various layers [23]

conductivity due to friction [7]. The conducting polymers reflect less and largely absorb electro-magnetic radiation [7]. It has been reported that textile fabrics coated with polypyrrole by chemical polymerization are suitable for EMI Shielding [1, 53, 93, 100, 109]. The desired value of electrical resistivity of an electromagnetic shield is less than 100 Ω/cm^2 [42, 131, 173]. Polyaniline coated polyester fabric of resistivity 5 kΩ/ϒ has been found suitable for EMI shielding [111].

5.5.1 Conductive Polymer Coated Woven Fabrics for Electromagnetic Shielding

The polymerization of aniline pyrrole and thiophene on insulating fabrics such as polyester, glass and high silica cloth is carried out by in situ chemical polymerization [42]. In the radio frequency range from 100 to 1000 MHz, conducting polyaniline-coated polyester fabrics show a shielding effectiveness in the range 30–40 dB as shown in Fig. 34. The reflectance studies show that the polyaniline-coated polyester fabrics can absorb about 98 % and can reflect only about 2 % of the energy. In case of polypyrrole coated polyester fabrics, 96 % of energy is absorbed and 4 % is reflected back, whereas in case of polythiophene coated polyester fabric, 82 % of the energy is absorbed and 18 % is reflected back [42]. An effective shield should absorb all the energy and nothing should reflect back. The shielding effectiveness of polypyrrole coated nylon/lycra fabric prepared with an anthraquinone-2-sulfonic acid (AQSA) dopant is reported as 89.9 % at 18 GHz [67]. In another study, polypyrrole is polymerized chemically and electrochemically in sequence on a polyester woven fabric, for preparation of a composite with very low electrical resistivity of 0.2 Ω cm. EMISE of this composite is about 36 dB over a wide frequency range up to 1.5 GHz as shown in Fig. 35. It can be seen that the composite restricts electro-magnetic waves in large by absorption and in small by reflection. Overall EMISE increases with increase of electrical conductivity [96]. PPy films with high conductivity and good adhesion are generated on the surface of

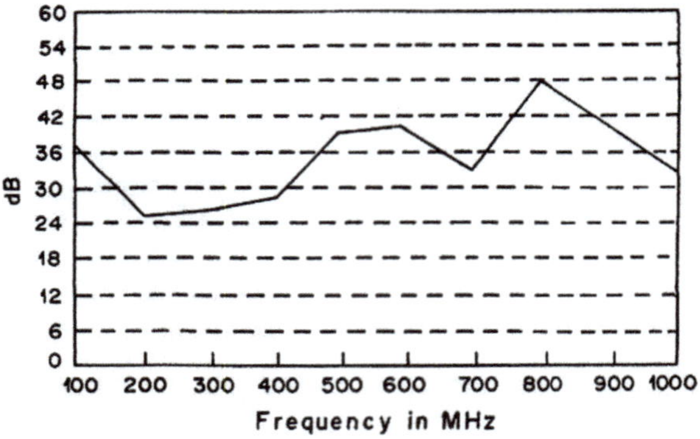

Fig. 34 Shielding effectiveness of polyaniline-coated fabrics in the frequency range 100–1000 MHz [42]

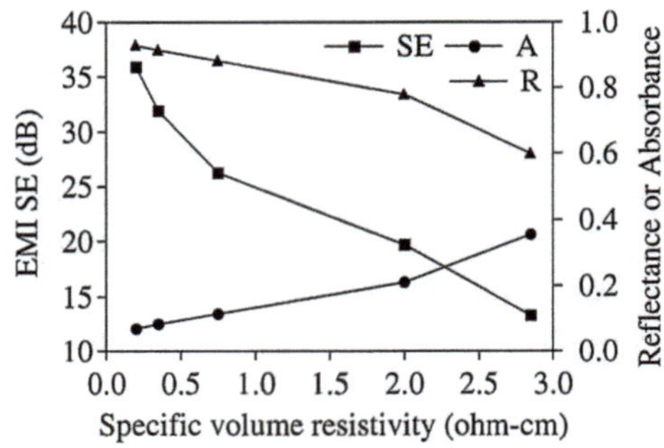

Fig. 35 EMISE absorbance (A), and reflectance (R) of PET fabric/PPy composites with various specific volume resistivity [100]

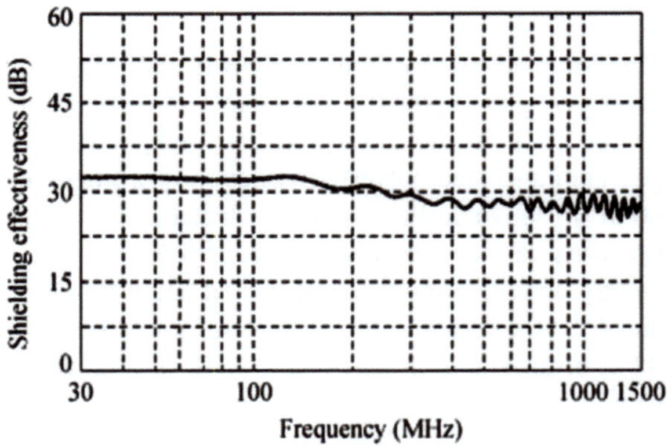

Fig. 36 Shielding effectiveness of PPy film [150]

insulating epoxy resin substrates using chemical polymerization. Those films show EMISE of about 30 dB over a wide frequency range from 30 MHz to 1500 MHz as shown in Fig. 36 [150]. PPy coated polyester and silica fabric show EMISE of 21.48 dB and 35.51 dB respectively at frequency of 101 GHz [99]. Polypyrrole coated glass fabrics having resistivity below 500 Ω/ϒ exhibit 98.67–99.23 % signal loss in the frequency range of 800–2,400 MHz [1]. These samples with low resistivity show high shielding effectiveness compared to those samples with high

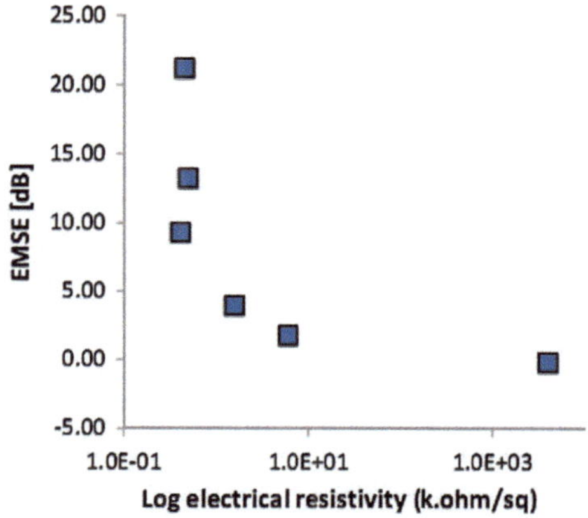

Fig. 37 EMSE of PPy coated glass fabric as a function of electrical resistivity [1]

resistivity as shown in Fig. 37. These fabrics are proposed for applications as shield for house hold appliances, FM/AM radio broadcast sets, wireless phones, cellular phones, computers, buildings, secret rooms and various electronic gadgets that operate up to 2.4 GHz frequency [1].

In another study, silver (Ag) is thermally vacuum evaporated on the surface of polypyrrole–polyester complexes. The EMISE of fabric complexes increases as the area of Ag evaporation layer increases. When the Ag is partially (~37 %) evaporated on one side of fabric complexes, the EMISE is 29 dB at 0.5 GHz, and while Ag is evaporated on total area of the sample the EMISE obtained is 33 dB at the same frequency range [75]. Also, PPy/Al_2O_3 textile nanocomposite is found suitable for effective EMI shield in the range of 8–12 GHz frequency and could be able to absorb more than 53 % of microwave radiation [190].

5.5.2 Conductive Polymer Coated Nonwoven Fabrics for Electromagnetic Shielding

EMISE of PPy coated hydro-entangled polyester nonwoven textiles is studied in the frequency range 100–800 MHz [6]. A positive correlation between the SE and the surface conductivity of those nonwoven textiles is found. EMISE of 37 dB was found for the sample with the lowest surface was resistivity of 3 Ω/square [92]. In another study, PPy coated polyester nonwoven composite fabrics shows effective EMISE of 20 dB in 1 GHz frequency [116]. For these nonwoven fabrics coating with Ag particles enhances the EMISE up to 55 dB and for multilayer complexes of those fabrics had EMISE of about 80 dB. PPy coated nonwoven fabrics are proposed for excellent radio frequency and microwave absorber because of high absorbance and low reflectance [197, 198]. It is reported that polypyrrole coated textile fabrics of high conductivity show reflection dominant EMISE and that of low conductivity showed absorption dominant EMISE [99, 101]. EMISE of some conductive textile materials proposed as effective shield, as cited in literature, are tabulated in Table 2.

Table 2 Shielding effectiveness of various conductive textiles

Material	Resistivity	Testing frequency	EMSE [dB]	References
Metallized fabrics	–	100 kHz–1 GHz	67, 75–80	Koprowska et al. [108]
PPy-coated fabrics	Surface resistivity 3 Ω/ϒ	800 MHz	37.02	Avloni et al. [6]
PPy-coated polyester fabrics	Volume resistivity 0.2 Ω cm	1500 MHz	36.6	Kim et al. [100]
PPy-coated polyester fabrics	Volume resistivity 0.3 Ω cm	1500 MHz	35 dB	Kim et al. [99]
PPy or poly (3,4-ethylenedioxythiopene coated polyester woven fabric	Volume resistivity 0.3 Ω cm	1500 MHz	36	Abbasi and Militky [1]
PPy coated glass fabrics	Surface resistivity 460 Ω/ϒ	800 MHz	18.75	Abbasi and Militky [1]
PPy coated glass fabrics	Surface resistivity 460 Ω/ϒ	2.4 GHz	21.16	Abbasi and Militky [1]
Ag/PPy coated fabric complex	–		80	Lee et al. [115]
PPy coated cotton fabric	Surface resistivity 1.18 MΩ/ϒ	2500 MHz	01	Yildiz et al. [210]
Polyaniline-coated polyester fabrics	Volume resistivity 10–60 Ω cm	101 GHz	21.48	Dhawan et al. [42]
Polyaniline-coated silica fabrics	Volume resistivity 10–28 Ω cm	101 GHz	35.61	Dhawan et al. [42]
Co-weaved/knitted fabrics	–	100 kHz–1 GHz	40–50	Chen et al. [22]
Metalized nylon	Surface resistivity 0.09 Ω/ϒ	800 MHz	67.04	Avloni et al. [6]
Twill PPy coated	Surface resistivity 40 Ω/ϒ	800 MHz	16.8	Avloni et al. [6]
PPy/chitosan composite film	Linear resistivity 0.0145 Ω cm^{-1}	1200 MHz	32.15–35.66	Abdi et al. [2]
PPy blended hot melt adhesives	Linear resistivity 0.2 Ω cm^{-1}	300 MHz	30	Pomposo et al. [146]
Boron and carbon fabrics	–	1200 MHz	30	Mistik et al. [130]

6 Multi-scale Reinforcements and Composites

Like all other materials, nanotechnology has been extensively applied in composite materials with the aim to overcome some existing drawbacks or to introduce some special features. Laminated composites are the most widely used type of composite materials which are applied in various advanced applications including aerospace, automobiles, civil engineering, etc. Common problems of these composites are the poor z-directional properties and delamination between the different layers under loading conditions. The approaches that have been explored to overcome these problems are ply stitching or the use of 3 dimensional fabrics in which all the fabric

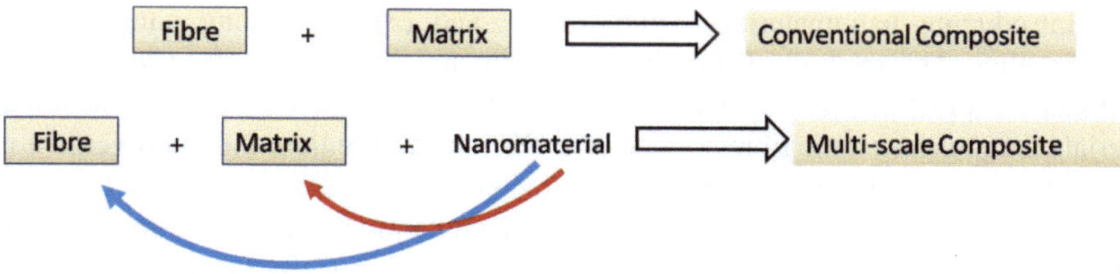

Fig. 38 Concept of developing multi-scale composites

layers are integrated using through-the-thickness yarns. However, these methods usually lead to reduced properties in in-plane direction. The most recent approach to solve this ever existing problem of laminated composites is to apply nanomaterials between the different plies of composites. The deposition of nanofibres and nanotubes between the plies could greatly enhance the inter-laminar shear strength of composites [80, 195]. Further, techniques to grow nanotubes (CNT) vertically between the plies have also been developed in order to stich the different layers with nanotubes [52]. Nanostructures, specially nanofibres and nanotubes have nano-diameters, high aspect ratio, huge surface area and extraordinary mechanical properties and therefore, stitching plies with nanomaterials is far more effective than conventional stitching methods. Moreover, nanomaterial addition does not affect the in-plane mechanical properties of composites. Besides nanomaterial deposition between plies, they can be also introduced within the surface of the fibres or within the matrix of composites [14, 95, 155, 156, 158–161, 197]. In both cases, significant improvement in the inter-laminar shear strength has been noticed, along with improvements in in-plane mechanical strength and stiffness and toughness Moreover, when nanomaterials like CNTs are introduced, they composite possesses some additional special features such as electrical and thermal conductivity, self-sensing behaviour, electromagnetic shielding and so on. These special features are highly important for advanced technical applications.

The incorporation of nanomaterials within conventional composites originated a special class of composite, known as multi-scale composites. Similarly, when nanomaterials are introduced within the fibres prior to composite reinforcement, the fibre systems are known as multi-scale reinforcements. These materials are called multi-scale as they are composed of materials from two different length scales. So, these composites are a special class of hybrid composites and sometimes, also called hierarchical composites. The concept of multi-scale composite is presented in Fig. 38.

Although different types of nanomaterials have been explored to develop multi-scale composites such as nano TiO_2, nano Al_2O_3, nano clay, polymeric nanofibres, CNF, CNT, etc., most of the research studies have been focused on CNF and CNT due to their exceptional mechanical, thermal and electrical properties. However, these nanomaterials with remarkable properties also led to a number of processing difficulties, mainly their agglomeration problems.

As discussed above, multi-scale composites can be developed by mainly three ways: (1) introducing nanomaterials within fibres (i.e. multi-scale reinforcement),

(2) introducing nanomaterials within matrix and (3) introducing nanomaterials within plies. The method of introducing nanomaterials on the surface of the fibres avoids the difficulty in dispersing them within the matrix. However, as nanomaterials are present on the fibre surface or at the interface region, the improvement in the matrix dominated properties such as fracture toughness is less likely to occur. One of the most studied approaches to introduce nanotubes on the fibre surface is through their growth directly on the fibre surface using CVD technique. For this purpose, the fibre substrate is first impregnated into the catalyst solution prior to introducing into the CVD chamber. A hydrocarbon gas is then passed through the CVD reactor maintained at very high temperature ($\sim$1000 °C), which breaks down the hydrocarbon gas and initiates the growth of CNTs on the fibre substrate containing catalyst. CNTs can also be grown vertically aligned through this process. One major problem of this process is the high temperature, which the conventional polymeric fibres do not sustain. So, another approach to avoid this problem is to first grow the nanotubes on a metal fabric such as alumina cloth, next to windup the cloth on a roller which, subsequently, transfers the nanotubes on a tacky fibre prepreg through application of pressure (this process is known as transfer printing). Coating of fibre surface with nanomaterial solution or simple spraying of the nanomaterial solution on to the fibre surface are other simple approaches followed for developing multi-scale reinforcements. Electrophoretic deposition of CNTs on to the carbon fibre surface has also been reported. In this process functionalized CNTs with negative charge are deposited on to carbon fibres which are used as the positive electrodes in the electrophoresis process. Grafting chemically functionalized CNTs on to surface activated carbon fibres is another reported technique.

The other most common approach of fabricating multi-scale composite is through dispersion of nanomaterials within the composite matrix. This process is very challenging due to very high agglomeration tendency of nanomaterials, especially for CNTs [139, 155, 157]. A number of physical and chemical techniques have been used for dispersing nanomaterials in various matrices. The most common physical approach is the ultrasonication technique, which creates a number sock waves in the nanomaterial solution, separating the nanotubes and distributing within the matrix. Further, the use of surfactants assists in separating the nanomaterials and the stabilization of their dispersion through steric/electrostatic stabilization mechanisms. Other mechanical techniques which are used to disperse nanomaterials within thermosetting matrices are high speed mechanical stirring, calendaring, ball milling, etc. which have been also used, sometimes, in combination with ultrasonication to reduce the period of ultrasonic treatment and associated problems (nanomaterial damage, process time and cost). In case of thermoplastic polymers, the efficient route of dispersing nanomaterials is through twin screw extrusion process.

The process of fabricating multi-scale composites is similar to the conventional composites, once the nanomaterials have been incorporated within the fibre or matrix system. Commonly, vacuum assisted resin transfer moulding (VARTM) and compression moulding are used. However, the composite fabrication system should not allow much resin flow, as it favours nanomaterial re-agglomeration.

Multi-scale composites possess enhanced mechanical properties as compared to the conventional composites. Improvements in in-plane mechanical strength and stiffness, inter-laminar shear strength, fracture toughness, impact performance, fatigue resistance, dynamic mechanical performance, etc. have been reported. The main factors responsible for enhanced mechanical performance of multi-scale composites are improved fibre/matrix interface, reinforcing effect as well as other energy absorbing effects of nanomaterials such as crack-bridging, nanomaterial pull-outs, crack pinning, etc. Besides mechanical properties, some nanomaterials with excellent thermal and electrical conductivity like CNT and CNF can produce conducting composite materials. Electromagnetic shielding, piezoresistivity and sensing behaviour are other important benefits of multi-scale composites. Table 3 lists some recently developed multi-scale composites and their various properties.

Table 3 Different types of multi-scale composites and their properties

Type of nano filler and concentration	Fibre/matrix and composite fabrication method	Property improvement
Nano Al_2O_3 10 vol% [77]	Carbon fibre/epoxy, Filament winding technique	Young's modulus, flexural strength, interlaminar shear strength and fracture toughness improved significantly
Nanoclay 1–3 wt% [33]	Carbon fabric/epoxy, Vacuum assisted resin infusion molding	Maximum improvement of 9.3 and 13.7 % in flexural modulus and strength, 52.4 % in storage modulus (35 °C), 30.2 % in shear strength at 2 wt% nanoclay
Graphite nanoplatelets 3 and 5 wt% [30]	Carbon fibre/epoxy, vacuum assisted wet layup	Maximum improvement of 18 and 11 % in in-plane shear modulus and strength and 16 % in longitudinal compressive strength at 5 wt%
Amino functionalized DWCNT, 0.1 and 0.3 wt%, dispersed in resin [64]	Glass fabric/epoxy, Resin transfer molding	No significant change in Young's modulus and tensile strength. Improvement of 20 % in interlaminar shear strength
MWCNT, thin-MWCNT, amine functionalized double walled CNT 0.5 wt%, dispersed in resin [62]	Carbon fibre/epoxy, Preparation of prepreg in drum winder and laminates in vacuum bag	No major improvement in modulus and strength. Fracture toughness improved by 80 % for MWCNTs and modifying epoxy by compatibilizer
MWCNT functionalized and non-functionalized 1 wt%, dispersed in resin [151]	Glass fibre/epoxy, vacuum assisted resin transfer molding	Improvement of 14 % in tensile strength, 20 % in Young's modulus, 5 % in shear strength
Silane functionalized MWCNTs, 1 wt%, Dispersed in resin [104].	Basalt/epoxy, VARTM	Flexural modulus and strength increased by approximately 54 and 34 %, respectively

(continued)

Table 3 (continued)

Type of nano filler and concentration	Fibre/matrix and composite fabrication method	Property improvement
Silane and acid functionalized MWCNTs, 1 wt% [103]	Basalt/epoxy, autoclave processing	Flexural modulus, strength and fracture toughness of silane-treated CNT based composites were 10, 14 and 40 % greater than those of acid-treated CNT based composites
DWCNT-NH_2, 0.025–0.1 wt%, dispersed in resin [55]	Carbon fibre/epoxy, Vacuum Infusion Technique	Enhancement in flexural modulus by up to 35, 5 % improvement in flexural strength, 6 % improvement in absorbed impact energy, and 23 % decrease in the mode I interlaminar toughness
Amine functionalized MWCNTs, 1 wt%, dispersed in resin [172]	Carbon fibre/epoxy, Hand Layup-Vacuum bag processing	Increase in Young's modulus, inter-laminar shear strength, and flexural modulus by 51.46, 39.62, and 38.04 %, respectively
SWNT, 0.1 wt%, dispersed in matrix [155]	Carbon fabric/epoxy, Hand layup- compression moulding	Improvements of 95 % in Young's modulus, 31 % in tensile strength, 76 % in compressive modulus and 41 % in compressive strength
SWNT, 0.1 wt%, sprayed on to fibres in midplane ply [80]	Glass/vinyl ester, VARTM	Up to 45 % increase in shear strength over control samples
Vertically aligned CNT on prepreg surface for ply stitching, 1 vol%, transfer printing method [52]	Carbon fibre/epoxy, Autoclave processing	Increase in fracture toughness by 1.5–2.5 times in Mode I, and 3 times in Mode II
MWCNT, 1–2 vol% in composite, grown on surface of the fibres [180]	Alumina fibre/epoxy, Vacuum bag processing	Improvement of steady-state toughness by 76 %, in-plane tension-bearing stiffness by 19 %, critical strength by 9 %, and ultimate strength by 5 %
CNT, 1–3 vol% in composite, grown on fibre surface [60]	Alumina fibre/epoxy, Vacuum bag processing	Enhancement of 69 % in interlaminar shear strength
MWCNTs, 0.25 wt% on fibre surface, electrophoretic deposition [12]	Carbon fibre/epoxy, VARTM technique	Enhancement of the interlaminar shear strength by 27 %

7 Conclusions

The initial development of the composite was a result of the quest for a strong at the same time light material. However, with change and advancement of technology, the material requirements have also changed drastically. This has lead to the development of a new breed of composite having various functional properties. These new composites with special properties have potential for many critical and demanding applications. Some of these special composites have already become a reality and are being used commercially either in high end or routine applications, but many are still in growing stage. The advent of nano materials and nano technology has opened up new vistas for the development of composites with special properties. The challenge is in the selection and processing of these materials to tailor make a specialised composite structure. The biggest challenge is the technology transfer and commercialisation of process to benefit the mankind and society at large.

A dream is the bearer of a new possibility, the enlarged horizon, the great hope
—Howard Thurman

References

1. Abbasi AMR, Militky J (2013) EMI shielding effectiveness of polypyrrole coated glass fabric. J Chem Chem Eng 7:256–259
2. Abdi MM, Kassim AB, Mahmud H et al (2009) Electromagnetic interference shielding effectiveness of new conducting polymer composite. J Macrom Sci, Part A: Pure Appl Chem 47:71–75
3. Abyaneh MK, Ekar S, Kulkarni SK (2012) Piezoresistivity and mechanical behavior of metal-polymer composites under uniaxial pressure. J Mater Sci Res 1:50–58
4. Aghaei M, Hanum Y, Thayoob M, et al (2012) A review on the impact of the electromagnetic radiation (EMR) on the human's health. In: Proceedings National graduate conference (NatGrad2012), Universiti Tenaga Nasional, Putrajaya Campus, 8–10 Nov 2012
5. Aıssa B, Tagziria K, Haddad E, Jamroz W, Loiseau J, Higgins A (2012) The self healing capability of carbon fibre composite structures subjected to hypervelocity impacts simulating orbital space debris. Int Schol Res Net ISRN Nanomater Article ID 351205:1–16. doi:10.5402/2012/351205
6. Avloni J, Lau R, Ouyang M, Florio L, Henn AR, Sparavigna A (2008) Polypyrrol e-coated nonwovens for electromagnetic shielding. J Ind Tex 38:55–68
7. Avloni J, Ouyang M, Florio L, Henn AR, Sparavigna A (2007) Shielding effectiveness evaluation of metallized and poly pyrrole-coated fabrics. J Therm Comp Mater 20:241–254
8. Baeza FJ, Galao O, Zornoza E, Garcés P (2013) Multifunctional cement composites strain and damage sensors applied on reinforced concrete (RC) structural elements. Materials 6:841–855
9. Ballou JW (1954) Static electricity in textiles, tex. Res J 24:146–155. doi:10.1177/004051755402400209
10. Baltopoulos A, Athanasopoulos N, Fotiou I, Vavouliotis A, Kostopoulos V (2013) Sensing strain and damage in polyurethane-MWCNT nano-composite foams using electrical measurements. eXPRESS Polym Lett 7:40–54. doi:10.3144/expresspolymlett.2013.4
11. Bayerl T, Duhovic M, Mitschang P, Bhattacharyya D (2014) The heating of polymer composites by electromagnetic induction—A review. Compos Part A: Appl Sci Manuf 57:27–40. doi:10.1016/j.compositesa.2013.10.024
12. Bekyarova E, Thostenson ET, Yu A, Kim H, Gao J, Tang J, Hahn HT, Chou TW, Itkis ME, Haddon RC (2007) Multiscale carbon nanotube-carbon fiber reinforcement for advanced epoxy composites. Langmuir 23(7):3970–3974
13. Bertuleit K (1991) Silver coated polyamide: a conductive fabric. J Indus Tex 20:211–215. doi:10.1177/152808379102000307

14. Bhattacharyya A, Rana S, Parveen S, Fangueiro R, Alagirusamy R, Joshi M (2013) Mechanical and thermal transmission properties of carbon nanofibre dispersed carbon/phenolic multi-scale composites. J Appl Polym Sci 129:2383–2392
15. Bleha M, Kudela V, Rosova E et al (1999) Synthesis and characterization of thin polypyrrole layers on polyethylene microporous films. Eur Polym J 35:613–620
16. Boger L, Viets C, Wichmann MHG, Schulte K (2009) Glass fibre reinforced composites with a carbon nanotube modified epoxy matrix as self sensing material. In: Proceedings of the 7th International workshop on structural health monitoring, Stanford, CA, USA, pp 973–978
17. Boiko YM, Guérin G, Marikhin VA, Prud'homme RE (2001) Healing of interfaces of amorphous and semi-crystalline poly(ethylene terephthalate) in the vicinity of the glass transition temperature. Polymer 42:8695–8702
18. Bonaldi RR, Siores E, Shah T (2010) Electromagnetic shielding characterisation of several conductive fabrics for medical applications. J Fiber Bioeng Inf 2:237–245. doi:10.3993/jfbi03201006
19. Bond IP, Trask RS, Williams HR (2008) self healing fiber-reinforced polymer composites. MRS Bull 33:770–774. doi:10.1557/mrs2008.164
20. Boschi A, Arosio C, Cucchi I et al (2008) Properties and performance of polypyrrole (PPy)-coated silk fibers. Fiber Polym 9:698–707
21. Boutrois JP, Jolly R, Petrescu C (1997) Process of polypyrrole deposit on textile: product characteristics and applications. Synth Met 85:1405–1406
22. Chen HC, Lee KC, Lin JH et al (2007) Comparison of electromagnetic shielding effectiveness properties of diverse conductive textiles via various measurement techniques. J Mater Process Technol 192–193:549–554
23. Chen HS, Lee KC, Lin JH, Koch M (2007) Fabrication of conductive woven fabric and analysis of electromagnetic shielding via measurement and empirical equation. J Mater Process Technol 184:124–130
24. Chen PW, Chung DDL (1995) Improving the electrical conductivity of composites comprised of short conductive fibres in a non-conducting matrix: the addition of nonconductive particulate filler. J Elecrtochem Mater 24:47–52
25. Chen X, Dam MA, Ono K, Mal A, Shen HB, Nutt SR, Sheran K, Wudl F (2002) A thermally re-mendable cross-linked polymeric material. Science 295:1698–1702
26. Chen X, Wudl F, Mal AK, Shen H, Nutt SR (2003) New thermally remendable highly cross-linked polymeric materials. Macromolecules 36:1802–1807
27. Cheng KB, Lee ML, Ramakrishna S (2001) Electromagnetic shielding effectiveness of stainless steel/polyester woven fabrics. Text Res J 7:42–49. doi:10.1177/004051750107100107
28. Cheng KB, Ramakrishna S, Lee KC (2000) Electromagnetic shielding effectiveness of copper/glass fiber knitted fabric reinforced polypropylene composites. Compos A 31:1039–1045
29. Chipara M, Wooley K (2005) Molecular self healing processes in polymers. Mater Res Soc Symp Proc 851:127–132
30. Cho H, Tabata I, Hisada K, Hirogaki K, Hori T (2013) Characterization of copper-plated conductive fibers after pretreatment with supercritical carbon dioxide and surface modification using Lyocell fiber. Tex Res J 83:780–793. doi:10.1177/0040517512467130
31. Cho JW, Jung H (1997) Electrically conducting high-strength aramid composite fibres prepared by vapour-phase polymerization of pyrrole. J Mater Sci 32:5371–5376
32. Cho J, Chen JY, Daniel IM (2007) Mechanical enhancement of carbon fiber/epoxy composites by graphite nanoplatelet reinforcement. Scripta Mater 56(8):685–688
33. Chowdhury FH, Hosur MV, Jeelani S (2006) Studies on the flexural and thermomechanical properties of woven carbon/nanoclay-epoxy laminates. Mater Sci Eng, A 421(1–2):298–306
34. Chung CM, Roh YS, Cho SY, Kim JG (2004) Crack healing in polymeric materials via photochemical [2 + 2] cycloaddition. Chem Mater 16:3982–3984
35. Chung DDL (2002) Electrical conduction behavior of cement–matrix composites. J Mater Eng Perform 11:194–204

36. Coates GW, Hustad PD, Reinartz S (2002) Catalysts for the living insertion polymerization of alkenes: access to new polyolefin architectures using Ziegler- Natta chemistry. Ang Chem Int 41:2236–2257
37. Cochrane C, Koncar V, Lewandowski M, Dufour C (2007) Design and development of a flexible strain sensor for textile structures based on a conductive polymer composite. Sensors 7:473–492
38. Coppola L, Buoso A, Corazza F (2011) Electrical properties of carbon nanotubes cement composites for monitoring stress conditions in cement structure. Appl Mech Mater 82:118–123. doi:10.4028/www.scientific.net/AMM.82.118
39. Dall'Acqua L, Tonin C, Varesano A, Canetti M, Porzio W, Catellani M (2006) Vapour phase polymerisation of pyrrole on cellulose-based textile substrates. Synth Met 156:379–386
40. De Gennes PG (1971) Reptation of a polymer chain in the presence of fixed obstacles. J Chem Phys 55:572–579
41. Dharap P, Li Z, Nagarajaiah S, Barrera EV (2004) Nanotube film based on single-wall carbon nanotubes for strain sensing. Nanotechnology 15:379–382
42. Dhawan SK, Singh N, Venkatachalam S (2002) Shielding behaviour of conducting polymer-coated fabrics in X-band, W-band and radio frequency range. Synth Met 129:261–267
43. Dietzel Y, Przyborowski W, Nocke G et al (2000) Investigation of PVD arc coatings on polyamide fabrics. Surf Coat Technol 135:75–81
44. Doi M, Edwards SF (1978) Dynamics of concentrated polymer systems. Faradays Transactions, J Chem Soc
45. Dry CM (1991) Alteration of matrix permeability and associated pore and crack structure by timed release of internal chemicals. Ceram Trans 16:729–768
46. Dry CM (1992) Passive tunable fibers and matrices Int J Mod Phys B 6:2763–2771
47. Dry CM (1993) Passive smart materials for sensing and actuation. J Intell Mater Sys Struct 4:420–425
48. Dry CM (1996) Procedures developed for self-repair of polymer matrix composite materials. Compos Struct 35:263–269
49. Easter MR (2005) Self healing cables. (Individual U) US: 2005136257-A1
50. Ehsan MN, Zaman MM, Mahabubuzzaman AKM (2010) enrichment of self healing material and advanced composite structures. J Innov Dev Strategy 4:28–32
51. Eisenberg A, Rinaudo M (1990) Polyelectrolytes and ionomers. Polym Bull 24:671
52. Enrique GJ, Wardle BL, Hart AJ (2008) Joining prepreg composite interfaces with aligned carbon nanotubes. Compos A Appl Sci Manuf 39(6):1065–1070
53. Erdoğan MK, Karakişla M, Saçak M (2012) Preparation, characterization and electromagnetic shielding effectiveness of conductive polythiophene/poly(ethterephthalate) composite fibers. J Macromol Sci Part A Pure Appl Chem 49:473–482
54. Esfandiari A (2008) PPy covered cellulosic and protein fibres using novel covering methods to improve the electrical property. World Appl Sci J 3:470–475
55. Fawad I, Wong DWY, Kuwata M, Peijs T (2010) Multiscale hybrid micro-nanocomposites based on carbon nanotubes and carbon fibers. J Nanomater 2010:12
56. Fink BK, McCullough RL, Gillespie WJ (1992) A local theory of heating in cross-ply carbon fiber thermoplastic composites by magnetic induction. Polym Eng Sci 32:357–369. doi:10.1002/pen.760320509
57. Fischer H (2010) Self-repairing material systems-a dream or a reality? Nat Sci 2:873–901. doi:10.4236/ns.2010.28110
58. Florio L, Sparavigna A (2004) Textiles for electromagnetic shielding, International Conference on Condensed Matter Physics, Genova, Abstract book, p 193
59. Gandhi OP (2002) Electromagnetic fields: human safety issues. Annu Rev Biomed Eng 4:211–234
60. Garcia EJ, Wardle BL, Hart AJ, Yamamoto N (2008) Fabrication and multifunctional properties of a hybrid laminate with aligned carbon nanotubes grown in situ. Compos Sci Technol 68(9):2034–2041

61. Ghorbel I, Akele N, Thominette F, Spiteri P, Verdu J (1995) Hydrolytic aging of polycarbonate. II. Hydrolysis kinetics, effect of static stresses. J Appl Polym Sci 55:173–179
62. Godara A, Mezzo L, Luizi F, Warrier A, Lomov SV, Van Vuure AW, Gorbatikh L, Moldenaers P, Verpoest I (2009) Influence of carbon nanotube reinforcement on the processing and the mechanical behaviour of carbon fiber/epoxy composites. Carbon 47(12):2914–2923
63. Goethals EJ, Du Prez F (2007) Carbocationic polymerizations. Prog Polym Sci 32:220–246
64. Gojny FH, Wichmann MHG, Fiedler B, Bauhofer W, Schulte K (2005) Influence of nano-modification on the mechanical and electrical properties of conventional fibre-reinforced composites. Compos Part A 36(11):1525–1535
65. Gupta N, Sharma S, Mir IA, Kumar D (2006) Advances in sensors based on conductive polymers. J Sci Ind Res 65:549–557
66. Håkansson E, Amiet A, Kaynak A (2006) Electromagnetic shielding properties of polypyrrole/polyester composites in the 1–18 GHz frequency range. Synth Met 156:917–925
67. Håkansson E, Amiet A, Nahavandi S, Kaynak A (2007) Electromagnetic interference shielding and radiation absorption in thin polypyrrole films. Europ Polym J 43(1):205–213
68. Hakansson E, Kaynak A, Lin T et al (2004) Characterization of conducting polymer coated synthetic fabrics for heat generation. Synth Met 144:21
69. Hamdani STA, Potluri P, Fernando A (2013) Thermo-mechanical behavior of textile heating fabric based on silver coated polymeric yarn. Materials 6:1072–1089. doi:10.3390/ma6031072
70. Han EG, Kim EA, Oh KW (2001) Electromagnetic interference shielding effectiveness of elecroless Cu-plated PET fabric. Synth Met 123:469–476
71. Harreld JH, Wong MS, Hansma PK, Morse DE, Stucky GD (2004) self healing organosiloxane materials containing reversable and energy-dispersive crosslinking domains. (University of California U) US patent: 2004007792-A1
72. Harris KM, Rajagopalan M (2003) Self healing polymers in sports equipment. (Acushnet Company U) US:2003032758-A1
73. Hasegawa M, Katsumata T, Ito Y, Saigo K, Iitaka Y (1988) Topochemical photoreactions of unsymmetrically substituted diolefins. 2. Photopolymerization of 4-(Alkoxycarbonyl)-2,5-distyrylpyrazines. Macromolecules 21:3134–3138
74. Homma D, Mihashi H, Nishiwaki T (2009) Self healing capability of fibre reinforced cementitious composites. J Adv Conc Tech 7:217–228
75. Hong YK, Lee CY, Jeong CK et al (2001) Electromagnetic interference shielding characteristics of fabric complexes coated with conductive polypyrrole and thermally evaporated Ag. Curr Appl Phys 1:439–442
76. Hou C, Huang T, Wang H, Yu H, Zhang Q, Li Y (2013) A strong and stretchable self healing film with self-activated pressure sensitivity for potential artificial skin applications. Sci Rep 3:3138–3144. doi:10.1038/srep03138
77. Hussain M, Nakahira A, Niihara K (1996) Mechanical property improvement of carbon fiber reinforced epoxy composites by Al_2O_3 filler dispersion. Mater Lett 26(3):185–191
78. Ichkitidze L, Podgaetsky V, Selishchev S, Blagov E, Galperin V, Shaman Y, Pavlov A, Kitsyuk E (2013) Electrically-conductive composite nanomaterial with multi-walled carbon nanotubes. Mater Sci Appl 4:1–7
79. Imaizumi K, Ohba T, Ikeda Y, Takeda K (2001) Self-repairing mechanism of polymer composite. Mater Sci Res Int (Japan) 7:249–253
80. Jiang Z, Imam A, Crane R, Lozano K, Khabashesku VN, Barrera EV (2007) Processing a glass fiber reinforced vinyl ester composite with nanotube enhancement of interlaminar shear strength. Compos Sci Technol 67(7–8):1509–1517
81. Johnson O, Gardner C, Seegmiller D, Mara N et al (2011) Multiscale model for the extreme piezoresistivity in silicone/nickel nanostrand nanocomposites. Metall Mater Trans A 42(13):3898–3906

82. Johnson OK, Gardner CJ, Fullwood DT, Adams BL, Hansen G, Hansen N (2010) The colossal piezoresistive effect in nickel nanostrand polymer composites and a quantum tunneling model. Comput Mater Continua 15(2):24
83. Johnson TM, Fullwood DT, Hansen G (2013) Strain monitoring of carbon fiber composite via embedded nickel nano-particles. Compos B Eng 43:1155–1163
84. Jones F, Hayes SA (2005) self healing composite material. (University of Sheffeld G) WO:2005066244-A2
85. Jud K, Kausch HH (1979) Load transfer through chain molecules after interpenetration at interfaces. Polym Bull 1:697–707
86. Jud K, Kausch HH, Williams JG (1981) Fracture-mechanics studies of crack healing and welding of polymers. J Mater Sci 16:204–210
87. Kalantari M, Dargahi J, Kövecses J, Ghanbari M, Nouri S (2012) A new approach for modeling piezoresistive force sensors based on semiconductive polymer composites. IEEE/ASME Trans Mechatron 17(3):572–581
88. Kalista SJ, Ward TC (2007) Thermal characteristics of the self healing response in poly (ethylene-co-methacrylic acid) copolymers. R Soc Interface 4:405–411
89. Kalista SJ, Ward TC, Oyetunji Z (2007) Self-healing of poly (ethylene-comethacrylic acid) opolymers following projectile puncture. Mech Adv Mater Struct 14:391–397
90. Katunin AW, Kostka Hufenbach P, Holeczek K (2010) Frequency dependence of the self-heating effect in polymer-based composites. J Achieve Mater Manuf Eng 41:9–15
91. Kausch HH, Jud K (1982) Molecular aspects of crack formation and healing in glassy polymers. Rubber Process Appl 2:265–268
92. Kaynak A, Foitzik R (2011) Methods of coating textiles with soluble conducting polymers. Res J Tex Appar 15:107–113
93. Kaynak A, Najar SS, Foitzik RC (2008) Conducting nylon, cotton and wool yarns by continuous vapor polymerization of pyrrole. Synth Met 158:1–5
94. Kaynak A, Håkansson E (2005) Generating heat from conducting polypyrrole-coated PET fabrics. Adv PolymTechnol 24:194–207
95. Khan S, Kim JK (2011) Impact and delamination failure of multiscale carbon nanotube-fiber reinforced polymer composites: a review. Int J Space Sci 12(2):115–133
96. Kim HK, Byun SW, Jeong SH, Lee et al (2002) Environmental Staility of EMI Shielding PET FabricPolypyrrole Composite. Mol Cryst Liq Cryst 377:369–372
97. Kim HK, Kim MS, Chun SY, Park YH, Jeon BS, Lee JY, Hong YK, Joo J, Kim SH (2003) Characteristics of electrically conducting polymer-coated textiles. Mol Cryst Liq Cryst 405 (1):161–169
98. Kim J, Sohn D, Sung Y, Kim E (2003) Fabrication and characterization of conductive polypyrrole thin film prepared by in situ vapor phase polymerization. Synth Met 132:309–313
99. Kim KH, Kim MS, Song K et al (2003) EMI shielding intrinsically conductive polymer/PET textile composites. Synth Met 135:105–106. doi:10.1016/S0379-6779(02)00876-7
100. Kim MS, Kim HK, Byun SW et al (2002) PET fabric/polypyrrole composite with high electrical conductivity for EMI shielding. Synth Met 126:233–239
101. Kim SH, Jang SH, Byun SW, Lee JY, Joo JS, Jeong SH, Park MJ (2003) Electrical properties and EMI shielding characteristics of polypyrrole-nylon 6 composite fabrics. J Appl Polym Sci 87:1969–1974. doi:10.1002/app.11566
102. Kim YH, Wool RP (1983) A theory of healing at a polymer-polymer interface. Macromolecules 16:1115–1120
103. Kim MT, Rhee KY, Park SJ, Hui D (2012) Effects of silane-modified carbon nanotubes on flexural and fracture behaviors of carbon nanotube-modified epoxy/basalt composites. Compos B Eng 43(5):2298–2302
104. Kim MT, Rhee KY (2011) Flexural behavior of carbon nanotube-modified epoxy/basalt composites. Carbon Lett 12(3):177–179
105. King RWP (2000) Electrical currents and fields induced in cells in the human brain by radiation from hand-held cellular telephones. J Appl Phys 87:893–900

106. Knittel D, Schollmeyer E (2009) electrically high-conductive textiles. Synth Met 159:1433–1437. doi:10.1016/j.synthmet.2009.03.021
107. Koecher MC, Pande JH, Merkley S, Henderson S, Fullwood DT, Bowden AE (2015) Piezoresistive in-situ strain sensing of composite laminate structures. Compos Part B 69:534–541. doi:10.1016/j.compositesb.2014.09.029
108. Koprowska J, Pietranik M, Stawski W (2004) New type of textiles with shielding properties. Fiber Tex Eastern Eur 12(3):47
109. Kuhn HH, Child AD, Kimbrell WC (1995) Toward real applications of conductive polymers. Synth Met 71:2139–2142
110. Kumar A, Stephenson LD (2006) Self healing coatings using microcapsules. (Individual U) US: 2006042504-A1
111. Kumar MN, Thilagavathi G (2012) Surface resistivity and EMI shielding effectiveness of polyaniline coated polyester fabric. J Tex Appar Tech Manag 7:1–6
112. Lak A (2012) Human health effects from radiofrequency and microwave fields. J Basic Appl Sci Res 2:12302–12305
113. Lakshmi K, John H, Mathew KT et al (2009) Microwave absorption, reflection and EMI shielding of PU–PANI composite. Acta Mater 57:371–375
114. Lao LL, Ramanujan RV (2004) Magnetic and hydrogel composite materials for hypothermia application. J Mater Sci: Mater Med 15:1061–1064
115. Lee CY, Lee DE, Jeong CK et al (2002) Electromagnetic interference shielding by using conductive polypyrrole and metal compound coated fabrics. Polym Adv Technol 13:577–583. doi:10.1002/pat.227
116. Lee CY, Lee J, Joo MS, Kim JY et al (2001) Conductivity and EMI shielding of polypyrrole and metal compounds coated on nonwoven fabrics. Synth Met 119:429–430
117. Lee JY (2003) Polypyrrole-coated woven fabric as a flexible surface-heating element. Macromol Res 11:481–487
118. LePing L, Wei Z, Yi X, HongMei W, Yang Z, WuJun L (2011) Preparation and characterization of microcapsule containing epoxy resin and its self-healing performance of anticorrosion covering material. Chinese Sci Bull 56:439–443. doi:10.1007/s11434-010-4133-0
119. Levin ZS, Robert C, Feller JF, Castro M, Grunlan JC (2013) Flexible latex—polyaniline segregated network composite coating capable of measuring large strain on epoxy. Smart Mater Struct 22:015008–015009. doi:10.1088/0964-1726/22/1/015008
120. Lin CB, Lee SB, Liu KS (1990) Methanol-induced crack healing in poly(methyl methacrylate). Polym Eng Sci 30:1399–1406
121. Lin VWJ, Lia M, Lynch JP, Li VC (2011) Mechanical and electrical characterization of self-sensing carbon black ECC. Proc SPIE 7983:1–12. doi:10.1117/12.880178
122. Loh KJ, Lynch JP, Kotov NA (2008) Inductively coupled nanocomposite wireless strain and pH sensors. Smart Struct Syst 4:531–548
123. Macasaquit AC, Binag CA (2010) Preparation of conducting polyester textile by in situ polymerization of pyrrole. Philippine J Sci 139:189–196
124. Mahajan A, Singh M (2012) Human health and electromagnetic radiations. Inter J Eng Innov Technol 1:95–97
125. Maity S, Chatterjee A (2013) Preparation and characterization of electro-conductive rotor yarn by in situ chemical polymerization of pyrrole. Fibre Polym 14(8):1407–1413. doi:10.1007/s12221-013-1407-6
126. Maity S, Chatterjee A, Singh B, Singh AP (2014) Polypyrrole based elelctro-conductive textiles for heat generation. J Tex Inst 105(8):887–893. doi:10.1080/00405000.2013.861149
127. Maity S, Singha K, Debnath P, Singha M (2013) Textiles in electromagnetic Radiation Protection. J Safety Eng 2:11–19. doi:10.5923/j.safety.20130202.01
128. Malhotra U, Maity S, Chatterjee A (2015) Polypyrrole-silk electro-conductive composite fabric by in situ chemical polymerization. J Appl Polym Sci 132(4):41336. doi:10.1002/app.41336

129. McNeill IC, Rincon A (1993) Thermal-degradation of polycarbonates—reaction conditions and reaction-mechanisms. Polym Degrad Stabil 39:13–19
130. Mistik SI, Sancak E, Usta IE et al (2012) Investigation of electromagnetic shielding properties of boron and carbon fibre woven fabrics and their polymer composites, In: RMUTP International conference: textiles & fashion, Bangkok, Thailand, 3–4 July 2012
131. Miyasaka K (1986) Mechanism of electrical conduction in electrically-conductive filler-polymer composites. Int Polym Sci Technol 13:41–48
132. Muthukumar N, Thilagavathi G (2012) Development and characterization of electrically conductive polyaniline coated fabrics. Indian J Chem Technol 19:423–441
133. Najar SS, Kaynak A, Foitzik RC (2007) Conductive wool yarns by continuous vapour phase polymerization of pyrrole. Synth Met 157:1–4
134. Nauman S, Cristian I, Koncar V (2011) Simultaneous application of fibrous piezoresistive sensors for compression and traction detection in glass laminate composites. Sensors 11:9478–9498. doi:10.3390/s111009478
135. Negru D, Buda CT, Avram D (2012) Electrical conductivity of woven fabrics coated with carbon black particles. Fiber Tex Eastern Eur 20:53–56
136. Paczkowski J (1996) Polymeric Materials Encyclopedia. In: Salamone JC. Boca Raton, FL: CRC Press, p. 5142
137. Palamutcu S, Özek A, Karpuz C, Dağ N (2010) Electrically conductive textile surfaces and their electromagnetic shielding efficiency measurement. Tekstil Ve Konfeksiyon 3:199–207
138. Pang JWC, Bond IP (2005) A hollow fibre reinforced polymer composite encompassing self-healing and enhanced damage visibility. Compos Sci Technol 65:1791–1799. doi:10.1016/j.compscitech.2005.03.008
139. Parveen S, Rana S, Fangueiro R (2013) A review on nanomaterial dispersion, microstructure and mechanical properties of carbon nanotube and nanofiber based cement composites. J Nanomater 2013(2013):1–19
140. Patel AJ, Sottos NR, Wetzel ED, White SR (2010) Autonomic healing of low-velocity impact damage in fiber-reinforced composites. Compos A 41:360–368. doi:10.1016/j.compositesa.2009.11.002
141. Patil A, Deogaonkar S (2012) A novel method of in situ chemical polymerization of polyaniline for synthesis of electrically conductive cotton fabrics. Tex Res J 82:1517–1530
142. Patil SJ, Duragkar N, Rao VR (2014) An ultra-sensitive piezoresistive polymer nano-composite microcantilever sensor electronic nose platform for explosivevapor detection. Sens Actuators B 192:444–451. doi:10.1016/j.snb.2013.10.111
143. Perumalraj R, Dasaradan BS (2011) Ellectroless nikel plated composite textile material for electromagnet compatibility. Indian J Fibre Text Res 36:35–41
144. Perumalraj R, Nalankilli G, Balasaravanan TR et al (2010) Electromagnetic shielding tester for conductive textile materials. Indian J Fibre Tex Res 35:361–365
145. Pingkarawat K, Bhat T, Craze DA, Wang CH, Varley RJ (2013) Mouritz A P (2013) Healing of carbon fibre–epoxy composites using thermoplastic additives. Polym Chem 4:5007–5015. doi:10.1039/C3PY00459G
146. Pomposo JA, Rodrıguez J, Grande H (1999) Polypyrrole-based conducting hot melt adhesives for EMI shielding applications. Synth Met 104:107–111
147. Power EJ, Dias T (2003) Knitting of Electroconductive Yarns. In: Eurowearable, 2003 The Institution of Electrical Engineers. Birmingham, UK, 4–5 Sept 2003, pp 55–60. doi:10.1049/ic:20030147
148. Prager S, Tirrell M (1981) The healing process at polymer-polymer interfaces. J Chem Phys 75:5194–5198
149. Pryde CA, Hellman MY (1980) Solid state hydrolysis of bisphenol-A polycarbonate I. Effect of phenolic end groups 1980. J Appl Polym Sci 25:2573–2587
150. Qiao Y, Shen L, Dou Y (2010) Polymerization and characterization of high conductivity and good adhesion polypyrrole films for electromagnetic interference shielding. Chin J Polym Sci 28:923–930

151. Qiu J, Zhang C, Wang B, Liang R (2007) Carbon nanotube integrated multifunctional multiscale composites. Nanotechnology 18(27):5708
152. Raghavan A, Kessler SS, Dunn CT, Barber D, Wicks S, Wardle BL (2009) Structural health monitoring using carbon nanotube (CNT) enhanced composites. In: Proceedings of the 7th International workshop on structural health monitoring, Stanford, CA, USA, pp 1034–1041
153. Rainieri C, Fabbrocino G, Song Y, Shanov V (2011) CNT composites for shm: a literature review. In: International workshop smart materials, structures & NDT in aerospace, Quebec, Canada, 2–4 Nov 2011
154. Rana S, Zdraveva E, Rosado K, Patinha S, Cunha F, Fangueiro R (2012) strain and damage sensing behaviour of core reinforced braided composite rods. In: ECCM15—15th European conference on composite materials, Venice, Italy, 24–28 June 2012
155. Rana S, Alagirusamy R, Joshi M (2011) Single-walled carbon nanotube incorporated novel three phase carbon/epoxy composite with enhanced properties. J Nanosci Nanotechnol 11(8):7033–7036
156. Rana S, Alagirusamy R, Joshi M (2009) A review on carbon epoxy nanocomposites. J Reinf Plastics Compos 28:461–487
157. Rana S, Alagirusamy R, Joshi M (2010) Mechanical properties of epoxy reinforced with homogeneously dispersed carbon nanofibre. Int J Plastics Technol 14(2):224–233
158. Rana S, Alagirusamy R, Joshi M (2011) Development of carbon nanofibre incorporated three phase carbon/epoxy composites with enhanced mechanical, electrical and thermal properties. Compos A Appl Sci Manuf 42(5):439–445
159. Rana S, Alagirusamy R, Joshi M (2011) Effect of carbon nanofibre dispersion on the tensile properties of epoxy nanocomposites. J Compos Mater 45(21):2247–2256
160. Rana S, Alagirusamy R, Joshi M (2012) Carbon nanomaterial based three phase multi-functional composites. Lap Lambert Academic Publishing Gmbh & Co, KG, Germany
161. Rana Sohel, Bhattacharyya Amitava, Parveen Shama, Fangueiro Raul, Alagirusamy Ramasamy, Joshi Mangala (2013) Processing and performance of carbon/epoxy multi-scale composites containing carbon nanofibres and single walled carbon nanotubes. J Polym Res 20(12):1–11
162. Ranade R, Zhang J, Lynch JP, Li VC (2014) Influence of micro-cracking on the composite resistivity of engineered cementitious composites. Cement Conc Res 58:1–12
163. Rodriguez J, Otero TF, Grande H, Moliton JP, Moliton A, Trigaud T (1996) Optimization of the electrical conductivity of polypyrrole films electrogenerated on aluminium electrodes. Synth Met 76(1–3):301–303
164. Roh JS, Chi YS, Kang TJ, Nam SW (2008) Electromagnetic shielding effectiveness of multifunctional metal composite fabrics. Tex Res J 78:825–835
165. Rui M, Marco S, Renato G, Gerardo R, Miguel NJ, Helder C, Pedro S, Senentxu LM (2014) Processing and electrical response of fully polymer piezoelectric filaments for e-textiles applications. J Text Eng 60(2):27–34
166. Sanders ML, Rowlands SF, Coombs PG (1998) Self healing UV barrier coating for flexible polymer substrate. (Optical Coating Laboratory Inc U) US: 5790304
167. Schnabel W, Kiwi J (1978) Photodegradation. Jellinek HHG. Degradation and stablization of polymers. Elsevier Science, Amsterdam, pp 195–246
168. Schwarz A, Kazani I, Cuny L et al (2011) Comparative study on the mechanical properties of elastic, electro-conductive hybrid yarns & their input materials. Tex Res J 81:1713–1724. doi:10.1177/0040517511410109
169. Schwarz A, Kazani I, Cuny L et al (2011) Electro-conductive & elastic hybrid yarns—the effects of stretching, cyclic straining & washing on their electro-conductive properties. Mater Des 32:4247–4256. doi:10.1016/j.matdes.2011.04.021
170. Seena V, Fernandes A, Pant P, Mukherji S, Rao VR (2011) Polymer nanocomposite nanomechanical cantilever sensors: material characterization, device development and application in explosive vapour detection. Nanotechnology 22:295501–295511. doi:10.1088/0957-4484/22/29/295501

171. Seshadri DT Bhat NV (2005) Synthesis & properties of cotton fabrics modifies with polypyrrole. BTRA SCAN, 1–8 Dec 2005
172. Sharma K, Shukla M (2014) Three-phase carbon fiber amine functionalized carbon nanotubes epoxy composite: processing, characterisation, and multiscale modeling. J Nanomater 2014(2014):1–10
173. Simon RM, Stutz D (1983) Test methods for shielding materials. EMC Technol 2:39–48
174. Singh AK (2013) Carbon nanotube based nanocomposite for electromagnetic absorption and dymnamic structural strain sensing. Indian J Pure Appl Phys 51:439–443
175. Soyaslan D, Çomlekçi S, Göktepe O (2010) Determination of electromagnetic shielding performance of plain knitting and 1X 1rib structures with coaxial test fixture relating to ASTMD 4935. J Tex Inst 101:890–897
176. Sparavigna AC, Florio L, Avloni J et al (2010) Polypyrrole coated PET fabrics for thermal applications. Mater Sci Appl 1:253–259. doi:10.4236/msa.2010.14037
177. Stampfer C, Helbling T, Obergfell D, Schöberle B, Tripp MK, Jungen A, Roth S, Bright VM, Hierold C (2006) Fabrication of single-walled carbon-nanotube-based pressure sensors. Nano Lett 6:233–237
178. Stankute R, Grinrvicicteiūte D, Gutauskas M et al (2010) Evaluation of electrostatic properties of fiber-forming polymers. Mater Sci (MEDŽIAGOTYRA) 16:72–75
179. Sundaresan VB, Morgan A, Castellucci M (2013) Self-healing of ionomeric polymers with carbon fibers from medium-velocity impact and resistive heating. Smart Mater Res Article ID 271546:12
180. Sunny WS, Villoria RG, Wardle BL (2010) Interlaminar and intralaminar reinforcement of composite laminates with aligned carbon nanotubes. Compos Sci Technol 70(1):20–28
181. Szwarc M (1956) Living polymers. Nature 178:1168–1169
182. Takeda K, Tanahashi M, Unno H (2003) Self-repairing mechanism of plastics. Sci Technol Adv Mater 4:435–444
183. Takeda K, Unno H, Zhang M (2004) Polymer reaction in polycarbonate with Na_2CO_3. J Appl Polym Sci 93:920–926
184. Tan KS, Hinberg I, Wadhwani J (2001) Electromagnetic interference in medical devices: health Canada's past and current perspectives and activities. IEEE Int Symp Electr Comp 2:1283–1288
185. Tanaka H, Tsunawaki K (1981) Electrically conductive fiber and method for producing the same. Patent US4267233A, USA
186. Todoroki A (2010) Self-sensing composites and optimization of composite structures in Japan. Int J Aeronaut Space Sci 11(3):155–166. doi:10.5139/IJASS.2010.11.3.155
187. Todoroki A, Haruyama D, Mizutani Y, Suzuki Y, Yasuoka T (2014) Electrical resistance change of carbon/epoxy composite laminates under cyclic loading under damage initiation limit. Open J Compos Mater 4:22–31. doi:10.4236/ojcm.2014.41003
188. Trask RS, Bond IP (2006) Biomimetic self-healing of advanced composite structures using hollow glass fibres. Smart Mater Struct 15:704–710
189. Trifigny N, Kelly FM, Cochrane C, Boussu F, Koncar V, Soulat D (2013) PEDOT: PSS-based piezo-resistive sensors applied to reinforcement glass fibres for in situ measurement during the composite material weaving process. Sensors 13:10749–10764. doi:10.3390/s130810749
190. Trung VO, Tung DN, Huyen GN (2009) Polypyrrole/Al2O3 nanocomposites: preparation, characyerization and electromagnetic shielding properties. J Experiment Nanosci 4:213–219. doi:10.1080/174580809031
191. Tyagi S, Lee JY, Buxton GA, Balazs AC (2004) Using nanocomposite coatings to heal surface defects. Macromolecules 37:9160–9168
192. Varghese S, Lele A, Mashelkar R (2006) Metal-ion-mediated healing of gels. J Polym Sci Part A-Polym Chem 44:666–670
193. Varghese S, Lele AK, Srinivas D, Mashelkar RA (2001) Role of hydrophobicity on structure of polymer-metal complexes. J Phys Chem B 105:5368–5373

194. Varnaite S (2010) The use of conductive yarns in woven fabric for protection against electrostatic field. Mater Sci (Medžiagotyra) 16:133–137
195. Veedu VP, Cao A, Li X, Ma K, Soldano C, Kar S, Ajayan PM, Nejhad MNG (2006) Multifunctional composites using reinforced laminate with carbon-nanotube forests. Nat Mater 5(6):457–462
196. Volodina AA, Belmesova AA, Murzina VB, Fursikova PV, Zolotarenkob AD, Tarasova BP (2013) Electroconductive composites based on titania and carbon nanotubes. Neorganic Mater 49:702–708
197. Wang JP, Xue P, Tao XM (2011) Strain sensing behavior of electrically conductive fibers under large deformation. Mater Sci Eng A 528:2863–2869. doi:10.1016/j.msea.2010.12.057
198. Wang Y, Xu Z, Chen L, Jiao Y, Wu X (2011) Multi-scale hybrid composites-based carbon nanotubes. Polym Compos 32(2):159–167
199. Wen S, Chung DDL (2001) Uniaxial compression in carbon fiber-reinforced cement, sensed by electrical resistivity measurement in longitudinal and transverse directions. Cement Concr Res 31:297–301
200. White SR, Sottos NR, Geubelle PH, Moore JS, Kessler MR, Sriram SR, Brown EN, Viswanathan S (2001) Autonomic healing of polymer composites. Nature 409:794–797
201. Wool RP, O'Connor KM (1981) A theory of crack healing in polymers. J Appl Phys 52:5953–5963
202. Wu DY, Meure S, Solomon D (2008) Self-healing polymeric materials: a review of recent developments. Prog Polym Sci doi:10.1016/j.progpolymsci.2008.02.001
203. Wu XF, Yarin AL (2013) Recent progress in interfacial toughening and damage self-healing of polymer composites based on electrospun and solution-blown nanofibers: an overview. doi:10.1002/app.39282
204. Wudl F, Chen X (2004) Thermally re-mendable cross-linked polymers. US:2004014933-A1. University of California
205. Xue P, Tao XM, Kwok K, Leung MY (2004) Electromechanical behavior of fibers coated with an electrically conductive polymer. Tex Res J 74:929–936
206. Yamaguchi M, Ono S, Terano M (2007) Self-repairing property of polymer network with dangling chains. Mater Lett 61:1396–1399
207. Yang F, Pitchumani R (2002) Healing of thermoplastic polymers at an interface under nonisothermal conditions. Macromolecules 35:3213–3224
208. Yavuz O, Ram MK, Aldissi M et al (2005) Polypyrrole composites for shielding applications. Synth Met 151:211–217
209. Yildiz Z, Usta I, Gungor A (2012) Electrical properties and electromagnetic shielding effectiveness of polyester yarns with polypyrrole deposition. Tex Res J 82:2137–2148
210. Yildiz Z, Usta I, Gungor A (2013) Investigation of the electrical properties and electromagnetic shielding effectiveness of polypyrrole coated cotton yarns. Fibre Tex Eastern Eur 98:32–37
211. Yuan YC, Yin T, Rong MZ, Zhang MQ (2008) Self healing in polymers and polymer composites. Concepts, realization and outlook: a review. eXPRESS Polym Let 2:238–250. doi:10.3144/expresspolymlett.2008.29
212. Zako M, Takano N (1999) Intelligent material systems using epoxy particles to repair microcracks and delamination damage in GFRP. J Intell Mater Sys Struct 10:836–841
213. Zhang H, Tao X, Wang S, Yu T (2005) Electro-mechanical properties of knitted fabric made from conductive multi-filament yarn under unidirectional extension. Tex Res J 75:598–606. doi:10.1177/0040517505056870
214. Zhao DL, Zhang HL, Zeng XW, Xia QS, Tang JT (2006) Inductive heat property of Fe_3O_4/polymer composite nanoparticles in an ac magnetic field for localized hyperthermia. Biomed Mater 1:198–201. doi:10.1088/1748-6041/1/4/004

215. Zhu X, Li X, Sun B (2012) Study on electromagnetic shielding efficacy of knitting clothing. Przeglad Elektrotechniczny (Electr Rev) 88:42–43
216. Zhu YF, Zhang L, Natsuki T et al (2012) Synthesis of hollow poly(aniline-co-pyrrole)–Fe3O4composite nanospheres and their microwave absorption behavior. Synth Met 162:337–343

Comparison of Performance, Cost-Effectiveness and Sustainability

Jack Howarth

Abstract This concluding chapter presents the cost-performance comparison of different fibres as well as their environmental impacts and sustainability. Future directions of research and developments on the use of fibres in composite materials have been pointed out and sources for further reading and information are listed.

1 Cost-Performance Comparison of Different Fibres

Textile based engineering composites exhibit a highly diverse range of properties and are hence utilised in many applications. The number of variables available to designers and manufacturers through choice of reinforcement, matrix, filler and manufacturing process leads to numerous combinations of characteristics in terms of mechanical performance, surface finish, corrosion resistance, operating temperature and processability. This section focuses on the cost-performance relationship between different reinforcements, i.e. fibres. The choice of reinforcement is critical in any design stage, and an analysis of the cost-performance relationship is necessary to aid in meeting both product performance and budget targets.

Reinforcing fibres are utilised in composite manufacture in both their as manufactured (tow) form and as fabrics, which requires processing of the tows similar to textile manufacture. Both forms will be considered.

1.1 Reinforcement Type

Fibrous reinforcements utilised in engineering composite materials are usually one of three types: glass, carbon or aromatic polyamide (aramid). Within each type there

J. Howarth (✉)
University of Manchester, Manchester, UK
e-mail: jack.howarth@manchester.ac.uk

© Springer Science+Business Media Singapore 2016

S. Rana and R. Fangueiro (eds.), *Fibrous and Textile Materials for Composite Applications*, Textile Science and Clothing Technology,
DOI 10.1007/978-981-10-0234-2_11

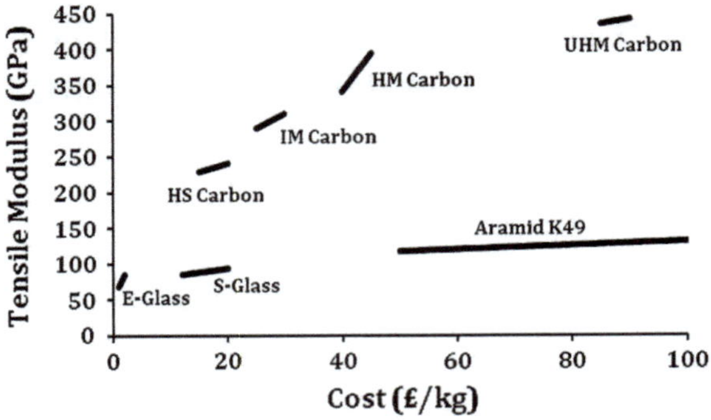

Fig. 1 Cost versus tensile modulus for common reinforcing fibres

can be many different grades and it is thus difficult to specify a fixed unit cost per fibre type. Figures 1 and 2 show the relationship between cost and tensile modulus and cost and tensile strength respectively, of the most common reinforcement types. Each individual entry does not necessarily imply a linear relationship between cost and performance, but is intended to represent the range of each within that fibre type. Cost is affected by many contributing factors such as raw material costs, varying processing conditions and ancillary equipment, oil price and exchange rate, and the data is intended as a guide only.

It is clear from Figs. 1 and 2 that there is a wide variety of cost-performance characteristics for these reinforcements. E-glass fibres represent the cheapest option, with corresponding inferior performance. E-glass fibres are by far the most common reinforcement used in the composites industry. Good mechanical performance combined with very low cost compared to other reinforcements has facilitated their use in a wide variety of applications. Although inferior to carbon fibres, their specific (per unit weight) strength and stiffness compares well with many metals. S-glass fibres exhibit higher strength and stiffness than E-glass fibres but are disproportionately more expensive and consequently relatively rare. When considering the cost-performance relationship in terms of modulus (Fig. 1), there is little motivation to replace E-glass with S-glass, however S-glass fibres have superior tensile strength, comparable with high strength (HS) carbon fibres (Fig. 2). S-glass

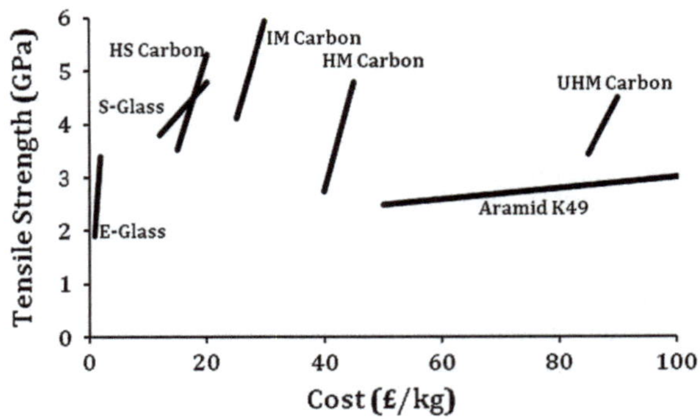

Fig. 2 Cost versus tensile strength for common reinforcing fibres

fibres are thus used in some military and aerospace applications; however the vast majority of glass fibres utilised in composites are E-glass.

Carbon fibres are classified according to their tensile modulus, and are referred to as: high strength (HS); intermediate modulus (IM); high modulus (HM) and ultra high modulus (UHM). Higher modulus fibres are generally more expensive as they require a higher graphitisation temperature. Thus the cost-performance relationship for stiffness is that cost increases with increasing fibre stiffness (Fig. 1). Tensile strength varies within each classification however there is a general trend towards reduced strength with increasing stiffness (Fig. 2) and thus these 2 properties cannot be optimised co-operatively. Carbon fibres are significantly more expensive than glass fibres and therefore not as commonly used. They are mainly utilised in weight-critical applications such as aerospace, as their high specific stiffness and strength gives them a considerable advantage over both glass fibres and metals.

Aramid fibres are generally the most expensive (along with UHM carbon) despite having lower strength and stiffness than most carbon fibres. This is due to the demands of the manufacturing process. Aramid fibres are however extremely tough, and find applications in for example ballistic protection and cabling where use of carbon fibres would be impractical due to their brittle nature.

Generally speaking, glass fibres are the reinforcement of choice where cost is the critical factor, carbon fibres for where specific strength and/or stiffness are required and aramid fibres where toughness is paramount. There are numerous examples of material selection models on cost-performance optimisation in the literature which are beyond the scope of this chapter. The reader is directed to Sect. 5.1 for recommendations on further reading.

1.2 Fibre Tows

Fibrous reinforcements are generally manufactured in continuous filament bundles known as tows. The number of fibres in a tow can be varied, and commercial fibre tows generally consist of between 3,000 and 80,000 (3–80 k) fibres. A greater number of fibres in a tow make the process cheaper, but there is less control over the number of fibres [1]. Thus in performance critical applications, lower tow counts are generally used. Table 1 lists the commercial prices of some common carbon fibre tows sold in the UK. These are retail prices based on small quantities, and are for comparison only.

Table 1 Relationship between number of fibres per tow and cost of 3 common carbon fibre tows sold in the UK

No. of fibres (k)	Price (£/kg)
3	79–80
6	55–70
12	44–53

It can be seen from Table 1 that the price of the material decreases with the increasing number of fibres in the tow bundles. This is also the case with reinforcements other than carbon. For example, 3220 decitex (3.22 kg per 10,000 m) aramid fibre tows are approximately 25 % cheaper than their 1610 decitex equivalents. Fibre tows are the precursor material to textile reinforcements, although they are used in their raw form in processes such as pultrusion and filament winding.

1.3 Fabric Reinforcements

The manufacture of textile fabrics used in composite applications is similar to conventional production of textiles. The main factors that affect the cost and performance of textiles in composite applications are:

- Reinforcement type (discussed in Sect. 1.1)
- Number of fibres per tow (discussed in Sect. 1.2)
- Fabric aerial density
- Fabric weave architecture

Fabrics produced from fibre tows are supplied in a variety of aerial densities (weight per unit area) and weave architectures. Aerial densities are usually quoted in grams per square metre (gsm). The higher the density of the fabric, the cheaper it is per unit weight. This is because textile production costs are closely related to the unit area of production, thus the more fabric is manufactured per unit area, the cheaper per unit weight it becomes. This is illustrated in Figs. 3 (for glass fibre) and 4 (for carbon fibre) which show a general decrease in cost per unit weight with increasing aerial density.

The weave architecture is defined by the interaction between the warp and weft fibres in a reinforcing fabric. The most common types of weave architecture are plain, twill and satin. Unidirectional (UD) fabrics are also produced, held together with a supporting net on one side which typically accounts for less than 1 % of the total fabric weight. Weave architecture is less cost critical than the aerial density.

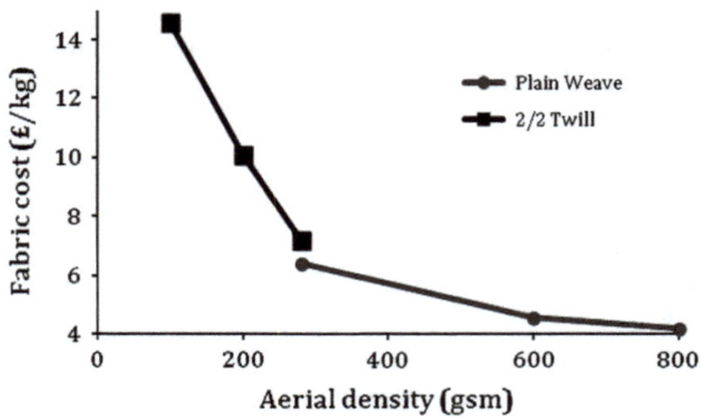

Fig. 3 Relationship between fabric aerial density and cost per unit weight of common glass fibre fabrics

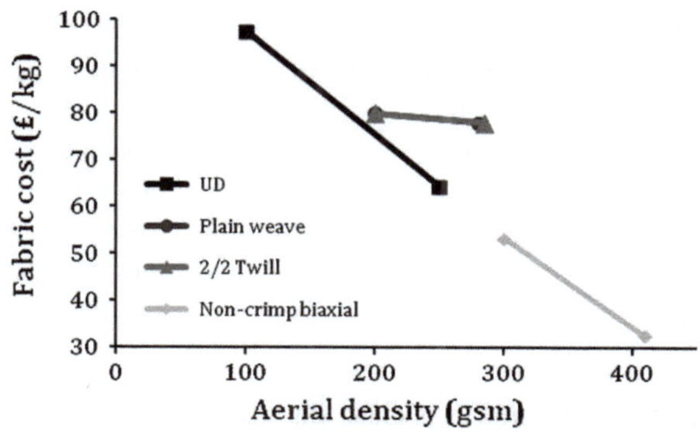

Fig. 4 Relationship between fabric aerial density and cost per unit weight of common carbon fibre fabrics

This is illustrated for glass fibre in Fig. 3 and carbon fibre in Fig. 4, where plain and twill weave fabrics have similar cost where the aerial densities are equivalent.

The choice of weave architecture depends on the application. UD and other non-crimp fabrics (such as biaxial fabrics) have greater compression properties. Plain weave fabrics have greater stability and symmetry than twill and satin weave fabrics. However the reduced crimp of twill, and particularly satin weave fabrics, give them superior resin wettability than plain weave fabrics. Satin weave fabrics have a high degree of drape (ability to conform to shapes) which makes them ideal for moulding more complex geometries, although their relatively poor symmetry means care must be taken when assembling multiple layers.

In summary the single largest contributor to cost and performance is the nature of the fibre. The least critical factor is the weave architecture of the fabric, choice of which is largely dependent on application only. The number of fibres per tow and fabric aerial density do affect cost, but are however not as significant as the nature of the fibre. More consistent properties are generally achieved using tows with a fewer number of fibres as control of the number of fibres is higher, and fabrics with smaller aerial densities have greater wettability than those with higher aerial densities.

1.4 Natural Fibre Composites

Looking beyond traditional reinforcement fibres such as glass, carbon and aramid are those extracted from plants—so called ‘natural fibres’. Advantages of natural fibres include that they are a renewable resource, are safe to handle, biodegradable and have excellent cost-specific properties. Disadvantages include their tendency to moisture absorption, compatibility issues with polymer matrices and inferior mechanical performance compared to synthetic fibres. Table 2 lists some common natural fibres, their mechanical properties and unit cost. Table 2 shows that these natural fibres are cheaper than all of the synthetic fibres depicted in Figs. 1 and 2 with the exception of E-glass, the price range of which has overlaps with flax, hemp

Table 2 Mechanical properties and unit cost of common natural fibres [2]

Fibre	Tensile modulus (GPa)	Tensile strength (MPa)	Price (€/kg)
Flax	27–80	343–1035	2.29–11.47
Hemp	3–90	580–1100	0.57–1.73
Jute	3–55	187–773	0.12–0.35
Kenaf	22–53	295–930	0.53–0.61
Coir	3–6	106–270	0.24–0.48
Cotton	5–13	287–597	1.61–4.59

and cotton fibres. It is clear that natural fibres cannot compete with synthetic fibres in terms of strength, although there is some overlap of their stiffness properties with that of E-glass fibres. Thus their potential applications as the reinforcement phase in polymer matrix composites where synthetic fibres are traditionally used are largely limited to the low end of the performance spectrum, although their low specific gravity compared to glass fibres offers an advantage in weight saving for automotive applications.

2 Environmental Impacts and Sustainability

2.1 Life Cycle Assessment

When considering the environmental impacts of composite products (or indeed any product) it is useful to separate the contributions into their constituent life cycle stages, of which there four: raw material acquisition; product manufacture; use and end of life disposal. There may also be additional stages in between, such as transportation, operation and maintenance, depending on the product. The life cycle assessment (LCA) methodology can be used to quantify the environmental impact of a product at each stage. There are many categories of environmental impact, including but not limited to: global warming potential; cumulative energy demand; ozone layer depletion; acidification and eutrophication. There are many impact assessment methods in LCA software packages that combine the data from each impact category into a single measure or value. However such steps are highly subjective as there is no standardised method for normalising and weighting the impacts of different categories, and thus this practice does not conform to the ISO standards on LCA [3, 4].

However, one impact category that transcends research fields and the international community is a commitment to energy efficiency and a reduction in greenhouse gas emissions (GHGE). Measured or modelled GHGE values are represented by the global warming potential impact assessment category. As energy consumption and global warming potential are closely related, reducing one generally reduces the other [5]. This section will therefore address environmental impacts according to energy demand for ease of comparison.

2.1.1 Raw Material Production

The energy intensity (sometimes referred to as embodied energy) of a material represents the total energy demand associated with its production. When performing an LCA on a composite product with the aim of analysing the cumulative energy demand, the sum of the energy intensities of the raw materials represents the cumulative energy demand of the raw material life cycle phase. LCAs could also be conducted on each individual raw material, where the energy intensity would be subdivided into each of its life cycle stages. However for the purpose of analysing the composite, the total energy intensities of common raw materials are listed in Table 3.

The first thing to note from Table 3 is that the energy intensity of carbon fibre is approximately an order of magnitude higher than that of glass fibre, and we can thus assume that the global warming potential will be proportionately higher. The wide range of values for both fibre types is due to several factors such as process differences across manufacturers and plant size effects (higher production rates are generally more energy efficient). Also for carbon fibres, the energy intensity changes with the modulus classification (Sect. 1.1). The energy intensities of epoxy and polyester resins are intermediate between glass and carbon fibres. Consequently, for carbon fibre composites, the fibre phase contributes most to the environmental impact of the raw material phase, whereas for glass fibre composites the matrix phase dominates the impact. Exact contributions to environmental impact depend on the relative weight fractions of each constituent phase.

It is important to note that the energy intensities of the fibres listed in Table 3 refer to the manufacture of the tows. The production of fabrics requires additional energy input, however this step is a relatively small contribution to the energy intensity of the fabric. Electrical energy values of 0.4 MJ/m^2 for glass and 0.77 MJ/m^2 carbon fabrics have been reported [7]. Values for multi-axial fabrics (such as biaxial and UD) are lower, 0.14 MJ/m^2 for glass, 0.19 MJ/m^2 for carbon and 0.1 MJ/m^2 for aramid. The high value for carbon fibres is due to their brittle nature [7].

Fabrics can be further processed by impregnating with resin and partially curing to form 'pre-pregs'. These materials are used in performance critical applications such as aerospace. The resin content can be tightly controlled by the speed of the process and resin viscosity. The process is much more energy intensive than fabric production, around 40 MJ/kg [8].

Table 3 Energy intensity of common raw materials used in composite manufacture [6]

Type	Material	Energy intensity (MJ/kg)
Fibres	Carbon	183–286
	Glass	13–32
Matrices	Epoxy	76–80
	Polyester	63–78

Table 4 Energy intensities of common composite manufacturing processes [6]

Process	Energy intensity (MJ/kg)
Autoclave moulding	21.9
Spray up	14.9
Resin transfer moulding (RTM)	12.8
Cold press	11.8
Vacuum assisted resin infusion (VARI)	10.2
Pre-form matched die	10.1
Sheet moulding compound (SMC)	3.5
Pultrusion	3.1
Filament winding	2.7

2.1.2 Product Manufacture

After raw material production, the next stage of a products' life cycle is its' manufacture. There are many techniques that can be used to manufacture composites, and their energy intensities are varied (Table 4). Fibre tows can be used directly in pultrusion and filament winding. Automated processes such as this and sheet mould compounding (SMC) have relatively low energy intensities. Higher energy processes employ pressure in addition to heat to achieve cure. Autoclave moulding is the most energy intensive process, and is almost exclusively used in moulding pre-pregs. As these products are used in high end, performance critical applications, cost is not the primary concern.[1]

The energy intensity of composite manufacturing processes are mostly lower than the energy intensities of the constituent materials. There are of course exceptions (such as a pultruded glass fibre composite) however in a cradle to gate LCA (where only raw material acquisition and product manufacture are considered) it is generally the raw material phase that dominates the environmental impact. This is especially so for carbon fibre composites.

2.1.3 The Use Phase

For the purpose of environmental impact assessment, the application of composites defined in the use phase can be categorised into 2 groups: passive and non-passive. In passive applications, there are no inputs (resources) or outputs (emissions) associated with the use phase, although in some cases there may be impacts associated with operation, maintenance and repair. In non-passive applications, resource consumption and resultant emissions contribute to the environmental impact.

[1]Cost minimisation is of course important; however the requirements of the product preclude this being achieved through use of inferior materials or manufacturing processes

Examples of passive applications include roofing panels, tooling moulds, bath tubs, and surfboards. In these products there are no impacts associated with the use phase, and the life cycle impact is generally dominated by the raw material phase. The most important non-passive applications in terms of environmental impact are automotive and aviation components. These products consume energy during their use phase in the form of fuel and emit GHGs. Additional impacts arise from operation and maintenance, although their contribution is minor compared to fuel consumption. In automobiles, there is a very close relationship between total weight and fuel consumption [9]. The use of composites in non-passive applications reduces costs and emissions compared to metals in the use phase because of their superior specific strength and stiffness.

As carbon fibres have significantly higher energy intensity than steel or aluminium, the benefit of replacing these metals is not immediately clear. However when accounting for the relationship between impacts and product or component weight, the point in the life cycle where use of composites becomes more efficient can be defined. In such non-passive applications the use phase is dominant in terms of environmental impact.

2.1.4 End of Life Disposal

Thermoset based composites have traditionally been disposed of in landfill at the end of their serviceable life. The heterogeneous nature of the material makes recycling a considerable challenge. Recycling is much easier for thermoplastic based composites, which can be melted and re-moulded. Impacts in terms of energy demand and global warming potential associated with landfill are negligible to the entire life cycle contribution. However tightening legislation on this disposal method is forcing research and industry to find viable alternatives. Incineration with energy recovery can be used as an alternative to landfill, exploiting the calorific value of thermoset resins; however the presence of fibres and fillers limits the amount of energy recovered and gives rise to the problem of disposal of the resulting ash. These unsustainable end of life scenarios are incompatible with international efforts to reduce resource consumption and GHGEs. The next section will discuss the drivers towards better waste management and the state of the art in composite recycling.

2.2 Composite Recycling

The global market for composite materials is steadily growing. The global demand for carbon fibre composites in 2012 was around 65,000 tonnes, and is projected to reach over 200,000 tonnes by 2020 [10]. The demand for glass fibre composites was estimated at 2.4 million tonnes in 2001, increasing to 3.8 million tonnes in 2008 (prior to the global financial crisis) at a cumulative annual growth rate (CAGR) of 6.8 %, and despite a drop in 2009, the market recovered to a demand of

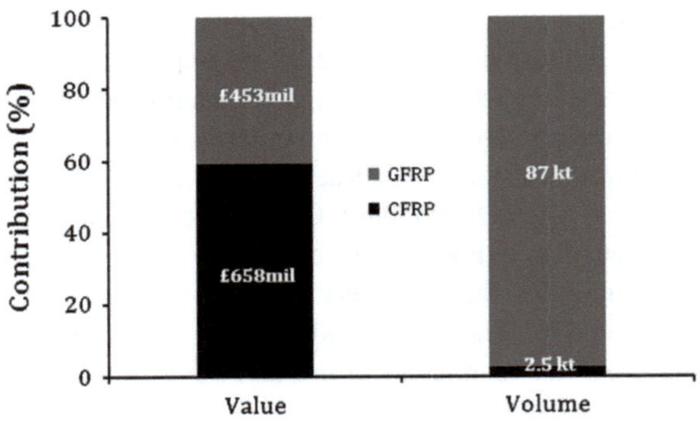

Fig. 5 Relative contributions to value and volume of glass fibre reinforced plastics (GFRP) and carbon fibre reinforced plastics (CFRP) in the UK, 2010 [12]

4.1 million tonnes in 2011 [11]. These figures highlight the legacy problem for these materials in terms of waste, particularly for glass fibre composites, which dominate the market in terms of volume. This is in contrast, however, to the value contributions.

Figure 5 shows that despite accounting for over 97 % of production volume, glass fibre composites represented just 41 % of the value in the UK in 2010. This discrepancy highlights the contrasting motivators for recycling: for carbon fibre composites, the incentive is to recover high value carbon fibres; for glass fibre composites, it is to comply with legislation and resolve the bulk of the waste problem.

2.2.1 Legislation

The EU Waste Hierarchy (Fig. 6) defines 5 levels of waste management ranked from most desirable; prevention, down to least desirable; disposal (for example landfill). Recycling ranks 3rd, and is the most applicable to composites that have reached the end of their serviceable life.

Specific EU legislation that affects composites includes: The Waste Landfill Directive (1999), the End of Life Vehicle Directive (2000) and the Directive on Waste Electrical and Electronic Equipment (2002). Such legislation, combined with the waste legacy problem, has driven research and development into composite recycling technologies.

2.2.2 Recycling Technologies

The state of the art in composite recycling has been extensively discussed elsewhere by Pickering [14] and Oliveux et al. [15], and particularly for carbon fibre composites by Pimenta and Pinho [16]. A brief outline will be given here.

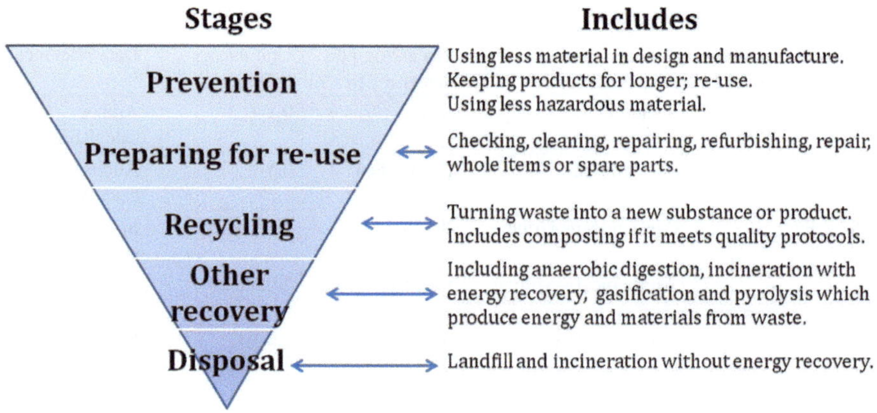

Fig. 6 The EU Waste Hierarchy adapted from [13]

Recycling processes fall into one of three main categories based on the nature of the process; mechanical, thermal and chemical. In mechanical recycling, the waste is gradually size reduced in processes such as grinding and milling, resulting in powderous material consisting of both fibre and matrix. The matrix contamination coupled with reduction in fibre length results in recyclates with little reinforcement capacity. As a result there is a significant drop in the value of the material, and with the exception of a few isolated cases (there are roofing companies who recycle their trimmings in this way and re-incorporate the filler into their production line), mechanical recycling is not practiced on an industrial scale. Its relatively low energy input and high throughput capacity [17] however, highlight its potential for processing the high volumes of glass fibre waste. In the UK, the trade association Composites UK and the Knowledge Transfer Network (KTN), in collaboration with research council funded projects such as EXHUME, are actively seeking out markets and the establishment of supply chains for mechanical recyclate.

Thermal recycling utilises heat to separate the fibre and matrix phases. Thermal recycling by pyrolysis (combustion in the absence of oxygen) is practiced on an industrial scale in the recovery of carbon fibres. For example, ELG Carbon Fibre in the UK, process in excess 2,000 tonnes of waste per annum [18]. The process is more energy intensive (Table 5) than mechanical recycling, and therefore not currently economically feasible for recovery of glass fibres. Also, the mechanical property degradation is more severe for glass fibres. There are studies however, which are trying to address the problem and make thermal recycling feasible for glass fibre composites [19, 20]. The potential of retaining the value of the glass fibres could make thermal recycling more economical than mechanical recycling, which is limited to recovering non-reinforcing fillers. Other than pyrolysis, thermal recycling by a fluidised bed process has been demonstrated at the laboratory scale [21]. Chemical recycling processes are currently only practiced on the laboratory scale, however it has the potential to recover the matrix phase as feedstock solution

Table 5 Processing rate and unit energy values of composite end of life scenarios

Scenario according to waste hierarchy	Process	Processing rate (kg/h)	Unit energy (MJ/kg)	Data reference
Disposal	Landfill	N/A	0.1–0.2	Authors calculation
Other recovery	Incineration	Unspecified	0.5–1	Authors calculation
Recycling	Mechanical—Grinding	28–29	4.85–4.65	Srivastava et al. [23]
	Mechanical—Milling	10–150	2.03–0.27	Howarth et al. [17]
	Thermal—Pyrolysis	200*	30–50	Witik et al. [24]

*Authors estimate based on processed volume of industrial process

as an additional source of value. The state of the art in chemical recycling was reviewed by Morin et al. [22].

2.2.3 Impacts of Recycling

There is little available data in the literature regarding the environmental impacts of recycling. The available data is presented in Table 5 for one thermal and 2 mechanical process, along with landfill and incineration for comparison. It is clear that the energy consumption in mechanical recycling is significantly lower than for thermal recycling, but produces recyclates of inferior quality. This explains why thermal recycling is only feasible for carbon fibre recovery at present. There is also no environmental motivation to recover glass fibres by pyrolysis, as the unit energy falls within the range of that of virgin glass fibres (Table 3). Carbon fibres recovered by pyrolysis retail at approximately £15/kg. This is significantly cheaper than virgin carbon fibre and provides a viable low-cost alternative in applications where short and milled fibres are utilised. There is a corresponding environmental benefit too, as the unit energy of pyrolysis is significantly lower than that of virgin carbon fibres (Table 3).

It is important to note that the unit energies of recycling processes will vary according to the machinery used. Also, the dependence on processing rate, where the unit energy consumption reduces with increasing processing rate [17] has implications for the static values quoted for unit energy in many manufacturing processes (including those in Table 4), which do not account for the processing rate.

Table 5 shows that in most cases the recycling scenarios have greater energy demand than both landfill and incineration. It is important to note therefore that recycling achieves reduced environmental burden through avoided material production. This occurs when the recyclate material is incorporated into a new product, displacing virgin material, production of which is avoided. For engineering materials such as thermoplastics and metals, their recyclates form a significant proportion of the supply chain. This is not yet the case for recovered fibres, though the benefits of such closed-loop (direct substitution of virgin fibres by recyclate fibres) recycling are clear when considering their respective embodied energies. The heterogeneous nature of the material and difficulty in separating the phases however results in recyclates that are vastly different to the original material, and thus there is a reduction in value. Therefore cross-sector (as opposed to closed loop) recycling may be more applicable to composites, where the recyclate is utilised in different product systems to the original material.

2.2.4 Life Cycle Assessment Case Study

The environmental impacts associated with composite materials are better illustrated through examples. Although there is no data the author is aware of in any software package on carbon fibre, there is data on glass fibre, thermoset and thermoplastic matrices. This example will assess the impacts of an automotive panel (a non-passive application) made of glass fibre composite. A non-passive application was chosen so that impacts will occur in the use phase. The goal was to assess the relative contribution of each life cycle phase (raw material acquisition, product manufacture, use and end of life disposal).

Considering just one component in a product system, in this case a single panel, part of the larger assembly of an automobile, highlights the problem of allocation in LCA. Allocation is the procedure of assigning impacts to different parts of a product system. This is usually done in one of two ways: by value or by weight. Value allocation is used where a process produces more than one product (by-products), and impacts are assigned according to the relative value of each product or by-product. Weight allocation is useful where impact data is only available for the entire product system (automobile), and the contribution of the component is assigned according to the weight fraction on the whole. This is a sensible approach, as we have seen previously that there is a close to linear relationship between vehicle weight and fuel consumption [9]. The impact assessment category used was global warming potential. The assumptions used in the analysis are listed below:

- Panel consists of equal weight fractions of glass fibre and polyester matrix totalling 5 kg.
- Panel was manufactured by SMC.
- Total vehicle weight was 1,000 kg, thus the panel accounts for 0.5 % of this and for allocation procedures, 0.5 % of the global warming potential in the use phase.
- Life of the vehicle was assumed to be 160,000 km.
- End of life disposal was in inert material landfill.

The analysis was performed using SimaPro (version 8.02) developed by PRe consultants of the Netherlands, (except in the case of the manufacturing phase, where no data was available for SMC. Here the CO_2 footprint per unit energy estimation was used). Figure 7 shows the global warming potential of each life

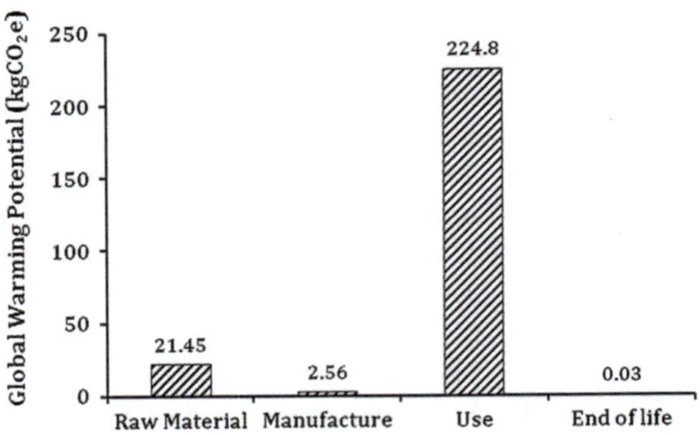

Fig. 7 Global warming potential of each life cycle phase of 5 kg composite automotive panel

cycle phase. It is clear that the use phase dominates the impact, accounting for over 90 % of the global warming potential.

Although there is no data in the software on carbon fibres, we can assume that as their embodied energy is significantly higher than glass fibres, if the glass fibres in this product were substituted for carbon, the impacts associated with the reinforcement (and thus the raw material phase) would increase considerably. However what this case study highlights is the weight dependence on impacts of non-passive applications that consume fuel. Increased impacts in other life cycle phases that reduce component and product weight (which can be achieved by substituting steel for composites or aluminium) are acceptable in these cases as the savings made in the use phase far outweigh the additional raw material impacts.

When considering a component made of the same materials and manufactured by the same process that is used in a passive application, the situation changes completely. In such products there are no impacts associated with the use phase (except in cases where operation and/or maintenance is required). The raw material phase then becomes dominant, and the most viable target for emission reductions. As discussed in Sect. 2.2.3, the impacts of recycling processes (where there is available data) although mostly greater than traditional methods of disposal, can result in reduced impacts through avoided material production by incorporating the recyclates into new products. LCA is a tool that through quantifying environmental impacts at each life cycle stage, can aid the decision making process when targeting reductions.

3 Future Directions

3.1 Out of Autoclave Processing

Although most composite manufacture is performed out of autoclave (OOA), the term used in the literature refers to those parts that are traditionally autoclave cured. Table 4 shows that autoclave curing is the most energy intensive of all composite manufacturing processes. The trade-off benefit is that this method produces parts of superior quality, but this has not stopped innovation into more efficient solutions. For manufacturers, energy consumption and its associate cost is not the primary motivator in moving to OOA processing. Brosius [25] highlighted that energy consumption is only a small fraction of total cost, and is perhaps as low as 5 %, and that the primary motivators are shorter cycle times and reduced capital and tooling investments.

One alternative is the use of vacuum bag only (VBO) pre-pregs. Autoclave cure produces superior quality parts by applying an external pressure onto the vacuum bagged part, which reduces void content. VBO pre-pregs can accommodate for this lack of pressure as they exhibit better flow. Brosius [25] points out that VBO pre-pregs are edging closer to the mechanical performance required of primary aircraft structures, but that problems such as long debulk and oven cure times need to be overcome for them to succeed in high volume production.

Other alternatives are variations on resin infusion. Table 4 shows that RTM and VARI are only around half as energy intensive as autoclave moulding. Resin infusion however produces components with inferior properties to autoclave cure. There are concerns regarding the brittle nature of infusion resins used in such techniques, and the resulting low damage tolerance of the components [26]. Addition of thermoplastic tougheners can mitigate this, however this increases the resin viscosity and therefore results in flow problems for infusion processes. As most pre-preg resin systems would be unsuitable for use in infusion processes because of their higher viscosity, re-certification for use in performance critical applications may be required. One way of overcoming this is the application of resin films. Here resins are cast directly onto films resulting in a material that could be described as pre-preg without the reinforcement. These films are then placed directly under the fibre pre-form in a process known as resin film infusion (RFI).

Going beyond OOA is out of oven (OOO) cure. OOO has the potential for further energy and cost savings by avoiding the heat (as well as the pressure) associated with autoclave cure, although in some cases the heat is supplied by other means. One example applicable to carbon fibres is electrical cure, which utilises the conductivity of carbon fibres which become heating elements in the cure of the surrounding resin. Hayes et al. [27] demonstrated similar degrees of cure compared to conventionally cured pre-pregs. Although at an early stage, the demonstrated potential of OOO processes such as electrical cure could lead to further significant reductions in energy consumption and manufacturing costs in high performance, high volume sectors. Suggested further reading on OOA techniques can be found in Sect. 5.5

3.2 3D Fabrics

Composites from traditional fabric reinforcements are usually multi-layer laminates of 2D in-plane fibre fabrics impregnated with resin. The in-plane specific strength and stiffness are excellent compared to metals however the through thickness properties such as impact resistance are extremely poor compared to non-laminated isotropic materials. Out of plane properties are so poor in composite laminates due to the lack of through thickness reinforcement resulting in low resistance to delamination [28]. In order to improve the through-thickness properties of 2D laminates, 3D fabrics have been developed by a variety of processes (weaving, knitting, stitching and braiding). A further important advantage of 3D fabrics is that preforms for complex geometries can be made to near net shape [29]. However there are limitations. The addition of fibres to the through thickness axis reduces the in plane specific strength and stiffness. Machinery to produce 3D fabrics is largely in its infancy, however further research into automation and further performance characterisation could propagate the growth of 3D fabrics as they further penetrate the market.

3.3 Meeting the Demand for Carbon Fibres

The increasing demand for carbon fibres and corresponding growth of the carbon fibre composites market is putting huge pressure on manufacturers. Continued growth could see demand soon outstrip supply. There is an urgent need to source alternative precursors to petroleum-based polyacrylonitrile (PAN). Lignin offers a potential alternative to PAN and holds advantages such as its detachment from oil price fluctuations and sustainability. Traditionally sourced from wood pulping, lignin availability is increasing as it is a by-product of ethanol production in bio-refineries. Progress is being made where Swerea SICOMP are leading a program to develop a supply chain for lignin based carbon fibre in Sweden. As discussed in Sect. 2.2, industrial recycling of carbon fibre composites could also contribute to the supply chain. Developments in technology and processing capacity will help reduce the burden on PAN-based virgin carbon fibres in short fibre applications where mechanical performance is less critical. Further processing of recyclate fibres into non-woven textiles have been demonstrated in applications such as battery fuel cells and electromagnetic interference shielding components [30].

3.4 Compliance with Legislation

Section 2.2 discussed current and emerging technologies in composites recycling that are driven partly by the need to comply with waste legislation. This is especially true for glass fibre composites, as there is little incentive to recover such a relatively low value material compared to carbon fibres. However there is another aspect to legislation that affects the use of composites in the automotive industry. Weight saving through incorporation of composites results in lower fuel consumption, which aids in meeting targets such as that of the EU; CO_2 emissions to be reduced to 130 g/km by 2015 and 95 g/km by 2020. However, the vehicles then have more difficulty in complying with the End of Life Vehicle Directive (2000), as the recycling of heterogeneous composites is much more difficult than for metals such as steel and aluminium. This puts further pressure on research and industry to develop resource efficient recycling technology.

Interestingly, one market where composites are unlikely to gain a foothold is in large commercial shipping, where weight and fuel savings from the automotive and aerospace sectors would appear to be transferable. This is due to combustibility regulations, which the presence of polymer matrices causes composites to fail [31].

4 Conclusions

The type of reinforcement in textile composite materials can be divided into 4 categories: fibre type; number of fibres per tow; fabric weave architecture and fabric aerial density. Each characteristic affects the performance of the composite in different ways and to different extents, with the fibre type generally being the most important. Although not strictly a linear relationship, the trend towards increased performance is generally accompanied by increased cost. Optimisation of reinforcement characteristics are product specific, and can still vary within a given product depending on the relative importance to the designer of cost versus performance. Further reading is suggested in Sect. 5.1.

The environmental impacts and sustainability of composite materials is becoming an increasing issue as demand for these materials grows. Although the production volumes are still relatively low compared to metals and unreinforced plastics, waste management is largely limited to landfill disposal. There has been some success in recycling of CFRPs for carbon fibre recovery as the inherent value of the fibres makes their recycling economically feasible. This is not the case for glass fibres, which represent the bulk of the waste problem, although work is ongoing towards achieving feasibility of glass fibre recovery through recycling.

Glass fibres exhibit a much lower environmental burden than carbon fibres per unit weight produced, however they are produced in much greater volumes than carbon fibres and thus have a higher overall environmental footprint. The case for utilising GFRPs and CFRPs in non-passive applications is clear, as their specific strength and stiffness allow for weight savings to be made for greater fuel efficiency in the use phase of such products.

Future challenges for the industry lie in cascading the value of the reinforcement materials through appropriate waste management, requiring the development of economically and environmentally feasible transformation technologies. Progress in efficient manufacturing, through for example increased automation and OOA processing, reduces environmental burden per unit weight of material produced, however such savings are mainly offset by increasing demand. Establishment of markets and supply chains for recyclate materials is a key issue that needs to be addressed alongside recycling technology development.

5 Recommendations for Further Reading

5.1 *Cost-Performance Optimisation*

- Bader M.G. Selection of composite materials and manufacturing routes for cost-effective performance. Composites Part A: Applied Science and Manufacturing. 33. 2002. pp 913–934.
- Park C.H., Saouab A., Breard J., Han W.S. and Vautrin A. Integrated optimization for weight, performance and cost of composite structures. Information Control Problems in Manufacturing 2006, a Proceedings Volume from the 12th IFAC Conference. Saint-Etienne. France. 2006.

5.2 *Natural Fibre Composites*

- Natural Fibre Composites. eds. Hodzic A. and Shanks R. Woodhead Publishing Ltd. 2014.
- Properties and Performance of Natural Fibre Composites. ed. Pickering K. Woodhead Publishing Ltd. 2008.

5.3 *Life Cycle Assessment*

- Letterier Y. Life Cycle Engineering of Composites. in Comprehensive Composite Materials. eds. Kelly A. and Zweben C.H. Elsevier. 2000. pp 1073–1102.

5.4 *Composite Recycling*

- Management, Recycling and Reuse of Waste Composites. ed. Goodship V. Elsevier. 2009.
- Oliveux G., Dandy L.O. and Leeke G.A. Current status of recycling of fibre reinforced polymers. Progress in Materials Science. 72. 2015. pp 61–99.

5.5 *Out of Autoclave (OOA) Processing*

- Centea T., Grunenfelder L.K. and Nutt S.R. A review of out-of-autoclave pre-pregs—Material properties, process phenomena and manufacturing considerations. Composites Part A: Applied Science and Manufacturing. 70. 2015. pp 132–154.

- Schlimbach J. and Ogale A. Out-of-autoclave curing process in polymer matrix composites. in Manufacturing Techniques for Polymer Matrix Composites. eds. Advani S. and Hsiao K-T. 2012. Elsevier. pp 435–480.

5.6 3D Fabrics

- 3D textile reinforcements in composite materials. ed. Miravete A. Woodhead Publishing Ltd. 1999.

References

1. Williamson A in Concise Encyclopedia of Composite Materials. ed. Mortensen A. Elsevier,2006, p 9
2. Biagiotti J, Puglia D, Kenny JM (2004) A review of natural-fibre based composites-part I. J Nat Fibres 1:37–68
3. ISO 14040:2006. Environmental management—Life cycle assessment—Principles and framework
4. ISO 14044:2006. Environmental management—Life cycle assessment—Requirements and guidelines
5. Ashby M, Ball N, Bream C The CES EduPack Eco Audit tool. A White Paper. Granta Design. Undated
6. Song YS, Youn JR, Gutowski TG (2009) Life cycle energy analysis of fiber-reinforced composites. Compos A Appl Sci Manuf 40:1257–1265
7. Stiller H (1999) Material intensity of advanced composite materials. Results of a study for the Verbundwerkstofflabor Bremen e.V. Wuppertal Papers. 90
8. Suzuki T, Takahashi J (2005) Prediction of energy intensity of carbon fiber reinforced plastics for mass-produced passenger car. In: 9th Japan International SAMPE Symposium
9. Duflou JR, De Moor J, Verpoest I, Dewulf W (2009) Environmental impact analysis of composite use in car manufacturing. CIRP Ann—Manuf Technol 58:9–12
10. Witten E, Jahn B (2013) Composites market report 2013. AVK
11. Koncept Analytics (2012) Global fibreglass composite market report: 2012 Edition
12. Smith F (2010) UK composites supply chain study. UK Trade and Investment
13. Government Review of Waste Policy in England (2011) Department for environment, food and rural affairs (DEFRA)
14. Pickering SJ (2006) Recycling technologies for thermoset composite materials—current status. Compos A Appl Sci Manuf 37:1206–1215
15. Oliveux G, Dandy LO, Leeke GA (2015) Current status of recycling of fibre reinforced polymers. Prog Mater Sci 72:61–99
16. Pimenta S, Pinho ST (2011) Recycling carbon fibre reinforced polymers for structural applications. Waste Manag 31:378–392
17. Howarth J, Mareddy SSR, Mativenga PT (2014) Energy intensity and environmental analysis of mechanical recycling of carbon fibre composite. J Clean Prod 81:46–50
18. ELG Carbon Fibre Ltd. www.elgcf.com. Accessed April 2015
19. Thomason JL, Yang L, Meier R (2014) The properties of glass fibres after conditioning at composite recycling temperatures. Compos A Appl Sci Manuf 61:201–208

20. Yang L, Saez ER, Nagel U, Thomason JL Can thermally degraded glass fibre be regenerated for closed-loop recycling of thermosetting composites? Compos A Appl Sci Manuf 72 (201):167–174
21. Pickering SJ, Kelly RM, Kennerley JR, Rudd CD (2000) A fluidised bed process for the recovery of glass fibres from scrap thermoset composites. Compos Sci Technol 60:509–523
22. Morin C, Loppinet-Serani A, Cansell F, Aymonier C (2012) Near- and supercritical solvolysis of carbon fibre reinforced polymers (CFRPs) for recycling carbon fibers as a valuable resource: state of the art. J Supercrit Fluids 66:232–240
23. Srivastava A, Bull S, Ord G (2012) Application of mechanically recycled waste material: Glass fibre reinforced plastic (GFRP). Compos Eng Conf. Conf Proc NetComposite:90–101
24. Witik RA, Teuscher R, Michaud V, Ludwig C, Manson J-A.E (2013) Carbon fibre reinforced composite waste: an environmental assessment of recycling, energy recovery and landfilling. Compos A Appl Sci Manuf 49: 89–99
25. Brosius D (2014) Out-of-autoclave manufacturing: the green solution. www.compositesworld.com/articles/out-of-autoclave-manufacturing-the-green-solution. Accessed April 2015
26. Garschke C, Weimer C, Parlevliet PP, Fox BL (2012) Out-of-autoclave cure cycle study of a resin film infusion process using in situ process monitoring. Compos A Appl Sci Manuf 43:935–944
27. Hayes SA, Lafferty AD, Altinkurt G, Wilson PR, Collinson M, Duchene P (2015) Direct electrical cure of carbon fiber composites. Adv Manuf Polym Compos Sci 1:112–119
28. Chou TW (1992) Microstructural design of fibre composites. Cambridge University Press, Cambridge
29. Mouritz AP, Bannister MK, Falzon PJ, leong KH (1999) Review of applications for advanced three-dimensional fibre textile composites. Compos A Appl Sci Manuf 30:1445–1461
30. Howarth J, Jeschke M (2009) Advanced nonwoven materials from recycled carbon fibre. Carbon fibre recycling and reuse conference. Intertech Pira, Hamburg, Germany
31. Job S (2015) Why not composites in ships? www.materialstoday.com/carbon-fiber/features/why-not-composites-in-ships/. Accessed April 2015

MIX
Papier aus verantwortungsvollen Quellen
Paper from responsible sources
FSC® C105338

If you have any concerns about our products,
you can contact us on
ProductSafety@springernature.com

In case Publisher is established outside the EU,
the EU authorized representative is:
Springer Nature Customer Service Center GmbH
Europaplatz 3, 69115 Heidelberg, Germany

Printed by Libri Plureos GmbH
in Hamburg, Germany